SCHAUM'S OUTLINE OF

THEORY AND PROBLEMS

of

COLLEGE ALGEBRA

Second Edition

MURRAY R. SPIEGEL, Ph.D.

Former Professor and Chairman
Mathematics Department
Rensselaer Polytechnic Institute
Hartford Graduate Center

ROBERT E. MOYER, Ph.D.

Professor of Mathematics
Fort Valley State University

SCHAUM'S OUTLINE SERIES

McGRAW-HILL

New York San Francisco Washington, D.C. Auckland Bogotá Caracus Lisbon
London Madrid Mexico City Milan Montreal New Dehli
San Juan Singapore Sydney Tokyo Toronto

MURRAY R. SPIEGEL received the M.S. degree in Physics and the Ph.D. in Mathematics from Cornell University. He had positions at Harvard University, Columbia University, Oak Ridge, and Rensselaer Polytechnic Institute, and had served as a mathematical consultant at several large companies. His last position was Professor and Chairman of Mathematics at the Rensselaer Polytechnic Institute, Hartford Graduate Center. He was interested in most branches of mathematics, especially those which involved applications to physics and engineering problems. He was the author of numerous journal articles and 14 books on various topics in mathematics.

ROBERT E. MOYER has been teaching mathematics at Fort Valley State University in Fort Valley, Georgia, since 1985. He served as head of the Department of Mathematics and Physics from 1992 to 1994. Before joining the FVSU faculty, he served as the mathematics consultant for a five-county public school cooperative. His experiences include 12 years of high school mathematics teaching in Illinois. He received his Doctor of Philosophy in Mathematics Education from the University of Illinois in 1974. From Southern Illinois University he received his Master of Science in 1967 and his Bachelor of Science in 1964, both in Mathematics Education.

Schaum's Outline of Theory and Problems of
COLLEGE ALGEBRA

10 11 12 13 14 15 16 17 18 19 20 PRS PRS 05 04 03

ISBN 0-07-060266-2

Sponsoring Editor: Barbara Gilson
Production Supervisor: Tina Cameron
Editing Supervisor: Maureen B. Walker

Library of Congress Cataloging-in-Publication Data

Spiegel, Murray R.
Schaum's outline of theory and problems of college algebra/
Murray R. Spiegel, Robert E. Moyer. -- 2nd ed.
p. cm. -- (Schaum's outline series)
Includes index.
ISBN 0-07-060266-2
1. Algebra -- Problems, excercises, etc. 2. Algebra -- Outlines, syllabi, etc. I. Moyer, Robert E. II. Title.
QA157.S725 1997
512.9'076--dc21 97-31223
CIP

McGraw-Hill
A Division of The **McGraw-Hill** *Companies* 

Preface

In revising this book, the strengths of the first edition were retained while reflecting the changes in the study of algebra since the first edition was written. Many of the changes were based on changes in terminology and notation. This edition focuses only on college algebra, includes the use of a graphing calculator, provides tables for both common and natural logarithms, expands the numbers of graphs and the material on graphing, adds material on matrices, and expands the material on analytic geometry.

The book is complete in itself and can be used equally well by those who are studying college algebra for the first time as well as those who wish to review the fundamental principles and procedures of college algebra. Students who are studying advanced algebra in high school will be able to use the book as a source of additional examples, explanations, and problems. The thorough treatment of the topics of algebra allows an instructor to use the book as the textbook for a course, as a resource for material on a specific topic, or as a source for additional problems.

Each chapter contains a summary of the necessary definitions and theorems followed by a set of solved problems. These solved problems include the proofs of theorems and the derivations of formulas. The chapters end with a set of supplementary problems and their answers.

The choice of whether to use a calculator or not is left to the student. A calculator is not required, but its use is explained. Problem-solving procedures for using logarithmic tables and for using a calculator are included. Procedures for graphing expressions are discussed both using a graphing calculator and manually.

ROBERT E. MOYER
Professor of Mathematics
Fort Valley State University

Contents

PROBLEMS ALSO FOUND IN THE COMPANION *SCHAUM'S ELECTRONIC TUTOR*

Some of the problems in this book have software components in the companion *Schaum's Electronic Tutor.* The Mathcad Engine, which "drives" the *Electronic Tutor*, allows every number, formula, and graph chosen to be completely live and interactive. To identify those items that are available in the *Electronic Tutor* software, please look for the Mathcad icons, , placed adjacent to a problem number. A complete list of these Mathcad entries follows below. For more information about the software, including the sample screens, see Appendix C on page 385.

Problem 1.2
Problem 1.15
Problem 1.16
Problem 1.17
Problem 1.19
Problem 1.21
Problem 1.23
Problem 1.25
Problem 1.26
Problem 2.1
Problem 2.2
Problem 2.11
Problem 2.12
Problem 2.13
Problem 2.14
Problem 2.16
Problem 2.17
Problem 3.6
Problem 3.7
Problem 4.3
Problem 4.5
Problem 4.7
Problem 5.13
Problem 5.15
Problem 5.16
Problem 5.19
Problem 6.6
Problem 6.7
Problem 6.8
Problem 6.10
Problem 6.11
Problem 7.1
Problem 7.3
Problem 7.6
Problem 7.19
Problem 7.20
Problem 8.11
Problem 8.12
Problem 8.13
Problem 8.14
Problem 9.6
Problem 9.8
Problem 10.10
Problem 10.12
Problem 10.13
Problem 10.14
Problem 10.15
Problem 11.19
Problem 11.21
Problem 11.29
Problem 11.33
Problem 11.34
Problem 12.29
Problem 12.31
Problem 12.33
Problem 12.34
Problem 12.35
Problem 12.36
Problem 12.39
Problem 12.40
Problem 12.42
Problem 12.43
Problem 12.44
Problem 12.45
Problem 12.47
Problem 12.50
Problem 12.51
Problem 12.56
Problem 13.34
Problem 13.35
Problem 13.39
Problem 13.41
Problem 13.42
Problem 14.10
Problem 14.11
Problem 14.14
Problem 14.15
Problem 14.16
Problem 14.17
Problem 15.26
Problem 15.28
Problem 15.29
Problem 15.33
Problem 15.34
Problem 15.35
Problem 15.36
Problem 16.30
Problem 16.31
Problem 16.32
Problem 16.33
Problem 16.34
Problem 16.35
Problem 16.36
Problem 16.37
Problem 16.40
Problem 16.41
Problem 16.43
Problem 16.45
Problem 16.48
Problem 17.1
Problem 17.12
Problem 17.14
Problem 17.15
Problem 17.18
Problem 17.19
Problem 18.17
Problem 18.18
Problem 18.19
Problem 19.21
Problem 19.23
Problem 19.30
Problem 19.32
Problem 19.34
Problem 19.35
Problem 20.43
Problem 20.44
Problem 20.47
Problem 20.51
Problem 20.55
Problem 20.60
Problem 20.72
Problem 21.5
Problem 21.6
Problem 21.8
Problem 21.9
Problem 22.1
Problem 22.2
Problem 22.58
Problem 22.60
Problem 22.62
Problem 22.72
Problem 22.78
Problem 22.91
Problem 22.97
Problem 23.26
Problem 23.27
Problem 23.28
Problem 23.29
Problem 23.33
Problem 23.34
Problem 23.37
Problem 23.42
Problem 23.43
Problem 24.22
Problem 24.23
Problem 24.24
Problem 24.27
Problem 24.32
Problem 24.37
Problem 24.42
Problem 24.44
Problem 24.46
Problem 25.49
Problem 25.50
Problem 25.51
Problem 25.69
Problem 25.70
Problem 25.75
Problem 25.82
Problem 25.83
Problem 25.87
Problem 25.92
Problem 26.27
Problem 26.28
Problem 26.30
Problem 27.29
Problem 27.30
Problem 27.31
Problem 27.35
Problem 27.38
Problem 27.39
Problem 27.40
Problem 27.45
Problem 28.11
Problem 28.12
Problem 28.13
Problem 28.14
Problem 28.15
Problem 30.9
Problem 30.11
Problem 30.12
Problem 30.13
Problem 30.14
Problem 31.9
Problem 31.12
Problem 31.17
Problem 31.21
Problem 32.2
Problem 32.3
Problem 32.4
Problem 32.5
Problem 32.8
Problem 32.16
Problem 32.18
Problem 32.21

Chapter 1

Fundamental Operations with Numbers

1.1 FOUR OPERATIONS

Four operations are fundamental in algebra, as in arithmetic. These are addition, subtraction, multiplication, and division

When two numbers a and b are added, their sum is indicated by $a + b$. Thus $3 + 2 = 5$.

When a number b is subtracted from a number a, the difference is indicated by $a - b$. Thus $6 - 2 = 4$.

Subtraction may be defined in terms of addition. That is, we may define $a - b$ to represent that number x such that x added to b yields a, or $x + b = a$. For example, $8 - 3$ is that number x which when added to 3 yields 8, i.e., $x + 3 = 8$; thus $8 - 3 = 5$.

The product of two numbers a and b is a number c such that $a \times b = c$. The operation of multiplication may be indicated by a cross, a dot or parentheses. Thus $5 \times 3 = 5 \cdot 3 = 5(3) = (5)(3) = 15$, where the factors are 5 and 3 and the product is 15. When letters are used, as in algebra, the notation $p \times q$ is usually avoided since $\times$ may be confused with a letter representing a number.

When a number a is divided by a number b, the quotient obtained is written

$$a \div b \text{ or } \frac{a}{b} \text{ or } a/b,$$

where a is called the dividend and b the divisor. The expression a/b is also called a fraction, having numerator a and denominator b.

Division by zero is not defined. See Problems 1.1(b) and (e).

Division may be defined in terms of multiplication. That is, we may consider a/b as that number x which upon multiplication by b yields a, or $bx = a$. For example, $6/3$ is that number x such that 3 multiplied by x yields 6, or $3x = 6$; thus $6/3 = 2$.

1.2 SYSTEM OF REAL NUMBERS

The system of real numbers as we know it today is a result of gradual progress, as the following indicates.

(1) *Natural numbers* 1, 2, 3, 4, . . . (three dots mean "and so on") used in counting are also known as the positive integers. If two such numbers are added or multiplied, the result is always a natural number.

(2) *Positive rational numbers* or positive fractions are the quotients of two positive integers, such as 2/3, 8/5, 121/17. The positive rational numbers include the set of natural numbers. Thus the rational number 3/1 is the natural number 3.

(3) *Positive irrational numbers* are numbers which are not rational, such as $\sqrt{2}$, π.

(4) *Zero*, written 0, arose in order to enlarge the number system so as to permit such operations as $6 - 6$ or $10 - 10$. Zero has the property that any number multiplied by zero is zero. Zero divided by any number $\neq 0$ (i.e., not equal to zero) is zero.

(5) *Negative* integers, negative rational numbers and negative irrational numbers such as -3, $-2/3$, and $-\sqrt{2}$, arose in order to enlarge the number system so as to permit such operations as $2 - 8$, $\pi - 3\pi$ or $2 - 2\sqrt{2}$.

When no sign is placed before a number, a plus sign is understood. Thus 5 is +5, $\sqrt{2}$ is $+\sqrt{2}$. Zero is considered a rational number without sign.

The real number system consists of the collection of positive and negative rational and irrational numbers and zero.

Note. The word "real" is used in contradiction to still other numbers involving $\sqrt{-1}$, which will be taken up later and which are known as *imaginary*, although they are very useful in mathematics and the sciences. Unless otherwise specified we shall deal with real numbers.

1.3 GRAPHICAL REPRESENTATION OF REAL NUMBERS

It is often useful to represent real numbers by points on a line. To do this, we choose a point on the line to represent the real number zero and call this point the origin. The positive integers +1, +2, +3, . . . are then associated with points on the line at distances 1, 2, 3, . . . units respectively to the *right* of the origin (see Fig. 1-1), while the negative integers −1, −2, −3, . . . are associated with points on the line at distances 1, 2, 3, . . . units respectively to the *left* of the origin.

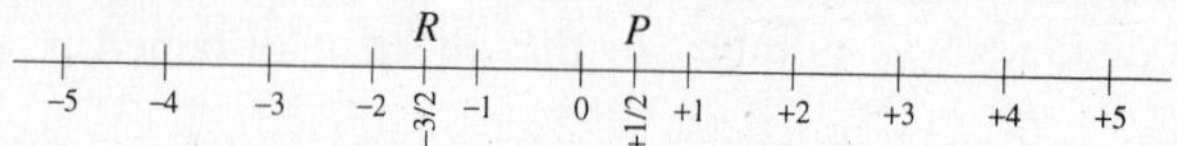

Fig. 1-1

The rational number 1/2 is represented on this scale by a point P halfway between 0 and +1. The negative number −3/2 or $-1\frac{1}{2}$ is represented by a point R $1\frac{1}{2}$ units to the left of the origin.

It can be proved that corresponding to each real number there is one and only one point on the line; and conversely, to every point on the line there corresponds one and only one real number.

The position of real numbers on a line establishes an order to the real number system. If a point A lies to the right of another point B on the line we say that the number corresponding to A is *greater* or *larger* than the number corresponding to B, or that the number corresponding to B is *less* or *smaller* than the number corresponding to A. The symbols for "greater than" and "less than" are $>$ and $<$ respectively. These symbols are called "inequality signs."

Thus since 5 is to the right of 3, 5 is greater than 3 or $5>3$; we may also say 3 is less than 5 and write $3<5$. Similarly, since −6 is to the left of −4, −6 is smaller than −4, i.e., $-6<-4$; we may also write $-4>-6$.

By the absolute value or numerical value of a number is meant the distance of the number from the origin on a number line. Absolute value is indicated by two vertical lines surrounding the number. Thus $|-6| = 6$, $|+4| = 4$, $|-3/4| = 3/4$.

1.4 PROPERTIES OF ADDITION AND MULTIPLICATION OF REAL NUMBERS

(1) *Commutative property for addition* The order of addition of two numbers does not affect the result.

Thus $$a+b=b+a, \qquad 5+3=3+5=8.$$

(2) *Associative property for addition* The terms of a sum may be grouped in any manner without affecting the result.

$$a+b+c=a+(b+c)=(a+b)+c, \qquad 3+4+1=3+(4+1)=(3+4)+1=8$$

(3) *Commutative property for multiplication* The order of the factors of a product does not affect the result.

$$a \cdot b = b \cdot a, \qquad 2 \cdot 5 = 5 \cdot 2 = 10$$

(4) *Associative property for multiplication* The factors of a product may be grouped in any manner without affecting the result.

$$abc = a(bc) = (ab)c, \qquad 3 \cdot 4 \cdot 6 = 3(4 \cdot 6) = (3 \cdot 4)6 = 72$$

(5) *Distributive property for multiplication over addition* The product of a number a by the sum of two numbers $(b + c)$ is equal to the sum of the products ab and ac.

$$a(b + c) = ab + ac, \qquad 4(3 + 2) = 4 \cdot 3 + 4 \cdot 2 = 20$$

Extensions of these laws may be made. Thus we may add the numbers a, b, c, d, e by grouping in any order, as $(a + b) + c + (d + e)$, $a + (b + c) + (d + e)$, etc. Similarly, in multiplication we may write $(ab)c(de)$ or $a(bc)(de)$, the result being independent of order or grouping.

1.5 RULES OF SIGNS

(1) To add two numbers with like signs, add their absolute values and prefix the common sign. Thus $3 + 4 = 7$, $(-3) + (-4) = -7$.

(2) To add two numbers with unlike signs, find the difference between their absolute values and prefix the sign of the number with greater absolute value.

EXAMPLES 1.1. $17 + (-8) = 9, \quad (-6) + 4 = -2, \quad (-18) + 15 = -3$

(3) To subtract one number b from another number a, change the operation to addition and replace b by its opposite, $-b$.

EXAMPLES 1.2. $12 - (7) = 12 + (-7) = 5, \quad (-9) - (4) = -9 + (-4) = -13, \quad 2 - (-8) = 2 + 8 = 10$

(4) To multiply (or divide) two numbers having like signs, mutiply (or divide) their absolute values and prefix a plus sign (or no sign).

EXAMPLES 1.3. $(5)(3) = 15, \quad (-5)(-3) = 15, \quad \dfrac{-6}{-3} = 2$

(5) To multiply (or divide) two numbers having unlike signs, multiply (or divide) their absolute values and prefix a minus sign.

EXAMPLES 1.4. $(-3)(6) = -18, \quad (3)(-6) = -18, \quad \dfrac{-12}{4} = -3$

1.6 EXPONENTS AND POWERS

When a number a is multiplied by itself n times, the product $a \cdot a \cdot a \cdots a$ (n times) is indicated by the symbol a^n which is referred to as "the nth power of a" or "a to the nth power" or "a to the nth."

EXAMPLES 1.5. $2 \cdot 2 \cdot 2 \cdot 2 \cdot 2 = 2^5 = 32, \quad (-5)^3 = (-5)(-5)(-5) = -125$

$2 \cdot x \cdot x \cdot x = 2x^3, \quad a \cdot a \cdot a \cdot b \cdot b = a^3b^2, \quad (a - b)(a - b)(a - b) = (a - b)^3$

In a^n, the number a is called the *base* and the positive integer n is the *exponent*.

If p and q are positive integers, then the following are laws of exponents.

(1) $a^p \cdot a^q = a^{p+q}$ Thus: $2^3 \cdot 2^4 = 2^{3+4} = 2^7$

(2) $\dfrac{a^p}{a^q} = a^{p-q} = \dfrac{1}{a^{q-p}}$ if $a \neq 0$ $\qquad \dfrac{3^5}{3^2} = 3^{5-2} = 3^3, \quad \dfrac{3^4}{3^6} = \dfrac{1}{3^{6-4}} = \dfrac{1}{3^2}$

(3) $(a^p)^q = a^{pq}$ $\qquad (4^2)^3 = 4^6, \quad (3^4)^2 = 3^8$

(4) $(ab)^p = a^p b^p, \quad \left(\dfrac{a}{b}\right)^p = \dfrac{a^p}{b^p}$ if $b \neq 0$ $\qquad (4 \cdot 5)^2 = 4^2 \cdot 5^2, \quad \left(\dfrac{5}{2}\right)^3 = \dfrac{5^3}{2^3}$

1.7 OPERATIONS WITH FRACTIONS

Operations with fractions may be performed according to the following rules.

(1) The value of a fraction remains the same if its numerator and denominator are both multiplied or divided by the same number provided the number is not zero.

EXAMPLES 1.6. $\dfrac{3}{4} = \dfrac{3 \cdot 2}{4 \cdot 2} = \dfrac{6}{8}, \quad \dfrac{15}{18} = \dfrac{15 \div 3}{18 \div 3} = \dfrac{5}{6}$

(2) Changing the sign of the numerator or denominator of a fraction changes the sign of the fraction.

EXAMPLE 1.7. $\dfrac{-3}{5} = -\dfrac{3}{5} = \dfrac{3}{-5}$

(3) Adding two fractions with a common denominator yields a fraction whose numerator is the sum of the numerators of the given fractions and whose denominator is the common denominator.

EXAMPLE 1.8. $\dfrac{3}{5} + \dfrac{4}{5} = \dfrac{3+4}{5} = \dfrac{7}{5}$

(4) The sum or difference of two fractions having different denominators may be found by writing the fractions with a common denominator.

EXAMPLE 1.9. $\dfrac{1}{4} + \dfrac{2}{3} = \dfrac{3}{12} + \dfrac{8}{12} = \dfrac{11}{12}$

(5) The product of two fractions is a fraction whose numerator is the product of the numerators of the given fractions and whose denominator is the product of the denominators of the fractions.

EXAMPLES 1.10. $\dfrac{2}{3} \cdot \dfrac{4}{5} = \dfrac{2 \cdot 4}{3 \cdot 5} = \dfrac{8}{15}, \quad \dfrac{3}{4} \cdot \dfrac{8}{9} = \dfrac{3 \cdot 8}{4 \cdot 9} = \dfrac{24}{36} = \dfrac{2}{3}$

(6) The reciprocal of a fraction is a fraction whose numerator is the denominator of the given fraction and whose denominator is the numerator of the given fraction. Thus the reciprocal of 3 (i.e., 3/1) is 1/3. Similarly the reciprocals of 5/8 and $-4/3$ are 8/5 and $3/-4$ or $-3/4$, respectively.

(7) To divide two fractions, multiply the first by the reciprocal of the second.

EXAMPLES 1.11. $\frac{a}{b} \div \frac{c}{d} = \frac{a}{b} \cdot \frac{d}{c} = \frac{ad}{bc}$, $\quad \frac{2}{3} \div \frac{4}{5} = \frac{2}{3} \cdot \frac{5}{4} = \frac{10}{12} = \frac{5}{6}$

This result may be established as follows:

$$\frac{a}{b} \div \frac{c}{d} = \frac{a/b}{c/d} = \frac{a/b \cdot bd}{c/d \cdot bd} = \frac{ad}{bc}.$$

Solved Problems

1.1 Write the sum S, difference D, product P, and quotient Q of each of the following pairs of numbers: (*a*) 48, 12; (*b*) 8, 0; (*c*) 0, 12; (*d*) 10, 20; (*e*) 0, 0.

SOLUTION

(*a*) $S = 48 + 12 = 60$, $D = 48 - 12 = 36$, $P = 48(12) = 576$, $Q = 48 \div 12 = \frac{48}{12} = 4$

(*b*) $S = 8 + 0 = 8$, $D = 8 - 0 = 8$, $P = 8(0) = 0$, $Q = 8 \div 0$ or $8/0$.
But by definition $8/0$ is that number x (if it exists) such that $x(0) = 8$. Clearly there is no such number, since any number multiplied by 0 must yield 0.

(*c*) $S = 0 + 12 = 12$, $D = 0 - 12 = -12$, $P = 0(12) = 0$, $Q = \frac{0}{12} = 0$

(*d*) $S = 10 + 20 = 30$, $D = 10 - 20 = -10$, $P = 10(20) = 200$, $Q = 10 \div 20 = \frac{10}{20} = \frac{1}{2}$

(*e*) $S = 0 + 0 = 0$, $D = 0 - 0 = 0$, $P = 0(0) = 0$. $Q = 0 \div 0$ or $0/0$ is by definition that number x (if it exists) such that $x(0) = 0$. Since this is true for *all* numbers x there is no one number which $0/0$ represents.

From (*b*) and (*e*) it is seen that division by zero is an undefined operation.

1.2 Perform each of the indicated operations.

(*a*) $42 + 23$, $23 + 42$

(*b*) $27 + (48 + 12)$, $(27 + 48) + 12$

(*c*) $125 - (38 + 27)$

(*d*) $6 \cdot 8$, $8 \cdot 6$

(*e*) $4(7 \cdot 6)$, $(4 \cdot 7)6$

(*f*) $35 \cdot 28$

(*g*) $756 \div 21$

(*h*) $\frac{(40 + 21)(72 - 38)}{(32 - 15)}$

(*i*) $72 \div 24 + 64 \div 16$

(*j*) $4 \div 2 + 6 \div 3 - 2 \div 2 + 3 \cdot 4$

(*k*) $128 \div (2 \cdot 4)$, $(128 \div 2) \cdot 4$

SOLUTION

(*a*) $42 + 23 = 65$, $23 + 42 = 65$. Thus $42 + 23 = 23 + 42$.
This illustrates the commutative law for addition.

(*b*) $27 + (48 + 12) = 27 + 60 = 87$, $(27 + 48) + 12 = 75 + 12 = 87$. Thus $27 + (48 + 12) = (27 + 48) + 12$.
This illustrates the associative law for addition.

(*c*) $125 - (38 + 27) = 125 - 65 = 60$

(*d*) $6 \cdot 8 = 48$, $8 \cdot 6 = 48$. Thus $6 \cdot 8 = 8 \cdot 6$, illustrating the commutative law for multiplication.

(*e*) $4(7 \cdot 6) = 4(42) = 168$, $(4 \cdot 7)6 = (28)6 = 168$. Thus $4(7 \cdot 6) = (4 \cdot 7)6$.
This illustrates the associative law for multiplication.

(*f*) $(35)(28) = 35(20 + 8) = 35(20) + 35(8) = 700 + 280 = 980$ by the distributive law for multiplication.

(*g*) $\frac{756}{21} = 36$ Check: $21 \cdot 36 = 756$

(*h*) $\frac{(40+21)(72-38)}{(32-15)} = \frac{(61)(34)}{17} = \frac{61 \cdot \overset{2}{\cancel{34}}}{\underset{1}{\cancel{17}}} = 61 \cdot 2 = 122$

(*i*) Computations in arithmetic, by convention, obey the following rule: Operations of multiplication and division precede operations of addition and subtraction.
Thus $72 \div 24 + 64 \div 16 = 3 + 4 = 7$.

(*j*) The rule of (*i*) is applied here. Thus $4 \div 2 + 6 \div 3 - 2 \div 2 + 3 \cdot 4 = 2 + 2 - 1 + 12 = 15$.

(*k*) $128 \div (2 \cdot 4) = 128 \div 8 = 16$, $(128 \div 2) \cdot 4 = 64 \cdot 4 = 256$
Hence if one wrote $128 \div 2 \cdot 4$ without parentheses we would do the operations of multiplication and division in the order they occur from left to right, so $128 \div 2 \cdot 4 = 64 \cdot 4 = 256$.

1.3 Classify each of the following numbers according to the categories: real number, positive integer, negative integer, rational number, irrational number, none of the foregoing.

$$-5,\ 3/5,\ 3\pi,\ 2,\ -1/4,\ 6.3,\ 0,\ \sqrt{5},\ \sqrt{-1},\ 0.3782,\ \sqrt{4},\ -18/7$$

SOLUTION

If the number belongs to one or more categories this is indicated by a check mark.

	Real number	Positive integer	Negative integer	Rational number	Irrational number	None of foregoing
-5	√		√	√		
$3/5$	√			√		
3π	√				√	
2	√	√		√		
$-1/4$	√			√		
6.3	√			√		
0	√			√		
$\sqrt{5}$	√				√	
$\sqrt{-1}$						√
0.3782	√			√		
$\sqrt{4}$	√	√		√		
$-18/7$	√			√		

1.4 Represent (approximately) by a point on a graphical scale each of the real numbers in Problem 1.3.

Note: 3π is approximately $3(3.14) = 9.42$, so that the corresponding point is between $+9$ and $+10$ as indicated. $\sqrt{5}$ is between 2 and 3, its value to three decimal places being 2.236.

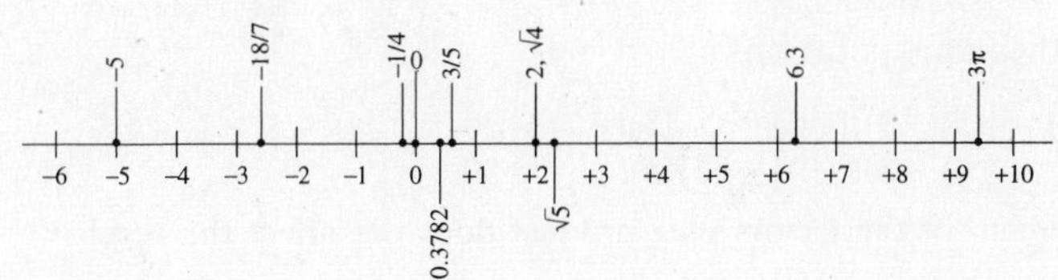

1.5 Place an appropriate inequality symbol (< or >) between each pair of real numbers.

(*a*) 2, 5 (*b*) 0, 2 (*c*) 3, −1 (*d*) −4, +2 (*e*) −4, −3 (*f*) π, 3 (*g*) $\sqrt{7}$, 3 (*h*) $-\sqrt{2}$, −1 (*i*) −3/5, −1/2

SOLUTION

(*a*) $2<5$ (or $5>2$), i.e., 2 is *less than* 5 (or 5 is *greater than* 2)
(*b*) $0<2$ (or $2>0$)
(*c*) $3>-1$ (or $-1<3$)
(*d*) $-4<+2$ (or $+2>-4$)
(*e*) $-4<-3$ (or $-3>-4$)
(*f*) $\pi>3$ (or $3<\pi$)
(*g*) $3>\sqrt{7}$ (or $\sqrt{7}<3$)
(*h*) $-\sqrt{2}<-1$ $(-1>-\sqrt{2})$
(*i*) $-3/5<-1/2$ since $-.6<-.5$

1.6 Arrange each of the following groups of real numbers in ascending order of magnitude.

(*a*) -3, $22/7$, $\sqrt{5}$, -3.2, 0 (*b*) $-\sqrt{2}$, $-\sqrt{3}$, -1.6, $-3/2$.

SOLUTION

(*a*) $-3.2<-3<0<\sqrt{5}<22/7$ (*b*) $-\sqrt{3}<-1.6<-3/2<-\sqrt{2}$

1.7 Write the absolute value of each of the following real numbers.

$$-1,\ +3,\ 2/5,\ -\sqrt{2},\ -3.14,\ 2.83,\ -3/8,\ -\pi,\ +5/7$$

SOLUTION

We may write the absolute values of these numbers as

$$|-1|,\ |+3|,\ |2/5|,\ |-\sqrt{2}|,\ |-3.14|,\ |2.83|,\ |-3/8|,\ |-\pi|,\ |+5/7|$$

which in turn may be written 1, 3, 2/5, $\sqrt{2}$, 3.14, 2.83, 3/8, π, 5/7 respectively.

1.8 The following illustrate addition and subtraction of real numbers.

(*a*) $(-3)+(-8)=-11$
(*b*) $(-2)+3=1$
(*c*) $(-6)+3=-3$
(*d*) $-2+5=3$
(*e*) $-15+8=-7$
(*f*) $(-32)+48+(-10)=6$
(*g*) $50-23-27=0$
(*h*) $-3-(-4)=-3+4=1$
(*i*) $-(-14)+(-2)=14-2=12$

1.9 Write the sum S, difference D, product P, and quotient Q of each of the following pairs of real numbers: (*a*) −2, 2; (*b*) −3, 6; (*c*) 0, −5; (*d*) −5, 0.

SOLUTION

(*a*) $S=-2+2=0$, $D=(-2)-2=-4$, $P=(-2)(2)=-4$, $Q=-2/2=-1$
(*b*) $S=(-3)+6=3$, $D=(-3)-6=-9$, $P=(-3)(6)=-18$, $Q=-3/6=-1/2$
(*c*) $S=0+(-5)=-5$, $D=0-(-5)=5$, $P=(0)(-5)=0$, $Q=0/-5=0$
(*d*) $S=(-5)+0=-5$, $D=(-5)-0=-5$, $P=(-5)(0)=0$, $Q=-5/0$ (an undefined operation, so it is not a number)

1.10 Perform the indicated operations.

(*a*) $(5)(-3)(-2) = [(5)(-3)](-2) = (-15)(-2) = 30$
$= (5)[(-3)(-2)] = (5)(6) \quad = 30$

The arrangement of the factors of a product does not affect the result.

(*b*) $8(-3)(10) = -240$

(*c*) $\dfrac{8(-2)}{-4} + \dfrac{(-4)(-2)}{2} = \dfrac{-16}{-4} + \dfrac{8}{2} = 4 + 4 = 8$

(*d*) $\dfrac{12(-40)(-12)}{5(-3) - 3(-3)} = \dfrac{12(-40)(-12)}{-15 - (-9)} = \dfrac{12(-40)(-12)}{-6} = -960$

1.11 Evaluate the following.

(*a*) $2^3 = 2 \cdot 2 \cdot 2 = 8$

(*b*) $5(3)^2 = 5 \cdot 3 \cdot 3 = 45$

(*c*) $2^4 \cdot 2^6 = 2^{4+6} = 2^{10} = 1024$

(*d*) $2^5 \cdot 5^2 = (32)(25) = 800$

(*e*) $\dfrac{3^4 \cdot 3^3}{3^2} = \dfrac{3^7}{3^2} = 3^{7-2} = 3^5 = 243$

(*f*) $\dfrac{5^2 \cdot 5^3}{5^7} = \dfrac{5^5}{5^7} = \dfrac{1}{5^{7-5}} = \dfrac{1}{5^2} = \dfrac{1}{25}$

(*g*) $(2^3)^2 = 2^{3 \cdot 2} = 2^6 = 64$

(*h*) $\left(\dfrac{2}{3}\right)^4 = \dfrac{2^4}{3^4} = \dfrac{16}{81}$

(*i*) $\dfrac{(3^4)^3 \cdot (3^2)^4}{(-3)^{15} \cdot 3^4} = \dfrac{3^{12} \cdot 3^8}{-3^{15} \cdot 3^4} = -\dfrac{3^{20}}{3^{19}} = -3^1 = -3$

(*j*) $\dfrac{3^8}{3^5} - \dfrac{4^2 \cdot 2^4}{2^6} + 3(-2)^3 = 3^3 - \dfrac{4^2}{2^2} + 3(-8) = 27 - 4 - 24 = -1$

1.12 Write each of the following fractions as an equivalent fraction having the indicated denominator.

(*a*) 1/3; 6 (*b*) 3/4; 20 (*c*) 5/8; 48 (*d*) −3/7; 63 (*e*) −12/5; 75

SOLUTION

(*a*) To obtain the denominator 6, multiply numerator and denominator of the fraction 1/3 by 2.

Then $\dfrac{1}{3} = \dfrac{1}{3} \cdot \dfrac{2}{2} = \dfrac{2}{6}$.

(*b*) $\dfrac{3}{4} = \dfrac{3 \cdot 5}{4 \cdot 5} = \dfrac{15}{20}$

(*c*) $\dfrac{5}{8} = \dfrac{5 \cdot 6}{8 \cdot 6} = \dfrac{30}{48}$

(*d*) $-\dfrac{3}{7} = -\dfrac{3 \cdot 9}{7\ 9} = -\dfrac{27}{63}$

(*e*) $-\dfrac{12}{5} = -\dfrac{12 \cdot 15}{5 \cdot 15} = -\dfrac{180}{75}$

1.13 Find the sum S, difference D, product P, and quotient Q of each of the following pairs of rational numbers: (*a*) 1/3, 1/6; (*b*) 2/5, 3/4; (*c*) −4/15, −11/24.

SOLUTION

(*a*) 1/3 may be written as the equivalent fraction 2/6.

$$S = \frac{1}{3} + \frac{1}{6} = \frac{2}{6} + \frac{1}{6} = \frac{3}{6} = \frac{1}{2} \qquad P = \left(\frac{1}{3}\right)\left(\frac{1}{6}\right) = \frac{1}{18}$$

$$D = \frac{1}{3} - \frac{1}{6} = \frac{2}{6} - \frac{1}{6} = \frac{1}{6} \qquad Q = \frac{1/3}{1/6} = \frac{1}{3}\cdot\frac{6}{1} = \frac{6}{3} = 2$$

(*b*) 2/5 and 3/4 may be expressed with denominator 20: 2/5 = 8/20, 3/4 = 15/20.

$$S = \frac{2}{5} + \frac{3}{4} = \frac{8}{20} + \frac{15}{20} = \frac{23}{20} \qquad P = \left(\frac{2}{5}\right)\left(\frac{3}{4}\right) = \frac{6}{20} = \frac{3}{10}$$

$$D = \frac{2}{5} - \frac{3}{4} = \frac{8}{20} - \frac{15}{20} = -\frac{7}{20} \qquad Q = \frac{2/5}{3/4} = \frac{2}{5}\cdot\frac{4}{3} = \frac{8}{15}$$

(*c*) −4/15 and −11/24 have a least common denominator 120: −4/15 = −32/120, −11/24 = −55/120.

$$S = \left(-\frac{4}{15}\right) + \left(-\frac{11}{24}\right) = -\frac{32}{120} - \frac{55}{120} = -\frac{87}{120} = -\frac{29}{40} \qquad P = \left(-\frac{4}{15}\right)\left(-\frac{11}{24}\right) = \frac{11}{90}$$

$$D = \left(-\frac{4}{15}\right) - \left(-\frac{11}{24}\right) = -\frac{32}{120} + \frac{55}{120} = \frac{23}{120} \qquad Q = \frac{-4/15}{-11/24} = \left(-\frac{4}{15}\right)\left(-\frac{24}{11}\right) = \frac{32}{55}$$

1.14 Evaluate the following expressions, given $x = 2$, $y = -3$, $z = 5$, $a = 1/2$, $b = -2/3$.

(*a*) $2x + y = 2(2) + (-3) = 4 - 3 = 1$

(*b*) $3x - 2y - 4z = 3(2) - 2(-3) - 4(5) = 6 + 6 - 20 = -8$

(*c*) $4x^2y = 4(2)^2(-3) = 4\cdot 4\cdot(-3) = -48$

(*d*) $\dfrac{x^3 + 4y}{2a - 3b} = \dfrac{2^3 + 4(-3)}{2(1/2) - 3(-2/3)} = \dfrac{8 - 12}{1 + 2} = -\dfrac{4}{3}$

(*e*) $\left(\dfrac{x}{y}\right)^2 - 3\left(\dfrac{b}{a}\right)^3 = \left(\dfrac{2}{-3}\right)^2 - 3\left(\dfrac{-2/3}{1/2}\right)^3 = \left(-\dfrac{2}{3}\right)^2 - 3\left(-\dfrac{4}{3}\right)^3 = \dfrac{4}{9} - 3\left(-\dfrac{64}{27}\right) = \dfrac{4}{9} + \dfrac{64}{9} = \dfrac{68}{9}$

Supplementary Problems

1.15 Write the sum S, difference D, product P, and quotient Q of each of the following pairs of numbers: (*a*) 54, 18; (*b*) 4, 0; (*c*) 0, 4; (*d*) 12, 24; (*e*) 50, 75.

1.16 Perform each of the indicated operations.

(*a*) $38 + 57,\ 57 + 38$

(*b*) $15 + (33 + 8),\ (15 + 33) + 8$

(*c*) $(23 + 64) - (41 + 12)$

(*d*) $12\cdot 8,\ 8\cdot 12$

(*e*) $6(4\cdot 8),\ (6\cdot 4)8$

(*f*) $42\cdot 68$

(*g*) $1296 \div 36$

(*h*) $\dfrac{(35 - 23)(28 + 17)}{43 - 25}$

(*i*) $45 \div 15 + 84 \div 12$

(*j*) $10 \div 5 - 4 \div 2 + 15 \div 3 + 2\cdot 5$

(*k*) $112 \div (4\cdot 7),\ (112 \div 4)\cdot 7$

(*l*) $\dfrac{15 + 3\cdot 2}{9 - 4 \div 2}$

1.17 Place an appropriate inequality symbol (< or >) between each of the following pairs of real numbers.

(*a*) 4, 3 (*b*) −2, 0 (*c*) −1, 2 (*d*) 3, −2 (*e*) −8, −7 (*f*) 1, $\sqrt{2}$ (*g*) −3, $-\sqrt{11}$ (*h*) −1/3, −2/5

1.18 Arrange each of the following groups of real numbers in ascending order of magnitude.

(*a*) $-\sqrt{3}$, -2, $\sqrt{6}$, -2.8, 4, 7/2 (*b*) 2π, -6, $\sqrt{8}$, -3π, 4.8, 19/3

1.19 Write the absolute value of each of the following real numbers: 2, $-3/2$, $-\sqrt{6}$, $+3.14$, 0, 5/3, $\sqrt{4}$, -0.001, $-\pi-1$.

1.20 Evaluate.

(*a*) $6+5$ (*d*) $6+(-4)$ (*g*) $(-18)+(-3)+22$ (*j*) $-(-16)-(-12)+(-5)-15$
(*b*) $(-4)+(-6)$ (*e*) $-8+4$ (*h*) $40-12+4$
(*c*) $(-4)+3$ (*f*) $-4+8$ (*i*) $-12-(-8)$

1.21 Write the sum S, difference D, product P, and quotient Q of each of the following pairs of real numbers: (*a*) 12, 4; (*b*) -6, -3; (*c*) -8, 4; (*d*) 0, -4; (*e*) 3, -2.

1.22 Perform the indicated operations.

(*a*) $(-3)(2)(-6)$ (*c*) $4(-1)(5)+(-3)(2)(-4)$ (*e*) $(-8)\div(-4)+(-3)(2)$
(*b*) $(6)(-8)(-2)$ (*d*) $\dfrac{(-4)(6)}{-3}+\dfrac{(-16)(-9)}{12}$ (*f*) $\dfrac{(-3)(8)(-2)}{(-4)(-6)-(2)(-12)}$

1.23 Evaluate.

(*a*) 3^3 (*f*) $\dfrac{3^4\cdot 3^8}{3^6\cdot 3^5}$ (*j*) $\dfrac{(-2)^3\cdot(2)^3}{3(2^2)^2}$
(*b*) $3(4)^2$
(*c*) $2^4\cdot 2^3$ (*g*) $\dfrac{7^5}{7^3\cdot 7^4}$ (*k*) $\dfrac{3(-3)^2+4(-2)^3}{2^3-3^2}$
(*d*) $4^2\cdot 3^2$
(*e*) $\dfrac{5^6\cdot 5^3}{5^5}$ (*h*) $(3^2)^3$ (*l*) $\dfrac{5^7}{5^4}+\dfrac{2^{10}}{8^2\cdot(-2)^3}-4(-3)^4$
(*i*) $(\frac{1}{2})^6\cdot 2^5$

1.24 Write each of the following fractions as an equivalent fraction having the indicated denominator.

(*a*) 2/5; 15 (*c*) 5/16; 64 (*e*) 11/12; 132
(*b*) $-4/7$; 28 (*d*) $-10/3$; 42 (*f*) 17/18; 90

1.25 Find the sum S, difference D, product P, and quotient Q of each of the following pairs of rational numbers: (*a*) 1/4, 3/8; (*b*) 1/3, 2/5; (*c*) -4, 2/3; (*d*) $-2/3$, $-3/2$.

1.26 Evaluate the following expressions, given $x=-2$, $y=4$, $z=1/3$, $a=-1$, $b=1/2$.

(*a*) $3x-2y+6z$ (*e*) $\dfrac{x^2y(x+y)}{3x+4y}$
(*b*) $2xy+6az$
(*c*) $4b^2x^3$ (*f*) $\left(\dfrac{y}{x}\right)^3-4\left(\dfrac{a}{b}\right)^2-\dfrac{xy}{z^2}$
(*d*) $\dfrac{3y^2-4x}{ax+by}$

ANSWERS TO SUPPLEMENTARY PROBLEMS

1.15 (*a*) $S=72$, $D=36$, $P=972$, $Q=3$ (*d*) $S=36$, $D=-12$, $P=288$, $Q=1/2$
(*b*) $S=4$, $D=4$, $P=0$, Q undefined (*e*) $S=125$, $D=-25$, $P=3750$, $Q=2/3$
(*c*) $S=4$, $D=-4$, $P=0$, $Q=0$

1.16 (*a*) 95, 95 (*c*) 34 (*e*) 192, 192 (*g*) 36 (*i*) 10 (*k*) 4, 196
(*b*) 56, 56 (*d*) 96, 96 (*f*) 2856 (*h*) 30 (*j*) 15 (*l*) 3

1.17 (*a*) $3<4$ or $4>3$ (*d*) $-2<3$ or $3>-2$ (*g*) $-\sqrt{11}<-3$ or $-3>-\sqrt{11}$
(*b*) $-2<0$ or $0>-2$ (*e*) $-8<-7$ or $-7>-8$ (*h*) $-2/5<-1/3$ or $-1/3>-2/5$
(*c*) $-1<2$ or $2>-1$ (*f*) $1<\sqrt{2}$ or $\sqrt{2}>1$

1.18 (*a*) $-2.8<-2<-\sqrt{3}<\sqrt{6}<7/2<4$ (*b*) $-3\pi<-6<\sqrt{8}<4.8<2\pi<19/3$

1.19 2, 3/2, $\sqrt{6}$, 3.14, 0, 5/3, $\sqrt{4}$, 0.001, $\pi+1$

1.20 (*a*) 11 (*c*) -1 (*e*) -4 (*g*) 1 (*i*) -4
(*b*) -10 (*d*) 2 (*f*) 4 (*h*) 32 (*j*) 8

1.21 (*a*) $S=16$, $D=8$, $P=48$, $Q=3$ (*d*) $S=-4$, $D=4$, $P=0$, $Q=0$
(*b*) $S=-9$, $D=-3$, $P=18$, $Q=2$ (*e*) $S=1$, $D=5$, $P=-6$, $Q=-3/2$
(*c*) $S=-4$, $D=-12$, $P=-32$, $Q=-2$

1.22 (*a*) 36 (*b*) 96 (*c*) 4 (*d*) 20 (*e*) -4 (*f*) 1

1.23 (*a*) 27 (*c*) 128 (*e*) $5^4=625$ (*g*) 1/49 (*i*) 1/2 (*k*) 5
(*b*) 48 (*d*) 144 (*f*) 3 (*h*) $3^6=729$ (*j*) $-4/3$ (*l*) -201

1.24 (*a*) 6/15 (*b*) $-16/28$ (*c*) 20/64 (*d*) $-140/42$ (*e*) 121/132 (*f*) 85/90

1.25 (*a*) $S=5/8$, $D=-1/8$, $P=3/32$, $Q=2/3$
(*b*) $S=11/15$, $D=-1/15$, $P=2/15$, $Q=5/6$
(*c*) $S=-10/3$, $D=-14/3$, $P=-8/3$, $Q=-6$
(*d*) $S=-13/6$, $D=5/6$, $P=1$, $Q=4/9$

1.26 (*a*) -12 (*b*) -18 (*c*) -8 (*d*) 14 (*e*) 16/5 (*f*) 48

Chapter 2

Fundamental Operations with Algebraic Expressions

2.1 ALGEBRAIC EXPRESSIONS

An algebraic expression is a combination of ordinary numbers and letters which represent numbers.

Thus $$3x^2 - 5xy + 2y^4, \qquad 2a^3b^5, \qquad \frac{5xy + 3z}{2a^3 - c^2}$$

are algebraic expressions.

A term consists of products and quotients of ordinary numbers and letters which represent numbers. Thus $6x^2y^3$, $5x/3y^4$, $-3x^7$ are terms.

However, $6x^2 + 7xy$ is an algebraic expression consisting of two terms.

A monomial is an algebraic expression consisting of only one term. Thus $7x^3y^4$, $3xyz^2$, $4x^2/y$ are monomials.

Because of this definition, monomials are sometimes simply called terms.

A binomial is an algebraic expression consisting of two terms. Thus $2x + 4y$, $3x^4 - 4xyz^3$ are binomials.

A trinomial is an algebraic expression consisting of three terms. Thus $3x^2 - 5x + 2$, $2x + 6y - 3z$, $x^3 - 3xy/z - 2x^3z^7$ are trinomials.

A multinomial is an algebraic expression consisting of more than one term. Thus $7x + 6y$, $3x^3 + 6x^2y - 7xy + 6$, $7x + 5x^2/y - 3x^3/16$ are multinomials.

2.2 TERMS

One factor of a term is said to be the coefficient of the rest of the term. Thus in the term $5x^3y^2$, $5x^3$ is the coefficient of y^2, $5y^2$ is the coefficient of x^3, and 5 is the coefficient of x^3y^2.

If a term consists of the product of an ordinary number and one or more letters, we call the number the numerical coefficient (or simply the coefficient) of the term. Thus in $-5x^3y^2$, -5 is the numerical coefficient or simply the coefficient.

Like terms, or similar terms, are terms which differ only in numerical coefficients. For example, $7xy$ and $-2xy$ are like terms; $3x^2y^4$ and $-\frac{1}{2}x^2y^4$ are like terms; however, $-2a^2b^3$ and $-3a^2b^7$ are unlike terms.

Two or more like terms in an algebraic expression may be combined into one term. Thus $7x^2y - 4x^2y + 2x^2y$ may be combined and written $5x^2y$.

A term is integral and rational in certain literals (letters which represent numbers) if the term consists of

(*a*) positive integer powers of the variables multiplied by a factor not containing any variable, or

(*b*) no variables at all.

For example, the terms $6x^2y^3$, $-5y^4$, 7, $-4x$, and $\sqrt{3}x^3y^6$ are integral and rational in the variables present. However, $3\sqrt{x}$ is not rational in x, $4/x$ is not integral in x.

A polynomial is a monomial or multinomial in which every term is integral and rational.

For example, $3x^2y^3 - 5x^4y + 2$, $2x^4 - 7x^3 + 3x^2 - 5x + 2$, $4xy + z$, and $3x^2$ are polynomials. However, $3x^2 - 4/x$ and $4\sqrt{y} + 3$ are not polynomials.

2.3 DEGREE

The degree of a monomial is the sum of all the exponents in the variables in the term. Thus the degree of $4x^3y^2z$ is $3 + 2 + 1 = 6$. The degree of a constant, such as 6, 0, $-\sqrt{3}$, or π, is zero.

The degree of a polynomial is the same as that of the term having highest degree and non-zero coefficient. Thus $7x^3y^2 - 4xz^5 + 2x^3y$ has terms of degree 5, 6, and 4 respectively; hence the degree of the polynomial is 6.

2.4 GROUPING

A symbol of grouping such as parentheses ( ), brackets [ ], or braces { } is often used to show that the terms contained in them are considered as a single quantity.

For example, the sum of two algebraic expressions $5x^2 - 3x + y$ and $2x - 3y$ may be written $(5x^2 - 3x + y) + (2x - 3y)$. The difference of these may be written $(5x^2 - 3x + y) - (2x - 3y)$, and their product $(5x^2 - 3x + y)(2x - 3y)$.

Removal of symbols of grouping is governed by the following laws.

(1) If a + sign precedes a symbol of grouping, this symbol of grouping may be removed without affecting the terms contained.

Thus $$(3x + 7y) + (4xy - 3x^3) = 3x + 7y + 4xy - 3x^3.$$

(2) If a − sign precedes a symbol of grouping, this symbol of grouping may be removed if each sign of the terms contained is changed.

Thus $$(3x + 7y) - (4xy - 3x^3) = 3x + 7y - 4xy + 3x^3.$$

(3) If more than one symbol of grouping is present, the inner ones are to be removed first.

Thus $$2x - \{4x^3 - (3x^2 - 5y)\} = 2x - \{4x^3 - 3x^2 + 5y\} = 2x - 4x^3 + 3x^2 - 5y.$$

2.5 COMPUTATION WITH ALGEBRAIC EXPRESSIONS

Addition of algebraic expressions is achieved by combining like terms. In order to accomplish this addition, the expressions may be arranged in rows with like terms in the same column; these columns are then added.

EXAMPLE 2.1. Add $7x + 3y^3 - 4xy$, $3x - 2y^3 + 7xy$, and $2xy - 5x - 6y^3$.

Write:	$7x$	$3y^3$	$-4xy$
	$3x$	$-2y^3$	$7xy$
	$-5x$	$-6y^3$	$2xy$
Addition:	$5x$	$-5y^3$	$5xy$.

Hence the result is $5x - 5y^3 + 5xy$.

Subtraction of two algebraic expressions is achieved by changing the sign of every term in the expression which is being subtracted (sometimes called the subtrahend) and adding this result to the other expression (called the minuend).

EXAMPLE 2.2. Subtract $2x^2 - 3xy + 5y^2$ from $10x^2 - 2xy - 3y^2$.

$$\begin{array}{rr} & 10x^2 - 2xy - 3y^2 \\ & 2x^2 - 3xy + 5y^2 \\ \hline \text{Subtraction:} & 8x^2 + xy - 8y^2 \end{array}$$

We may also write $(10x^2 - 2xy - 3y^2) - (2x^2 - 3xy + 5y^2) = 10x^2 - 2xy - 3y^2 - 2x^2 + 3xy - 5y^2 = 8x^2 + xy - 8y^2$.

Multiplication of algebraic expressions is achieved by multiplying the terms in the factors of the expressions.

(1) To multiply two or more monomials: Use the laws of exponents, the rules of signs, and the commutative and associative properties of multiplication.

EXAMPLE 2.3. Multiply $-3x^2y^3z$, $2x^4y$, and $-4xy^4z^2$.

Write $$(-3x^2y^3z)(2x^4y)(-4xy^4z^2).$$

Arranging according to the commutative and associative laws,

$$\{(-3)(2)(-4)\}\{(x^2)(x^4)(x)\}\{(y^3)(y)(y^4)\}\{(z)(z^2)\}. \qquad (1)$$

Combine using rules of signs and laws of exponents to obtain

$$24x^7y^8z^3.$$

Step (1) may be done mentally when experience is acquired.

(2) To multiply a polynomial by a monomial: Multiply each term of the polynomial by the monomial and combine results.

EXAMPLE 2.4. Multiply $3xy - 4x^3 + 2xy^2$ by $5x^2y^4$.

Write
$$\begin{aligned} &(5x^2y^4)(3xy - 4x^3 + 2xy^2) \\ &= (5x^2y^4)(3xy) + (5x^2y^4)(-4x^3) + (5x^2y^4)(2xy^2) \\ &= 15x^3y^5 - 20x^5y^4 + 10x^3y^6. \end{aligned}$$

(3) To multiply a polynomial by a polynomial: Multiply each of the terms of one polynomial by each of the terms of the other polynomial and combine results.

It is very often useful to arrange the polynomials according to ascending (or descending) powers of one of the letters involved.

EXAMPLE 2.5. Multiply $-3x + 9 + x^2$ by $3 - x$.
Arranging in descending powers of x,

$$\begin{array}{rl} & x^2 - 3x + 9 \qquad (2) \\ & -x + 3 \\ \hline \text{Multiplying (2) by } -x, & -x^3 + 3x^2 - 9x \\ \text{Multiplying (2) by 3,} & \qquad 3x^2 - 9x + 27 \\ \hline \text{Adding,} & -x^3 + 6x^2 - 18x + 27 \end{array}$$

Division of algebraic expressions is achieved by using the division laws of exponents.

(1) To divide a monomial by a monomial: Find the quotient of the numerical coefficients, find the quotients of the variables, and multiply these quotients.

EXAMPLE 2.6. Divide $24x^4y^2z^3$ by $-3x^3y^4z$.

Write $$\frac{24x^4y^2z^3}{-3x^3y^4z} = \left(\frac{24}{-3}\right)\left(\frac{x^4}{x^3}\right)\left(\frac{y^2}{y^4}\right)\left(\frac{z^3}{z}\right) = (-8)(x)\left(\frac{1}{y^2}\right)(z^2) = -\frac{8xz^2}{y^2}.$$

(2) To divide a polynomial by a polynomial:

(*a*) Arrange the terms of both polynomials in descending (or ascending) powers of one of the variables common to both polynomials.

(*b*) Divide the first term in the dividend by the first term in the divisor. This gives the first term of the quotient.

(*c*) Multiply the first term of the quotient by the divisor and subtract from the dividend, thus obtaining a new dividend.

(*d*) Use the dividend obtained in (*c*) to repeat steps (*b*) and (*c*) until a remainder is obtained which is either of degree lower than the degree of the divisor or zero.

(*e*) The result is written:

$$\frac{\text{dividend}}{\text{divisor}} = \text{quotient} + \frac{\text{remainder}}{\text{divisor}}.$$

EXAMPLE 2.7. Divide $x^2 + 2x^4 - 3x^3 + x - 2$ by $x^2 - 3x + 2$.

Write the polynomials in descending powers of x and arrange the work as follows.

$$\begin{array}{r|l}
 & 2x^2 + 3x + 6 \\
\hline
x^2 - 3x + 2 & 2x^4 - 3x^3 + x^2 + x - 2 \\
 & \underline{2x^4 - 6x^3 + 4x^2} \\
 & 3x^3 - 3x^2 + x - 2 \\
 & \underline{3x^3 - 9x^2 + 6x} \\
 & 6x^2 - 5x - 2 \\
 & \underline{6x^2 - 18x + 12} \\
 & 13x - 14
\end{array}$$

Hence $$\frac{2x^4 - 3x^3 + x^2 + x - 2}{x^2 - 3x + 2} = 2x^2 + 3x + 6 + \frac{13x - 14}{x^2 - 3x + 2}.$$

Solved Problems

2.1 Evaluate each of the following algebraic expressions, given that $x = 2$, $y = -1$, $z = 3$, $a = 0$, $b = 4$, $c = 1/3$.

(*a*) $2x^2 - 3yz = 2(2)^2 - 3(-1)(3) = 8 + 9 = 17$

(*b*) $2z^4 - 3z^3 + 4z^2 - 2z + 3 = 2(3)^4 - 3(3)^3 + 4(3)^2 - 2(3) + 3 = 162 - 81 + 36 - 6 + 3 = 114$

(*c*) $4a^2 - 3ab + 6c = 4(0)^2 - 3(0)(4) + 6(1/3) = 0 - 0 + 2 = 2$

(*d*) $\dfrac{5xy + 3z}{2a^3 - c^2} = \dfrac{5(2)(-1) + 3(3)}{2(0)^3 - (1/3)^2} = \dfrac{-10 + 9}{-1/9} = \dfrac{-1}{-1/9} = 9$

(*e*) $\dfrac{3x^2y}{z} - \dfrac{bc}{x+1} = \dfrac{3(2)^2(-1)}{3} - \dfrac{4(1/3)}{3} = -4 - 4/9 = -40/9$

(*f*) $\dfrac{4x^2y(z-1)}{a + b - 3c} = \dfrac{4(2)^2(-1)(3-1)}{0 + 4 - 3(1/3)} = \dfrac{4(4)(-1)(2)}{4 - 1} = -\dfrac{32}{3}$

2.2 Classify each of the following algebraic expressions according to the categories: term or monomial, binomial, trinomial, multinomial, polynomial.

(*a*) $x^3 + 3y^2z$ (*d*) $y + 3$ (*g*) $\sqrt{x^2 + y^2 + z^2}$
(*b*) $2x^2 - 5x + 3$ (*e*) $4z^2 + 3z - 2\sqrt{z}$ (*h*) $\sqrt{y} + \sqrt{z}$
(*c*) $4x^2y/z$ (*f*) $5x^3 + 4/y$ (*i*) $a^3 + b^3 + c^3 - 3abc$

SOLUTION

If the expression belongs to one or more categories, this is indicated by a check mark.

	Term or monomial	Binomial	Trinomial	Multinomial	Polynomial
$x^3 + 3y^2z$		√		√	√
$2x^2 - 5x + 3$			√	√	√
$4x^2y/z$	√				
$y + 3$		√		√	√
$4z^2 + 3z - 2\sqrt{z}$			√	√	
$5x^3 + 4/y$		√		√	
$\sqrt{x^2 + y^2 + z^2}$	√				
$\sqrt{y} + \sqrt{z}$		√		√	
$a^3 + b^3 + c^3 - 3abc$				√	√

2.3 Find the degree of each of the following polynomials.

(*a*) $2x^3y + 4xyz^4$. The degree of $2x^3y$ is 4 and that of $4xyz^4$ is 6; hence the polynomial is of degree 6.

(*b*) $x^2 + 3x^3 - 4$. The degree of x^2 is 2, of $3x^3$ is 3, and of -4 is 0; hence the degree of the polynomial is 3.

(*c*) $y^3 - 3y^2 + 4y - 2$ is of degree 3.

(*d*) $xz^3 + 3x^2z^2 - 4x^3z + x^4$. Each term is of degree 4; hence the polynomial is of degree 4.

(*e*) $x^2 - 10^5$ is of degree 2. (The degree of the constant 10^5 is zero.)

2.4 Remove the symbols of grouping in each of the following and simplify the resulting expressions by combining like terms.

(*a*) $3x^2 + (y^2 - 4z) - (2x - 3y + 4z) = 3x^2 + y^2 - 4z - 2x + 3y - 4z = 3x^2 + y^2 - 2x + 3y - 8z$

(*b*) $2(4xy + 3z) + 3(x - 2xy) - 4(z - 2xy) = 8xy + 6z + 3x - 6xy - 4z + 8xy = 10xy + 3x + 2z$

(*c*) $x - 3 - 2\{2 - 3(x - y)\} = x - 3 - 2\{2 - 3x + 3y\} = x - 3 - 4 + 6x - 6y = 7x - 6y - 7$

(*d*) $4x^2 - \{3x^2 - 2[y - 3(x^2 - y)] + 4\} = 4x^2 - \{3x^2 - 2[y - 3x^2 + 3y] + 4\}$
$= 4x^2 - \{3x^2 - 2y + 6x^2 - 6y + 4\} = 4x^2 - \{9x^2 - 8y + 4\}$
$= 4x^2 - 9x^2 + 8y - 4 = -5x^2 + 8y - 4$

2.5 Add the algebraic expressions in each of the following groups.

(*a*) $x^2+y^2-z^2+2xy-2yz$, $\quad y^2+z^2-x^2+2yz-2zx$, $\quad z^2+x^2-y^2+2zx-2xy$, $1-x^2-y^2-z^2$

SOLUTION

Arranging,
$$\begin{array}{l} x^2+y^2-z^2+2xy-2yz \\ -x^2+y^2+z^2 \qquad +2yz-2zx \\ x^2-y^2+z^2-2xy \qquad +2zx \\ -x^2-y^2-z^2 \qquad\qquad +1 \\ \hline \end{array}$$

Adding, $\quad 0+0+0+0+0+0+1 \quad$ The result of the addition is 1.

(*b*) $5x^3y-4ab+c^2$, $\quad 3c^2+2ab-3x^2y$, $\quad x^3y+x^2y-4c^2-3ab$, $\quad 4c^2-2x^2y+ab^2-3ab$

SOLUTION

Arranging,
$$\begin{array}{l} \qquad\quad 5x^3y-4ab+c^2 \\ -3x^2y \qquad\quad +2ab+3c^2 \\ x^2y+x^3y-3ab-4c^2 \\ -2x^2y \qquad\quad -3ab+4c^2+ab^2 \\ \hline \end{array}$$

Adding, $\quad -4x^2y+6x^3y-8ab+4c^2+ab^2$

2.6 Subtract the second of each of the following expressions from the first.

(*a*) $a-b+c-d$, $\quad c-a+d-b$.

SOLUTION

Write
$$\begin{array}{l} a-b+c-d \\ -a-b+c+d \\ \hline \end{array}$$

Subtracting, $\quad 2a+0+0-2d \quad$ The result is $2a-2d$.

Otherwise: $(a-b+c-d)-(c-a+d-b)=a-b+c-d-c+a-d+b=2a-2d$

(*b*) $4x^2y-3ab+2a^2-xy$, $\quad 4xy+ab^2-3a^2+2ab$.

SOLUTION

Write
$$\begin{array}{l} 4x^2y-3ab+2a^2-xy \\ \qquad\quad 2ab-3a^2+4xy+ab^2 \\ \hline \end{array}$$

Subtracting, $\quad 4x^2y-5ab+5a^2-5xy-ab^2$

Otherwise: $(4x^2y-3ab+2a^2-xy)-(4xy+ab^2-3a^2+2ab)$
$$= 4x^2y-3ab+2a^2-xy-4xy-ab^2+3a^2-2ab$$
$$= 4x^2y-5ab+5a^2-5xy-ab^2$$

2.7 In each of the following find the indicated product of the algebraic expressions.

(*a*) $(-2ab^3)(4a^2b^5)$
(*b*) $(-3x^2y)(4xy^2)(-2x^3y^4)$
(*c*) $(3ab^2)(2ab+b^2)$
(*d*) $(x^2-3xy+y^2)(4xy^2)$
(*e*) $(x^2-3x+9)(x+3)$
(*f*) $(x^4+x^3y+x^2y^2+xy^3+y^4)(x-y)$
(*g*) $(x^2-xy+y^2)(x^2+xy+y^2)$
(*h*) $(2x+y-z)(3x-z+y)$

SOLUTION

(*a*) $(-2ab^3)(4a^2b^5) = \{(-2)(4)\}\{(a)(a^2)\}\{(b^3)(b^5)\} = -8a^3b^8$

(*b*) $(-3x^2y)(4xy^2)(-2x^3y^4) = \{(-3)(4)(-2)\}\{(x^2)(x)(x^3)\}\{(y)(y^2)(y^4)\} = 24x^6y^7$

(*c*) $(3ab^2)(2ab + b^2) = (3ab^2)(2ab) + (3ab^2)(b^2) = 6a^2b^3 + 3ab^4$

(*d*) $(x^2 - 3xy + y^2)(4xy^2) = (x^2)(4xy^2) + (-3xy)(4xy^2) + (y^2)(4xy^2) = 4x^3y^2 - 12x^2y^3 + 4xy^4$

(*e*)
$$\begin{array}{r}
x^2 - 3x + 9 \\
x + 3 \\
\hline
x^3 - 3x^2 + 9x \\
3x^2 - 9x + 27 \\
\hline
x^3 + 0 + 0 + 27
\end{array}$$
Ans. $x^3 + 27$

(*f*)
$$\begin{array}{r}
x^4 + x^3y + x^2y^2 + xy^3 + y^4 \\
x - y \\
\hline
x^5 + x^4y + x^3y^2 + x^2y^3 + xy^4 \\
- x^4y - x^3y^2 - x^2y^3 - xy^4 - y^5 \\
\hline
x^5 + 0 + 0 + 0 + 0 - y^5
\end{array}$$
Ans. $x^5 - y^5$

(*g*)
$$\begin{array}{r}
x^2 - xy + y^2 \\
x^2 + xy + y^2 \\
\hline
x^4 - x^3y + x^2y^2 \\
x^3y - x^2y^2 + xy^3 \\
x^2y^2 - xy^3 + y^4 \\
\hline
x^4 + 0 + x^2y^2 + 0 + y^4
\end{array}$$
Ans. $x^4 + x^2y^2 + y^4$

(*h*)
$$\begin{array}{r}
2x + y - z \\
3x + y - z \\
\hline
6x^2 + 3xy - 3xz \\
2xy + y^2 - yz \\
- 2xz - yz + z^2 \\
\hline
6x^2 + 5xy - 5xz + y^2 - 2yz + z^2
\end{array}$$

2.8 Perform the indicated divisions.

(*a*) $\dfrac{24x^3y^2z}{4xyz^2} = \left(\dfrac{24}{4}\right)\left(\dfrac{x^3}{x}\right)\left(\dfrac{y^2}{y}\right)\left(\dfrac{z}{z^2}\right) = (6)(x^2)(y)\left(\dfrac{1}{z}\right) = \dfrac{6x^2y}{z}$

(*b*) $\dfrac{-16a^4b^6}{-8ab^2c} = \left(\dfrac{-16}{-8}\right)\left(\dfrac{a^4}{a}\right)\left(\dfrac{b^6}{b^2}\right)\left(\dfrac{1}{c}\right) = \dfrac{2a^3b^4}{c}$

(*c*) $\dfrac{3x^3y + 16xy^2 - 12x^4yz^4}{2x^2yz} = \left(\dfrac{3x^3y}{2x^2yz}\right) + \left(\dfrac{16xy^2}{2x^2yz}\right) + \left(\dfrac{-12x^4yz^4}{2x^2yz}\right) = \dfrac{3x}{2z} + \dfrac{8y}{xz} - 6x^2z^3$

(*d*) $\dfrac{4a^3b^2 + 16ab - 4a^2}{-2a^2b} = \left(\dfrac{4a^3b^2}{-2a^2b}\right) + \left(\dfrac{16ab}{-2a^2b}\right) + \left(\dfrac{-4a^2}{-2a^2b}\right) = -2ab - \dfrac{8}{a} + \dfrac{2}{b}$

(*e*) $\dfrac{2x^4 + 3x^3 - x^2 - 1}{x - 2}$

$$\begin{array}{r|l}
 & 2x^3 + 7x^2 + 13x + 26 \\
\hline
x - 2 & 2x^4 + 3x^3 - x^2 \qquad - 1 \\
 & 2x^4 - 4x^3 \\
\hline
 & 7x^3 - x^2 \qquad - 1 \\
 & 7x^3 - 14x^2 \\
\hline
 & 13x^2 \qquad - 1 \\
 & 13x^2 - 26x \\
\hline
 & 26x - 1 \\
 & 26x - 52 \\
\hline
 & 51
\end{array}$$

(*f*) $\dfrac{16y^4 - 1}{2y - 1}$

$$\begin{array}{r|l}
 & 8y^3 + 4y^2 + 2y + 1 \\
\hline
2y - 1 & 16y^4 \qquad - 1 \\
 & 16y^4 - 8y^3 \\
\hline
 & 8y^3 \qquad - 1 \\
 & 8y^3 - 4y^2 \\
\hline
 & 4y^2 \qquad - 1 \\
 & 4y^2 - 2y \\
\hline
 & 2y - 1 \\
 & 2y - 1 \\
\hline
 & 0
\end{array}$$

Thus $\dfrac{2x^4 + 3x^3 - x^2 - 1}{x - 2} = 2x^3 + 7x^2 + 13x + 26 + \dfrac{51}{x - 2}$ and $\dfrac{16y^4 - 1}{2y - 1} = 8y^3 + 4y^2 + 2y + 1.$

(*g*) $\dfrac{2x^6 + 5x^4 - x^3 + 1}{-x^2 + x + 1}.$

Arrange in descending powers of x.

$$
\begin{array}{r|l}
 & -2x^4 - 2x^3 - 9x^2 - 10x - 19 \\
\hline
-x^2 + x + 1 & 2x^6 \quad + 5x^4 - x^3 \qquad\qquad + 1 \\
 & 2x^6 - 2x^5 - 2x^4 \\
\hline
 & 2x^5 + 7x^4 - x^3 \qquad\qquad + 1 \\
 & 2x^5 - 2x^4 - 2x^3 \\
\hline
 & 9x^4 + x^3 \qquad\qquad + 1 \\
 & 9x^4 - 9x^3 - 9x^2 \\
\hline
 & 10x^3 + 9x^2 \qquad + 1 \\
 & 10x^3 - 10x^2 - 10x \\
\hline
 & 19x^2 + 10x + 1 \\
 & 19x^2 - 19x - 19 \\
\hline
 & 29x + 20
\end{array}
$$

Thus $\dfrac{2x^6 + 5x^4 - x^3 + 1}{-x^2 + x + 1} = -2x^4 - 2x^3 - 9x^2 - 10x - 19 + \dfrac{29x + 20}{-x^2 + x + 1}$.

(*h*) $\dfrac{x^4 - x^3y + x^2y^2 + 2x^2y - 2xy^2 + 2y^3}{x^2 - xy + y^2}$.

Arrange in descending powers of a letter, say x.

$$
\begin{array}{r|l}
 & x^2 + 2y \\
\hline
x^2 - xy + y^2 & x^4 - x^3y + x^2y^2 + 2x^2y - 2xy^2 + 2y^3 \\
 & x^4 - x^3y + x^2y^2 \\
\hline
 & 2x^2y - 2xy^2 + 2y^3 \\
 & 2x^2y - 2xy^2 + 2y^3 \\
\hline
 & 0
\end{array}
$$

Thus $\dfrac{x^4 - x^3y + x^2y^2 + 2x^2y - 2xy^2 + 2y^3}{x^2 - xy + y^2} = x^2 + 2y$.

2.9 Check the work in Problems 2.7(*h*) and 2.8(*g*) by using the values $x = 1$, $y = -1$, $z = 2$.

SOLUTION

From Problem 2.7(*h*), $(2x + y - z)(3x - z + y) = 6x^2 + 5xy - 5xz - 2yz + z^2 + y^2$.
Substitute $x = 1$, $y = -1$, $z = 2$ and obtain

$$[2(1) + (-1) - 2][3(1) - (2) - 1] = 6(1)^2 + 5(1)(-1) - 5(1)(2) - 2(-1)(2) + (2)^2 + (-1)^2$$

or

$$[-1][0] = 6 - 5 - 10 + 4 + 4 + 1, \text{ i.e. } 0 = 0.$$

From Problem 2.8(*g*),

$$\frac{2x^6 + 5x^4 - x^3 + 1}{-x^2 + x + 1} = -2x^4 - 2x^3 - 9x^2 - 10x - 19 + \frac{29x + 20}{-x^2 + x + 1}.$$

Put $x = 1$ and obtain

$$\frac{2 + 5 - 1 + 1}{-1 + 1 + 1} = -2 - 2 - 9 - 10 - 19 + \frac{29 + 20}{-1 + 1 + 1} \quad \text{or} \quad 7 = 7.$$

Although a check by substitution of numbers for variables is not conclusive, it can be used to indicate possible errors.

Supplementary Problems

2.10 Evaluate each algebraic expression, given that $x = -1$, $y = 3$, $z = 2$, $a = 1/2$, $b = -2/3$.

(a) $4x^3y^2 - 3xz^2$

(b) $(x-y)(y-z)(z-x)$

(c) $9ab^2 + 6ab - 4a^2$

(d) $\dfrac{xy^2 - 3z}{a+b}$

(e) $\dfrac{z(x+y)}{8a^2} - \dfrac{3ab}{y-x+1}$

(f) $\dfrac{(x-y)^2 + 2z}{ax+by}$

(g) $\dfrac{1}{x} + \dfrac{1}{y} + \dfrac{1}{z}$

(h) $\dfrac{(x-1)(y-1)(z-1)}{(a-1)(b-1)}$

2.11 Determine the degree of each of the following polynomials.

(a) $3x^4 - 2x^3 + x^2 - 5$

(b) $4xy^4 - 3x^3y^3$

(c) $x^5 + y^5 + z^5 - 5xyz$

(d) $\sqrt{3}xyz - 5$

(e) -10^3

(f) $y^2 - 3y^5 - y + 2y^3 - 4$

2.12 Remove the symbols of grouping and simplify the resulting expressions by combining like terms.

(a) $(x + 3y - z) - (2y - x + 3z) + (4z - 3x + 2y)$

(b) $3(x^2 - 2yz + y^2) - 4(x^2 - y^2 - 3yz) + x^2 + y^2$

(c) $3x + 4y + 3\{x - 2(y - x) - y\}$

(d) $3 - \{2x - [1 - (x + y)] + [x - 2y]\}$

2.13 Add the algebraic expressions in each of the following groups.

(a) $2x^2 + y^2 - x + y$, $3y^2 + x - x^2$, $x - 2y + x^2 - 4y^2$

(b) $a^2 - ab + 2bc + 3c^2$, $2ab + b^2 - 3bc - 4c^2$, $ab - 4bc + c^2 - a^2$, $a^2 + 2c^2 + 5bc - 2ab$

(c) $2a^2bc - 2acb^2 + 5c^2ab$, $4b^2ac + 4bca^2 - 7ac^2b$, $4abc^2 - 3a^2bc - 3ab^2c$, $b^2ac - abc^2 - 3a^2bc$

2.14 Subtract the second of each of the following expressions from the first.

(a) $3xy - 2yz + 4zx$, $3zx + yz - 2xy$

(b) $4x^2 + 3y^2 - 6x + 4y - 2$, $2x - y^2 + 3x^2 - 4y + 3$

(c) $r^3 - 3r^2s + 4rs^2 - s^3$, $2s^3 + 3s^2r - 2sr^2 - 3r^3$

2.15 Subtract $xy - 3yz + 4xz$ from twice the sum of the following expressions: $3xy - 4yz + 2xz$ and $3yz - 4zx - 2xy$.

2.16 Obtain the product of the algebraic expressions in each of the following groups.

(a) $4x^2y^5$, $-3x^3y^2$

(b) $3abc^2$, $-2a^3b^2c^4$, $6a^2b^2$

(c) $-4x^2y$, $3xy^2 - 4xy$

(d) $r^2s + 3rs^3 - 4rs + s^3$, $2r^2s^4$

(e) $y - 4$, $y + 3$

(f) $y^2 - 4y + 16$, $y + 4$

(g) $x^3 + x^2y + xy^2 + y^3$, $x - y$

(h) $x^2 + 4x + 8$, $x^2 - 4x + 8$

(i) $3r - s - t^2$, $2s + r + 3t^2$

(j) $3 - x - y$, $2x + y + 1$, $x - y$

2.17 Perform the indicated divisions.

(a) $\dfrac{-12x^4yz^3}{3x^2y^4z}$ (b) $\dfrac{-18r^3s^2t}{-4r^5st^2}$ (c) $\dfrac{4ab^3 - 3a^2bc + 12a^3b^2c^4}{-2ab^2c^3}$ (d) $\dfrac{4x^3 - 5x^2 + 3x - 2}{x+1}$

2.18 Perform the indicated divisions.

(a) $\dfrac{27s^3 - 64}{3s - 4}$ (b) $\dfrac{1 - x^2 + x^4}{1 - x}$ (c) $\dfrac{2y^3 + y^5 - 3y - 2}{y^2 - 3y + 1}$ (d) $\dfrac{4x^3y + 5x^2y^2 + x^4 + 2xy^3}{x^2 + 2y^2 + 3xy}$

2.19 Perform the indicated operations and check by using the values $x = 1$, $y = 2$.

(a) $(x^4 + x^2y^2 + y^4)(y^4 - x^2y^2 + x^4)$ (b) $\dfrac{x^4 + xy^3 + x^3y + 2x^2y^2 + y^4}{xy + x^2 + y^2}$

ANSWERS TO SUPPLEMENTARY PROBLEMS

2.10 (a) -24 (b) -12 (c) -1 (d) 90 (e) $11/5$ (f) -8 (g) $-1/6$ (h) $-24/5$

2.11 (a) 4 (b) 6 (c) 5 (d) 3 (e) 0 (f) 5

2.12 (a) $3y - x$ (b) $8y^2 + 6yz$ (c) $12x - 5y$ (d) $y - 4x + 4$

2.13 (a) $2x^2 + x - y$ (b) $a^2 + b^2 + 2c^2$ (c) abc^2

2.14 (a) $5xy - 3yz + zx$ (b) $x^2 + 4y^2 - 8x + 8y - 5$ (c) $4r^3 - r^2s + rs^2 - 3s^3$

2.15 $xy + yz - 8xz$

2.16 (a) $-12x^5y^7$ (f) $y^3 + 64$
(b) $-36a^6b^5c^6$ (g) $x^4 - y^4$
(c) $-12x^3y^3 + 16x^3y^2$ (h) $x^4 + 64$
(d) $2r^4s^5 + 6r^3s^7 - 8r^3s^5 + 2r^2s^7$ (i) $3r^2 + 5rs + 8rt^2 - 2s^2 - 5st^2 - 3t^4$
(e) $y^2 - y - 12$ (j) $y^3 - 2y^2 - 3y + 3x + 5x^2 - 3xy - 2x^3 - x^2y + 2xy^2$

2.17 (a) $-\dfrac{4x^2z^2}{y^3}$ (b) $\dfrac{9s}{2r^2t}$ (c) $-\dfrac{2b}{c^3} + \dfrac{3a}{2bc^2} - 6a^2c$ (d) $4x^2 - 9x + 12 + \dfrac{-14}{x+1}$

2.18 (a) $9s^2 + 12s + 16$ (b) $-x^3 - x^2 + \dfrac{1}{1-x}$ (c) $y^3 + 3y^2 + 10y + 27 + \dfrac{68y - 29}{y^2 - 3y + 1}$ (d) $x^2 + xy$

2.19 (a) $x^8 + x^4y^4 + y^8$. Check: $21(13) = 273$. (b) $x^2 + y^2$. Check: $35/7 = 5$.

Chapter 3

Properties of Numbers

3.1 SETS OF NUMBERS

The set of counting (or natural) numbers is the set of numbers: 1, 2, 3, 4, 5, ...

The set of whole numbers is the set of counting numbers and zero: 0, 1, 2, 3, 4, ...

The set of integers is the set of counting numbers, zero, and the opposites of the counting numbers: . . ., −5, −4, −3, −2, −1, 0, 1, 2, 3, 4, 5, ...

The set of real numbers is the set of all numbers that correspond to the points on a number line. The real numbers can be separated into two distinct subsets: the rational numbers and the irrational numbers.

The set of rational numbers is the set of real numbers that can be written in the form a/b, where a and b are integers and b is not zero. The rational numbers can be thought of as the set of integers and the common fractions. The numbers −4, 2/3, 50/7, $\sqrt{9}$, 10/5, −1/2, 0, 145, and 15/1 are examples of rational numbers.

The set of irrational numbers is the set of real numbers that are not rational numbers. The numbers $\sqrt{2}$, $\sqrt[3]{5}$, $\sqrt[4]{10}$, $\sqrt{3}+4$, $\sqrt[3]{6}-5$, and the mathematical constants π and e are examples of irrational numbers.

3.2 PROPERTIES

A set has closure under an operation if the result of performing the operation with two elements of the set is also an element of the set. The set X is closed under the operation * if for all elements a and b in set X, the result $a*b$ is in set X.

A set has an identity under an operation if there is an element in the set that when combined with each element in the set leaves that element unchanged. The set X has an identity under the operation * if there is an element j in set X such that $j*a = a*j = a$ for all elements a in set X.

A set has inverses under an operation if for each element of the set there is an element of the set such that when these two elements are combined using the operation, the result is the identity for the set under the operation. If a set does not have an identity under an operation, it cannot have the inverse property for the operation. If X is a set that has identity j under operation *, then it has inverses if for each element a in set X there is an element a' in set X such that $a*a' = j$ and $a'*a = j$.

Sets under an operation may also have the associative property and the commutative property, as described in Section 1.4. If there are two operations on the set then the set could have the distributive property, also described in Section 1.4.

EXAMPLE 3.1. Which properties are true for the counting numbers, whole numbers, integers, rational numbers, irrational numbers, and real numbers under the operation of addition?

+	Counting	Whole	Integers	Rational	Irrational	Real
Closure	Yes	Yes	Yes	Yes	No	Yes
Identity	No	Yes	Yes	Yes	No	Yes
Inverse	No	No	Yes	Yes	No	Yes
Associativity	Yes	Yes	Yes	Yes	Yes	Yes
Commutativity	Yes	Yes	Yes	Yes	Yes	Yes

3.3 ADDITIONAL PROPERTIES

There are some properties that sets of numbers have that do not depend on an operation to be true. Three such properties are order, density, and completeness.

A set of numbers has order if given two distinct elements in the set one element is greater than the other.

A set of numbers has density if between any two elements of the set there is another element of the set.

A set of numbers has completeness if the points using its elements as coordinates completely fill a line or plane.

EXAMPLE 3.2. Which properties are true for the counting numbers, whole numbers, integers, rational numbers, irrational numbers, and real numbers?

	Counting	Whole	Integers	Rational	Irrational	Real
Order	Yes	Yes	Yes	Yes	Yes	Yes
Density	No	No	No	Yes	Yes	Yes
Completeness	No	No	No	No	No	Yes

Solved Problems

3.1 Which of the properties closure, identity, and inverse does the set of even integers have under addition?

SOLUTION

Since even integers are of the form $2n$ where n is an integer, we let $2m$ and $2k$ be any two even integers. The sum of two even integers is $2m + 2k = 2(m + k)$. From Example 3.1 we know that $m + k$ is an integer, since m and k are integers. Thus, $2(m + k)$ is 2 times an integer and is even, so $2m + 2k$ is even. Therefore, the even integers are closed under addition.

Zero is an even integer since $2(0) = 0$. $2m + 0 = 2m + 2(0) = 2(m + 0) = 2m$. Thus, 0 is the identity for the even integers under addition.

For the even integer $2m$, the inverse is $-2m$. Since m is an integer, $-m$ is an integer. Thus, $-2m = 2(-m)$ is an even integer. Also, $2m + (-2m) = 2(m + (-m)) = 2(0) = 0$. Therefore, each even integer has an inverse.

3.2 Which of the properties closure, identity, inverse, associativity, and commutativity are true under multiplication for the sets counting numbers, whole numbers, integers, rational numbers, irrational numbers, and real numbers?

SOLUTION

	Counting	Whole	Integers	Rational	Irrational	Real
Closure	Yes	Yes	Yes	Yes	No	Yes
Identity	Yes	Yes	Yes	Yes	No	Yes
Inverse	No	No	No	Yes	No	Yes
Associativity	Yes	Yes	Yes	Yes	Yes	Yes
Commutativity	Yes	Yes	Yes	Yes	Yes	Yes

3.3 Which of the properties closure, identity, and inverse does the set of odd integers have under multiplication?

SOLUTION

Since the odd integers are of the form $2n + 1$ where n is an integer, we let $2m + 1$ and $2k + 1$ be any two odd integers. The product of two odd integers is represented by $(2m + 1)(2k + 1) = 4mk + 2m + 2k + 1 = 2(2mk + m + k) + 1$. Since the integers are closed under both addition and multiplication, $(2mk + m + k)$ is an integer and the product $(2m + 1)(2k + 1)$ is equal to 2 times an integer plus 1. Thus, the product is an odd integer. Therefore, the odd integers are closed under multiplication.

One is an odd integer, since $2(0) + 1 = 0 + 1 = 1$. Also $(2m + 1)(1) = (2m)(1) + (1)(1) = 2m + 1$. Thus, 1 is the identity for the odd integers under multiplication.

Seven is an odd integer, since $2(3) + 1 = 7$. Also, $7(1/7) = 1$, but $1/7$ is not an odd integer. Thus, 7 does not have an inverse under multiplication. Since there is at least one odd integer that does not have an inverse under multiplication, the set of odd integers under multiplication does not have the inverse property.

3.4 Does the set of even integers have the order, density, and completeness properties?

SOLUTION

Given two distinct even integers $2m$ and $2k$ where m and k are integers, we know that either $m > k$ or $k > m$. If $m > k$, then $2m > 2k$, but if $k > m$, then $2k > 2m$. Thus, the set of even integers has the order property, since for two distinct even integers $2m$ and $2k$ either $2m > 2k$ or $2k > 2m$.

The numbers $2m$ and $2m + 2$ are even integers. There is no even integer between $2m$ and $2m + 2$, since $2m + 2 = 2(m + 1)$ and there is no integer between m and $m + 1$. Thus, the even integers do not have the density property.

Between the two even integers 8 and 10 is the odd integer 9. Thus, the even integers do not represent the coordinates for all points on a number line. Therefore the even integers do not have the completeness property.

3.5 Let $K = \{-1, 1\}$. (*a*) Is K closed under multiplication? (*b*) Does K have an identity under multiplication? (*c*) Does K have inverses under multiplication?

SOLUTION

(*a*) $(1)(1) = 1$, $(-1)(-1) = 1$, $(1)(-1) = -1$, and $(-1)(1) = -1$. For all possible products of two elements in K, the result is in K. Thus, K is closed under multiplication.

(*b*) 1 is in K, $(1)(1) = 1$, and $(1)(-1) = -1$. Thus 1 is the identity for K under multiplication.

(*c*) Since $(1)(1) = 1$ and $(-1)(-1) = 1$, each element of K is its own inverse.

Supplementary Problems

3.6 Which of the properties closure, identity, and inverse does the set of even integers have under multiplication?

3.7 Which of the properties closure, identity, and inverse does the set of odd integers have under addition?

3.8 Does the set of odd integers have the order, density, and completeness properties?

3.9 Which of the properties closure, identity, inverse, associativity, and commutativity are true under subtraction for the sets of counting numbers, whole numbers, integers, rational numbers, irrational numbers, and real numbers?

3.10 Which of the properties closure, identity, inverse, associativity, and commutativity are true under non-zero division for the set of counting numbers, whole numbers, integers, rational numbers, irrational numbers, and real numbers?

3.11 Which of the properties closure, identity, inverse, associativity, and commutativity are true for the set of zero, $\{0\}$, under (*a*) addition, (*b*) subtraction, and (*c*) multiplication?

3.12 Which of the properties closure, identity, inverse, associativity, and commutativity are true for the set of one, $\{1\}$, under (*a*) addition, (*b*) subtraction, (*c*) multiplication, and (*d*) division?

ANSWERS TO SUPPLEMENTARY PROBLEMS

3.6 Closure: yes; identity: no; inverse: no.

3.7 Closure: no; identity: no; inverse: no.

3.8 Order: yes; density: no; completeness: no.

3.9

–	Counting	Whole	Integers	Rational	Irrational	Real
Closure	No	No	Yes	Yes	No	Yes
Identity	No	No	No	No	No	No
Inverse	No	No	No	No	No	No
Associativity	No	No	No	No	No	No
Commutativity	No	No	No	No	No	No

3.10

÷	Counting	Whole	Integers	Rational	Irrational	Real
Closure	No	No	No	Yes	No	Yes
Identity	No	No	No	No	No	No
Inverse	No	No	No	No	No	No
Associativity	No	No	No	No	No	No
Commutativity	No	No	No	No	No	No

3.11

	Closure	Identity	Inverse	Associativity	Commutativity
(*a*) +	Yes	Yes	Yes	Yes	Yes
(*b*) −	Yes	Yes	Yes	Yes	Yes
(*c*) ·	Yes	Yes	Yes	Yes	Yes

3.12

	Closure	Identity	Inverse	Associativity	Commutativity
(*a*) +	No	No	No	Yes	Yes
(*b*) −	No	No	No	No	Yes
(*c*) ·	Yes	Yes	Yes	Yes	Yes
(*d*) ÷	Yes	Yes	Yes	Yes	Yes

Chapter 4

Special Products

4.1 SPECIAL PRODUCTS

The following are some of the products which occur frequently in mathematics, and the student should become familiar with them as soon as possible. Proofs of these results may be obtained by performing the multiplications.

I. Product of a monomial and a binomial

$$a(c+d) = ac + ad$$

II. Product of the sum and the difference of two terms

$$(a+b)(a-b) = a^2 - b^2$$

III. Square of a binomial

$$(a+b)^2 = a^2 + 2ab + b^2$$
$$(a-b)^2 = a^2 - 2ab + b^2$$

IV. Product of two binomials

$$(x+a)(x+b) = x^2 + (a+b)x + ab$$
$$(ax+b)(cx+d) = acx^2 + (ad+bc)x + bd$$
$$(a+b)(c+d) = ac + bc + ad + bd$$

V. Cube of a binomial

$$(a+b)^3 = a^3 + 3a^2b + 3ab^2 + b^3$$
$$(a-b)^3 = a^3 - 3a^2b + 3ab^2 - b^3$$

VI. Square of a trinomial

$$(a+b+c)^2 = a^2 + b^2 + c^2 + 2ab + 2ac + 2bc$$

4.2 PRODUCTS YIELDING ANSWERS OF THE FORM $a^n \pm b^n$

It may be verified by multiplication that

$$(a-b)(a^2+ab+b^2) = a^3 - b^3$$
$$(a-b)(a^3+a^2b+ab^2+b^3) = a^4 - b^4$$
$$(a-b)(a^4+a^3b+a^2b^2+ab^3+b^4) = a^5 - b^5$$
$$(a-b)(a^5+a^4b+a^3b^2+a^2b^3+ab^4+b^5) = a^6 - b^6$$

etc., the rule being clear. These may be summarized by

VII. $$(a-b)(a^{n-1}+a^{n-2}b+a^{n-3}b^2+\cdots+ab^{n-2}+b^{n-1}) = a^n - b^n$$

where n is *any positive integer* $(1, 2, 3, 4, \ldots)$.

Similarly, it may be verified that

$$(a+b)(a^2-ab+b^2) = a^3 + b^3$$
$$(a+b)(a^4-a^3b+a^2b^2-ab^3+b^4) = a^5 + b^5$$
$$(a+b)(a^6-a^5b+a^4b^2-a^3b^3+a^2b^4-ab^5+b^6) = a^7 + b^7$$

etc., the rule being clear. These may be summarized by

VIII. $$(a+b)(a^{n-1}-a^{n-2}b+a^{n-3}b^2-\cdots-ab^{n-2}+b^{n-1})=a^n+b^n$$

where n is *any positive odd integer* $(1,3,5,7,\ldots)$.

Solved Problems

Find each of the following products.

4.1 (*a*) $3x(2x+3y)=(3x)(2x)+(3x)(3y)=6x^2+9xy$, using I with $a=3x$, $c=2x$, $d=3y$.

(*b*) $x^2y(3x^3-2y+4)=(x^2y)(3x^3)+(x^2y)(-2y)+(x^2y)(4)=3x^5y-2x^2y^2+4x^2y$

(*c*) $(3x^3y^2+2xy-5)(x^2y^3)=(3x^3y^2)(x^2y^3)+(2xy)(x^2y^3)+(-5)(x^2y^3)$
$=3x^5y^5+2x^3y^4-5x^2y^3$

(*d*) $(2x+3y)(2x-3y)=(2x)^2-(3y)^2=4x^2-9y^2$, using II with $a=2x$, $b=3y$.

(*e*) $(1-5x^3)(1+5x^3)=(1)^2-(5x^3)^2=1-25x^6$

(*f*) $(5x+x^3y^2)(5x-x^3y^2)=(5x)^2-(x^3y^2)^2=25x^2-x^6y^4$

(*g*) $(3x+5y)^2=(3x)^2+2(3x)(5y)+(5y)^2=9x^2+30xy+25y^2$, using III with $a=3x$, $b=5y$.

(*h*) $(x+2)^2=x^2+2(x)(2)+2^2=x^2+4x+4$

(*i*) $(7x^2-2xy)^2=(7x^2)^2-2(7x^2)(2xy)+(2xy)^2$
$=49x^4-28x^3y+4x^2y^2$, using III with $a=7x^2$, $b=2xy$.

(*j*) $(ax-2by)^2=(ax)^2-2(ax)(2by)+(2by)^2=a^2x^2-4axby+4b^2y^2$

(*k*) $(x^4+6)^2=(x^4)^2+2(x^4)(6)+(6)^2=x^8+12x^4+36$

(*l*) $(3y^2-2)^2=(3y^2)^2-2(3y^2)(2)+(2)^2=9y^4-12y^2+4$

(*m*) $(x+3)(x+5)=x^2+(3+5)x+(3)(5)=x^2+8x+15$, using IV with $a=3$, $b=5$.

(*n*) $(x-2)(x+8)=x^2+(-2+8)x+(-2)(8)=x^2+6x-16$

(*o*) $(x+2)(x-8)=x^2+(2-8)x+(2)(-8)=x^2-6x-16$

(*p*) $(t^2+10)(t^2-12)=(t^2)^2+(10-12)t^2+(10)(-12)=t^4-2t^2-120$

(*q*) $(3x+4)(2x-3)=(3)(2)x^2+[(3)(-3)+(4)(2)]x+(4)(-3)$
$=6x^2-x-12$, using IV with $a=3$, $b=4$, $c=2$, $d=-3$.

(*r*) $(2x+5)(4x-1)=(2)(4)x^2+[(2)(-1)+(5)(4)]x+(5)(-1)=8x^2+18x-5$

(*s*) $(3x+y)(4x-2y)=(3x)(4x)+(y)(4x)+(3x)(-2y)+(y)(-2y)$
$=12x^2-2xy-2y^2$, using IV with $a=3x$, $b=y$, $c=4x$, $d=-2y$.

(*t*) $(3t^2s-2)(4t-3s)=(3t^2s)(4t)+(-2)(4t)+(3t^2s)(-3s)+(-2)(-3s)$
$=12t^3s-8t-9t^2s^2+6s$

(*u*) $(3xy+1)(2x^2-3y)=(3xy)(2x^2)+(3xy)(-3y)+(1)(2x^2)+(1)(-3y)$
$=6x^3y-9xy^2+2x^2-3y$

(*v*) $(x+y+3)(x+y-3)=(x+y)^2-3^2=x^2+2xy+y^2-9$

(*w*) $(2x-y-1)(2x-y+1)=(2x-y)^2-(1)^2=4x^2-4xy+y^2-1$

(*x*) $(x^2+2xy+y^2)(x^2-2xy+y^2)=(x^2+y^2+2xy)(x^2+y^2-2xy)$
$=(x^2+y^2)^2-(2xy)^2=x^4+2x^2y^2+y^4-4x^2y^2=x^4-2x^2y^2+y^4$

(*y*) $(x^3+2+xy)(x^3-2+xy)=(x^3+xy+2)(x^3+xy-2)$
$=(x^3+xy)^2-2^2=x^6+2(x^3)(xy)+(xy)^2-4=x^6+2x^4y+x^2y^2-4$

4.2 (a) $(x+2y)^3 = x^3 + 3(x)^2(2y) + 3(x)(2y)^2 + (2y)^3$
$= x^3 + 6x^2y + 12xy^2 + 8y^3$, using V with $a = x$, $b = 2y$.

(b) $(3x+2)^3 = (3x)^3 + 3(3x)^2(2) + 3(3x)(2)^2 + (2)^3 = 27x^3 + 54x^2 + 36x + 8$

(c) $(2y-5)^3 = (2y)^3 - 3(y)^2(5) + 3(2y)(5)^2 - (5)^3$
$= 8y^3 - 60y^2 + 150y - 125$, using IV with $a = 2y$, $b = 5$.

(d) $(xy-2)^3 = (xy)^3 - 3(xy)^2(2) + 3(xy)(2)^2 - (2)^3 = x^3y^3 - 6x^2y^2 + 12xy - 8$

(e) $(x^2y - y^2)^3 = (x^2y)^3 - 3(x^2y)^2(y^2) + 3(x^2y)(y^2)^2 - (y^2)^3 = x^6y^3 - 3x^4y^4 + 3x^2y^5 - y^6$

(f) $(x-1)(x^2+x+1) = x^3 - 1$, using VII with $a = x$, $b = 1$.
If the form is not recognized, multiply as follows.
$(x-1)(x^2+x+1) = x(x^2+x+1) - 1(x^2+x+1) = x^3 + x^2 + x - x^2 - x - 1 = x^3 - 1$

(g) $(x-2y)(x^2+2xy+4y^2) = x^3 - (2y)^3 = x^3 - 8y^3$, using VII with $a = x$, $b = 2y$.

(h) $(xy+2)(x^2y^2 - 2xy + 4) = (xy)^3 + (2)^3 = x^3y^3 + 8$, using VIII with $a = xy$, $b = 2$.

(i) $(2x+1)(4x^2 - 2x + 1) = (2x)^3 + 1 = 8x^3 + 1$

(j) $(2x+3y+z)^2 = (2x)^2 + (3y)^2 + (z)^2 + 2(2x)(3y) + 2(2x)(z) + 2(3y)(z)$
$= 4x^2 + 9y^2 + z^2 + 12xy + 4xz + 6yz$, using VI with $a = 2x$, $b = 3y$, $c = z$.

(k) $(u^3 - v^2 + 2w)^2 = (u^3)^2 + (-v^2)^2 + (2w)^2 + 2(u^3)(-v^2) + 2(u^3)(2w) + 2(-v^2)(2w)$
$= u^6 + v^4 + 4w^2 - 2u^3v^2 + 4u^3w - 4v^2w$

4.3 (a) $(x-1)(x^5+x^4+x^3+x^2+x+1) = x^6 - 1$, using VII with $a = x$, $b = 1$, $n = 6$.

(b) $(x-2y)(x^4 + 2x^3y + 4x^2y^2 + 8xy^3 + 16y^4) = x^5 - (2y)^5$
$= x^5 - 32y^5$, using VII with $a = x$, $b = 2y$.

(c) $(3y+x)(81y^4 - 27y^3x + 9y^2x^2 - 3yx^3 + x^4) = (3y)^5 + x^5$
$= 243y^5 + x^5$, using VIII with $a = 3y$, $b = x$.

4.4 (a) $(x+y+z)(x+y-z)(x-y+z)(x-y-z)$. The first two factors may be written as

$$(x+y+z)(x+y-z) = (x+y)^2 - z^2 = x^2 + 2xy + y^2 - z^2,$$

and the second two factors as

$$(x-y+z)(x-y-z) = (x-y)^2 - z^2 = x^2 - 2xy + y^2 - z^2.$$

The result may be written
$(x^2+y^2-z^2+2xy)(x^2+y^2-z^2-2xy) = (x^2+y^2-z^2)^2 - (2xy)^2$
$= (x^2)^2 + (y^2)^2 + (-z^2)^2 + 2(x^2)(y^2) + 2(x^2)(-z^2) + 2(y^2)(-z^2) - 4x^2y^2$
$= x^4 + y^4 + z^4 - 2x^2y^2 - 2x^2z^2 - 2y^2z^2$

(b) $(x+y+z+1)^2 = [(x+y)+(z+1)]^2 = (x+y)^2 + 2(x+y)(z+1) + (z+1)^2$
$= x^2 + 2xy + y^2 + 2xz + 2x + 2yz + 2y + z^2 + 2z + 1$

(c) $(u-v)^3(u+v)^3 = [(u-v)(u+v)]^3 = (u^2-v^2)^3$
$= (u^2)^3 - 3(u^2)^2v^2 + 3(u^2)(v^2)^2 - (v^2)^3 = u^6 - 3u^4v^2 + 3u^2v^4 - v^6$

(d) $(x^2-x+1)^2(x^2+x+1)^2 = [(x^2-x+1)(x^2+x+1)]^2 = [(x^2+1-x)(x^2+1+x)]^2$
$= [(x^2+1)^2 - x^2]^2 = [x^4 + 2x^2 + 1 - x^2]^2 = (x^4 + x^2 + 1)^2$
$= (x^4)^2 + (x^2)^2 + 1^2 + 2(x^4)(x^2) + 2(x^4)(1) + 2(x^2)(1)$
$= x^8 + x^4 + 1 + 2x^6 + 2x^4 + 2x^2 = x^8 + 2x^6 + 3x^4 + 2x^2 + 1$

(e) $(e^y+1)(e^y-1)(e^{2y}+1)(e^{4y}+1)(e^{8y}+1) = (e^{2y}-1)(e^{2y}+1)(e^{4y}+1)(e^{8y}+1)$
$= (e^{4y}-1)(e^{4y}+1)(e^{8y}+1) = (e^{8y}-1)(e^{8y}+1) = e^{16y} - 1$

Supplementary Problems

Find each of the following products.

4.5 (*a*) $2xy(3x^2y - 4y^3) = 6x^3y^2 - 8xy^4$

(*b*) $3x^2y^3(2xy - x - 2y) = 6x^3y^4 - 3x^3y^3 - 6x^2y^4$

(*c*) $(2st^3 - 4rs^2 + 3s^3t)(5rst^2) = 10rs^2t^5 - 20r^2s^3t^2 + 15rs^4t^3$

(*d*) $(3a + 5b)(3a - 5b) = 9a^2 - 25b^2$

(*e*) $(5xy + 4)(5xy - 4) = 25x^2y^2 - 16$

(*f*) $(2 - 5y^2)(2 + 5y^2) = 4 - 25y^4$

(*g*) $(3a + 5a^2b)(3a - 5a^2b) = 9a^2 - 25a^4b^2$

(*h*) $(x + 6)^2 = x^2 + 12x + 36$

(*i*) $(y + 3x)^2 = y^2 + 6xy + 9x^2$

(*j*) $(z - 4)^2 = z^2 - 8z + 16$

(*k*) $(3 - 2x^2)^2 = 9 - 12x^2 + 4x^4$

(*l*) $(x^2y - 2z)^2 = x^4y^2 - 4x^2yz + 4z^2$

(*m*) $(x + 2)(x + 4) = x^2 + 6x + 8$

(*n*) $(x - 4)(x + 7) = x^2 + 3x - 28$

(*o*) $(y + 3)(y - 5) = y^2 - 2y - 15$

(*p*) $(xy + 6)(xy - 4) = x^2y^2 + 2xy - 24$

(*q*) $(2x - 3)(4x + 1) = 8x^2 - 10x - 3$

(*r*) $(4 + 3r)(2 - r) = 8 + 2r - 3r^2$

(*s*) $(5x + 3y)(2x - 3y) = 10x^2 - 9xy - 9y^2$

(*t*) $(2t^2 + s)(3t^2 + 4s) = 6t^4 + 11t^2s + 4s^2$

(*u*) $(x^2 + 4y)(2x^2y - y^2) = 2x^4y + 7x^2y^2 - 4y^3$

(*v*) $x(2x - 3)(3x + 4) = 6x^3 - x^2 - 12x$

(*w*) $(r + s - 1)(r + s + 1) = r^2 + 2rs + s^2 - 1$

(*x*) $(x - 2y + z)(x - 2y - z) = x^2 - 4xy + 4y^2 - z^2$

(*y*) $(x^2 + 2x + 4)(x^2 - 2x + 4) = x^4 + 4x^2 + 16$

4.6 (*a*) $(2x + 1)^3 = 8x^3 + 12x^2 + 6x + 1$

(*b*) $(3x + 2y)^3 = 27x^3 + 54x^2y + 36xy^2 + 8y^3$

(*c*) $(r - 2s)^3 = r^3 - 6r^2s + 12rs^2 - 8s^3$

(*d*) $(x^2 - 1)^3 = x^6 - 3x^4 + 3x^2 - 1$

(*e*) $(ab^2 - 2b)^3 = a^3b^6 - 6a^2b^5 + 12ab^4 - 8b^3$

(*f*) $(t - 2)(t^2 + 2t + 4) = t^3 - 8$

(*g*) $(z - x)(x^2 + xz + z^2) = z^3 - x^3$

(*h*) $(x + 3y)(x^2 - 3xy + 9y^2) = x^3 + 27y^3$

4.7 (*a*) $(x - 2y + z)^2 = x^2 - 4xy + 4y^2 + 2zx - 4zy + z^2$

(*b*) $(s - 1)(s^3 + s^2 + s + 1) = s^4 - 1$

(*c*) $(1 + t^2)(1 - t^2 + t^4 - t^6) = 1 - t^8$

(*d*) $(3x + 2y)^2(3x - 2y)^2 = 81x^4 - 72x^2y^2 + 16y^4$

(*e*) $(x^2 + 2x + 1)^2(x^2 - 2x + 1)^2 = x^8 - 4x^6 + 6x^4 - 4x^2 + 1$

(*f*) $(y - 1)^3(y + 1)^3 = y^6 - 3y^4 + 3y^2 - 1$

(*g*) $(u + 2)(u - 2)(u^2 + 4)(u^4 + 16) = u^8 - 256$

Chapter 5

Factoring

5.1 FACTORING

The factors of a given algebraic expression consist of two or more algebraic expressions which when multiplied together produce the given expression.

EXAMPLES 5.1. Factor each algebraic expression.

(*a*) $x^2 - 7x + 6 = (x-1)(x-6)$

(*b*) $x^2 + 8x = x(x+8)$

(*c*) $6x^2 - 7x - 5 = (3x-5)(2x+1)$

(*d*) $x^2 + 2xy - 8y^2 = (x+4y)(x-2y)$

The factorization process is generally restricted to finding factors of polynomials with integer coefficients in each of its terms. In such cases it is required that the factors also be polynomials with integer coefficients. Unless otherwise stated we shall adhere to this limitation.

Thus we shall not consider $(x-1)$ as being factorable into $(\sqrt{x}+1)(\sqrt{x}-1)$ because these factors are not polynomials. Similarly, we shall not consider (x^2-3y^2) as being factorable into $(x-\sqrt{3}y)(x+\sqrt{3}y)$ because these factors are not polynomials with integer coefficients. Also, even though $3x+2y$ could be written $3(x+\frac{2}{3}y)$ we shall not consider this to be a factored form because $x+\frac{2}{3}y$ is not a polynomial with integer coefficients.

A given polynomial with integer coefficients is said to be *prime* if it cannot itself be factored in accordance with the above restrictions. Thus $x^2-7x+6=(x-1)(x-6)$ has been expressed as a product of the prime factors $x-1$ and $x-6$.

A polynomial is said to be factored completely when it is expressed as a product of prime factors.

Note 1. In factoring we shall allow trivial changes in sign. Thus x^2-7x+6 can be factored either as $(x-1)(x-6)$ or $(1-x)(6-x)$. It can be shown that factorization into prime factors, apart from the trivial changes in sign and arrangement of factors, is possible in one and only one way. This is often referred to as the Unique Factorization Theorem.

Note 2. Sometimes the following definition of prime is used. A polynomial is said to be prime if it has no factors other than plus or minus itself and ± 1. This is in analogy with the definition of a prime number or integer such as $2, 3, 5, 7, 11, \ldots$ and may be seen to be equivalent to the previous definition.

Note 3. Occasionally we may factor polynomials with rational coefficients, e.g., $x^2 - 9/4 = (x+3/2)(x-3/2)$. In such cases the factors should be polynomials with rational coefficients.

Note 4. There are times when we want to factor an expression over a specific set of numbers, e.g., $x^2-2=(x+\sqrt{2})(x-\sqrt{2})$ over the set of real numbers, but it is prime over the set of rational numbers. Unless the set of numbers to use for the coefficients of the factors is specified it is assumed to be the set of integers.

5.2 FACTORIZATION PROCEDURES

In factoring, formulas I–VIII of Chapter 4 are very useful. Just as when read from left to right they helped to obtain *products*, so when read from right to left they help to find *factors*.

The following procedures in factoring are very useful.

A. Common monomial factor. Type: $ac + ad = a(c + d)$

EXAMPLES 5.2. (*a*) $6x^2y - 2x^3 = 2x^2(3y - x)$

(*b*) $2x^3y - xy^2 + 3x^2y = xy(2x^2 - y + 3x)$

B. Difference of two squares. Type: $a^2 - b^2 = (a + b)(a - b)$

EXAMPLES 5.3. (*a*) $x^2 - 25 = x^2 - 5^2 = (x + 5)(x - 5)$ where $a = x$, $b = 5$

(*b*) $4x^2 - 9y^2 = (2x)^2 - (3y)^2 = (2x + 3y)(2x - 3y)$ where $a = 2x$, $b = 3y$

C. Perfect square trinomials. Types: $a^2 + 2ab + b^2 = (a + b)^2$

$$a^2 - 2ab + b^2 = (a - b)^2$$

It follows that a trinomial is a perfect square if two terms are perfect squares and the third term is numerically twice the product of the square roots of the other two terms.

EXAMPLES 5.4. (*a*) $x^2 + 6x + 9 = (x + 3)^2$

(*b*) $9x^2 - 12xy + 4y^2 = (3x - 2y)^2$

D. Other trinomials. Types: $x^2 + (a + b)x + ab = (x + a)(x + b)$

$$acx^2 + (ad + bc)x + bd = (ax + b)(cx + d)$$

EXAMPLES 5.5. (*a*) $x^2 - 5x + 4 = (x - 4)(x - 1)$ where $a = -4$, $b = -1$ so that their sum $(a + b) = -5$ and their product $ab = 4$.

(*b*) $x^2 + xy - 12y^2 = (x - 3y)(x + 4y)$ where $a = -3y$, $b = 4y$

(*c*) $3x^2 - 5x - 2 = (x - 2)(3x + 1)$. Here $ac = 3$, $bd = -2$, $ad + bc = -5$; and we find by trial that $a = 1$, $c = 3$, $b = -2$, $d = 1$ satisfies $ad + bc = -5$.

(*d*) $6x^2 + x - 12 = (3x - 4)(2x + 3)$

(*e*) $8 - 14x + 5x^2 = (4 - 5x)(2 - x)$

E. Sum, difference of two cubes. Types: $a^3 + b^3 = (a + b)(a^2 - ab + b^2)$

$$a^3 - b^3 = (a - b)(a^2 + ab + b^2)$$

EXAMPLES 5.6. (*a*)
$$\begin{aligned} 8x^3 + 27y^3 &= (2x)^3 + (3y)^3 \\ &= (2x + 3y)[(2x)^2 - (2x)(3y) + (3y)^2] \\ &= (2x + 3y)(4x^2 - 6xy + 9y^2) \end{aligned}$$

(*b*) $8x^3y^3 - 1 = (2xy)^3 - 1^3 = (2xy - 1)(4x^2y^2 + 2xy + 1)$

F. Grouping of terms. Type: $ac + bc + ad + bd = c(a + b) + d(a + b) = (a + b)(c + d)$

EXAMPLE 5.7. $2ax - 4bx + ay - 2by = 2x(a - 2b) + y(a - 2b) = (a - 2b)(2x + y)$

G. Factors of $a^n \pm b^n$. Here we use formulas VII and VIII of Chapter 4.

EXAMPLES 5.8. (*a*)
$$\begin{aligned} 32x^5 + 1 = (2x)^5 + 1^5 &= (2x + 1)[(2x)^4 - (2x)^3 + (2x)^2 - 2x + 1] \\ &= (2x + 1)(16x^4 - 8x^3 + 4x^2 - 2x + 1) \end{aligned}$$

(*b*) $x^7 - 1 = (x - 1)(x^6 + x^5 + x^4 + x^3 + x^2 + x + 1)$

H. Addition and subtraction of suitable terms.

EXAMPLE 5.9. Factor $x^4 + 4$.

Adding and subtracting $4x^2$ (twice the product of the square roots of x^4 and 4), we have

$$\begin{aligned} x^4 + 4 &= (x^4 + 4x^2 + 4) - 4x^2 = (x^2 + 2)^2 - (2x)^2 \\ &= (x^2 + 2 + 2x)(x^2 + 2 - 2x) = (x^2 + 2x + 2)(x^2 - 2x + 2) \end{aligned}$$

I. Miscellaneous combinations of previous methods.

EXAMPLES 5.10. (*a*)
$$\begin{aligned} x^4 - xy^3 - x^3y + y^4 &= (x^4 - xy^3) - (x^3y - y^4) \\ &= x(x^3 - y^3) - y(x^3 - y^3) \\ &= (x^3 - y^3)(x - y) = (x - y)(x^2 + xy + y^2)(x - y) \\ &= (x - y)^2(x^2 + xy + y^2) \end{aligned}$$

(*b*)
$$\begin{aligned} x^2y - 3x^2 - y + 3 &= (x^2y - 3x^2) + (-y + 3) \\ &= x^2(y - 3) - (y - 3) \\ &= (y - 3)(x^2 - 1) \\ &= (y - 3)(x + 1)(x - 1) \end{aligned}$$

(*c*)
$$\begin{aligned} x^2 + 6x + 9 - y^2 &= (x^2 + 6x + 9) - y^2 \\ &= (x + 3)^2 - y^2 \\ &= [(x + 3) + y][(x + 3) - y] \\ &= (x + y + 3)(x - y + 3) \end{aligned}$$

5.3 GREATEST COMMON FACTOR

The greatest common factor (GCF) of two or more given polynomials is the polynomial of highest degree and largest numerical coefficients (apart from trivial changes in sign) which is a factor of all the given polynomials.

The following method is suggested for finding the GCF of several polynomials. (*a*) Write each polynomial as a product of prime factors. (*b*) The GCF is the product obtained by taking each factor to the *lowest* power to which it occurs in any of the polynomials.

EXAMPLE 5.11. The GCF of $2^3 3^2(x - y)^3(x + 2y)^2$, $2^2 3^3(x - y)^2(x + 2y)^3$, $3^2(x - y)^2(x + 2y)$ is $3^2(x - y)^2(x + 2y)$.

Two or more polynomials are *relatively prime* if their GCF is 1.

5.4 LEAST COMMON MULTIPLE

The least common multiple (LCM) of two or more given polynomials is the polynomial of lowest degree and smallest numerical coefficients (apart from trivial changes in sign) for which each of the given polynomials will be a factor.

The following procedure is suggested for determining the LCM of several polynomials. (*a*) Write each polynomial as a product of prime factors. (*b*) The LCM is the product obtained by taking each factor to the *highest* power to which it occurs.

EXAMPLE 5.12. The LCM of $2^3 3^2(x - y)^3(x + 2y)^2$, $2^2 3^3(x - y)^2(x + 2y)^3$, $3^2(x - y)^2(x + 2y)$ is $2^3 3^3(x - y)^3(x + 2y)^3$.

Solved Problems

Common Monomial Factor

Type: $ac + ad = a(c + d)$

5.1 (*a*) $2x^2 - 3xy = x(2x - 3y)$

(*b*) $4x + 8y + 12z = 4(x + 2y + 3z)$

(*c*) $3x^2 + 6x^3 + 12x^4 = 3x^2(1 + 2x + 4x^2)$

(*d*) $9s^3t + 15s^2t^3 - 3s^2t^2 = 3s^2t(3s + 5t^2 - t)$

(*e*) $10a^2b^3c^4 - 15a^3b^2c^4 + 30a^4b^3c^2 = 5a^2b^2c^2(2bc^2 - 3ac^2 + 6a^2b)$

(*f*) $4a^{n+1} - 8a^{2n} = 4a^{n+1}(1 - 2a^{n-1})$

Difference of Two Squares

Type: $a^2 - b^2 = (a + b)(a - b)$

5.2 (*a*) $x^2 - 9 = x^2 - 3^2 = (x + 3)(x - 3)$

(*b*) $25x^2 - 4y^2 = (5x)^2 - (2y)^2 = (5x + 2y)(5x - 2y)$

(*c*) $9x^2y^2 - 16a^2 = (3xy)^2 - (4a)^2 = (3xy + 4a)(3xy - 4a)$

(*d*) $1 - m^2n^4 = 1^2 - (mn^2)^2 = (1 + mn^2)(1 - mn^2)$

(*e*) $3x^2 - 12 = 3(x^2 - 4) = 3(x + 2)(x - 2)$

(*f*) $x^2y^2 - 36y^4 = y^2[x^2 - (6y)^2] = y^2(x + 6y)(x - 6y)$

(*g*) $x^4 - y^4 = (x^2)^2 - (y^2)^2 = (x^2 + y^2)(x^2 - y^2) = (x^2 + y^2)(x + y)(x - y)$

(*h*) $1 - x^8 = (1 + x^4)(1 - x^4) = (1 + x^4)(1 + x^2)(1 - x^2) = (1 + x^4)(1 + x^2)(1 + x)(1 - x)$

(*i*) $32a^4b - 162b^5 = 2b(16a^4 - 81b^4) = 2b(4a^2 + 9b^2)(4a^2 - 9b^2)$
$= 2b(4a^2 + 9b^2)(2a + 3b)(2a - 3b)$

(*j*) $x^3y - y^3x = xy(x^2 - y^2) = xy(x + y)(x - y)$

(*k*) $(x + 1)^2 - 36y^2 = [(x + 1) + (6y)][(x + 1) - (6y)] = (x + 6y + 1)(x - 6y + 1)$

(*l*) $(5x + 2y)^2 - (3x - 7y)^2 = [(5x + 2y) + (3x - 7y)][(5x + 2y) - (3x - 7y)]$
$= (8x - 5y)(2x + 9y)$

Perfect Square Trinomials

Types: $a^2 + 2ab + b^2 = (a + b)^2$
$a^2 - 2ab + b^2 = (a - b)^2$

5.3 (*a*) $x^2 + 8x + 16 = x^2 + 2(x)(4) + 4^2 = (x + 4)^2$

(*b*) $1 + 4y + 4y^2 = (1 + 2y)^2$

(*c*) $t^2 - 4t + 4 = t^2 - 2(t)(2) + 2^2 = (t - 2)^2$

(*d*) $x^2 - 16xy + 64y^2 = (x - 8y)^2$

(*e*) $25x^2 + 60xy + 36y^2 = (5x + 6y)^2$

(*f*) $16m^2 - 40mn + 25n^2 = (4m - 5n)^2$

(*g*) $9x^4 - 24x^2y + 16y^2 = (3x^2 - 4y)^2$

(*h*) $2x^3y^3 + 16x^2y^4 + 32xy^5 = 2xy^3(x^2 + 8xy + 16y^2) = 2xy^3(x + 4y)^2$

(*i*) $16a^4 - 72a^2b^2 + 81b^4 = (4a^2 - 9b^2)^2 = [(2a + 3b)(2a - 3b)]^2 = (2a + 3b)^2(2a - 3b)^2$

(*j*) $(x + 2y)^2 + 10(x + 2y) + 25 = (x + 2y + 5)^2$

(*k*) $a^2x^2 - 2abxy + b^2y^2 = (ax - by)^2$

(*l*) $4m^6n^6 + 32m^4n^4 + 64m^2n^2 = 4m^2n^2(m^4n^4 + 8m^2n^2 + 16) = 4m^2n^2(m^2n^2 + 4)^2$

Other Trinomials

Types: $x^2 + (a+b)x + ab = (x+a)(x+b)$

$acx^2 + (ad+bc)x + bd = (ax+b)(cx+d)$

5.4 (a) $x^2 + 6x + 8 = (x+4)(x+2)$

(b) $x^2 - 6x + 8 = (x-4)(x-2)$

(c) $x^2 + 2x - 8 = (x+4)(x-2)$

(d) $x^2 - 2x - 8 = (x-4)(x+2)$

(e) $x^2 - 7xy + 12y^2 = (x-3y)(x-4y)$

(f) $x^2 + xy - 12y^2 = (x+4y)(x-3y)$

(g) $16 - 10x + x^2 = (8-x)(2-x)$

(h) $20 - x - x^2 = (5+x)(4-x)$

(i) $3x^3 - 3x^2 - 18x = 3x(x^2 - x - 6) = 3x(x-3)(x+2)$

(j) $y^4 + 7y^2 + 12 = (y^2+4)(y^2+3)$

(k) $m^4 + m^2 - 2 = (m^2+2)(m^2-1) = (m^2+2)(m+1)(m-1)$

(l) $(x+1)^2 + 3(x+1) + 2 = [(x+1)+2][(x+1)+1] = (x+3)(x+2)$

(m) $s^2t^2 - 2st^3 - 63t^4 = t^2(s^2 - 2st - 63t^2) = t^2(s-9t)(s+7t)$

(n) $z^4 - 10z^2 + 9 = (z^2-1)(z^2-9) = (z+1)(z-1)(z+3)(z-3)$

(o) $2x^6y - 6x^4y^3 - 8x^2y^5 = 2x^2y(x^4 - 3x^2y^2 - 4y^4)$

$= 2x^2y(x^2+y^2)(x^2-4y^2) = 2x^2y(x^2+y^2)(x+2y)(x-2y)$

(p) $x^2 - 2xy + y^2 + 10(x-y) + 9 = (x-y)^2 + 10(x-y) + 9$

$= [(x-y)+1][(x-y)+9] = (x-y+1)(x-y+9)$

(q) $4x^8y^{10} - 40x^5y^7 + 84x^2y^4 = 4x^2y^4(x^6y^6 - 10x^3y^3 + 21) = 4x^2y^4(x^3y^3 - 7)(x^3y^3 - 3)$

(r) $x^{2a} - x^a - 30 = (x^a - 6)(x^a + 5)$

(s) $x^{m+2n} + 7x^{m+n} + 10x^m = x^m(x^{2n} + 7x^n + 10) = x^m(x^n+2)(x^n+5)$

(t) $a^{2(y-1)} - 5a^{y-1} + 6 = (a^{y-1} - 3)(a^{y-1} - 2)$

5.5 (a) $3x^2 + 10x + 3 = (3x+1)(x+3)$

(b) $2x^2 - 7x + 3 = (2x-1)(x-3)$

(c) $2y^2 - y - 6 = (2y+3)(y-2)$

(d) $10s^2 + 11s - 6 = (5s-2)(2s+3)$

(e) $6x^2 - xy - 12y^2 = (3x+4y)(2x-3y)$

(f) $10 - x - 3x^2 = (5-3x)(2+x)$

(g) $4z^4 - 9z^2 + 2 = (z^2-2)(4z^2-1) = (z^2-2)(2z+1)(2z-1)$

(h) $16x^3y + 28x^2y^2 - 30xy^3 = 2xy(8x^2 + 14xy - 15y^2) = 2xy(4x-3y)(2x+5y)$

(i) $12(x+y)^2 + 8(x+y) - 15 = [6(x+y)-5][2(x+y)+3] = (6x+6y-5)(2x+2y+3)$

(j) $6b^{2n+1} + 5b^{n+1} - 6b = b(6b^{2n} + 5b^n - 6) = b(2b^n+3)(3b^n-2)$

(k) $18x^{4p+m} - 66x^{2p+m}y^2 - 24x^my^4 = 6x^m(3x^{4p} - 11x^{2p}y^2 - 4y^4) = 6x^m(3x^{2p}+y^2)(x^{2p}-4y^2)$

$= 6x^m(3x^{2p}+y^2)(x^p+2y)(x^p-2y)$

(l) $64x^{12}y^3 - 68x^8y^7 + 4x^4y^{11} = 4x^4y^3(16x^8 - 17x^4y^4 + y^8) = 4x^4y^3(16x^4 - y^4)(x^4 - y^4)$

$= 4x^4y^3(4x^2+y^2)(4x^2-y^2)(x^2+y^2)(x^2-y^2)$

$= 4x^4y^3(4x^2+y^2)(2x+y)(2x-y)(x^2+y^2)(x+y)(x-y)$

Sum of Difference of Two Cubes

Types: $a^3 + b^3 = (a+b)(a^2 - ab + b^2)$

$a^3 - b^3 = (a-b)(a^2 + ab + b^2)$

5.6 (a) $x^3+8 = x^3+2^3 = (x+2)(x^2-2x+2^2) = (x+2)(x^2-2x+4)$

(b) $a^3-27 = a^3-3^3 = (a-3)(a^2+3a+3^2) = (a-3)(a^2+3a+9)$

(c) $a^6+b^6 = (a^2)^3+(b^2)^3 = (a^2+b^2)[(a^2)^2-a^2b^2+(b^2)^2]$
$= (a^2+b^2)(a^4-a^2b^2+b^4)$

(d) $a^6-b^6 = (a^3+b^3)(a^3-b^3) = (a+b)(a^2-ab+b^2)(a-b)(a^2+ab+b^2)$

(e) $a^9+b^9 = (a^3)^3+(b^3)^3 = (a^3+b^3)[(a^3)^2-a^3b^3+(b^3)^2]$
$= (a+b)(a^2-ab+b^2)(a^6-a^3b^3+b^6)$

(f) $a^{12}+b^{12} = (a^4)^3+(b^4)^3 = (a^4+b^4)(a^8-a^4b^4+b^8)$

(g) $64x^3+125y^3 = (4x)^3+(5y)^3 = (4x+5y)[(4x)^2-(4x)(5y)+(5y)^2]$
$= (4x+5y)(16x^2-20xy+25y^2)$

(h) $(x+y)^3-z^3 = (x+y-z)[(x+y)^2+(x+y)z+z^2]$
$= (x+y-z)(x^2+2xy+y^2+xz+yz+z^2)$

(i) $(x-2)^3+8y^3 = (x-2)^3+(2y)^3 = (x-2+2y)[(x-2)^2-(x-2)(2y)+(2y)^2]$
$= (x-2+2y)(x^2-4x+4-2xy+4y+4y^2)$

(j) $x^6-7x^3-8 = (x^3-8)(x^3+1)$
$= (x^3-2^3)(x^3+1) = (x-2)(x^2+2x+4)(x+1)(x^2-x+1)$

(k) $x^8y-64x^2y^7 = x^2y(x^6-64y^6) = x^2y(x^3+8y^3)(x^3-8y^3) = x^2y[x^3+(2y)^3][x^3-(2y)^3]$
$= x^2y(x+2y)(x^2-2xy+4y^2)(x-2y)(x^2+2xy+4y^2)$

(l) $54x^6y^2-38x^3y^2-16y^2 = 2y^2(27x^6-19x^3-8) = 2y^2(27x^3+8)(x^3-1)$
$= 2y^2[(3x)^3+2^3](x^3-1) = 2y^2(3x+2)(9x^2-6x+4)(x-1)(x^2+x+1)$

Grouping of Terms

Type: $ac+bc+ad+bd = c(a+b)+d(a+b) = (a+b)(c+d)$

5.7 (a) $bx-ab+x^2-ax = b(x-a)+x(x-a) = (x-a)(b+x) = (x-a)(x+b)$

(b) $3ax-ay-3bx+by = a(3x-y)-b(3x-y) = (3x-y)(a-b)$

(c) $6x^2-4ax-9bx+6ab = 2x(3x-2a)-3b(3x-2a) = (3x-2a)(2x-3b)$

(d) $ax+ay+x+y = a(x+y)+(x+y) = (x+y)(a+1)$

(e) $x^2-4y^2+x+2y = (x+2y)(x-2y)+(x+2y) = (x+2y)(x-2y+1)$

(f) $x^3+x^2y+xy^2+y^3 = x^2(x+y)+y^2(x+y) = (x+y)(x^2+y^2)$

(g) $x^7+27x^4-x^3-27 = x^4(x^3+27)-(x^3+27) = (x^3+27)(x^4-1)$
$= (x^3+3^3)(x^2+1)(x^2-1) = (x+3)(x^2-3x+9)(x^2+1)(x+1)(x-1)$

(h) $x^3y^3-y^3+8x^3-8 = y^3(x^3-1)+8(x^3-1) = (x^3-1)(y^3+8)$
$= (x-1)(x^2+x+1)(y+2)(y^2-2y+4)$

(i) $a^6+b^6-a^2b^4-a^4b^2 = a^6-a^2b^4+b^6-a^4b^2 = a^2(a^4-b^4)-b^2(a^4-b^4)$
$= (a^4-b^4)(a^2-b^2) = (a^2+b^2)(a^2-b^2)(a+b)(a-b)$
$= (a^2+b^2)(a+b)(a-b)(a+b)(a-b) = (a^2+b^2)(a+b)^2(a-b)^2$

(j) $a^3+3a^2-5ab+2b^2-b^3 = (a^3-b^3)+(3a^2-5ab+2b^2)$
$= (a-b)(a^2+ab+b^2)+(a-b)(3a-2b)$
$= (a-b)(a^2+ab+b^2+3a-2b)$

Factors of $a^n \pm b^n$

5.8 a^n+b^n has $a+b$ as a factor if and only if n is a positive odd integer. Then

$$a^n+b^n = (a+b)(a^{n-1}-a^{n-2}b+a^{n-3}b^2-\cdots-ab^{n-2}+b^{n-1}).$$

(a) $a^3+b^3 = (a+b)(a^2-ab+b^2)$

(b) $64+y^3 = 4^3+y^3 = (4+y)(4^2-4y+y^2) = (4+y)(16-4y+y^2)$

(c) $x^3 + 8y^6 = x^3 + (2y^2)^3 = (x + 2y^2)[x^2 - x(2y^2) + (2y^2)^2]$
$= (x + 2y^2)(x^2 - 2xy^2 + 4y^4)$

(d) $a^5 + b^5 = (a + b)(a^4 - a^3b + a^2b^2 - ab^3 + b^4)$

(e) $1 + x^5y^5 = 1^5 + (xy)^5 = (1 + xy)(1 - xy + x^2y^2 - x^3y^3 + x^4y^4)$

(f) $z^5 + 32 = z^5 + 2^5 = (z + 2)(z^4 - 2z^3 + 2^2z^2 - 2^3z + 2^4)$
$= (z + 2)(z^4 - 2z^3 + 4z^2 - 8z + 16)$

(g) $a^{10} + x^{10} = (a^2)^5 + (x^2)^5 = (a^2 + x^2)[(a^2)^4 - (a^2)^3x^2 + (a^2)^2(x^2)^2 - (a^2)(x^2)^3 + (x^2)^4]$
$= (a^2 + x^2)(a^8 - a^6x^2 + a^4x^4 - a^2x^6 + x^8)$

(h) $u^7 + v^7 = (u + v)(u^6 - u^5v + u^4v^2 - u^3v^3 + u^2v^4 - uv^5 + v^6)$

(i) $x^9 + 1 = (x^3)^3 + 1^3 = (x^3 + 1)(x^6 - x^3 + 1) = (x + 1)(x^2 - x + 1)(x^6 - x^3 + 1)$

5.9 $a^n - b^n$ has $a - b$ as a factor if n is any positive integer. Then

$$a^n - b^n = (a - b)(a^{n-1} + a^{n-2}b + a^{n-3}b^2 + \cdots + ab^{n-2} + b^{n-1}).$$

If n is an even positive integer, $a^n - b^n$ also has $a + b$ as factor.

(a) $a^2 - b^2 = (a - b)(a + b)$

(b) $a^3 - b^3 = (a - b)(a^2 + ab + b^2)$

(c) $27x^3 - y^3 = (3x)^3 - y^3 = (3x - y)[(3x)^2 + (3x)y + y^2] = (3x - y)(9x^2 + 3xy + y^2)$

(d) $1 - x^3 = (1 - x)(1^2 + 1x + x^2) = (1 - x)(1 + x + x^2)$

(e) $a^5 - 32 = a^5 - 2^5 = (a - 2)(a^4 + a^3 \cdot 2 + a^2 \cdot 2^2 + a \cdot 2^3 + 2^4)$
$= (a - 2)(a^4 + 2a^3 + 4a^2 + 8a + 16)$

(f) $y^7 - z^7 = (y - z)(y^6 + y^5z + y^4z^2 + y^3z^3 + y^2z^4 + yz^5 + z^6)$

(g) $x^6 - a^6 = (x^3 + a^3)(x^3 - a^3) = (x + a)(x^2 - ax + a^2)(x - a)(x^2 + ax + a^2)$

(h) $u^8 - v^8 = (u^4 + v^4)(u^4 - v^4) = (u^4 + v^4)(u^2 + v^2)(u^2 - v^2)$
$= (u^4 + v^4)(u^2 + v^2)(u + v)(u - v)$

(i) $x^9 - 1 = (x^3)^3 - 1 = (x^3 - 1)(x^6 + x^3 + 1) = (x - 1)(x^2 + x + 1)(x^6 + x^3 + 1)$

(j) $x^{10} - y^{10} = (x^5 + y^5)(x^5 - y^5)$
$= (x + y)(x^4 - x^3y + x^2y^2 - xy^3 + y^4)(x - y)(x^4 + x^3y + x^2y^2 + xy^3 + y^4)$

Addition and Subtraction of Suitable Terms

5.10 (a) $a^4 + a^2b^2 + b^4$ (adding and subtracting a^2b^2)
$= (a^4 + 2a^2b^2 + b^4) - a^2b^2 = (a^2 + b^2)^2 - (ab)^2$
$= (a^2 + b^2 + ab)(a^2 + b^2 - ab)$

(b) $36x^4 + 15x^2 + 4$ (adding and subtracting $9x^2$)
$= (36x^4 + 24x^2 + 4) - 9x^2 = (6x^2 + 2)^2 - (3x)^2$
$= [(6x^2 + 2) + 3x][(6x^2 + 2) - 3x] = (6x^2 + 3x + 2)(6x^2 - 3x + 2)$

(c) $64x^4 + y^4$ (adding and subtracting $16x^2y^2$)
$= (64x^4 + 16x^2y^2 + y^4) - 16x^2y^2 = (8x^2 + y^2)^2 - (4xy)^2$
$= (8x^2 + y^2 + 4xy)(8x^2 + y^2 - 4xy)$

(d) $u^8 - 14u^4 + 25$ (adding and subtracting $4u^4$)
$= (u^8 - 10u^4 + 25) - 4u^4 = (u^4 - 5)^2 - (2u^2)^2$
$= (u^4 - 5 + 2u^2)(u^4 - 5 - 2u^2) = (u^4 + 2u^2 - 5)(u^4 - 2u^2 - 5)$

Miscellaneous Problems

5.11 (a) $x^2 - 4z^2 + 9y^2 - 6xy = (x^2 - 6xy + 9y^2) - 4z^2$
$= (x - 3y)^2 - (2z)^2 = (x - 3y + 2z)(x - 3y - 2z)$

(b) $16a^2 + 10bc - 25c^2 - b^2 = 16a^2 - (b^2 - 10bc + 25c^2)$
$= (4a)^2 - (b - 5c)^2 = (4a + b - 5c)(4a - b + 5c)$

(c) $x^2 + 7x + y^2 - 7y - 2xy - 8 = (x^2 - 2xy + y^2) + 7(x - y) - 8$
$= (x - y)^2 + 7(x - y) - 8 = (x - y + 8)(x - y - 1)$

(d) $a^2 - 8ab - 2ac + 16b^2 + 8bc - 15c^2 = (a^2 - 8ab + 16b^2) - (2ac - 8bc) - 15c^2$
$= (a - 4b)^2 - 2c(a - 4b) - 15c^2 = (a - 4b - 5c)(a - 4b + 3c)$

(e) $m^4 - n^4 + m^3 - mn^3 - n^3 + m^3n = (m^4 - mn^3) + (m^3n - n^4) + (m^3 - n^3)$
$= m(m^3 - n^3) + n(m^3 - n^3) + (m^3 - n^3)$
$= (m^3 - n^3)(m + n + 1) = (m - n)(m^2 + mn + n^2)(m + n + 1)$

Greatest Common Factor and Least Common Multiple

5.12 (a) $9x^4y^2 = 3^2x^4y^2, \quad 12x^3y^3 = 2^2 \cdot 3x^3y^3$
$\text{GCF} = 3x^3y^2, \quad \text{LCM} = 2^2 \cdot 3^2x^4y^3 = 36x^4y^3$

(b) $48r^3t^4 = 2^4 \cdot 3r^3t^4, \quad 54r^2t^6 = 2 \cdot 3^3r^2t^6, \quad 60r^4t^2 = 2^2 \cdot 3 \cdot 5r^4t^2$
$\text{GCF} = 2 \cdot 3r^2t^2 = 6r^2t^2, \quad \text{LCM} = 2^4 \cdot 3^3 \cdot 5r^4t^6 = 2160r^4t^6$

(c) $6x - 6y = 2 \cdot 3(x - y), \quad 4x^2 - 4y^2 = 2^2(x^2 - y^2) = 2^2(x + y)(x - y)$
$\text{GCF} = 2(x - y), \quad \text{LCM} = 2^2 \cdot 3(x + y)(x - y)$

(d) $y^4 - 16 = (y^2 + 4)(y + 2)(y - 2), \quad y^2 - 4 = (y + 2)(y - 2), \quad y^2 - 3y + 2 = (y - 1)(y - 2)$
$\text{GCF} = y - 2, \quad \text{LCM} = (y^2 + 4)(y + 2)(y - 2)(y - 1)$

(e) $3 \cdot 5^2(x + 3y)^2(2x - y)^4, \quad 2^3 \cdot 3^2 \cdot 5(x + 3y)^3(2x - y)^2, \quad 2^2 \cdot 3 \cdot 5(x + 3y)^4(2x - y)^5$
$\text{GCF} = 3 \cdot 5(x + 3y)^2(2x - y)^2, \quad \text{LCM} = 2^3 \cdot 3^2 \cdot 5^2(x + 3y)^4(2x - y)^5$

Supplementary Problems

Factor each expression.

5.13 (a) $3x^2y^4 + 6x^3y^3$
(b) $12s^2t^2 - 6s^5t^4 + 4s^4t$
(c) $2x^2yz - 4xyz^2 + 8xy^2z^3$
(d) $4y^2 - 100$
(e) $1 - a^4$
(f) $64x - x^3$
(g) $8x^4 - 128$
(h) $18x^3y - 8xy^3$
(i) $(2x + y)^2 - (3y - z)^2$
(j) $4(x + 3y)^2 - 9(2x - y)^2$
(k) $x^2 + 4x + 4$
(l) $4 - 12y + 9y^2$
(m) $x^2y^2 - 8xy + 16$
(n) $4x^3y + 12x^2y^2 + 9xy^3$
(o) $3a^4 + 6a^2b^2 + 3b^4$
(p) $(m^2 - n^2)^2 + 8(m^2 - n^2) + 16$
(q) $x^2 + 7x + 12$
(r) $y^2 - 4y - 5$
(s) $x^2 - 8xy + 15y^2$
(t) $2z^3 + 10z^2 - 28z$
(u) $15 + 2x - x^2$

5.14 (a) $m^4 - 4m^2 - 21$
(b) $a^4 - 20a^2 + 64$
(c) $4s^4t - 4s^3t^2 - 24s^2t^3$
(d) $x^{2m+4} + 5x^{m+4} - 50x^4$
(e) $2x^2 + 3x + 1$
(f) $3y^2 - 11y + 6$
(g) $5m^3 - 3m^2 - 2m$
(h) $6x^2 + 5xy - 6y^2$
(i) $36z^6 - 13z^4 + z^2$
(j) $12(x - y)^2 + 7(x - y) - 12$
(k) $4x^{2n+2} - 4x^{n+2} - 3x^2$

5.15 (a) $y^3 + 27$
(b) $x^3 - 1$
(c) $x^3y^3 + 8$
(d) $8z^4 - 27z^7$
(e) $8x^4y - 64xy^4$
(f) $m^9 - n^9$
(g) $y^6 + 1$
(h) $(x - 2)^3 + (y + 1)^3$
(i) $8x^6 + 7x^3 - 1$

5.16 (a) $xy + 3y - 2x - 6$
(b) $2pr - ps + 6qr - 3qs$
(c) $ax^2 + bx - ax - b$
(d) $x^3 - xy^2 - x^2y + y^3$
(e) $z^7 - 2z^6 + z^4 - 2z^3$
(f) $m^3 - mn^2 + m^2n - n^3 + m^2 - n^2$

5.17 (a) z^5+1 (b) x^5+32y^5 (c) $32-u^5$ (d) $m^{10}-1$ (e) $1-z^7$

5.18 (a) z^4+64 (b) $4x^4+3x^2y^2+y^4$ (c) x^8-12x^4+16 (d) $m^2-4p^2+4mn+4n^2$ (e) $6ab+4-a^2-9b^2$ (f) $9x^2-x^2y^2+4y^2+12xy$ (g) $x^2+y^2-4z^2+2xy+3xz+3yz$

5.19 Find the GCF and LCM of each group of polynomials.

(a) $16y^2z^4$, $24y^3z^2$

(b) $9r^3s^2t^5$, $12r^2s^4t^3$, $21r^5s^2$

(c) $x^2-3xy+2y^2$, $4x^2-16xy+16y^2$

(d) $6y^3+12y^2z$, $6y^2-24z^2$, $4y^2-4yz-24z^2$

(e) x^5-x, x^5-x^2, x^5-x^3

ANSWERS TO SUPPLEMENTARY PROBLEMS

5.13 (a) $3x^2y^3(y+2x)$ (b) $2s^2t(6t-3s^3t^3+2s^2)$ (c) $2xyz(x-2z+4yz^2)$ (d) $4(y+5)(y-5)$ (e) $(1+a^2)(1+a)(1-a)$ (f) $x(8+x)(8-x)$ (g) $8(x^2+4)(x+2)(x-2)$ (h) $2xy(3x+2y)(3x-2y)$ (i) $(2x+4y-z)(2x-2y+z)$ (j) $(8x+3y)(9y-4x)$ (k) $(x+2)^2$ (l) $(2-3y)^2$ (m) $(xy-4)^2$ (n) $xy(2x+3y)^2$ (o) $3(a^2+b^2)^2$ (p) $(m^2-n^2+4)^2$ (q) $(x+3)(x+4)$ (r) $(y-5)(y+1)$ (s) $(x-3y)(x-5y)$ (t) $2z(z+7)(z-2)$ (u) $(5-x)(3+x)$

5.14 (a) $(m^2-7)(m^2+3)$ (b) $(a+2)(a-2)(a+4)(a-4)$ (c) $4s^2t(s-3t)(s+2t)$ (d) $x^4(x^m-5)(x^m+10)$ (e) $(2x+1)(x+1)$ (f) $(3y-2)(y-3)$ (g) $m(5m+2)(m-1)$ (h) $(2x+3y)(3x-2y)$ (i) $z^2(2z+1)(2z-1)(3z+1)(3z-1)$ (j) $(4x-4y-3)(3x-3y+4)$ (k) $x^2(2x^n+1)(2x^n-3)$

5.15 (a) $(y+3)(y^2-3y+9)$ (b) $(x-1)(x^2+x+1)$ (c) $(xy+2)(x^2y^2-2xy+4)$ (d) $z^4(2-3z)(4+6z+9z^2)$ (e) $8xy(x-2y)(x^2+2xy+4y^2)$ (f) $(m-n)(m^2+mn+n^2)(m^6+m^3n^3+n^6)$ (g) $(y^2+1)(y^4-y^2+1)$ (h) $(x+y-1)(x^2-xy+y^2-5x+4y+7)$ (i) $(2x-1)(4x^2+2x+1)(x+1)(x^2-x+1)$

5.16 (a) $(x+3)(y-2)$ (b) $(2r-s)(p+3q)$ (c) $(ax+b)(x-1)$ (d) $(x-y)^2(x+y)$ (e) $z^3(z-2)(z+1)(z^2-z+1)$ (f) $(m+n)(m-n)(m+n+1)$

5.17 (a) $(z+1)(z^4-z^3+z-z+1)$

(b) $(x+2y)(x^4-2x^3y+4x^2y^2-8xy^3+16y^4)$

(c) $(2-u)(16+8u+4u^2+2u^3+u^4)$

(d) $(m+1)(m^4-m^3+m^2-m+1)(m-1)(m^4+m^3+m^2+m+1)$

(e) $(1-z)(1+z+z^2+z^3+z^4+z^5+z^6)$

5.18 (a) $(z^2+4z+8)(z^2-4z+8)$ (b) $(2x^2+xy+y^2)(2x^2-xy+y^2)$ (c) $(x^4+2x^2-4)(x^4-2x^2-4)$ (d) $(m+2n+2p)(m+2n-2p)$ (e) $(2+a-3b)(2-a+3b)$ (f) $(3x+xy+2y)(3x-xy+2y)$ (g) $(x+y+4z)(x+y-z)$

5.19 (*a*) GCF $= 2^3y^2z^2 = 8y^2z^2$, LCM $= 2^4 \cdot 3y^3z^4 = 48y^3z^4$

(*b*) GCF $= 3r^2s^2$, LCM $= 252r^5s^4t^5$

(*c*) GCF $= x - 2y$, LCM $= 4(x - y)(x - 2y)^2$

(*d*) GCF $= 2(y + 2z)$, LCM $= 12y^2(y + 2z)(y - 2z)(y - 3z)$

(*e*) GCF $= x(x - 1)$, LCM $= x^3(x + 1)(x - 1)(x^2 + 1)(x^2 + x + 1)$

Chapter 6

Fractions

6.1 RATIONAL ALGEBRAIC FRACTIONS

A rational algebraic fraction is an expression which can be written as the quotient of two polynomials, P/Q. P is called the numerator and Q the denominator of the fraction. Thus

$$\frac{3x-4}{x^2-6x+8} \quad \text{and} \quad \frac{x^3+2y^2}{x^4-3xy+2y^3}$$

are rational algebraic fractions.

Rules for manipulation of algebraic fractions are the same as for fractions in arithmetic. One such fundamental rule is: The value of a fraction is unchanged if its numerator and denominator are both multiplied by the same quantity or both divided by the same quantity, provided only that this quantity is not zero. In such case we call the fractions *equivalent*.

For example, if we multiply the numerator and denominator of $(x+2)/(x-3)$ by $(x-1)$ we obtain the equivalent fraction

$$\frac{(x+2)(x-1)}{(x-3)(x-1)} = \frac{x^2+x-2}{x^2-4x+3}$$

provided $(x-1)$ is not zero, i.e. provided $x \neq 1$.

Similarly, given the fraction $(x^2+3x+2)/(x^2+4x+3)$ we may write it as

$$\frac{(x+2)(x+1)}{(x+3)(x+1)}$$

and divide numerator and denominator by $(x+1)$ to obtain $(x+2)/(x+3)$ provided $(x+1)$ is not zero, i.e., provided $x \neq -1$. The operation of dividing out common factors of the numerator and denominator is called *cancellation* and may be indicated by a sloped line thus:

$$\frac{(x+2)\cancel{(x+1)}}{(x+3)\cancel{(x+1)}} = \frac{x+2}{x+3}.$$

To simplify a given fraction is to convert it into an equivalent form in which numerator and denominator have no common factor (except ± 1). In such case we say that the fraction is *reduced to lowest terms*. This reduction is achieved by factoring numerator and denominator and canceling common factors assuming they are not equal to zero.

Thus $\quad \dfrac{x^2-4xy+3y^2}{x^2-y^2} = \dfrac{(x-3y)\cancel{(x-y)}}{(x+y)\cancel{(x-y)}} = \dfrac{x-3y}{x+y} \quad$ provided $(x-y) \neq 0$.

Three signs are associated with a fraction: the sign of the numerator, of the denominator, and of the entire fraction. Any two of these signs may be changed without changing the value of the fraction. If there is no sign before a fraction, a plus sign is implied.

EXAMPLES 6.1.

$$\frac{-a}{b} = \frac{a}{-b} = -\frac{a}{b}, \qquad \frac{-a}{-b} = \frac{a}{b}, \qquad -\left(\frac{-a}{-b}\right) = -\frac{a}{b}$$

Change of sign may often be of use in simplification. Thus

$$\frac{x^2-3x+2}{2-x}=\frac{(x-2)(x-1)}{2-x}=\frac{(x-2)(x-1)}{-(x-2)}=\frac{x-1}{-1}=1-x.$$

6.2 OPERATIONS WITH ALGEBRAIC FRACTIONS

The algebraic sum of fractions having a *common denominator* is a fraction whose numerator is the algebraic sum of the numerators of the given fractions and whose denominator is the common denominator.

EXAMPLES 6.2.

$$\frac{3}{5}-\frac{4}{5}-\frac{2}{5}+\frac{1}{5}=\frac{3-4-2+1}{5}=\frac{-2}{5}=-\frac{2}{5}$$

$$\frac{2}{x-3}-\frac{3x+4}{x-3}+\frac{x^2+5}{x-3}=\frac{2-(3x+4)+(x^2+5)}{x-3}=\frac{x^2-3x+3}{x-3}$$

To add and subtract fractions having *different denominators*, write each of the given fractions as equivalent fractions all having a common denominator.

The *least common denominator* (LCD) of a given set of fractions is the LCM of the denominators of the fractions.

Thus the LCD of $\frac{3}{4}$, $\frac{4}{5}$, and $\frac{7}{10}$ is the LCM of 4, 5, 10 which is 20, and the LCD of

$$\frac{2}{x^2}, \frac{3}{2x}, \frac{x}{7} \quad \text{is } 14x^2.$$

EXAMPLES 6.3.

$$\frac{3}{4}-\frac{4}{5}+\frac{7}{10}=\frac{15}{20}-\frac{16}{20}+\frac{14}{20}=\frac{15-16+14}{20}=\frac{13}{20}$$

$$\frac{2}{x^2}-\frac{3}{2x}-\frac{x}{7}=\frac{2(14)-3(7x)-x(2x^2)}{14x^2}=\frac{28-21x-2x^3}{14x^2}$$

$$\frac{2x+1}{x(x+2)}-\frac{3}{(x+2)(x-1)}=\frac{(2x+1)(x-1)-3x}{x(x+2)(x-1)}=\frac{2x^2-4x-1}{x(x+2)(x-1)}$$

The product of two or more given fractions produces a fraction whose numerator is the product of the numerators of the given fractions and whose denominator is the product of the denominators of the given fractions.

EXAMPLES 6.4.

$$\frac{2}{3}\cdot\frac{4}{5}\cdot\frac{15}{16}=\frac{2\cdot 4\cdot 15}{3\cdot 5\cdot 16}=\frac{1}{2}$$

$$\frac{x^2-9}{x^2-6x+5}\cdot\frac{x-5}{x+3}=\frac{(x+3)(x-3)}{(x-5)(x-1)}\cdot\frac{x-5}{x+3}$$

$$=\frac{\cancel{(x+3)}(x-3)\cancel{(x-5)}}{\cancel{(x-5)}(x-1)\cancel{(x+3)}}=\frac{x-3}{x-1}$$

The quotient of two given fractions is obtained by inverting the divisor and then multiplying.

EXAMPLES 6.5.

$$\frac{3}{8} \div \frac{5}{4} \quad \text{or} \quad \frac{3/8}{5/4} = \frac{3}{8} \cdot \frac{4}{5} = \frac{3}{10}$$

$$\frac{7}{x^2-4} \div \frac{xy}{x+2} = \frac{7}{(x+2)(x-2)} \cdot \frac{x+2}{xy} = \frac{7}{xy(x-2)}$$

6.3 COMPLEX FRACTIONS

A complex fraction is one which has one or more fractions in the numerator or denominator, or in both. To simplify a complex fraction:

Method 1

(1) Reduce the numerator and denominator to simple fractions.

(2) Divide the two resulting fractions.

EXAMPLE 6.6.

$$\frac{x - \dfrac{1}{x}}{1 + \dfrac{1}{x}} = \frac{\dfrac{x^2-1}{x}}{\dfrac{x+1}{x}} = \frac{x^2-1}{x} \cdot \frac{x}{x+1} = \frac{x^2-1}{x+1} = x-1$$

Method 2

(1) Multiply the numerator and denominator of the complex fraction by the LCM of all denominators of the fractions in the complex fraction.

(2) Reduce the resulting fraction to lowest terms.

EXAMPLE 6.7.

$$\frac{\dfrac{1}{x^2} - 4}{\dfrac{1}{x} - 2} = \frac{\left(\dfrac{1}{x^2} - 4\right)x^2}{\left(\dfrac{1}{x} - 2\right)x^2} = \frac{1-4x^2}{x-2x^2} = \frac{(1+2x)(1-2x)}{x(1-2x)}$$

$$= \frac{1+2x}{x}$$

Solved Problems

Reduction of Fractions to Lowest Terms

6.1 (a) $\dfrac{15x^2}{12xy} = \dfrac{3 \cdot 5 \cdot x \cdot x}{3 \cdot 4 \cdot x \cdot y} = \dfrac{5x}{4y}$ (c) $\dfrac{14a^3b^3c^2}{-7a^2b^4c^2} = -\dfrac{2a}{b}$

(b) $\dfrac{4x^2y}{18xy^3} = \dfrac{2 \cdot 2 \cdot x \cdot x \cdot y}{2 \cdot 9 \cdot x \cdot y \cdot y^2} = \dfrac{2x}{9y^2}$ (d) $\dfrac{8x-8y}{16x-16y} = \dfrac{8\cancel{(x-y)}}{16\cancel{(x-y)}} = \dfrac{1}{2}$ (where $x - y \neq 0$)

(e) $\dfrac{x^3y - y^3x}{x^2y - xy^2} = \dfrac{xy(x^2-y^2)}{xy(x-y)} = \dfrac{\cancel{xy(x-y)}(x+y)}{\cancel{(xy(x-y)}} = x + y$

(*f*) $\dfrac{x^2-4xy+3y^2}{y^2-x^2}=\dfrac{(x-3y)(x-y)}{(y-x)(y+x)}=-\dfrac{(x-3y)\cancel{(x-y)}}{\cancel{(x-y)}(y+x)}=-\dfrac{x-3y}{y+x}=\dfrac{3y-x}{y+x}$

(*g*) $\dfrac{6x^2-3xy}{-4x^2y+2xy^2}=\dfrac{3x(2x-y)}{2xy(y-2x)}=-\dfrac{3x\cancel{(2x-y)}}{2xy\cancel{(2x-y)}}=-\dfrac{3}{2y}$

(*h*) $\dfrac{r^3s+3r^2s+9rs}{r^3-27}=\dfrac{rs(r^2+3r+9)}{r^3-3^3}=\dfrac{rs(r^2+3r+9)}{(r-3)(r^2+3r+9)}=\dfrac{rs}{r-3}$

(*i*) $\dfrac{(8xy+4y^2)^2}{8x^3y+y^4}=\dfrac{(4y[2x+y])^2}{y(8x^3+y^3)}=\dfrac{16y^2(2x+y)^2}{y(2x+y)(4x^2-2xy+y^2)}=\dfrac{16y(2x+y)}{4x^2-2xy+y^2}$

(*j*) $\dfrac{x^{2n+1}-x^{2n}y}{x^{n+3}-x^ny^3}=\dfrac{x^{2n}(x-y)}{x^n(x^3-y^3)}=\dfrac{x^{2n}(x-y)}{x^n(x-y)(x^2+xy+y^2)}=\dfrac{x^n}{x^2+xy+y^2}$

Multiplication of Fractions

6.2 (*a*) $\dfrac{2x}{3y^2}\cdot\dfrac{6y}{x^2}=\dfrac{12xy}{3x^2y^2}=\dfrac{4}{xy}$ (*b*) $\dfrac{9}{3x+3}\cdot\dfrac{x^2-1}{6}=\dfrac{9}{3(x+1)}\cdot\dfrac{(x+1)(x-1)}{6}=\dfrac{x-1}{2}$

(*c*) $\dfrac{x^2-4}{xy^2}\cdot\dfrac{2xy}{x^2-4x+4}=\dfrac{(x+2)(x-2)}{xy^2}\cdot\dfrac{2xy}{(x-2)^2}=\dfrac{2(x+2)}{y(x-2)}$

(*d*) $\dfrac{6x-12}{4xy+4x}\cdot\dfrac{y^2-1}{2-3x+x^2}=\dfrac{6(x-2)}{4x(y+1)}\cdot\dfrac{(y+1)(y-1)}{(2-x)(1-x)}$

$=-\dfrac{6\cancel{(x-2)}\cancel{(y+1)}(y-1)}{4x\cancel{(y+1)}\cancel{(x-2)}(1-x)}=-\dfrac{3(y-1)}{2x(1-x)}=\dfrac{3(y-1)}{2x(x-1)}$

(*e*) $\left(\dfrac{ax+ab+cx+bc}{a^2-x^2}\right)\left(\dfrac{x^2-2ax+a^2}{x^2+(b+a)x+ab}\right)=\dfrac{(a+c)(x+b)}{(a-x)(a+x)}\cdot\dfrac{(x-a)(x-a)}{(x+a)(x+b)}$

$=-\dfrac{(a+c)(x+b)}{(x-a)(a+x)}\cdot\dfrac{(x-a)(x-a)}{(x+a)(x+b)}=-\dfrac{(a+c)(x-a)}{(x+a)^2}=\dfrac{(a+c)(a-x)}{(x+a)^2}$

Division of Fractions

6.3 (*a*) $\dfrac{5}{4}\div\dfrac{3}{11}=\dfrac{5}{4}\cdot\dfrac{11}{3}=\dfrac{55}{12}$

(*b*) $\dfrac{9}{7}\div\dfrac{4}{7}=\dfrac{9}{7}\cdot\dfrac{7}{4}=\dfrac{9}{4}$

(*c*) $\dfrac{3x}{2}\div\dfrac{6x^2}{4}=\dfrac{3x}{2}\cdot\dfrac{4}{6x^2}=\dfrac{1}{x}$

(*d*) $\dfrac{10xy^2}{3z}\div\dfrac{5xy}{6z^3}=\dfrac{10xy^2}{3z}\cdot\dfrac{6z^3}{5xy}=4yz^2$

(*e*) $\dfrac{x+2xy}{3x^2}\div\dfrac{2y+1}{6x}=\dfrac{x+2xy}{3x^2}\cdot\dfrac{6x}{2y+1}=\dfrac{x(1+2y)}{3x^2}\cdot\dfrac{6x}{(2y+1)}=2$

(*f*) $\dfrac{9-x^2}{x^4+6x^3}\div\dfrac{x^3-2x^2-3x}{x^2+7x+6}=\dfrac{9-x^2}{x^4+6x^3}\cdot\dfrac{x^2+7x+6}{x^3-2x^2-3x}$

$=\dfrac{(3-x)(3+x)}{x^3(x+6)}\cdot\dfrac{(x+1)(x+6)}{x(x-3)(x+1)}=-\dfrac{3+x}{x^4}$

(*g*) $\dfrac{2x^2-5x+2}{\left(\dfrac{2x-1}{3}\right)}=(2x^2-5x+2)\cdot\dfrac{3}{2x-1}=(2x-1)(x-2)\cdot\dfrac{3}{2x-1}=3(x-2)$

(*h*) $\dfrac{\left(\dfrac{x^2-5x+6}{x^2+7x-8}\right)}{\left(\dfrac{9-x^2}{64-x^2}\right)} = \dfrac{x^2-5x+6}{x^2+7x-8}\cdot\dfrac{64-x^2}{9-x^2} = \dfrac{(x-3)(x-2)}{(x+8)(x-1)}\cdot\dfrac{(8-x)(8+x)}{(3-x)(3+x)} = -\dfrac{(x-2)(8-x)}{(x-1)(3+x)}$

Addition and Subtraction of Fractions

6.4 (*a*) $\dfrac{1}{3}+\dfrac{1}{6}=\dfrac{2}{6}+\dfrac{1}{6}=\dfrac{3}{6}=\dfrac{1}{2}$

(*b*) $\dfrac{5}{18}+\dfrac{7}{24}=\dfrac{5(4)}{72}+\dfrac{7(3)}{72}=\dfrac{41}{72}$

(*c*) $\dfrac{x}{6}+\dfrac{5x}{21}=\dfrac{x(7)+5x(2)}{42}=\dfrac{17x}{42}$

(*d*) $\dfrac{3t^2}{5}-\dfrac{4t^2}{15}=\dfrac{3t^2(3)-4t^2(1)}{15}=\dfrac{5t^2}{15}=\dfrac{t^2}{3}$

(*e*) $\dfrac{1}{x}+\dfrac{1}{y}=\dfrac{y+x}{xy}$

(*f*) $\dfrac{3}{x}+\dfrac{4}{3y}=\dfrac{3(3y)+4(x)}{3xy}=\dfrac{9y+4x}{3xy}$

(*g*) $\dfrac{5}{2x}-\dfrac{3}{4x^2}=\dfrac{5(2x)-3(1)}{4x^2}=\dfrac{10x-3}{4x^2}$

(*h*) $\dfrac{3a}{bc}+\dfrac{2b}{ac}=\dfrac{3a(a)+2b(b)}{abc}=\dfrac{3a^2+2b^2}{abc}$

(*i*) $\dfrac{3t-1}{10}+\dfrac{5-2t}{15}=\dfrac{(3t-1)3+(5-2t)2}{30}=\dfrac{9t-3+10-4t}{30}=\dfrac{5t+7}{30}$

(*j*) $\dfrac{3}{x}-\dfrac{2}{x+1}+\dfrac{2}{x^2}=\dfrac{3x(x+1)-2x^2+2(x+1)}{x^2(x+1)}=\dfrac{x^2+5x+2}{x^2(x+1)}$

(*k*) $5-\dfrac{5}{x+3}+\dfrac{10}{x^2-9}=\dfrac{5(x^2-9)-5(x-3)+10}{x^2-9}=\dfrac{5(x^2-x-4)}{x^2-9}$

(*l*) $\dfrac{3}{y-2}-\dfrac{2}{y+2}-\dfrac{y}{y^2-4}=\dfrac{3(y+2)-2(y-2)-y}{y^2-4}=\dfrac{10}{y^2-4}$

(*m*) $\dfrac{5}{2s+4}-\dfrac{3}{s^2+3s+2}+\dfrac{s}{s^2-s-2}=\dfrac{5}{2(s+2)}-\dfrac{3}{(s+1)(s+2)}+\dfrac{s}{(s-2)(s+1)}$

$=\dfrac{5(s+1)(s-2)-3(2)(s-2)+s(2)(s+2)}{2(s+2)(s+1)(s-2)}=\dfrac{7s^2-7s+2}{2(s+2)(s+1)(s-2)}$

(*n*) $\dfrac{3x-6}{4x^2+12x-16}-\dfrac{2x-5}{6x^2-6}+\dfrac{3x^2+3}{8x^2+40x+32}=\dfrac{3x-6}{4(x+4)(x-1)}-\dfrac{2x-5}{6(x+1)(x-1)}+\dfrac{3x^2+3}{8(x+4)(x+1)}$

$=\dfrac{(3x-6)(6)(x+1)-(2x-5)(4)(x+4)+(3x^2+3)(3)(x-1)}{24(x+4)(x-1)(x+1)}=\dfrac{9x^3+x^2-21x+35}{24(x+4)(x-1)(x+1)}$

Complex Fractions

6.5 (*a*) $\dfrac{5/7}{3/4}=\dfrac{5}{7}\cdot\dfrac{4}{3}=\dfrac{20}{21}$ (*b*) $\dfrac{2/3}{7}=\dfrac{2}{3}\cdot\dfrac{1}{7}=\dfrac{2}{21}$ (*c*) $\dfrac{10}{5/6}=10\cdot\dfrac{6}{5}=\dfrac{60}{5}=12$

(*d*) $\dfrac{\left(\dfrac{2}{3}+\dfrac{5}{6}\right)}{\dfrac{3}{8}}=\dfrac{\left(\dfrac{4}{6}+\dfrac{5}{6}\right)}{\dfrac{3}{8}}=\dfrac{\dfrac{9}{6}}{\dfrac{3}{8}}=\dfrac{9}{6}\cdot\dfrac{8}{3}=4$

(*e*) $\dfrac{\left(\dfrac{x+y}{3x^2}\right)}{\left(\dfrac{x-y}{x}\right)}=\dfrac{x+y}{3x^2}\cdot\dfrac{x}{x-y}=\dfrac{x+y}{3x(x-y)}$

(*f*) $\dfrac{\left(\dfrac{2}{a-b}\right)}{a-b}=\dfrac{2}{a-b}\cdot\dfrac{1}{a-b}=\dfrac{2}{(a-b)^2}$

(*g*) $\dfrac{2a}{\left(\dfrac{a}{x+1}\right)}=2a\cdot\dfrac{x+1}{a}=2(x+1)$

$$(h)\quad \frac{\left(\dfrac{a+b}{a-b}-\dfrac{a-b}{a+b}\right)}{\left(1+\dfrac{a-b}{a+b}\right)}=\frac{\left(\dfrac{(a+b)^2-(a-b)^2}{(a-b)(a+b)}\right)}{\left(\dfrac{(a+b)+(a-b)}{a+b}\right)}=\frac{\left(\dfrac{4ab}{(a-b)(a+b)}\right)}{\left(\dfrac{2a}{a+b}\right)}=\frac{4ab}{(a-b)(a+b)}\cdot\frac{a+b}{2a}=\frac{2b}{a-b}$$

$$(i)\quad \frac{\left(\dfrac{2}{x+h-3}-\dfrac{2}{x-3}\right)}{h}=\frac{\left(\dfrac{2(x-3)-2(x+h-3)}{(x+h-3)(x-3)}\right)}{h}=\frac{\left(\dfrac{-2h}{(x+h-3)(x-3)}\right)}{h}=\frac{-2}{(x+h-3)(x-3)}$$

$$(j)\quad 3y+\frac{\left(1+\dfrac{2}{y}\right)}{\left(\dfrac{y+2}{y-2}\right)}=3y+\frac{\left(\dfrac{y+2}{y}\right)}{\left(\dfrac{y+2}{y-2}\right)}=3y+\left(\frac{y+2}{y}\right)\left(\frac{y-2}{y+2}\right)=3y+\frac{y-2}{y}=\frac{3y^2+y-2}{y}$$

$$(k)\quad \frac{1}{1-\dfrac{1}{\left(1+\dfrac{1}{x}\right)}}=\frac{1}{1-\dfrac{1}{\left(\dfrac{x+1}{x}\right)}}=\frac{1}{1-\dfrac{x}{x+1}}=\frac{1}{\left(\dfrac{x+1-x}{x+1}\right)}=\frac{1}{\left(\dfrac{1}{x+1}\right)}=x+1$$

$$(l)\quad \frac{a}{a-b+\dfrac{a+b}{\left(\dfrac{a}{b}-\dfrac{b}{a}\right)}}=\frac{a}{a-b+\dfrac{a+b}{\left(\dfrac{a^2-b^2}{ab}\right)}}=\frac{a}{a-b+\dfrac{ab(a+b)}{(a+b)(a-b)}}=\frac{a}{(a-b)+\dfrac{ab}{a-b}}$$

$$=\frac{a}{\left(\dfrac{(a-b)^2+ab}{a-b}\right)}=\frac{a}{\left(\dfrac{a^2-ab+b^2}{a-b}\right)}=\frac{a(a-b)}{a^2-ab+b^2}$$

$$(m)\quad 1-\frac{1}{2-\dfrac{1}{\left(3-\dfrac{2a-1}{2a+1}\right)}}=1-\frac{1}{2-\dfrac{1}{\left(\dfrac{3(2a+1)-(2a-1)}{2a+1}\right)}}=1-\frac{1}{2-\left(\dfrac{2a+1}{4a+4}\right)}$$

$$=1-\frac{1}{\left(\dfrac{2(4a+4)-(2a+1)}{4a+4}\right)}=1-\frac{4a+4}{6a+7}=\frac{6a+7-(4a+4)}{6a+7}=\frac{2a+3}{6a+7}$$

$$(n)\quad \frac{\left(\dfrac{1}{x}+1\right)}{\left(\dfrac{1}{x}-1\right)}=\frac{\left(\dfrac{1}{x}+1\right)x}{\left(\dfrac{1}{x}-1\right)x}=\frac{1+x}{1-x}$$

$$(o)\quad \frac{\left(\dfrac{a+6}{a^2-9}\right)}{\left(\dfrac{a}{a-3}-\dfrac{a+4}{a+3}\right)}=\frac{\left(\dfrac{a+6}{(a+3)(a-3)}\right)(a+3)(a-3)}{\left(\dfrac{a}{a-3}-\dfrac{a+4}{a+3}\right)(a+3)(a-3)}=\frac{a+6}{a(a+3)-(a+4)(a-3)}$$

$$=\frac{a+6}{a^2+3a-(a^2+a-12)}=\frac{a+6}{a^2+3a-a^2-a+12}=\frac{a+6}{2a+12}=\frac{a+6}{2(a+6)}=\frac{1}{2}$$

$$(p)\quad \frac{\left(\dfrac{2}{6ab}-\dfrac{1}{4a}\right)}{\left(\dfrac{1}{3b^2}+\dfrac{1}{2a}\right)}=\frac{\left(\dfrac{2}{6ab}-\dfrac{1}{4a}\right)12ab^2}{\left(\dfrac{1}{3b^2}+\dfrac{1}{2a}\right)12ab^2}=\frac{2(2b)-1(3b^2)}{1(4a)+1(6b^2)}=\frac{4b-3b^2}{4a+6b^2}$$

Supplementary Problems

Show that:

6.6 (a) $\dfrac{24x^3y^2}{18xy^3} = \dfrac{4x^2}{3y}$ (d) $\dfrac{4x^2-16}{x^2-2x} = \dfrac{4(x+2)}{x}$ (g) $\dfrac{ax^4-a^2x^3-6a^3x^2}{9a^4x-a^2x^3} = -\dfrac{x(x+2a)}{a(x+3a)}$

(b) $\dfrac{36xy^4z^2}{-15x^4y^3z} = \dfrac{-12yz}{5x^3}$ (e) $\dfrac{y^2-5y+6}{4-y^2} = \dfrac{3-y}{y+2}$ (h) $\dfrac{xy-y^2}{x^4y-xy^4} = \dfrac{1}{x(x^2+xy+y^2)}$

(c) $\dfrac{5a^2-10ab}{a-2b} = 5a$ (f) $\dfrac{(x^2+4x)^2}{x^2+6x+8} = \dfrac{x^2(x+4)}{x+2}$ (i) $\dfrac{3a^2}{4b^3}\cdot\dfrac{2b^4}{9a^3} = \dfrac{b}{6a}$

6.7 (a) $\dfrac{8xyz^2}{3x^3y^2z}\cdot\dfrac{9xy^2z}{4xz^5} = \dfrac{6y}{x^2z^3}$ (d) $\dfrac{x^2-4y^2}{3xy+3x}\cdot\dfrac{2y^2-2}{2y^2+xy-x^2} = -\dfrac{2(x+2y)(y-1)}{3x(x+y)}$

(b) $\dfrac{xy^2}{2x-2y}\cdot\dfrac{x^2-y^2}{x^3y^2} = \dfrac{x+y}{2x^2}$ (e) $\dfrac{y^2-y-6}{y^2-2y+1}\cdot\dfrac{y^2+3y-4}{9y-y^3} = -\dfrac{(y+2)(y+4)}{y(y-1)(y+3)}$

(c) $\dfrac{x^2+3x}{4x^2-4}\cdot\dfrac{2x^2+2x}{x^2-9}\cdot\dfrac{x^2-4x+3}{x^2} = \dfrac{1}{2}$ (f) $\dfrac{t^3+3t^2+t+3}{4t^2-16t+16}\cdot\dfrac{8-t^3}{t^3+t} = \dfrac{(t+3)(t^2+2t+4)}{4t(2-t)}$

6.8 (a) $\dfrac{3x}{8y} \div \dfrac{9x}{16y} = \dfrac{2}{3}$ (b) $\dfrac{24x^3y^2}{5z^2} \div \dfrac{8x^2y^3}{15z^4} = \dfrac{9xz^2}{y}$ (c) $\dfrac{x^2-4y^2}{x^2+xy} \div \dfrac{x^2-xy-6y^2}{y^2+xy} = \dfrac{y(x-2y)}{x(x-3y)}$

6.9 (a) $\dfrac{6x^2-x-2}{\left(\dfrac{3x-2}{2x+1}\right)} = (2x+1)^2$ (b) $\dfrac{\left(\dfrac{y^2-3y+2}{y^2+4y-21}\right)}{\left(\dfrac{4-4y+y^2}{9-y^2}\right)} = -\dfrac{(y-1)(y+3)}{(y-2)(y+7)}$ (c) $\dfrac{\left(\dfrac{x^2y+xy^2}{x-y}\right)}{x+y} = \dfrac{xy}{x-y}$

6.10 (a) $\dfrac{2x}{3} - \dfrac{x}{2} = \dfrac{x}{6}$ (e) $\dfrac{1}{x+2} + \dfrac{1}{x-2} - \dfrac{x}{x^2-4} = \dfrac{x}{x^2-4}$

(b) $\dfrac{4}{3x} - \dfrac{5}{4x} = \dfrac{1}{12x}$ (f) $\dfrac{r-1}{r^2+r-6} - \dfrac{r+2}{r^2+4r+3} + \dfrac{1}{3r-6} = \dfrac{r^2+4r+12}{3(r+3)(r-2)(r+1)}$

(c) $\dfrac{3}{2y^2} - \dfrac{8}{y} = \dfrac{3-16y}{2y^2}$ (g) $\dfrac{x}{2x^2+3xy+y^2} - \dfrac{x-y}{y^2-4x^2} + \dfrac{y}{2x^2+xy-y^2} = \dfrac{3x^2+xy}{(2x+y)(2x-y)(x+y)}$

(d) $\dfrac{x+y^2}{x^2} + \dfrac{x-1}{x} - 1 = \dfrac{y^2}{x^2}$ (h) $\dfrac{a}{(c-a)(a-b)} + \dfrac{b}{(a-b)(b-c)} + \dfrac{c}{(b-c)(c-a)} = 0$

6.11 (a) $\dfrac{x+y}{\dfrac{1}{x}+\dfrac{1}{y}} = xy$ (d) $\dfrac{\left(\dfrac{x+1}{x-1} - \dfrac{x-1}{x+1}\right)}{\left(\dfrac{1}{x+1} + \dfrac{1}{x-1}\right)} = 2$

(b) $\dfrac{2+\dfrac{1}{x}}{2x^2+x} = \dfrac{1}{x^2}$ (e) $\dfrac{x}{1-\left(\dfrac{1}{1+\dfrac{x}{y}}\right)} = x+y$

(c) $\dfrac{\left(y+\dfrac{2y}{y-2}\right)}{\left(1+\dfrac{4}{y^2-4}\right)} = y+2$ (f) $2 - \dfrac{2}{1-\left(\dfrac{2}{2-\dfrac{2}{x^2}}\right)} = 2x^2$

Chapter 7

Exponents

7.1 POSITIVE INTEGRAL EXPONENT

If n is a positive integer, a^n represents the product of n factors each of which is a. Thus $a^4 = a \cdot a \cdot a \cdot a$. In a^n, a is called the base and n the exponent or index. We may read a^n as the "nth power of a" or "a to the nth." If $n = 2$ we read a^2 as "a squared"; a^3 is read "a cubed."

EXAMPLES 7.1.

$$x^3 = x \cdot x \cdot x, \qquad 2^5 = 2 \cdot 2 \cdot 2 \cdot 2 \cdot 2 = 32, \qquad (-3)^3 = (-3)(-3)\,(-3) = -27$$

7.2 NEGATIVE INTEGRAL EXPONENT

If n is a positive integer, we define

$$a^{-n} = \frac{1}{a^n} \qquad \text{assuming } a \neq 0.$$

EXAMPLES 7.2.

$$2^{-4} = \frac{1}{2^4} = \frac{1}{16}, \qquad \frac{1}{3^{-3}} = 3^3 = 27, \qquad -4x^{-2} = \frac{-4}{x^2}, \qquad (a+b)^{-1} = \frac{1}{(a+b)}$$

7.3 ROOTS

If n is a positive integer and if a and b are such that $a^n = b$, then a is said to be an nth root of b.

If b is positive, there is only one positive number a such that $a^n = b$. We write this positive number $\sqrt[n]{b}$ and call it the *principal* nth root of b.

EXAMPLE 7.3. $\sqrt[4]{16}$ is that positive number which when raised to the 4th power yields 16. Clearly this is +2, so we write $\sqrt[4]{16} = +2$.

EXAMPLE 7.4. The number -2 when raised to the 4th power also yields 16. We call -2 a 4th root of 16 but not the principal 4th root of 16.

If b is negative, there is no positive nth root of b, but there is a negative nth root of b if n is odd. We call this negative number the principal nth root of b and we write it $\sqrt[n]{b}$.

EXAMPLE 7.5. $\sqrt[3]{-27}$ is that number which raised to the third power (or cubed) yields -27. Clearly this is -3 and so we write $\sqrt[3]{-27} = -3$ as the principal cube root of -27.

EXAMPLE 7.6. If n is even, as in $\sqrt[4]{-16}$, there is no principal nth root in terms of real numbers.

Note. In advanced mathematics it can be shown that there are exactly n values of a such that $a^n = b$, $b \neq 0$, provided we allow imaginary (or complex) numbers.

7.4 RATIONAL EXPONENTS

If m and n are positive integers we define

$$a^{m/n} = \sqrt[n]{a^m} \text{ (assume } a \geq 0 \text{ if } n \text{ is even)}$$

EXAMPLES 7.7.

$$4^{3/2} = \sqrt{4^3} = \sqrt{64} = 8, \qquad (27)^{2/3} = \sqrt[3]{(27)^2} = 9$$

If m and n are positive integers we define

$$a^{-m/n} = \frac{1}{a^{m/n}}$$

EXAMPLES 7.8.

$$8^{-2/3} = \frac{1}{8^{2/3}} = \frac{1}{\sqrt[3]{8^2}} = \frac{1}{\sqrt[3]{64}} = \frac{1}{4}, \qquad x^{-5/2} = \frac{1}{x^{5/2}} = \frac{1}{\sqrt{x^5}}$$

We define $a^0 = 1$ if $a \neq 0$.

EXAMPLES 7.9.

$$10^0 = 1, \qquad (-3)^0 = 1, \qquad (ax)^0 = 1 \qquad (\text{if } ax \neq 0)$$

7.5 GENERAL LAWS OF EXPONENTS

If p and q are real numbers, the following laws hold.

A. $a^p \cdot a^q = a^{p+q}$

EXAMPLES 7.10.

$$2^3 \cdot 2^2 = 2^{3+2} = 2^5, \qquad 5^{-3} \cdot 5^7 = 5^{-3+7} = 5^4, \qquad 2^{1/2} \cdot 2^{5/2} = 2^3 = 8$$

$$3^{1/3} \cdot 3^{1/6} = 3^{1/3+1/6} = 3^{1/2} = \sqrt{3}, \qquad 3^9 \cdot 3^{-2} \cdot 3^{-3} = 3^4 = 81$$

B. $(a^p)^q = a^{pq}$

EXAMPLES 7.11.

$$(2^4)^3 = 2^{12}, \qquad (5^{1/3})^{-3} = 5^{(1/3)(-3)} = 5^{-1} = 1/5, \qquad (3^2)^0 = 3^0 = 1$$

$$(x^5)^{-4} = x^{-20}, \qquad (a^{2/3})^{3/4} = a^{(2/3)(3/4)} = a^{1/2}$$

C. $\dfrac{a^p}{a^q} = a^{p-q} \qquad a \neq 0$

EXAMPLES 7.12.

$$\frac{2^6}{2^4} = 2^{6-4} = 2^2 = 4, \qquad \frac{3^{-2}}{3^4} = 3^{-2-4} = 3^{-6}, \qquad \frac{x^{1/2}}{x^{-1}} = x^{1/2-(-1)} = x^{3/2}$$

$$\frac{(x+15)^{4/3}}{(x+15)^{5/6}} = (x+15)^{4/3-5/6} = (x+15)^{1/2} = \sqrt{x+15}$$

D. $(ab)^p = a^p b^p$

EXAMPLES 7.13.

$$(2 \cdot 3)^4 = 2^4 \cdot 3^4, \qquad (2x)^3 = 2^3x^3 = 8x^3, \qquad (3a)^{-2} = 3^{-2}a^{-2} = \frac{1}{9a^2}$$

$$(4x)^{1/2} = 4^{1/2}x^{1/2} = 2x^{1/2} = 2\sqrt{x}$$

E. $\left(\frac{a}{b}\right)^p = \frac{a^p}{b^p} \qquad b \neq 0$

EXAMPLES 7.14.

$$\left(\frac{2}{3}\right)^5 = \frac{2^5}{3^5} = \frac{32}{243}, \qquad \left(\frac{x^2}{y^3}\right)^{-3} = \frac{(x^2)^{-3}}{(y^3)^{-3}} = \frac{x^{-6}}{y^{-9}} = \frac{y^9}{x^6}$$

$$\left(\frac{5^3}{2^6}\right)^{-1/3} = \frac{(5^3)^{-1/3}}{(2^6)^{-1/3}} = \frac{5^{-1}}{2^{-2}} = \frac{2^2}{5^1} = \frac{4}{5}$$

7.6 SCIENTIFIC NOTATION

Very large and very small numbers are often written in scientific notation when they are used in computation. A number is written in scientific notation by expressing it as a number N times a power of 10 where $1 \leq N < 10$ and N contains all of the significant digits on the number.

EXAMPLES 7.15. Write each number in scientific notation. (*a*) 5 834 000, (*b*) 0.028 031, (*c*) 45.6.

(*a*) $5\,834\,000 = 5.834 \times 10^6$

(*b*) $0.028\,031 = 2.8031 \times 10^{-2}$

(*c*) $45.6 = 4.56 \times 10^1$

We can enter the number 3.1416×10^3 into a calculator by using the EE key or the EXP key. When we enter 3.1416, press the EE key, and then enter 3 followed by pressing the ENTER key or the = key, we get a display of 3141.6. Similarly, we can enter 4.902×10^{-2} by entering 4.902, pressing the EE key, and then entering −2 followed by pressing the ENTER key to obtain a display of 0.049 02. The exponent can usually be any integer from −99 to 99. Depending on the number of digits in the number and the exponent used, a calculator may round off the number and/or leave the result in scientific notation. How many digits you can have in the number N varies from calculator to calculator, as does whether or not the calculator displays a particular result in scientific notation or in standard notation.

Calculators sometimes display the result in scientific notation, such as 3.69E-7 or 3.69^{-07}. In each case the answer is to be interpreted as 3.69×10^{-7}. Calculators display the significant digits in the result followed by the power of 10 to be used. When entering a number in scientific notation into your calculator as part of a computation, press the operation sign after each number until you are ready to compute the result.

EXAMPLE 7.16. Compute $(1.892 \times 10^8) \times (5.34 \times 10^{-3})$ using a calculator.

Enter 1.892, press the EE key, enter 8, press the × sign, enter 5.34, press the EE key, enter −3, and press the ENTER key and we get a display of 1 010 328.

$$(1.892 \times 10^8) \times (5.34 \times 10^{-3}) = 1\,010\,328$$

Solved Problems

Positive Integral Exponent

7.1 (*a*) $2^3 = 2\cdot 2\cdot 2 = 8$

(*b*) $(-3)^4 = (-3)(-3)(-3)(-3) = 81$

(*c*) $\left(\dfrac{2}{3}\right)^5 = \left(\dfrac{2}{3}\right)\left(\dfrac{2}{3}\right)\left(\dfrac{2}{3}\right)\left(\dfrac{2}{3}\right)\left(\dfrac{2}{3}\right) = \dfrac{32}{243}$

(*d*) $(3y)^2(2y)^3 = (3y)(3y)(2y)(2y)(2y) = 72y^5$

(*e*) $(-3xy^2)^3 = (-3xy^2)(-3xy^2)(-3xy^2) = -27x^3y^6$

Negative Integral Exponent

7.2 (*a*) $2^{-3} = \dfrac{1}{2^3} = \dfrac{1}{8}$

(*b*) $3^{-1} = \dfrac{1}{3^1} = \dfrac{1}{3}$

(*c*) $-4(4)^{-2} = -4\left(\dfrac{1}{4^2}\right) = -\dfrac{1}{4}$

(*d*) $-2b^{-2} = -2\left(\dfrac{1}{b^2}\right) = -\dfrac{2}{b^2}$

(*e*) $(-2b)^{-2} = \dfrac{1}{(-2b)^2} = \dfrac{1}{4b^2}$

(*f*) $5\cdot 10^{-3} = 5\left(\dfrac{1}{10^3}\right) = \dfrac{5}{1000} = \dfrac{1}{200}$

(*g*) $\dfrac{8}{10^{-2}} = 8\cdot 10^2 = 800$

(*h*) $\dfrac{4}{x^{-2}y^{-2}} = 4x^2y^2$

(*i*) $\left(\dfrac{3}{4}\right)^{-3} = \dfrac{1}{(3/4)^3} = \left(\dfrac{4}{3}\right)^3 = \dfrac{64}{27}$

(*j*) $\left(\dfrac{x}{y}\right)^{-3} = \dfrac{1}{(x/y)^3} = \left(\dfrac{y}{x}\right)^3 = \dfrac{y^3}{x^3}$

(*k*) $(0.02)^{-1} = \left(\dfrac{2}{100}\right)^{-1} = \dfrac{100}{2} = 50$

(*l*) $\dfrac{ab^{-4}}{a^{-2}b} = \dfrac{a\cdot a^2}{b\cdot b^4} = \dfrac{a^3}{b^5}$

(*m*) $\dfrac{x^{2n+1}}{y^{3n-1}} = x^{2n+1}y^{1-3n}$

(*n*) $\dfrac{(x-1)^{-2}(x+3)^{-1}}{(2x-4)^{-1}(x+5)^{-3}} = \dfrac{(2x-4)(x+5)^3}{(x-1)^2(x+3)}$

Rational Exponents

7.3 (*a*) $(8)^{2/3} = \sqrt[3]{8^2} = \sqrt[3]{64} = 4$

(*b*) $(-8)^{2/3} = \sqrt[3]{(-8)^2} = \sqrt[3]{64} = 4$

(*c*) $(-x^3)^{1/3} = \sqrt[3]{-x^3} = -x$

(*d*) $\left(\dfrac{1}{16}\right)^{1/2} = \sqrt{\dfrac{1}{16}} = \dfrac{1}{4}$

(*e*) $\left(-\dfrac{1}{8}\right)^{2/3} = \sqrt[3]{\left(-\dfrac{1}{8}\right)^2} = \sqrt[3]{\dfrac{1}{64}} = \dfrac{1}{4}$

7.4 (*a*) $x^{-1/3} = \dfrac{1}{x^{1/3}} = \dfrac{1}{\sqrt[3]{x}}$

(*b*) $(8)^{-2/3} = \dfrac{1}{8^{2/3}} = \dfrac{1}{\sqrt[3]{8^2}} = \dfrac{1}{4}$

(*c*) $(-8)^{-2/3} = \dfrac{1}{(-8)^{2/3}} = \dfrac{1}{\sqrt[3]{(-8)^2}} = \dfrac{1}{4}$

(*d*) $(-x^3)^{-1/3} = \dfrac{1}{(-x^3)^{1/3}} = \dfrac{1}{\sqrt[3]{-x^3}} = \dfrac{1}{-x} = -\dfrac{1}{x}$

(*e*) $(-1)^{-2/3} = \dfrac{1}{(-1)^{2/3}} = \dfrac{1}{\sqrt[3]{(-1)^2}} = 1$

(*f*) $-(1)^{-2/3} = -\dfrac{1}{1^{2/3}} = -1$

(*g*) $-(-1)^{-3/5} = -\dfrac{1}{(-1)^{3/5}} = -\dfrac{1}{\sqrt[5]{(-1)^3}} = -\dfrac{1}{\sqrt[5]{-1}} = -\dfrac{1}{-1} = 1$

7.5 (*a*) $7^0 = 1,\quad (-3)^0 = 1,\quad (-2/3)^0 = 1$

(*b*) $(x-y)^0 = 1,\quad$ if $\quad x - y \neq 0$

(*c*) $3x^0 = 3\cdot 1 = 3,\quad$ if $\quad x \neq 0$

(*d*) $(3x)^0 = 1,\quad$ if $\quad 3x \neq 0$, i.e. if $x \neq 0$

(*e*) $4\cdot 10^0 = 4\cdot 1 = 4$

(*f*) $(4\cdot 10)^0 = (40)^0 = 1$

(*g*) $-(1)^0 = -1$

(*h*) $(-1)^0 = 1$

(*i*) $(3x)^0(4y)^0 = 1\cdot 1 = 1,\quad$ if $\quad 3x \neq 0\quad$ and $\quad 4y \neq 0,\quad$ i.e. if $\quad x \neq 0,\quad y \neq 0$

(*j*) $-2(3x+2y-4)^0 = -2(1) = -2,\quad$ if $\quad 3x+2y-4 \neq 0$

(*k*) $\dfrac{(5x+3y)}{(5x+3y)^0} = \dfrac{5x+3y}{1} = 5x+3y,\quad$ if $\quad 5x+3y \neq 0$

(*l*) $4(x^2+y^2)(x^2+y^2)^0 = 4(x^2+y^2)(1) = 4(x^2+y^2),\quad$ if $\quad x^2+y^2 \neq 0$

General Laws of Exponents

Mathcad

7.6 (a) $a^p \cdot a^q = a^{p+q}$
(b) $a^3 \cdot a^5 = a^{3+5} = a^8$
(c) $3^4 \cdot 3^5 = 3^9$
(d) $a^{n+1} \cdot a^{n-2} = a^{2n-1}$
(e) $x^{1/2} \cdot x^{1/3} = x^{1/2+1/3} = x^{5/6}$
(f) $x^{1/2} \cdot x^{-1/3} = x^{1/2-1/3} = x^{1/6}$
(g) $10^7 \cdot 10^{-3} = 10^{7-3} = 10^4$
(h) $(4 \cdot 10^{-6})(2 \cdot 10^4) = 8 \cdot 10^{4-6} = 8 \cdot 10^{-2}$
(i) $a^x \cdot a^y \cdot a^{-z} = a^{x+y-z}$
(j) $(\sqrt{x+y})(x+y) = (x+y)^{1/2}\,(x+y)^1 = (x+y)^{3/2}$
(k) $10^{1.7} \cdot 10^{2.6} = 10^{4.3}$
(l) $10^{-4.1} \cdot 10^{3.5} \cdot 10^{-.1} = 10^{-4.1+3.5-.1} = 10^{-.7}$
(m) $\left(\frac{b}{a}\right)^{3/2} \cdot \left(\frac{b}{a}\right)^{-2/3} = \left(\frac{b}{a}\right)^{3/2-2/3} = \left(\frac{b}{a}\right)^{5/6}$
(n) $\left(\frac{x}{x+y}\right)^{-1}\left(\frac{x}{x+y}\right)^{1/2} = \left(\frac{x}{x+y}\right)^{-1/2} = \left(\frac{x+y}{x}\right)^{1/2} = \sqrt{\frac{x+y}{x}}$
(o) $(x^2+1)^{-5/2}(x^2+1)^0(x^2+1)^2 = (x^2+1)^{-5/2+0+2} = (x^2+1)^{-1/2} = \frac{1}{(x^2+1)^{1/2}} = \frac{1}{\sqrt{x^2+1}}$

7.7 (a) $(a^p)^q = a^{pq}$
(b) $(x^3)^4 = x^{3\cdot 4} = x^{12}$
(c) $(a^{m+2})^n = a^{(m+2)n} = a^{mn+2n}$
(d) $(10^3)^2 = 10^{3\cdot 2} = 10^6$
(e) $(10^{-3})^2 = 10^{-3\cdot 2} = 10^{-6}$
(f) $(49)^{3/2} = (7^2)^{3/2} = 7^{2\cdot 3/2} = 7^3 = 343$
(g) $(3^{-1/2})^{-2} = 3^1 = 3$
(h) $(u^{-2})^{-3} = u^{(-2)(-3)} = u^6$
(i) $(81)^{3/4} = (3^4)^{3/4} = 3^3 = 27$
(j) $(\sqrt{x+y})^5 = [(x+y)^{1/2}]^5 = (x+y)^{5/2}$
(k) $(\sqrt[3]{x^3+y^3})^6 = [(x^3+y^3)^{1/3}]^6 = (x^3+y^3)^{1/3\cdot 6} = (x^3+y^3)^2$
(l) $\sqrt[6]{\sqrt[3]{a^2}} = \sqrt[6]{a^{2/3}} = (a^{2/3})^{1/6} = a^{1/9} = \sqrt[9]{a}$

7.8 (a) $\frac{a^p}{a^q} = a^{p-q}$
(b) $\frac{a^5}{a^3} = a^{5-3} = a^2$
(c) $\frac{7^4}{7^3} = 7^{4-3} = 7^1 = 7$
(d) $\frac{p^{2n+3}}{p^{n+1}} = p^{(2n+3)-(n+1)} = p^{n+2}$
(e) $\frac{10^2}{10^5} = 10^{2-5} = 10^{-3}$
(f) $\frac{x^{m+3}}{x^{m-1}} = x^4$
(g) $\frac{y^{2/3}}{y^{1/3}} = y^{2/3-1/3} = y^{1/3}$
(h) $\frac{z^{1/2}}{z^{3/4}} = z^{1/2-3/4} = z^{-1/4}$
(i) $\frac{(x+y)^{3a+1}}{(x+y)^{2a+5}} = (x+y)^{a-4}$
(j) $\frac{8\cdot 10^2}{2\cdot 10^{-6}} = \frac{8}{2}\cdot 10^{2+6} = 4\cdot 10^8$
(k) $\frac{9\cdot 10^{-2}}{3\cdot 10^4} = \frac{9}{3}\cdot 10^{-2-4} = 3\cdot 10^{-6}$
(l) $\frac{a^3 b^{-1/2}}{ab^{-3/2}} = a^2 b^1 = a^2 b$
(m) $\frac{4x^3 y^{-2} z^{-3/2}}{2x^{-1/2} y^{-4} z} = \left(\frac{4}{2}\right) x^{3+1/2} y^{-2+4} z^{-3/2-1} = 2x^{7/2} y^2 z^{-5/2}$
(n) $\frac{8\sqrt[3]{x^2}\sqrt[4]{y}\sqrt{1/z}}{-2\sqrt[3]{x}\sqrt{y^5}\sqrt{z}} = \frac{8x^{2/3} y^{1/4} z^{-1/2}}{-2x^{1/3} y^{5/2} z^{1/2}} = -4x^{1/3} y^{-9/4} z^{-1}$

7.9 (a) $(ab)^p = a^p b^p$
(b) $(2a)^4 = 2^4 a^4 = 16a^4$
(c) $(3\cdot 10^2)^4 = 3^4 \cdot 10^8 = 81 \cdot 10^8$
(d) $(4x^8 y^4)^{1/2} = 4^{1/2}(x^8)^{1/2}(y^4)^{1/2} = 2x^4 y^2$
(e) $\sqrt[3]{64a^{12}b^6} = (64a^{12}b^6)^{1/3} = (64)^{1/3}(a^{12})^{1/3}(b^6)^{1/3} = 4a^4 b^2$
(f) $(x^{2n} y^{-1/2} z^{n-1})^2 = x^{4n} y^{-1} z^{2n-2}$
(g) $(27x^{3p} y^{6q} z^{12r})^{1/3} = (27)^{1/3}(x^{3p})^{1/3}(y^{6q})^{1/3}(z^{12r})^{1/3} = 3x^p y^{2q} z^{4r}$

7.10 (*a*) $\left(\dfrac{a}{b}\right)^p = \dfrac{a^p}{b^p}$

(*b*) $\left(\dfrac{2}{3}\right)^4 = \dfrac{2^4}{3^4} = \dfrac{16}{81}$

(*c*) $\left(\dfrac{3a}{4b}\right)^3 = \dfrac{(3a)^3}{(4b)^3} = \dfrac{27a^3}{64b^3}$

(*d*) $\left(\dfrac{x^2}{y^3}\right)^n = \dfrac{x^{2n}}{y^{3n}}$

(*e*) $\left(\dfrac{a^{m+1}}{b}\right)^m = \dfrac{a^{m^2+m}}{b^m}$

(*f*) $\left(\dfrac{a^2}{b^4}\right)^{3/2} = \dfrac{(a^2)^{3/2}}{(b^4)^{3/2}} = \dfrac{a^3}{b^6}$ (where $a \geq 0,\ b \neq 0$)

(*g*) $\left(\dfrac{2}{5}\right)^{-3} = \left(\dfrac{5}{2}\right)^3 = \dfrac{125}{8}$

(*h*) $\left(\dfrac{5^3}{2^6}\right)^{-1/3} = \left(\dfrac{2^6}{5^3}\right)^{1/3} = \dfrac{2^2}{5} = \dfrac{4}{5}$

(*i*) $\sqrt[3]{\dfrac{8x^{3n}}{27y^6}} = \left(\dfrac{8x^{3n}}{27y^6}\right)^{1/3} = \dfrac{(8x^{3n})^{1/3}}{(27y^6)^{1/3}} = \dfrac{8^{1/3}x^n}{27^{1/3}y^2} = \dfrac{2x^n}{3y^2}$

(*j*) $\left(\dfrac{a^{1/3}}{x^{1/3}}\right)^{3/2} = \dfrac{(a^{1/3})^{3/2}}{(x^{1/3})^{3/2}} = \dfrac{a^{1/2}}{x^{1/2}}$

(*k*) $\left(\dfrac{x^{-1/3}y^{-2}}{z^{-4}}\right)^{-3/2} = \dfrac{(x^{-1/3}y^{-2})^{-3/2}}{(z^{-4})^{-3/2}} = \dfrac{(x^{-1/3})^{-3/2}(y^{-2})^{-3/2}}{z^6} = \dfrac{x^{1/2}y^3}{z^6}$

(*l*) $\sqrt{\dfrac{\sqrt[5]{x}\,\sqrt[4]{y^3}}{\sqrt[3]{z^2}}} = \sqrt{\dfrac{x^{1/5}y^{3/4}}{z^{2/3}}} = \left(\dfrac{x^{1/5}y^{3/4}}{z^{2/3}}\right)^{1/2} = \dfrac{x^{1/10}y^{3/8}}{z^{1/3}}$

Miscellaneous Examples

7.11 (*a*) $2^3 + 2^2 + 2^1 + 2^0 + 2^{-1} + 2^{-2} + 2^{-3} = 8 + 4 + 2 + 1 + \dfrac{1}{2} + \dfrac{1}{4} + \dfrac{1}{8} = 15\dfrac{7}{8}$

(*b*) $4^{3/2} + 4^{1/2} + 4^{-1/2} + 4^{-3/2} = 8 + 2 + \dfrac{1}{2} + \dfrac{1}{8} = 10\dfrac{5}{8}$

(*c*) $\dfrac{4x^0}{2^{-4}} = 4(1)(2^4) = 4 \cdot 16 = 64$

(*d*) $10^4 + 10^3 + 10^2 + 10^1 + 10^0 + 10^{-1} + 10^{-2} = 10\,000 + 1000 + 100 + 10 + 1 + 0.1 + 0.01$
$= 11\,111.11$

(*e*) $3 \cdot 10^3 + 5 \cdot 10^2 + 2 \cdot 10^1 + 4 \cdot 10^0 = 3524$

(*f*) $\dfrac{4^{3n}}{2^n} = \dfrac{(2^2)^{3n}}{2^n} = \dfrac{2^{6n}}{2^n} = 2^{6n-n} = 2^{5n}$

(*g*) $(0.125)^{1/3}(0.25)^{-1/2} = \dfrac{\sqrt[3]{0.125}}{\sqrt{0.25}} = \dfrac{0.5}{0.5} = 1$

7.12 (*a*) Evaluate $4x^{-2/3} + 3x^{1/3} + 2x^0$ when $x = 8$.

$$4 \cdot 8^{-2/3} + 3 \cdot 8^{1/3} + 2 \cdot 8^0 = \frac{4}{8^{2/3}} + 3 \cdot 8^{1/3} + 2 \cdot 8^0 = \frac{4}{4} + 3 \cdot 2 + 2 \cdot 1 = 9$$

(*b*) Evaluate

$$\frac{(-3)^2(-2x)^{-3}}{(x+1)^{-2}}$$

when $x = 2$.

$$\frac{(-3)^2(-4)^{-3}}{3^{-2}} = \frac{9\left(\dfrac{1}{-4}\right)^3}{\dfrac{1}{3^2}} = 9\left(-\frac{1}{4}\right)^3(9) = -\frac{81}{64}$$

7.13 (*a*) $\dfrac{2^0 - 2^{-2}}{2 - 2(2)^{-2}} = \dfrac{1 - 1/2^2}{2 - 2/2^2} = \dfrac{1 - 1/4}{2 - 2/4} = \dfrac{3/4}{6/4} = \dfrac{1}{2}$

(*b*) $\dfrac{2a^{-1} + a^0}{a^{-2}} = \dfrac{\left(\dfrac{2}{a} + 1\right)}{\left(\dfrac{1}{a^2}\right)} = \dfrac{\left(\dfrac{2 + a}{a}\right)}{\left(\dfrac{1}{a^2}\right)} = \dfrac{2 + a}{a} \cdot a^2 = (2 + a)a = 2a + a^2$

(*c*) $\left(\dfrac{2^0}{8^{1/3}}\right)^{-1} = \left(\dfrac{1}{2}\right)^{-1} = \dfrac{1}{1/2} = 2$ or $\left(\dfrac{2^0}{8^{1/3}}\right)^{-1} = \left(\dfrac{8^{1/3}}{2^0}\right)^{1} = \dfrac{2}{1} = 2$

(*d*) $\left(\dfrac{1}{3}\right)^{-2} - (-3)^{-2} = (3)^2 - \left(-\dfrac{1}{3}\right)^2 = 9 - \dfrac{1}{9} = \dfrac{80}{9}$

(*e*) $\left(-\dfrac{1}{27}\right)^{-2/3} + \left(-\dfrac{1}{32}\right)^{2/5} = (-27)^{2/3} + \left(-\dfrac{1}{2^5}\right)^{2/5} = [(-3)^3]^{2/3} + \left[\left(-\dfrac{1}{2}\right)^5\right]^{2/5}$

$= (-3)^2 + \left(-\dfrac{1}{2}\right)^2 = \dfrac{37}{4}$

7.14 (*a*) $\dfrac{(-3a)^3 \cdot 3a^{-2/3}}{(2a)^{-2} \cdot a^{1/3}} = \dfrac{(-3a)^3 \cdot 3 \cdot (2a)^2}{a^{2/3} \cdot a^{1/3}} = \dfrac{-27a^3 \cdot 3 \cdot 4a^2}{a^{2/3+1/3}} = \dfrac{-324a^5}{a} = -324a^4$

(*b*) $\dfrac{(x^{-2})^{-3} \cdot (x^{-1/3})^9}{(x^{1/2})^{-3} \cdot (x^{-3/2})^5} = \dfrac{x^6 \cdot x^{-3}}{x^{-3/2} \cdot x^{-15/2}} = \dfrac{x^{6-3}}{x^{-3/2-15/2}} = \dfrac{x^3}{x^{-9}} = x^{12}$

(*c*) $\dfrac{\left(x + \dfrac{1}{y}\right)^m \cdot \left(x - \dfrac{1}{y}\right)^n}{\left(y + \dfrac{1}{x}\right)^m \cdot \left(y - \dfrac{1}{x}\right)^n} = \dfrac{\left(\dfrac{xy + 1}{y}\right)^m \cdot \left(\dfrac{xy - 1}{y}\right)^n}{\left(\dfrac{xy + 1}{x}\right)^m \cdot \left(\dfrac{xy - 1}{x}\right)^n} = \dfrac{\left(\dfrac{(xy + 1)^m}{y^m}\right) \cdot \left(\dfrac{(xy - 1)^n}{y^n}\right)}{\left(\dfrac{(xy + 1)^m}{x^m}\right) \cdot \left(\dfrac{(xy - 1)^n}{x^n}\right)}$

$= \dfrac{\left(\dfrac{(xy + 1)^m (xy - 1)^n}{y^{m+n}}\right)}{\left(\dfrac{(xy + 1)^m (xy - 1)^n}{x^{m+n}}\right)} = \dfrac{x^{m+n}}{y^{m+n}} = \left(\dfrac{x}{y}\right)^{m+n}$

(*d*) $\dfrac{3^{pq+q}}{3^{pq+p}} \cdot \dfrac{3^{2p}}{3^{2q}} = \dfrac{3^{pq+q+2p}}{3^{pq+p+2q}} = 3^{(pq+q+2p)-(pq+p+2q)} = 3^{p-q}$

7.15 (*a*) $\dfrac{(x^{3/4} \cdot x^{1/2})^{1/3}}{(y^{2/3} \cdot y^{4/3})^{1/2}} = \dfrac{(x^{5/4})^{1/3}}{(y^2)^{1/2}} = \dfrac{x^{5/12}}{y}$

(*b*) $\dfrac{(x^{3/4})^{2/3} - (y^{5/4})^{2/5}}{(x^{3/4})^{1/3} + (y^{2/3})^{3/8}} = \dfrac{x^{1/2} - y^{1/2}}{x^{1/4} + y^{1/4}} = \dfrac{(x^{1/4})^2 - (y^{1/4})^2}{x^{1/4} + y^{1/4}} = \dfrac{(x^{1/4} + y^{1/4})(x^{1/4} - y^{1/4})}{x^{1/4} + y^{1/4}} = x^{1/4} - y^{1/4}$

(*c*) $\dfrac{1}{1 + x^{p-q}} + \dfrac{1}{1 + x^{q-p}} = \dfrac{1}{1 + \left(\dfrac{x^p}{x^q}\right)} + \dfrac{1}{1 + \left(\dfrac{x^q}{x^p}\right)} = \dfrac{x^q}{x^q + x^p} + \dfrac{x^p}{x^p + x^q} = \dfrac{x^q + x^p}{x^q + x^p} = 1$

(*d*) $\dfrac{x^{3n} - y^{3n}}{x^n - y^n} = \dfrac{(x^n)^3 - (y^n)^3}{x^n - y^n} = \dfrac{(x^n - y^n)(x^{2n} + x^n y^n + y^{2n})}{x^n - y^n} = x^{2n} + x^n y^n + y^{2n}$

(*e*) $\sqrt[a]{a^{a^2-a}} = [a^{a(a-1)}]^{1/a} = a^{a-1}$

(*f*) $2^{2^{3^2}} = 2^{2^9} = 2^{512}$

7.16 (*a*) $(0.004)(30\,000)^2 = (4 \times 10^{-3})(3 \times 10^4)^2 = (4 \times 10^{-3})(3^2 \times 10^8) = 4 \cdot 3^2 \times 10^{-3+8}$
$= 36 \times 10^5$ or $3\,600\,000$

(*b*) $\dfrac{48\,000\,000}{1200} = \dfrac{48 \times 10^6}{12 \times 10^2} = 4 \times 10^{6-2} = 4 \times 10^4$ or $40\,000$

(c) $\dfrac{0.078}{0.00012} = \dfrac{78 \times 10^{-3}}{12 \times 10^{-5}} = 6.5 \times 10^{-3+5} = 6.5 \times 10^2 \quad \text{or} \quad 650$

(d) $\dfrac{(80\,000\,000)^2(0.000\,003)}{(600\,000)(0.0002)^4} = \dfrac{(8 \times 10^7)^2(3 \times 10^{-6})}{(6 \times 10^5)(2 \times 10^{-4})^4} = \dfrac{8^2 \cdot 3}{6 \cdot 2^4} \cdot \dfrac{10^{14} \cdot 10^{-6}}{10^5 \cdot 10^{-16}} = 2 \times 10^{19}$

(e) $\sqrt[3]{\dfrac{(0.004)^4(0.0036)}{(120\,000)^2}} = \sqrt[3]{\dfrac{(4 \times 10^{-3})^4(36 \times 10^{-4})}{(12 \times 10^4)^2}} = \sqrt[3]{\dfrac{256(36)}{144} \cdot \dfrac{10^{-12} \cdot 10^{-4}}{10^8}}$
$= \sqrt[3]{64 \times 10^{-24}} = 4 \times 10^{-8}$

7.17 For what real values of the variables involved will each of the following operations be valid and yield real numbers?

(a) $\sqrt{x^2} = (x^2)^{1/2} = x^1 = x$

(b) $\sqrt{a^2 + 2a + 1} = \sqrt{(a+1)^2} = a + 1$

(c) $\dfrac{a^{-2} - b^{-2}}{a^{-1} - b^{-1}} = \dfrac{(a^{-1})^2 - (b^{-1})^2}{a^{-1} - b^{-1}} = \dfrac{(a^{-1} + b^{-1})(a^{-1} - b^{-1})}{(a^{-1} - b^{-1})} = a^{-1} + b^{-1}$

(d) $\sqrt{x^4 + 2x^2 + 1} = \sqrt{(x^2+1)^2} = x^2 + 1$

(e) $\dfrac{x-1}{\sqrt{x-1}} = \dfrac{(x-1)^1}{(x-1)^{1/2}} = (x-1)^{1-1/2} = (x-1)^{1/2} = \sqrt{x-1}$

SOLUTION

(a) When x is a real number, $\sqrt{x^2}$ must be positive or zero. Assuming $\sqrt{x^2} = x$ were true for all x, then if $x = -1$ we would have $\sqrt{(-1)^2} = -1$ or $\sqrt{1} = -1$, i.e. $1 = -1$, a contradiction. Thus $\sqrt{x^2} = x$ cannot be true for all values of x. We will have $\sqrt{x^2} = x$ if $x \geq 0$. If $x \leq 0$ we have $\sqrt{x^2} = -x$. A result valid both for $x \geq 0$ and $x \leq 0$ may be $\sqrt{x^2} = |x|$ (absolute value of x).

(b) $\sqrt{a^2 + 2a + 1}$ must be positive or zero and thus will equal $a + 1$ if $a + 1 \geq 0$, i.e. if $a \geq -1$. A result valid for all real values of a is given by $\sqrt{a^2 + 2a + 1} = |a + 1|$.

(c) $(a^{-2} - b^{-2})/(a^{-1} - b^{-1})$ is not defined if a or b or both is equal to zero. Similarly it is not defined if the denominator $a^{-1} - b^{-1} = 0$, i.e. if $a^{-1} = b^{-1}$ or $a = b$. Hence the result $(a^{-2} - b^{-2})/(a^{-1} - b^{-1}) = a^{-1} + b^{-1}$ is valid if and only if $a \neq 0$, $b \neq 0$, and $a \neq b$.

(d) $\sqrt{x^4 + 2x^2 + 1}$ must be positive or zero and will equal $x^2 + 1$ if $x^2 + 1 \geq 0$. Since $x^2 + 1$ is greater than zero for all real numbers x, the result is valid for all real values of x.

(e) $\sqrt{x-1}$ will not be a real number if $x - 1 < 0$, i.e. if $x < 1$. Also, $(x-1)/(\sqrt{x-1})$ will not be defined if the denominator is zero, i.e. if $x = 1$. Hence $(x-1)/(\sqrt{x-1}) = \sqrt{x-1}$ if only if $x > 1$.

7.18 A student was asked to evaluate the expression $x + 2y + \sqrt{(x-2y)^2}$ for $x = 2$, $y = 4$. She wrote

$$x + 2y + \sqrt{(x-2y)^2} = x + 2y + x - 2y = 2x$$

and thus obtained the value $2x = 2(2) = 4$ for her answer. Was she correct?

SOLUTION

Putting $x = 2$, $y = 4$ in the given expression, we obtain

$$x + 2y + \sqrt{(x-2y)^2} = 2 + 2(4) + \sqrt{(2-8)^2} = 2 + 8 + \sqrt{36} = 2 + 8 + 6 = 16.$$

The student made the mistake of writing $\sqrt{(x-2y)^2} = x - 2y$ which is true only if $x \geq 2y$. If $x \leq 2y$, $\sqrt{(x-2y)^2} = 2y - x$. In all cases, $\sqrt{(x-2y)^2} = |x - 2y|$. The required simplification should have been $x + 2y + 2y - x = 4y$, which does give 16 when $y = 4$.

Supplementary Problems

Evaluate each of the following.

7.19
(*a*) 3^4
(*b*) $(-2x)^3$
(*c*) $\left(\frac{3y}{4}\right)^3$
(*d*) 4^{-3}
(*e*) $(-4x)^{-2}$
(*f*) $(2y^{-1})^{-1}$
(*g*) $\frac{3^{-1}x^2y^{-4}}{2^{-2}x^{-3}y^3}$
(*h*) $(16)^{1/4}$
(*i*) $\frac{8^{-2/3}(-8)^{2/3}}{8^{1/3}}$
(*j*) $(-a^3b^3)^{-2/3}$
(*k*) $-3(-1)^{-1/5}(4)^{-1/2}$
(*l*) $(10^3)^0$
(*m*) $(x-y)^0[(x-y)^4]^{-1/2}$
(*n*) $x^y \cdot x^{4y}$
(*o*) $3y^{2/3} \cdot y^{4/3}$
(*p*) $(4 \cdot 10^3)(3 \cdot 10^{-5})(6 \cdot 10^4)$

7.20
(*a*) $\frac{2^3 \cdot 2^{-2} \cdot 2^4}{2^{-1} \cdot 2^0 \cdot 2^{-3}}$
(*b*) $\frac{10^{x+y} \cdot 10^{y-x} \cdot 10^{y+1}}{10^{y+1} \cdot 10^{2y+1}}$
(*c*) $\frac{3^{1/2} \cdot 3^{-2/3}}{3^{-1/2} \cdot 3^{1/3}}$
(*d*) $\frac{(x+y)^{2/3}(x+y)^{-1/6}}{[(x+y)^2]^{1/4}}$
(*e*) $\frac{(10^2)^{-3}(10^3)^{1/6}}{\sqrt{10} \cdot (10^4)^{-1/2}}$
(*f*) $[(x^{-1})^{-2}]^{-3}$
(*g*) $\frac{4^{-1/2}a^{2/3}b^{-1/6}c^{-3/2}}{8^{2/3}a^{-1/3}b^{-2/3}c^{5/2}}$
(*h*) $\left(\frac{2^{-8} \cdot 3^4}{5^{-4}}\right)^{-1/4}$
(*i*) $\sqrt{\frac{\sqrt[4]{a^2}\sqrt[3]{b^5}}{c^{-2}d^2}}$

7.21
(*a*) $\sqrt{27^{-2/3}} + 5^{2/3} \cdot 5^{1/3}$
(*b*) $4\left(\frac{1}{2}\right)^0 + 2^{-1} - 16^{-1/2} \cdot 4 \cdot 3^0$
(*c*) $8^{2/3} + 3^{-2} - \frac{1}{9}(10)^0$
(*d*) $27^{2/3} - 3(3x)^0 + 25^{1/2}$
(*e*) $8^{2/3} \cdot 16^{-3/4} \cdot 2^0 - 8^{-2/3}$
(*f*) $\sqrt[3]{(x-2)^{-2}}$ when $x = -6$
(*g*) $x^{3/2} + 4x^{-1} - 5x^0$ when $x = 4$
(*h*) $y^{2/3} + 3y^{-1} - 2y^0$ when $y = 1/8$
(*i*) $64^{-2/3} \cdot 16^{5/4} \cdot 2^0 \cdot (\sqrt{3})^4$
(*j*) $\frac{\sqrt{a} \cdot a^{-2/3}}{\sqrt[6]{a^5}} + \frac{a^{-5/6}}{\sqrt[3]{a^2} \cdot a^{-1/2}}$
(*k*) $\left(\frac{\sqrt{72y^{2n}}}{3} \cdot 9^0\right)(2y^{n+2})^{-1}$

7.22
(*a*) $25^0 + 0.25^{1/2} - 8^{1/3} \cdot 4^{-1/2} + 0.027^{1/3}$
(*b*) $\frac{1}{8^{-2/3}} - 3a^0 + (3a)^0 + (27)^{-1/3} - 1^{3/2}$
(*c*) $\frac{3^{-2} + 5(2)^0}{3 - 4(3)^{-1}}$
(*d*) $\frac{3^0x + 4x^{-1}}{x^{-2/3}}$ if $x = 8$
(*e*) $\frac{2 + 2^{-1}}{5} + (-8)^0 - 4^{3/2}$
(*f*) $(64)^{-2/3} - 3(150)^0 + 12(2)^{-2}$
(*g*) $(0.125)^{-2/3} + \frac{3}{2 + 2^{-1}}$
(*h*) $\sqrt[n]{\frac{32}{2^{5+n}}}$
(*i*) $\frac{(60\,000)^3(0.00\,002)^4}{100^2(72\,000\,000)(0.0002)^5}$

7.23
(*a*) $\frac{(x^2+3x+4)^{1/3}[-\frac{1}{2}(5-x)^{-1/2}] - (5-x)^{1/2}[(x^2+3x+4)^{-2/3}(2x+3)/3]}{(x^2+3x+4)^{2/3}}$ if $x = 1$
(*b*) $\frac{(9x^2-5y)^{1/4}(2x) - x^2[\frac{1}{4}(9x^2-5y)^{-3/4}(18x)]}{(9x^2-5y)^{1/2}}$ if $x = 2,\quad y = 4$
(*c*) $\frac{(x+1)^{2/3}[\frac{1}{2}(x-1)^{-1/2}] - (x-1)^{1/2}[\frac{2}{3}(x+1)^{-1/3}]}{(x+1)^{4/3}}$
(*d*) $x - 1 + \sqrt{x^2 + 2x + 1}$
(*e*) $3x - 2y - \sqrt{4x^2 - 4xy + y^2}$

ANSWERS TO SUPPLEMENTARY PROBLEMS

7.19 (a) 81 (d) 1/64 (g) $\dfrac{4x^5}{3y^7}$ (j) $\dfrac{1}{a^2b^2}$ (m) $\dfrac{1}{(x-y)^2}$ (p) 7200

(b) $-8x^3$ (e) $\dfrac{1}{16x^2}$ (h) 2 (k) 3/2 (n) x^{5y}

(c) $\dfrac{27y^3}{64}$ (f) $y/2$ (i) 1/2 (l) 1 (o) $3y^2$

7.20 (a) 2^9 (b) 1/10 (c) 1 (d) 1 (e) 10^{-4} (f) x^{-6} (g) $\dfrac{a\sqrt{b}}{8c^4}$ (h) $\dfrac{4}{15}$ (i) $\dfrac{a^{1/4}b^{5/6}c}{d}$

7.21 (a) $\dfrac{16}{3}$ (b) $\dfrac{7}{2}$ (c) 4 (d) 11 (e) $\dfrac{1}{4}$ (f) $\dfrac{1}{4}$ (g) 4 (h) $\dfrac{89}{4}$ (i) 18 (j) $\dfrac{2}{a}$ (k) $\dfrac{\sqrt{2}}{y^2}$

7.22 (a) 0.8 (b) $\dfrac{4}{3}$ (c) $\dfrac{46}{15}$ (d) 34 (e) $-\dfrac{13}{2}$ (f) $\dfrac{1}{16}$ (g) $\dfrac{26}{5}$ (h) $\dfrac{1}{2}$ (i) 150

7.23 (a) $-\dfrac{1}{3}$

(b) $\dfrac{7}{8}$

(c) $\dfrac{7-x}{6(x-1)^{1/2}(x+1)^{5/3}}$

(d) $2x$ if $x \geq -1$, -2 if $x \leq -1$

(e) $x-y$ if $2x \geq y$, $5x-3y$ if $2x \leq y$

Chapter 8

Radicals

8.1 RADICAL EXPRESSIONS

A radical is an expression of the form $\sqrt[n]{a}$ which denotes the principal nth root of a. The positive integer n is the index, or order, of the radical and the number a is the radicand. The index is omitted if $n = 2$.

Thus $\sqrt[3]{5}$, $\sqrt[4]{7x^3 - 2y^2}$, $\sqrt{x+10}$ are radicals which have respectively indices 3, 4, and 2 and radicands 5, $7x^3 - 2y^2$, and $x + 10$.

8.2 LAWS FOR RADICALS

The laws for radicals are the same as the laws for exponents, since $\sqrt[n]{a} = a^{1/n}$. The following are the laws most frequently used. Note: If n is even, assume $a, b \geq 0$.

A. $(\sqrt[n]{a})^n = a$

EXAMPLES 8.1. $(\sqrt[3]{6})^3 = 6, \qquad (\sqrt[4]{x^2+y^2})^4 = x^2 + y^2$

B. $\sqrt[n]{ab} = \sqrt[n]{a}\sqrt[n]{b}$

EXAMPLES 8.2. $\sqrt[3]{54} = \sqrt[3]{27 \cdot 2} = \sqrt[3]{27} \cdot \sqrt[3]{2} = 3\sqrt[3]{2}, \qquad \sqrt[7]{x^2y^5} = \sqrt[7]{x^2}\,\sqrt[7]{y^5}$

C. $\sqrt[n]{\dfrac{a}{b}} = \dfrac{\sqrt[n]{a}}{\sqrt[n]{b}} \qquad b \neq 0$

EXAMPLES 8.3. $\sqrt[5]{\dfrac{5}{32}} = \dfrac{\sqrt[5]{5}}{\sqrt[5]{32}} = \dfrac{\sqrt[5]{5}}{2}, \qquad \sqrt[3]{\dfrac{(x+1)^3}{(y-2)^6}} = \dfrac{\sqrt[3]{(x+1)^3}}{\sqrt[3]{(y-2)^6}} = \dfrac{x+1}{(y-2)^2}$

D. $\sqrt[n]{a^m} = (\sqrt[n]{a})^m$

EXAMPLE 8.4. $\sqrt[3]{(27)^4} = (\sqrt[3]{27})^4 = 3^4 = 81$

E. $\sqrt[m]{\sqrt[n]{a}} = \sqrt[mn]{a}$

EXAMPLES 8.5. $\sqrt[3]{\sqrt{5}} = \sqrt[6]{5}, \qquad \sqrt[4]{\sqrt[3]{2}} = \sqrt[12]{2}, \qquad \sqrt[5]{\sqrt[3]{x^2}} = \sqrt[15]{x^2}$

8.3 SIMPLIFYING RADICALS

The form of a radical may be changed in the following ways.

(*a*) Removal of perfect nth powers from the radicand.

EXAMPLES 8.6. $\sqrt[3]{32} = \sqrt[3]{2^3(4)} = \sqrt[3]{2^3} \cdot \sqrt[3]{4} = 2\sqrt[3]{4}$

$\sqrt{8x^5y^7} = \sqrt{(4x^4y^6)(2xy)} = \sqrt{4x^4y^6}\,\sqrt{2xy} = 2x^2y^3\sqrt{2xy}$

(*b*) Reduction of the index of the radical.

EXAMPLES 8.7. $\sqrt[4]{64} = \sqrt[4]{2^6} = 2^{6/4} = 2^{3/2} = \sqrt{2^3} = \sqrt{8}$, where the index is reduced from 4 to 2.
$\sqrt[6]{25x^6} = \sqrt[6]{(5x^3)^2} = (5x^3)^{2/6} = (5x^3)^{1/3} = \sqrt[3]{5x^3} = x\sqrt[3]{5}$, where the index is reduced from 6 to 3.
Note. $\sqrt[4]{(-4)^2} = \sqrt[4]{16} = 2$. It is *incorrect* to write $\sqrt[4]{(-4)^2} = (-4)^{2/4} = (-4)^{1/2} = \sqrt{-4}$.

(*c*) Rationalization of the denominator in the radicand.

EXAMPLE 8.8. Rationalize the denominator of $\sqrt[3]{9/2}$.

Multiply the numerator and denominator of the radicand (9/2) by such a number as will make the denominator a perfect *n*th power (here $n = 3$) and then remove the denominator from under the radical sign. The number in this case is 2^2. Then

$$\sqrt[3]{\frac{9}{2}} = \sqrt[3]{\frac{9}{2}\left(\frac{2^2}{2^2}\right)} = \sqrt[3]{\frac{9(2^2)}{2^3}} = \frac{\sqrt[3]{36}}{2}.$$

EXAMPLE 8.9. Rationalize the denominator of $\sqrt[4]{\frac{7a^3y^2}{8b^6x^3}}$.

To make $8b^6x^3$ a perfect 4th power, multiply numerator and denominator by $2b^2x$. Then

$$\sqrt[4]{\frac{7a^3y^2}{8b^6x^3}} = \sqrt[4]{\frac{7a^3y^2}{8b^6x^3} \cdot \frac{2b^2x}{2b^2x}} = \sqrt[4]{\frac{14a^3y^2b^2x}{16b^8x^4}} = \frac{\sqrt[4]{14a^3y^2b^2x}}{2b^2x}$$

A radical is said to be in simplest form if:

(*a*) all perfect *n*th powers have been removed from the radical,
(*b*) the index of the radical is as small as possible,
(*c*) no fractions are present in the radicand, i.e., the denominator has been rationalized.

8.4 OPERATIONS WITH RADICALS

Two or more radicals are said to be similar if after being reduced to simplest form they have the same index and radicand.

Thus $\sqrt{32}$, $\sqrt{1/2}$, and $\sqrt{8}$ are similar since

$$\sqrt{32} = \sqrt{16 \cdot 2} = 4\sqrt{2}, \qquad \sqrt{\frac{1}{2}} = \sqrt{\frac{1}{2} \cdot \frac{2}{2}} = \frac{\sqrt{2}}{2}, \qquad \text{and} \qquad \sqrt{8} = \sqrt{4 \cdot 2} = 2\sqrt{2}.$$

Here each radicand is 2 and each index is 2. However, $\sqrt[3]{32}$ and $\sqrt[3]{2}$ are dissimilar since $\sqrt[3]{32} = \sqrt[3]{8 \cdot 4} = 2\sqrt[3]{4}$.

To add algebraically two or more radicals, reduce each radical to simplest form and combine terms with similar radicals. Thus:

$$\sqrt{32} - \sqrt{1/2} - \sqrt{8} = 4\sqrt{2} - \frac{\sqrt{2}}{2} - 2\sqrt{2} = \left(4 - \frac{1}{2} - 2\right)\sqrt{2} = \frac{3}{2}\sqrt{2}.$$

When multiplying two radicals, we choose the procedure to use based on whether or not the indices of the radicals are the same.

(*a*) To multiply two or more radicals having the *same index*, use Law *B*:

$$\sqrt[n]{a}\sqrt[n]{b} = \sqrt[n]{ab}.$$

EXAMPLES 8.10. $(2\sqrt[3]{4})(3\sqrt[3]{16}) = 2 \cdot 3\sqrt[3]{4}\,\sqrt[3]{16} = 6\sqrt[3]{64} = 6 \cdot 4 = 24$
$(3\sqrt[4]{x^2y})(\sqrt[4]{x^3y^2}) = 3\sqrt[4]{(x^2y)(x^3y^2)} = 3\sqrt[4]{x^5y^3} = 3x\sqrt[4]{xy^3}$.

(*b*) To multiply radicals with *different indices* it is convenient to use fractional exponents and the laws of exponents.

EXAMPLES 8.11. $\sqrt[3]{5}\sqrt{2} = 5^{1/3}\cdot 2^{1/2} = 5^{2/6}\cdot 2^{3/6} = (5^2\cdot 2^3)^{1/6} = (25\cdot 8)^{1/6} = \sqrt[6]{200}.$

$\sqrt[3]{4}\sqrt{2} = \sqrt[3]{2^2}\sqrt{2} = 2^{2/3}\cdot 2^{1/2} = 2^{4/6}\cdot 2^{3/6} = 2^{7/6} = \sqrt[6]{2^7} = 2\sqrt[6]{2}$

When dividing two radicals, we choose the procedure to use based on whether or not the indices of the radicals are the same.

(*a*) To divide two radicals having the *same index*, use Law C,

$$\frac{\sqrt[n]{a}}{\sqrt[n]{b}} = \sqrt[n]{\frac{a}{b}},$$

and simplify.

EXAMPLE 8.12.

$$\frac{\sqrt[3]{5}}{\sqrt[3]{3}} = \sqrt[3]{\frac{5}{3}} = \sqrt[3]{\frac{5}{3}\cdot\frac{3^2}{3^2}} = \sqrt[3]{\frac{45}{3^3}} = \frac{\sqrt[3]{45}}{3}$$

We may also rationalize the denominator directly, as follows.

$$\frac{\sqrt[3]{5}}{\sqrt[3]{3}} = \frac{\sqrt[3]{5}}{\sqrt[3]{3}}\cdot\frac{\sqrt[3]{3^2}}{\sqrt[3]{3^2}} = \frac{\sqrt[3]{5\cdot 3^2}}{\sqrt[3]{3^3}} = \frac{\sqrt[3]{45}}{3}$$

(*b*) To divide two radicals with *different indices* it is convenient to use fractional exponents and the laws of exponents.

EXAMPLES 8.13.

$$\frac{\sqrt{6}}{\sqrt[4]{2}} = \frac{6^{1/2}}{2^{1/4}} = \frac{6^{2/4}}{2^{1/4}} = \sqrt[4]{\frac{6^2}{2}} = \sqrt[4]{\frac{36}{2}} = \sqrt[4]{18}$$

$$\frac{\sqrt[3]{4}}{\sqrt{2}} = \frac{\sqrt[3]{2^2}}{\sqrt{2}} = \frac{2^{2/3}}{2^{1/2}} = \frac{2^{4/6}}{2^{3/6}} = 2^{1/6} = \sqrt[6]{2}$$

8.5 RATIONALIZING BINOMIAL DENOMINATORS

The binomial irrational numbers $\sqrt{a}+\sqrt{b}$ and $\sqrt{a}-\sqrt{b}$ are called conjugates of each other. Thus $2\sqrt{3}+\sqrt{2}$ and $2\sqrt{3}-\sqrt{2}$ are conjugates. The property of these conjugates that makes them useful is the fact that they are the sum and difference of the same two terms, so their product is the difference of the squares of these terms. Hence, $(\sqrt{a}+\sqrt{b})(\sqrt{a}-\sqrt{b}) = (\sqrt{a})^2 - (\sqrt{b})^2 = a - b$.

To rationalize a fraction whose denominator is a binomial with radicals of index 2, multiply numerator and denominator by the conjugate.

EXAMPLE 8.14.

$$\frac{5}{2\sqrt{3}+\sqrt{2}} = \frac{5}{2\sqrt{3}+\sqrt{2}}\cdot\frac{2\sqrt{3}-\sqrt{2}}{2\sqrt{3}-\sqrt{2}} = \frac{5(2\sqrt{3}-\sqrt{2})}{12-2} = \frac{2\sqrt{3}-\sqrt{2}}{2}$$

If the denominator of a fraction is $\sqrt[3]{a}+\sqrt[3]{b}$, we multiply the numerator and denominator of the fraction by $\sqrt[3]{a^2}-\sqrt[3]{ab}+\sqrt[3]{b^2}$ and get a denominator of $a+b$. If the original denominator has the form $\sqrt[3]{a}-\sqrt[3]{b}$, we multiply the numerator and denominator of the fraction by $\sqrt[3]{a^2}+\sqrt[3]{ab}+\sqrt[3]{b^2}$ and get a denominator of $a-b$. (See Section 4.2 for the special product rules.)

EXAMPLE 8.15.

$$\frac{3}{\sqrt[3]{5}-2}=\frac{3(\sqrt[3]{25}+2\sqrt[3]{5}+4)}{(\sqrt[3]{5}-2)(\sqrt[3]{25}+2\sqrt[3]{5}+4)}=\frac{3(\sqrt[3]{25}+2\sqrt[3]{5}+4)}{(\sqrt[3]{5})^3-(2)^3}$$
$$=\frac{3(\sqrt[3]{25}+2\sqrt[3]{5}+4)}{5-8}=\frac{3(\sqrt[3]{25}+2\sqrt[3]{5}+4)}{-3}=-\sqrt[3]{25}-2\sqrt[3]{5}-4$$

Solved Problems

Reduction of a Radical Expression to Simplest Form

8.1 (*a*) $\sqrt{18}=\sqrt{9\cdot 2}=\sqrt{3^2\cdot 2}=3\sqrt{2}$

(*b*) $\sqrt[3]{80}=\sqrt[3]{8\cdot 10}=\sqrt[3]{2^3\cdot 10}=2\sqrt[3]{10}$

(*c*) $5\sqrt[3]{243}=5\sqrt[3]{27\cdot 9}=5\sqrt[3]{3^3\cdot 9}=15\sqrt[3]{9}$

(*d*) $\sqrt[3]{648}=\sqrt[3]{8\cdot 27\cdot 3}=\sqrt[3]{2^3\cdot 3^3\cdot 3}=6\sqrt[3]{3}$

(*e*) $a\sqrt{9b^4c^3}=a\sqrt{3^2b^4c^2\cdot c}=3ab^2c\sqrt{c}$

(*f*) $\sqrt[6]{343}=\sqrt[6]{7^3}=7^{3/6}=7^{1/2}=\sqrt{7}$

(*g*) $\sqrt[6]{81a^2}=\sqrt[6]{3^4a^2}=3^{4/6}a^{2/6}=3^{2/3}a^{1/3}=\sqrt[3]{9a}$ Note that $a\geq 0$. See (*k*) below.

(*h*) $\sqrt[3]{64x^7y^{-6}}=\sqrt[3]{4^3x^6\cdot xy^{-6}}=4x^2y^{-2}\sqrt[3]{x}=\dfrac{4x^2}{y^2}\sqrt[3]{x}$

(*i*) $\sqrt[5]{(72)^4}=(72)^{4/5}=(8\cdot 9)^{4/5}=(2^3\cdot 3^2)^{4/5}=2^{12/5}\cdot 3^{8/5}$
$=(2^2\cdot 2^{2/5})(3\cdot 3^{3/5})=2^2\cdot 3\sqrt[5]{2^2\cdot 3^3}=12\sqrt[5]{108}$

(*j*) $(7\sqrt[3]{4ab})^2=49(4ab)^{2/3}=49\sqrt[3]{16a^2b^2}=98\sqrt[3]{2a^2b^2}$

(*k*) $2a\sqrt{a^2+6a+9}=2a\sqrt{(a+3)^2}=2a(a+3)$. The student is reminded that $\sqrt{(a+3)^2}$ is a positive number or zero; hence $\sqrt{(a+3)^2}=a+3$ only if $a+3\geq 0$. If values of a such that $a+3<0$ are included, we must write $\sqrt{(a+3)^2}=|a+3|$.

(*l*) $\dfrac{x-25}{\sqrt{x}+5}=\dfrac{(\sqrt{x}+5)(\sqrt{x}-5)}{\sqrt{x}+5}=\sqrt{x}-5$

(*m*) $\sqrt{12x^4-36x^2y^2+27y^4}=\sqrt{3(4x^4-12x^2y^2+9y^4)}=\sqrt{3(2x^2-3y^2)^2}=(2x^2-3y^2)\sqrt{3}$
Note that this is valid only if $2x^2\geq 3y^2$. See (*k*) above.

(*n*) $\sqrt[n]{a^nb^{2n}c^{3n+1}d^{n+2}}=(a^nb^{2n}c^{3n+1}d^{n+2})^{1/n}=ab^2c^3c^{1/n}dd^{2/n}=ab^2c^3d\sqrt[n]{cd^2}$

(*o*) $\sqrt[3]{\sqrt{256}}=\sqrt[3]{16}=\sqrt[3]{8\cdot 2}=2\sqrt[3]{2}$

(*p*) $\sqrt[4]{\sqrt[3]{6ab^2}}=[(6ab^2)^{1/3}]^{1/4}=(6ab^2)^{1/12}=\sqrt[12]{6ab^2}$

(*q*) $\sqrt[5]{729\sqrt{a^3}}=\sqrt[5]{729a^{3/2}}=(3^6a^{3/2})^{1/5}=3^{12/10}a^{3/10}=3\sqrt[10]{9a^3}$

Change in Form of a Radical

8.2 Express as radicals of the 12th order.

(*a*) $\sqrt[3]{5}=5^{1/3}=5^{4/12}=\sqrt[12]{5^4}=\sqrt[12]{625}$

(*b*) $\sqrt{ab}=(ab)^{1/2}=(ab)^{6/12}=\sqrt[12]{(ab)^6}=\sqrt[12]{a^6b^6}$

(*c*) $\sqrt[6]{x^n}=x^{n/6}=x^{2n/12}=\sqrt[12]{x^{2n}}$

8.3 Express in terms of radicals of least order.

(*a*) $\sqrt[4]{9}=9^{1/4}=(3^2)^{1/4}=3^{1/2}=\sqrt{3}$

(*b*) $\sqrt[12]{8x^3y^6}=\sqrt[12]{(2xy^2)^3}=(2xy^2)^{3/12}=(2xy^2)^{1/4}=\sqrt[4]{2xy^2}$

(*c*) $\sqrt[8]{a^2+2ab+b^2}=\sqrt[8]{(a+b)^2}=(a+b)^{2/8}=(a+b)^{1/4}=\sqrt[4]{a+b}$

8.4 Convert into entire radicals, i.e., radicals having coefficient 1.

(*a*) $6\sqrt{3} = \sqrt{36 \cdot 3} = \sqrt{108}$

(*b*) $4x^2\sqrt[3]{y^2} = \sqrt[3]{(4x^2)^3y^2} = \sqrt[3]{64x^6y^2}$

(*c*) $\dfrac{2x}{y}\sqrt[4]{\dfrac{2y}{x}} = \sqrt[4]{\left(\dfrac{2x}{y}\right)^4 \cdot \dfrac{2y}{x}} = \sqrt[4]{\dfrac{16x^4}{y^4} \cdot \dfrac{2y}{x}} = \sqrt[4]{\dfrac{32x^3}{y^3}}$

(*d*) $\dfrac{a-b}{a+b}\sqrt{\dfrac{a+b}{a-b}} = \sqrt{\dfrac{(a-b)^2}{(a+b)^2} \cdot \dfrac{a+b}{a-b}} = \sqrt{\dfrac{a-b}{a+b}}$

8.5 Determine which of the following irrational numbers is the greater:

(*a*) $\sqrt[3]{2}, \sqrt[4]{3}$ (*b*) $\sqrt{5}, \sqrt[3]{11}$ (*c*) $2\sqrt{5}, 3\sqrt{2}$

SOLUTION

(*a*) $\sqrt[3]{2} = 2^{1/3} = 2^{4/12} = (2^4)^{1/12} = (16)^{1/12}$; $\sqrt[4]{3} = 3^{1/4} = 3^{3/12} = (3^3)^{1/12} = (27)^{1/12}$.
Since $(27)^{1/12} > (16)^{1/12}$, $\sqrt[4]{3} > \sqrt[3]{2}$.

(*b*) $\sqrt{5} = 5^{1/2} = 5^{3/6} = (5^3)^{1/6} = (125)^{1/6}$; $\sqrt[3]{11} = (11)^{1/3} = (11)^{2/6} = (11^2)^{1/6} = (121)^{1/6}$.
Since $125 > 121$, $\sqrt{5} > \sqrt[3]{11}$.

(*c*) $2\sqrt{5} = \sqrt{2^2 \cdot 5} = \sqrt{20}$; $3\sqrt{2} = \sqrt{3^2 \cdot 2} = \sqrt{18}$.
Hence $2\sqrt{5} > 3\sqrt{2}$.

8.6 Rationalize the denominator.

(*a*) $\sqrt{\dfrac{2}{3}} = \sqrt{\dfrac{2}{3} \cdot \dfrac{3}{3}} = \sqrt{\dfrac{6}{3^2}} = \dfrac{1}{3}\sqrt{6}$

(*b*) $\dfrac{3}{\sqrt[3]{6}} = \dfrac{3}{\sqrt[3]{6}} \cdot \dfrac{\sqrt[3]{6^2}}{\sqrt[3]{6^2}} = \dfrac{3\sqrt[3]{6^2}}{\sqrt[3]{6 \cdot 6^2}} = \dfrac{3\sqrt[3]{36}}{6} = \dfrac{1}{2}\sqrt[3]{36}$

Another method: $\dfrac{3}{\sqrt[3]{6}} = \dfrac{3}{6^{1/3}} \cdot \dfrac{6^{2/3}}{6^{2/3}} = \dfrac{3 \cdot 6^{2/3}}{6^1} = \dfrac{3\sqrt[3]{6^2}}{6} = \dfrac{1}{2}\sqrt[3]{36}$

(*c*) $3x\sqrt[4]{\dfrac{y}{2x}} = 3x\sqrt[4]{\dfrac{y(2x)^3}{2x(2x)^3}} = 3x\sqrt[4]{\dfrac{y(8x^3)}{(2x)^4}} = \dfrac{3x}{2x}\sqrt[4]{8x^3y} = \dfrac{3}{2}\sqrt[4]{8x^3y}$

(*d*) $\sqrt{\dfrac{a-b}{a+b}} = \sqrt{\dfrac{a-b}{a+b} \cdot \dfrac{a+b}{a+b}} = \sqrt{\dfrac{a^2-b^2}{(a+b)^2}} = \dfrac{1}{a+b}\sqrt{a^2-b^2}$

(*e*) $\dfrac{4xy^2}{\sqrt[3]{2xy^2}} = \dfrac{4xy^2}{\sqrt[3]{2xy^2}} \cdot \dfrac{\sqrt[3]{(2xy^2)^2}}{\sqrt[3]{(2xy^2)^2}} = \dfrac{4xy^2\sqrt[3]{(2xy^2)^2}}{2xy^2} = 2\sqrt[3]{4x^2y^4} = 2y\sqrt[3]{4x^2y}$

Addition and Subtraction of Similar Radicals

8.7 (*a*) $\sqrt{18} + \sqrt{50} - \sqrt{72} = \sqrt{9 \cdot 2} + \sqrt{25 \cdot 2} - \sqrt{36 \cdot 2} = 3\sqrt{2} + 5\sqrt{2} - 6\sqrt{2} = (3+5-6)\sqrt{2} = 2\sqrt{2}$

(*b*) $2\sqrt{27} - 4\sqrt{12} = 2\sqrt{9 \cdot 3} - 4\sqrt{4 \cdot 3} = 2 \cdot 3\sqrt{3} - 4 \cdot 2\sqrt{3} = 6\sqrt{3} - 8\sqrt{3} = -2\sqrt{3}$

(*c*) $4\sqrt{75} + 3\sqrt{4/3} - 2\sqrt{48} = 4 \cdot 5\sqrt{3} + 3\sqrt{\dfrac{4}{3} \cdot \dfrac{3}{3}} - 2 \cdot 4\sqrt{3} = \left(20 + 3 \cdot \dfrac{2}{3} - 8\right)\sqrt{3} = 14\sqrt{3}$

(*d*) $\sqrt[3]{432} - \sqrt[3]{250} + \sqrt[3]{1/32} = \sqrt[3]{6^3 \cdot 2} - \sqrt[3]{5^3 \cdot 2} + \sqrt[3]{\dfrac{1}{2^5} \cdot \dfrac{2}{2}} = \left(6 - 5 + \dfrac{1}{4}\right)\sqrt[3]{2} = \dfrac{5}{4}\sqrt[3]{2}$

(*e*) $\sqrt{3} + \sqrt[3]{81} - \sqrt{27} + 5\sqrt[3]{3} = \sqrt{3} + \sqrt[3]{27 \cdot 3} - \sqrt{9 \cdot 3} + 5\sqrt[3]{3}$
$= \sqrt{3} + 3\sqrt[3]{3} - 3\sqrt{3} + 5\sqrt[3]{3} = -2\sqrt{3} + 8\sqrt[3]{3}$

(*f*) $2a\sqrt[3]{27x^3y} + 3b\sqrt[3]{8x^3y} - 6c\sqrt[3]{-x^3y} = 6ax\sqrt[3]{y} + 6bx\sqrt[3]{y} + 6cx\sqrt[3]{y} = 6x(a+b+c)\sqrt[3]{y}$

(g) $2\sqrt{\frac{2}{3}}+4\sqrt{\frac{3}{8}}-5\sqrt{\frac{1}{24}}=2\sqrt{\frac{2}{3}\cdot\frac{3}{3}}+4\sqrt{\frac{3}{8}\cdot\frac{2}{2}}-5\sqrt{\frac{1}{24}\cdot\frac{6}{6}}$

$=\left(\frac{2}{3}+4\cdot\frac{1}{4}-\frac{5}{12}\right)\sqrt{6}=\frac{5}{4}\sqrt{6}$

(h) $\frac{\sqrt{5}}{\sqrt{2}}+\frac{3}{\sqrt{0.1}}-\sqrt{1.6}=\sqrt{\frac{5}{2}\cdot\frac{2}{2}}+\frac{3}{\sqrt{1/10}}-\sqrt{(0.16)(10)}=\frac{1}{2}\sqrt{10}+3\sqrt{10}-0.4\sqrt{10}=3.1\sqrt{10}$

(i) $2\sqrt{\frac{a}{b}}-3\sqrt{\frac{b}{a}}+\frac{4}{\sqrt{ab}}=2\sqrt{\frac{a}{b}\cdot\frac{b}{b}}-3\sqrt{\frac{b}{a}\cdot\frac{a}{a}}+\frac{4}{\sqrt{ab}}\cdot\frac{\sqrt{ab}}{\sqrt{ab}}$

$=\frac{2}{b}\sqrt{ab}-\frac{3}{a}\sqrt{ab}+\frac{4}{ab}\sqrt{ab}=\left(\frac{2}{b}-\frac{3}{a}+\frac{4}{ab}\right)\sqrt{ab}=\left(\frac{2a-3b+4}{ab}\right)\sqrt{ab}$

Multiplication of Radicals

8.8 (a) $(2\sqrt{7})(3\sqrt{5})=(2\cdot3)\sqrt{7\cdot5}=6\sqrt{35}$

(b) $(3\sqrt[3]{2})(5\sqrt[3]{6})(8\sqrt[3]{4})=(3\cdot5\cdot8)\sqrt[3]{2\cdot6\cdot4}=120\sqrt[3]{48}=240\sqrt[3]{6}$

(c) $(\sqrt[3]{18x^2})(\sqrt[3]{2x})=\sqrt[3]{36x^3}=x\sqrt[3]{36}$

(d) $\sqrt[4]{ab^{-1}c^5}\cdot\sqrt[4]{a^3b^3c^{-1}}=\sqrt[4]{a^4b^2c^4}=\sqrt{a^2bc^2}=ac\sqrt{b}$

(e) $\sqrt{3}\cdot\sqrt[3]{2}=3^{1/2}\cdot2^{1/3}=3^{3/6}\cdot2^{2/6}=\sqrt[6]{3^3\cdot2^2}=\sqrt[6]{108}$

(f) $(\sqrt[3]{14})(\sqrt[4]{686})=(\sqrt[3]{7\cdot2})\ (\sqrt[4]{7^3\cdot2})=(7^{1/3}\cdot2^{1/3})\ (7^{3/4}\cdot2^{1/4})=(7^{4/12}\cdot2^{4/12})(7^{9/12}\cdot2^{3/12})$

$=7(7^{1/12}\cdot2^{7/12})=7\sqrt[12]{7\cdot2^7}=7\sqrt[12]{896}$

(g) $(-\sqrt{5}\sqrt[3]{x})^6=5^{6/2}x^{6/3}=5^3x^2=125x^2$

(h) $(\sqrt{4\times10^{-6}})(\sqrt{8.1\times10^3})(\sqrt{0.0016})=(\sqrt{4\times10^{-6}})(\sqrt{81\times10^2})(\sqrt{16\times10^{-4}})$

$=(2\times10^{-3})(9\times10)(4\times10^{-2})=72\times10^{-4}=0.0072$

(i) $(\sqrt{6}+\sqrt{3})(\sqrt{6}-2\sqrt{3})=\sqrt{6}\sqrt{6}+\sqrt{3}\sqrt{6}+(\sqrt{6})(-2\sqrt{3})\ +(\sqrt{3})(-2\sqrt{3})$

$=6+\sqrt{18}-2\sqrt{18}-2\cdot3=-\sqrt{18}=-3\sqrt{2}$

(j) $(\sqrt{5}+\sqrt{2})^2=(\sqrt{5})^2+\ 2(\sqrt{5})(\sqrt{2})+(\sqrt{2})^2=5+2\sqrt{10}+2=7+2\sqrt{10}$

(k) $(7\sqrt{5}-4\sqrt{3})^2=(7\sqrt{5})^2-\ 2(7\sqrt{5})(4\sqrt{3})+(4\sqrt{3})^2$

$=7^2\cdot5-2\cdot7\cdot4\sqrt{15}+4^2\cdot3=245-56\sqrt{15}+48=293-56\sqrt{15}$

(l) $(\sqrt{3}+1)(\sqrt{3}-1)=(\sqrt{3})^2-(1)^2=3-1=2$

(m) $(2\sqrt{3}-\sqrt{5})(2\sqrt{3}+\sqrt{5})=(2\sqrt{3})^2-(\sqrt{5})^2=4\cdot3-5=12-5=7$

(n) $(2\sqrt{5}-3\sqrt{2})(2\sqrt{5}+3\sqrt{2})=(2\sqrt{5})^2-(3\sqrt{2})^2=4\cdot5-9\cdot2=20-18=2$

(o) $(2+\sqrt[3]{3})(2-\sqrt[3]{3})=4-\sqrt[3]{9}$

(p) $(3\sqrt{2}+2\sqrt[3]{4})(3\sqrt{2}-2\sqrt[3]{4})\ =(3\sqrt{2})^2-(2\sqrt[3]{4})^2=18-4\sqrt[3]{16}=18\ -8\sqrt[3]{2}$

(q) $(3\sqrt{2}-4\sqrt{5})(2\sqrt{3}+3\sqrt{6})=(3\sqrt{2})(2\sqrt{3})-(4\sqrt{5})(2\sqrt{3})+(3\sqrt{2})(3\sqrt{6})-(4\sqrt{5})(3\sqrt{6})$

$=6\sqrt{6}-8\sqrt{15}+9\sqrt{12}-12\sqrt{30}\ =6\sqrt{6}-8\sqrt{15}+18\sqrt{3}-12\sqrt{30}$

(r) $(\sqrt{x+y}-z)(\sqrt{x+y}+z)\ =x+y-z^2$

(s) $(2\sqrt{x-1}-x\sqrt{2})(3\sqrt{x-1}\ +2x\sqrt{2})=6(x-1)-3x\sqrt{2(x-1)}+\ 4x\sqrt{2(x-1)}-4x^2$

$=6(x-1)+x\sqrt{2(x-1)}-4x^2$

8.9 (a) $(\sqrt{2}+\sqrt{3}+\sqrt{5})(\sqrt{2}+\sqrt{3}\ -\sqrt{5})$

$=[(\sqrt{2}+\sqrt{3})+\sqrt{5}][(\sqrt{2}+\sqrt{3})-\sqrt{5}]$

$=(\sqrt{2}+\sqrt{3})^2-(\sqrt{5})^2=2+2\sqrt{6}+3-5=2\sqrt{6}$

(b) $(2\sqrt{3}+3\sqrt{2}+1)(2\sqrt{3}-3\sqrt{2}-1)=[2\sqrt{3}+(3\sqrt{2}+1)][2\sqrt{3}-(3\sqrt{2}+1)]$

$=(2\sqrt{3})^2-(3\sqrt{2}+1)^2=12-(9\cdot2+6\sqrt{2}+1)=-7-6\sqrt{2}$

(c) $(\sqrt{2}+\sqrt{3}+\sqrt{5})^2=(\sqrt{2})^2+(\sqrt{3})^2+(\sqrt{5})^2+\ 2(\sqrt{2})(\sqrt{3})+2(\sqrt{3})(\sqrt{5})+2(\sqrt{2})(\sqrt{5})$

$=2+3+5+2\sqrt{6}+2\sqrt{15}+2\sqrt{10}=10+\ 2\sqrt{6}+2\sqrt{15}+2\sqrt{10}$

(d) $(\sqrt{6+3\sqrt{3}})=(\sqrt{6-3\sqrt{3}})=\sqrt{(6+3\sqrt{3})(6-3\sqrt{3})}=\sqrt{36-9\cdot3}=\sqrt{9}=3$

(e) $(\sqrt{a+b}-\sqrt{a-b})^2=a+b-2\sqrt{(a+b)(a-b)}+a-b=2a-2\sqrt{a^2-b^2}$

Division of Radicals. Rationalization of Denominators

8.10 (*a*) $\dfrac{10\sqrt{6}}{5\sqrt{2}} = \dfrac{10}{5}\sqrt{\dfrac{6}{2}} = 2\sqrt{3}$

(*b*) $\dfrac{2\sqrt[4]{30}}{3\sqrt[4]{5}} = \dfrac{2}{3}\sqrt[4]{\dfrac{30}{5}} = \dfrac{2}{3}\sqrt[4]{6}$

(*c*) $\dfrac{4x}{y} \cdot \dfrac{\sqrt[3]{x^2y^2}}{\sqrt[3]{xy}} = \dfrac{4x}{y}\sqrt[3]{\dfrac{x^2y^2}{xy}} = \dfrac{4x}{y}\sqrt[3]{xy}$

(*d*) $\dfrac{\sqrt{2}}{\sqrt{3}} = \sqrt{\dfrac{2}{3}} = \sqrt{\dfrac{2}{3}\cdot\dfrac{3}{3}} = \sqrt{\dfrac{6}{9}} = \dfrac{1}{3}\sqrt{6}$

(*e*) $\dfrac{\sqrt[3]{2}}{\sqrt[3]{3}} = \sqrt[3]{\dfrac{2}{3}} = \sqrt[3]{\dfrac{2}{3}\cdot\dfrac{9}{9}} = \sqrt[3]{\dfrac{18}{27}} = \dfrac{1}{3}\sqrt[3]{18}$

(*f*) $\sqrt[5]{\dfrac{1}{2}} = \sqrt[5]{\dfrac{1}{2}\cdot\dfrac{16}{16}} = \sqrt[5]{\dfrac{16}{32}} = \dfrac{1}{2}\sqrt[5]{16}$

(*g*) $\dfrac{\sqrt{3}+4\sqrt{2}-5\sqrt{8}}{\sqrt{2}} = \dfrac{\sqrt{3}+4\sqrt{2}-5\sqrt{8}}{\sqrt{2}} \cdot \dfrac{\sqrt{2}}{\sqrt{2}} = \dfrac{\sqrt{6}+4\cdot 2-5\sqrt{16}}{2} = \dfrac{\sqrt{6}-12}{2}$

(*h*) $\dfrac{3}{\sqrt{5}+\sqrt{2}} = \dfrac{3}{\sqrt{5}+\sqrt{2}} \cdot \dfrac{\sqrt{5}-\sqrt{2}}{\sqrt{5}-\sqrt{2}} = \dfrac{3(\sqrt{5}-\sqrt{2})}{5-2} = \sqrt{5}-\sqrt{2}$

(*i*) $\dfrac{1+\sqrt{2}}{1-\sqrt{2}} = \dfrac{1+\sqrt{2}}{1-\sqrt{2}} \cdot \dfrac{1+\sqrt{2}}{1+\sqrt{2}} = \dfrac{1+2\sqrt{2}+2}{1-2} = -(3+2\sqrt{2})$

(*j*) $\dfrac{1}{x-\sqrt{x^2-y^2}} - \dfrac{1}{x+\sqrt{x^2-y^2}} = \dfrac{(x+\sqrt{x^2-y^2})-(x-\sqrt{x^2-y^2})}{(x-\sqrt{x^2-y^2})(x+\sqrt{x^2-y^2})} = \dfrac{2\sqrt{x^2-y^2}}{y^2}$

(*k*) $\dfrac{3\sqrt{3}}{4\sqrt[3]{2}} = \dfrac{3\sqrt{3}}{4\sqrt[3]{2}} \cdot \dfrac{\sqrt[3]{4}}{\sqrt[3]{4}} = \dfrac{3(3^{3/6})(4^{2/6})}{4\sqrt[3]{8}} = \dfrac{3\sqrt[6]{3^3\cdot 4^2}}{8} = \dfrac{3}{8}\sqrt[6]{432}$

(*l*) $\dfrac{\sqrt{x-1}-\sqrt{x+1}}{\sqrt{x-1}-\sqrt{x+1}} \cdot \dfrac{\sqrt{x-1}-\sqrt{x+1}}{\sqrt{x-1}-\sqrt{x+1}} = \dfrac{(x-1)-2\sqrt{(x-1)(x+1)}+(x+1)}{(x-1)-(x+1)} = \sqrt{x^2-1}-x$

(*m*) $\dfrac{x+\sqrt{x}}{1+\sqrt{x}+x} = \dfrac{x+\sqrt{x}}{1+x+\sqrt{x}} \cdot \dfrac{1+x-\sqrt{x}}{1+x-\sqrt{x}} = \dfrac{x^2+\sqrt{x}}{(1+x)^2-x} = \dfrac{x^2+\sqrt{x}}{1+x+x^2}$

(*n*) $\dfrac{1}{\sqrt[3]{3}+\sqrt[3]{4}} = \dfrac{1}{3^{1/3}+4^{1/3}}$ Let $x = 3^{1/3}, y = 4^{1/3}$. Then

$\dfrac{1}{x+y} \cdot \dfrac{x^2-xy+y^2}{x^2-xy+y^2} = \dfrac{x^2-xy+y^2}{x^3+y^3} = \dfrac{3^{2/3}-3^{1/3}4^{1/3}+4^{2/3}}{(3^{1/3})^3+(4^{1/3})^3} = \dfrac{\sqrt[3]{9}-\sqrt[3]{12}+2\sqrt[3]{2}}{7}$

(*o*) $\dfrac{1}{\sqrt[3]{x}-\sqrt[3]{y}} = \dfrac{1(\sqrt[3]{x^2}+\sqrt[3]{xy}+\sqrt[3]{y^2})}{(\sqrt[3]{x}-\sqrt[3]{y})(\sqrt[3]{x^2}+\sqrt[3]{xy}+\sqrt[3]{y^2})} = \dfrac{\sqrt[3]{x^2}+\sqrt[3]{xy}+\sqrt[3]{y^2}}{x-y}$

(*p*) $\dfrac{m+n}{\sqrt[3]{m}+\sqrt[3]{n}} = \dfrac{(m+n)(\sqrt[3]{m^2}-\sqrt[3]{mn}+\sqrt[3]{n^2})}{(\sqrt[3]{m}+\sqrt[3]{n})(\sqrt[3]{m^2}-\sqrt[3]{mn}+\sqrt[3]{n^2})} = \dfrac{(m+n)(\sqrt[3]{m^2}-\sqrt[3]{mn}+\sqrt[3]{n^2})}{(m+n)}$

$= \sqrt[3]{m^2}-\sqrt[3]{mn}+\sqrt[3]{n^2}$

Supplementary Problems

Show that:

8.11 (*a*) $\sqrt{72} = 6\sqrt{2}$

(*b*) $\sqrt{27} = 3\sqrt{3}$

(*c*) $3\sqrt{20} = 6\sqrt{5}$

(*d*) $\dfrac{2}{5}\sqrt{50a^2} = 2a\sqrt{2}$ (assuming $a \ge 0$)

(*e*) $\dfrac{a}{b}\sqrt{75a^3b^2} = 5a^2\sqrt{3a}$

(f) $\frac{4}{ab}\sqrt{98a^2b^3} = 28\sqrt{2b}$

(g) $\sqrt[3]{640} = 4\sqrt[3]{10}$

(h) $\sqrt[3]{88x^3y^6z^5} = 2xy^2z\sqrt[3]{11z^2}$

(i) $\sqrt{a/b} = \frac{\sqrt{ab}}{b}$

(j) $14\sqrt{2/7} = 2\sqrt{14}$

(k) $3\sqrt[3]{2/3} = \sqrt[3]{18}$

(l) $\frac{3a}{4}\sqrt[3]{\frac{3}{2a}} = \frac{3}{8}\sqrt[3]{12a^2}$

(m) $xyz\sqrt{\frac{5}{2x^2yz}} = \frac{1}{2}\sqrt{10yz}$

(n) $60\sqrt{4/45} = 8\sqrt{5}$

(o) $3\sqrt[4]{4/9} = \sqrt{6}$

8.12 (a) $\sqrt{27} + \sqrt{48} - \sqrt{12} = 5\sqrt{3}$

(b) $5\sqrt{8} - 3\sqrt{18} = \sqrt{2}$

(c) $2\sqrt{150} - 4\sqrt{54} + 6\sqrt{48} = 24\sqrt{3} - 2\sqrt{6}$

(d) $5\sqrt{2} - 3\sqrt{50} + 7\sqrt{288} = 74\sqrt{2}$

(e) $\sqrt{16a^3 - 48a^2b} = 4a\sqrt{a - 3b}\quad (a \geq 0)$

(f) $3\sqrt[3]{16a^3} + 8\sqrt[3]{a^3/4} = 10a\sqrt[3]{2}$

(g) $\sqrt{\sqrt[3]{128}} = 2\sqrt[6]{2}$

(h) $\sqrt[n]{x^{n+1}y^{2n-1}z^{3n}} = xy^2z^3\sqrt[n]{x/y}$

(i) $3\sqrt[4]{9} - 2\sqrt[6]{27} = \sqrt{3}$

(j) $6\sqrt{8a^3/3} - 2\sqrt{24ab^2} + a\sqrt{54a} = (7a - 4b)\sqrt{6a}$
$a, b \geq 0$

(k) $(x + 1)\sqrt[3]{16x^2} - 4x\sqrt[3]{x^2/4} = 2\sqrt[3]{2x^2}$

(l) $2\sqrt{54} - 6\sqrt{2/3} - \sqrt{96} = 0$

(m) $4\sqrt{x/y} + \frac{3}{\sqrt[4]{x^2y^2}} - 5\sqrt[6]{y^3/x^3} = \frac{4x - 5y + 3}{xy}\sqrt{xy}$

8.13 (a) $(3\sqrt{8})(6\sqrt{5}) = 36\sqrt{10}$

(b) $\sqrt{48x^5}\,\sqrt{3x^3} = 12x^4$

(c) $\sqrt[3]{2}\,\sqrt[3]{32} = 4$

(d) $\sqrt{2}(\sqrt{2} + \sqrt{18}) = 8$

(e) $(5 + \sqrt{2})(5 - \sqrt{2}) = 23$

(f) $(x + \sqrt{y})(x + \sqrt{y}) = x^2 - y$

(g) $(2\sqrt{3} - \sqrt{6})(3\sqrt{3} + 3\sqrt{6}) = 9\sqrt{2}$

(h) $(3\sqrt{2} - 2\sqrt{3})(4\sqrt{2} + 3\sqrt{3}) = 6 + \sqrt{6}$

(i) $(\sqrt{2} - \sqrt{3})^2 + (\sqrt{2} + \sqrt{3})^2 = 10$

(j) $(2\sqrt{a} + 5\sqrt{a - b})(\sqrt{a} + \sqrt{a - b})$
$= 7a - 5b + 7\sqrt{a^2 - ab}$

(k) $(\sqrt{3} + \sqrt{5} + \sqrt{7})(\sqrt{3} + \sqrt{5} - \sqrt{7}) = 1 + 2\sqrt{15}$

(l) $\sqrt{8 - 2\sqrt{7}}\sqrt{8 + 2\sqrt{7}} = 6$

(m) $\frac{4 + \sqrt{8}}{2} = 2 + \sqrt{2}$

(n) $\frac{6 - \sqrt{18}}{3} = 2 - \sqrt{2}$

(o) $\frac{3}{\sqrt{2}} - \sqrt{\frac{1}{2}} = \sqrt{2}$

(p) $\frac{8 + 4\sqrt{48}}{8} = 1 + 2\sqrt{3}$

(q) $\frac{36 - 2\sqrt[3]{81}}{6} = 6 - \sqrt[3]{3}$

8.14 (a) $\frac{2\sqrt{24x^3}}{\sqrt{3x}} = 4x\sqrt{2}\quad (x > 0)$

(b) $\frac{a\sqrt{b}}{b\sqrt{a}} = \frac{\sqrt{ab}}{b}$

(c) $\frac{\sqrt{3} + \sqrt{2}}{\sqrt{2}} = 1 + \frac{1}{2}\sqrt{6}$

(d) $\frac{\sqrt{6} - \sqrt{10} - \sqrt{12}}{\sqrt{18}} = \frac{\sqrt{3} - \sqrt{5} - \sqrt{6}}{3}$

(e) $\sqrt[3]{\frac{9V^2}{16\pi^2}} = \frac{1}{4\pi}\sqrt[3]{36\pi V^2}$

(f) $\frac{6a}{\sqrt[3]{12}} = a\sqrt[3]{18}$

(g) $\frac{\sqrt[5]{3a^7b^6c^5}}{\sqrt[5]{24a^2bc}} = \frac{ab}{2}\sqrt[5]{4c^4}$

(h) $\sqrt[3]{\frac{x^{-2}y^{-3}z^{-1}}{4xyz^2}} = \frac{1}{2xy^2z}\sqrt[3]{2y^2}$

(i) $\frac{\sqrt{2}}{\sqrt[3]{3}} = \frac{\sqrt{2}\,\sqrt[3]{9}}{3} = \frac{\sqrt[6]{648}}{3}$

(j) $\frac{\sqrt[3]{20} - \sqrt[3]{18}}{\sqrt[3]{12}} = \frac{1}{3}\sqrt[3]{45} - \frac{1}{2}\sqrt[3]{12}$

(k) $\frac{1}{\sqrt{7} - 2} = \frac{\sqrt{7} + 2}{3}$

(l) $\frac{5}{3 + \sqrt{2}} = \frac{5}{7}(3 - \sqrt{2})$

(m) $\frac{-2}{2 - \sqrt{3}} = -4 - 2\sqrt{3}$

(n) $\frac{s\sqrt{3}}{\sqrt{3} - 1} = \frac{3s}{2} + \frac{s\sqrt{3}}{2}$

(o) $\frac{2\sqrt{3} - 1}{\sqrt{3} + 2} = 5\sqrt{3} - 8$

(p) $\frac{1 - \sqrt{x + 1}}{1 + \sqrt{x - 1}} = \frac{2\sqrt{x + 1} - x - 2}{x}$

8.15 (*a*) $\dfrac{4}{2+\sqrt{5}}+\dfrac{3}{5+2\sqrt{5}}=\dfrac{14}{5}\sqrt{5}-5$

(*b*) $\dfrac{3\sqrt{3}+2\sqrt{5}}{3\sqrt{3}-2\sqrt{5}}=\dfrac{47+12\sqrt{15}}{7}$

(*c*) $\dfrac{\sqrt{2}+\sqrt{3}+\sqrt{6}}{\sqrt{2}+\sqrt{3}}=1+3\sqrt{2}-2\sqrt{3}$

(*d*) $\dfrac{a\sqrt{b}-b\sqrt{a}}{a\sqrt{b}+b\sqrt{a}}=\dfrac{a+b-2\sqrt{ab}}{a-b}$

(*e*) $\dfrac{x+\sqrt{y}}{x-\sqrt{y}}+\dfrac{x-\sqrt{y}}{x+\sqrt{y}}=\dfrac{2x^2+2y}{x^2-y}$

(*f*) $\sqrt{\dfrac{x-y}{x^3y-2x^2y^2+xy^3}}=\dfrac{\sqrt{xy(x-y)}}{xy(x-y)}$

(*g*) $\dfrac{2+\sqrt{3}+\sqrt{5}}{2+\sqrt{3}-\sqrt{5}}=\dfrac{6+10\sqrt{3}+4\sqrt{5}+3\sqrt{15}}{11}$

(*h*) $\dfrac{1}{2+\sqrt[3]{2}}=\dfrac{4-2\sqrt[3]{2}+\sqrt[3]{4}}{10}$

(*i*) $\dfrac{3}{x+y-\sqrt{x^2-2xy+y^2}}=\dfrac{3}{2y}$ if $x\geq y$

$=\dfrac{3}{2x}$ if $x\leq y$

(*j*) $\dfrac{2}{x^2-\sqrt{x^4+2x^2+1}}=-2$

(*k*) $\dfrac{1}{1-\sqrt[3]{x}}=\dfrac{1+\sqrt[3]{x}+\sqrt[3]{x^2}}{1-x}$

(*l*) $\dfrac{4}{\sqrt[3]{m}+2}=\dfrac{4\sqrt[3]{m^2}-8\sqrt[3]{m}+16}{m+8}$

Chapter 9

Simple Operations with Complex Numbers

9.1 COMPLEX NUMBERS

The unit of imaginary numbers is $\sqrt{-1}$ and is designated by the letter i. Many laws which hold for real numbers hold for imaginary numbers as well.

Thus $\sqrt{-4} = \sqrt{(4)(-1)} = 2\sqrt{-1} = 2i$, $\sqrt{-18} = \sqrt{(18)(-1)} = \sqrt{18}\sqrt{-1} = 3\sqrt{2}\,i$. Also, since $i = \sqrt{-1}$, we have $i^2 = -1$, $i^3 = i^2 \cdot i = (-1)\,i = -i$, $i^4 = (i^2)^2 = (-1)^2 = 1$, $i^5 = i^4 \cdot i = 1 \cdot i = i$, and similarly for any integral power of i.

Note. One must be very careful in applying some of the laws which hold for real numbers. For example, one might be tempted to write

$$\sqrt{-4}\sqrt{-4} = \sqrt{(-4)(-4)} = \sqrt{16} = 4, \qquad \text{which is incorrect.}$$

To avoid such difficulties, always express $\sqrt{-m}$, where m is a positive number, as $\sqrt{m}\,i$; and use $i^2 = -1$ whenever it arises. Thus:

$$\sqrt{-4}\sqrt{-4} = (2i)(2i) = 4i^2 = -4, \qquad \text{which is correct.}$$

A complex number is an expression of the form $a + bi$, where a and b are real numbers and $i = \sqrt{-1}$. In the complex number $a + bi$, a is called the *real part* and bi the *imaginary part*. When $a = 0$, the complex number is called a *pure imaginary*. If $b = 0$, the complex number reduces to the real number a. Thus complex numbers include all real numbers and all pure imaginary numbers.

Two complex numbers $a + bi$ and $c + di$ are *equal* if and only if $a = c$ and $b = d$. Thus $a + bi = 0$ if and only if $a = 0$, $b = 0$. If $c + di = 3$, then $c = 3$, $d = 0$.

The conjugate of a complex number $a + bi$ is $a - bi$, and conversely. Thus $5 - 3i$ and $5 + 3i$ are conjugates.

9.2 GRAPHICAL REPRESENTATION OF COMPLEX NUMBERS

Employing rectangular coordinate axes, the complex number $x + yi$ is represented by, or corresponds to, the point whose coordinates are (x,y). See Fig. 9-1.

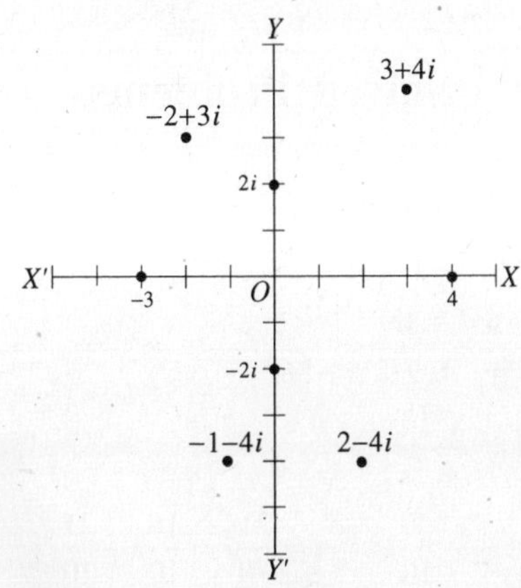

Fig. 9-1

To represent the complex number $3+4i$, measure off 3 units distance along $X'X$ and to the right of O, and then up 4 units distance.

To represent the number $-2+3i$, measure off 2 units distance along $X'X$ and to the left of O, and then up 3 units distance.

To represent the number $-1-4i$, measure off 1 unit distance along $X'X$ and to the left of O, and then down 4 units distance.

To represent the number $2-4i$, measure off 2 units distance along $X'X$ and to the right of O, and then down 4 units distance.

Pure imaginary numbers (such as $2i$ and $-2i$) are represented by points on the line $Y'Y$. Real numbers (such as 4 and -3) are represented by points on the line $X'X$.

9.3 ALGEBRAIC OPERATIONS WITH COMPLEX NUMBERS

To *add* two complex numbers, add the real parts and the imaginary parts separately. Thus

$$(a+bi)+(c+di)=(a+c)+(b+d)i$$
$$(5+4i)+(3+2i)=(5+3)+(4+2)i=8+6i$$
$$(-6+2i)+(4-5i)=(-6+4)+(2-5)\ i=-2-3i.$$

To subtract two complex numbers, subtract the real parts and the imaginary parts separately. Thus

$$(a+bi)-(c+di)=(a-c)+(b-d)i$$
$$(3+2i)-(5-3i)=(3-5)+(2+3)i=-2+5i$$
$$(-1+i)-(-3+2i)=(-1+\ 3)+(1-2)i=2-i.$$

To multiply two complex numbers, treat the numbers as ordinary binomials and replace i^2 by -1. Thus

$$(a+bi)(c+di)=ac+adi+bci+bdi^2=(ac-bd)+(ad+bc)i$$
$$(5+3i)(2-2i)=10-10i+6i-6i^2=10-4i-6(-1)=16-4i.$$

To divide two complex numbers, multiply the numerator and denominator of the fraction by the conjugate of the denominator, replacing i^2 by -1. Thus

$$\frac{2+i}{3-4i}=\frac{2+i}{3-4i}\cdot\frac{3+4i}{3+4i}=\frac{6+8i+3i+4i^2}{9-16i^2}=\frac{2+11i}{25}=\frac{2}{25}+\frac{11}{25}i.$$

Solved Problems

9.1 Express each of the following in terms of i.

(*a*) $\sqrt{-25}=\sqrt{(25)(-1)}=\sqrt{25}\sqrt{-1}=5i$

(*b*) $3\sqrt{-36}=3\sqrt{36}\sqrt{-1}=3\cdot 6\cdot i=18i$

(*c*) $-4\sqrt{-81}=-4\sqrt{81}\sqrt{-1}=-4\cdot 9\cdot i=-36i$

(*d*) $\sqrt{-\frac{1}{2}}=\sqrt{\frac{1}{2}}\sqrt{-1}=\sqrt{\frac{2}{4}}i=\frac{\sqrt{2}}{2}i$

(*e*) $2\sqrt{\frac{-16}{25}}-3\sqrt{\frac{-49}{100}}=2\cdot\frac{4}{5}i-3\cdot\frac{7}{10}i=\frac{8}{5}i=\frac{21}{10}i=\frac{16}{10}i-\frac{21}{10}i=-\frac{1}{2}i$

(*f*) $\sqrt{-12}-\sqrt{-3}=\sqrt{12}i-\sqrt{3}i=2\sqrt{3}i-\sqrt{3}i=\sqrt{3}i$

(g) $3\sqrt{-50} + 5\sqrt{-18} - 6\sqrt{-200} = 15\sqrt{2}i + 15\sqrt{2}i - 60\sqrt{2}i = -30\sqrt{2}i$

(h) $-2 + \sqrt{-4} = -2 + \sqrt{4}i = -2 + 2i$

(i) $6 - \sqrt{-50} = 6 - \sqrt{50}i = 6 - 5\sqrt{2}i$

(j) $\sqrt{8} + \sqrt{-8} = \sqrt{8} + \sqrt{8}i = 2\sqrt{2} + 2\sqrt{2}i$

(k) $\frac{1}{5}(-10 + \sqrt{-125}) = \frac{1}{5}(-10 + 5\sqrt{5}i) = -2 + \sqrt{5}i$

(l) $\frac{1}{4}(\sqrt{32} + \sqrt{-128}) = \frac{1}{4}(4\sqrt{2} + 8\sqrt{2}i) = \sqrt{2} + 2\sqrt{2}i$

(m) $\dfrac{\sqrt[3]{-8} + \sqrt{-8}}{2} = \dfrac{-2 + 2\sqrt{2}i}{2} = -1 + \sqrt{2}i$

9.2 Perform the indicated operations both algebraically and graphically:

(a) $(2 + 6i) + (5 + 3i)$, (b) $(-4 + 2i) - (3 + 5i)$.

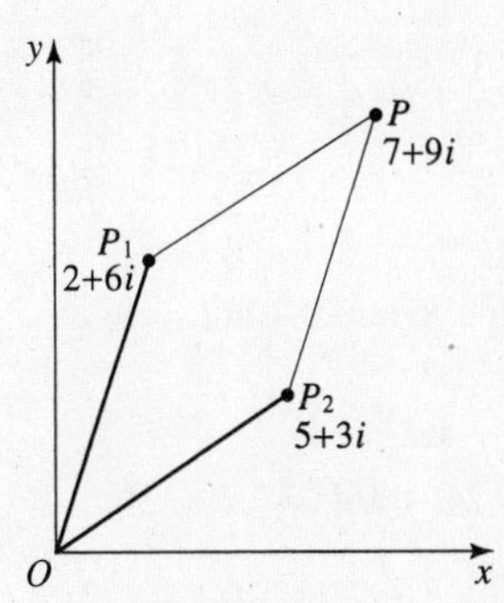

Fig. 9-2

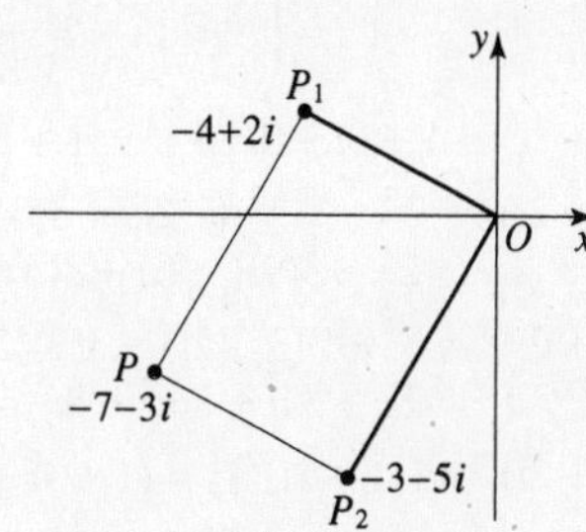

Fig. 9-3

SOLUTION

(a) Algebraically: $(2 + 6i) + (5 + 3i) = 7 + 9i$

Graphically: Represent the two complex numbers by the points P_1 and P_2 respectively, as shown in Fig. 9-2. Connect P_1 and P_2 with the origin O. Complete the parallelogram having OP_1 and OP_2 as adjacent sides. The vertex P (point $7 + 9i$) represents the sum of the two given complex numbers.

(b) Algebraically: $(-4 + 2i) - (3 + 5i) = -7 - 3i$

Graphically: $(-4 + 2i) - (3 + 5i) = (-4 + 2i) + (-3 - 5i)$. We now proceed to add $(-4 + 2i)$ with $(-3 - 5i)$.

Represent the two complex numbers $(-4 + 2i)$ and $(-3 - 5i)$ by the points P_1 and P_2 respectively, as shown in Fig. 9-3. Connect P_1 and P_2 with the origin O. Complete the parallelogram having OP_1 and OP_2 as adjacent sides. The vertex P (point $-7 - 3i$) represents the subtraction $(-4 + 2i) - (3 + 5i)$.

9.3 Perform the indicated operations and simplify.

(a) $(5 - 2i) + (6 + 3i) = 11 + i$

(b) $(6 + 3i) - (4 - 2i) = 6 + 3i - 4 + 2i = 2 + 5i$

(c) $(5 - 3i) - (-2 + 5i) = 5 - 3i + 2 - 5i = 7 - 8i$

(*d*) $\left(\frac{3}{2}+\frac{5}{8}i\right)+\left(-\frac{1}{4}+\frac{1}{4}i\right)=\frac{3}{2}-\frac{1}{4}+\left(\frac{5}{8}+\frac{1}{4}\right)i=\frac{5}{4}+\frac{7}{8}i$

(*e*) $(a+bi)+(a-bi)=2a$

(*f*) $(a+bi)-(a-bi)=a+bi-a+bi=2bi$

(*g*) $(5-\sqrt{-125})-(4-\sqrt{-20})=(5-5\sqrt{5}i)-(4-2\sqrt{5}i)=1-3\sqrt{5}i$

9.4 (*a*) $\sqrt{-2}\sqrt{-32}=(\sqrt{2}i)(\sqrt{32}i)=\sqrt{2}\sqrt{32}i^2=\sqrt{64}(-1)=-8$

(*b*) $-3\sqrt{-5}\sqrt{-20}=-3(\sqrt{5}i)(\sqrt{20}i)=-3(\sqrt{5}\sqrt{20})i^2=-3\sqrt{100}(-1)=30$

(*c*) $(4i)(-3i)=-12i^2=12$

(*d*) $(6i)^2=36i^2=-36$

(*e*) $(2\sqrt{-1})^3=(2i)^3=8i^3=8i(i^2)=-8i$

(*f*) $3i(i+2)=3i^2+6i=-3+6i$

(*g*) $(3-2i)(4+i)=3\cdot 4+3\cdot i-(2i)4-(2i)i=12+3i-8i+2=14-5i$

(*h*) $(5-3i)(i+2)=5i+10-3i^2-6i=5i+10+3-6i=13-i$

(*i*) $(5+3i)^2=5^2+2(5)3i+(3i)^2=25+30i+9i^2=16+30i$

(*j*) $(2-i)(3+2i)(1-4i)=(6+4i-3i-2i^2)(1-4i)=(8+i)(1-4i)$
$=8-32i+i-4i^2=12-31i$

(*k*) $\left(\frac{\sqrt{2}}{2}+\frac{\sqrt{2}}{2}i\right)^2=\left(\frac{\sqrt{2}}{2}\right)^2+2\left(\frac{\sqrt{2}}{2}\right)\left(\frac{\sqrt{2}}{2}i\right)+\left(\frac{\sqrt{2}}{2}i\right)^2=\frac{1}{2}+i-\frac{1}{2}=i$

(*l*) $(1+i)^3=1+3i+3i^2+i^3=1+3i-3-i=-2+2i$

(*m*) $(3-2i)^3=3^3+3(3^2)(-2i)+3(3)(-2i)^2+(-2i)^3$
$=27+3(9)(-2i)+3(3)(4i^2)-8i^3=27-54i-36+8i=-9-46i$

(*n*) $(3+2i)^3=3^3+3(3^2)(2i)+3(3)(2i)^2+(2i)^3$
$=27+54i+36i^2+8i^3=27+54i-36-8i=-9+46i$

(*o*) $(1+2i)^4=[(1+2i)^2]^2=(1+4i+4i^2)^2=(-3+4i)^2=9-24i+16i^2=-7-24i$

(*p*) $(-1+i)^8=[(-1+i^2]^4=(1-2i+i^2)^4=(-2i)^4=16i^4=16$

9.5 (*a*) $\frac{1+i}{3-i}=\frac{1+i}{3-i}\cdot\frac{3+i}{3+i}=\frac{3+3i+i+i^2}{3^2-i^2}=\frac{2+4i}{10}=\frac{1}{5}+\frac{2}{5}i$

(*b*) $\frac{1}{i}=\frac{1}{i}\left(\frac{-i}{-i}\right)=\frac{-i}{-i^2}=\frac{-i}{1}=-i$

(*c*) $\frac{2\sqrt{3}+\sqrt{2}i}{3\sqrt{2}-4\sqrt{3}i}=\frac{2\sqrt{3}+\sqrt{2}i}{3\sqrt{2}-4\sqrt{3}i}\cdot\frac{3\sqrt{2}+4\sqrt{3}i}{3\sqrt{2}+4\sqrt{3}i}=\frac{(2\sqrt{3}+\sqrt{2}i)(3\sqrt{2}+4\sqrt{3}i)}{(3\sqrt{2})^2-(4\sqrt{3})^2i^2}$
$=\frac{6\sqrt{6}+8\sqrt{9}i+3\sqrt{4}i+4\sqrt{6}i^2}{18+48}=\frac{2\sqrt{6}+30i}{66}=\frac{\sqrt{6}}{33}+\frac{5}{11}i$

Supplementary Problems

9.6 Express each of the following in terms of i.

(*a*) $2\sqrt{-49}$ (*d*) $4\sqrt{-1/8}$ (*g*) $\frac{-4+\sqrt{-4}}{2}$ (*i*) $4\sqrt{-81}-3\sqrt{-36}+4\sqrt{25}$

(*b*) $-4\sqrt{-64}$ (*e*) $3\sqrt{-25}-5\sqrt{-100}$ (*h*) $\frac{1}{6}(-12-\sqrt{-288})$ (*j*) $3\sqrt{12}-3\sqrt{-12}$

(*c*) $6\sqrt{-1/9}$ (*f*) $2\sqrt{-72}+3\sqrt{-32}$

9.7 Perform the indicated operations both algebraically and graphically.

(*a*) $(3+2i)+(2+3i)$ (*c*) $(4-3i)-(-2+i)$

(*b*) $(2-i)+(-4+5i)$ (*d*) $(-2+2i)-(-2-i)$

Perform each of the indicated operations and simplify.

9.8 (*a*) $(3+4i)+(-1-6i)$ (*g*) $(2i)^4$ (*m*) $(3-4i)^2$

(*b*) $(-2+5i)-(3-2i)$ (*h*) $(\frac{1}{2}\sqrt{-3})^6$ (*n*) $(1+i)(2+2i)(3-i)$

(*c*) $\left(\frac{2}{3}-\frac{1}{2}i\right)-\left(-\frac{1}{3}+\frac{1}{2}i\right)$ (*i*) $5i(2-i)$ (*o*) $(i-1)^3$

(*d*) $(3+\sqrt{-8})-(2-\sqrt{-32})$ (*j*) $(2+i)(2-i)$ (*p*) $(2+3i)^3$

(*e*) $\sqrt{-3}\sqrt{-12}$ (*k*) $(-3+4i)(-3-4i)$ (*q*) $(1-i)^4$

(*f*) $(-i\sqrt{2})(i\sqrt{2})$ (*l*) $(2-5i)(3+2i)$ (*r*) $(i+2)^5$

9.9 (*a*) $\dfrac{2-5i}{4+3i}$ (*d*) $\dfrac{3-\sqrt{2}i}{\sqrt{2}i}$ (*g*) $\dfrac{i+i^2+i^3+i^4}{1+i}$

(*b*) $\dfrac{-1}{2-2i}$ (*e*) $\dfrac{1}{1-2i}+\dfrac{3}{1+4i}$ (*h*) $\dfrac{i^{26}-i}{i-1}$

(*c*) $\dfrac{3\sqrt{2}+2\sqrt{3}i}{3\sqrt{2}-2\sqrt{3}i}$ (*f*) $\dfrac{5}{3-4i}+\dfrac{10}{4+3i}$ (*i*) $\left(\dfrac{4i^{11}-i}{1+2i}\right)^2$

ANSWERS TO SUPPLEMENTARY PROBLEMS

9.6 (*a*) $14i$ (*c*) $2i$ (*e*) $-35i$ (*g*) $-2+i$ (*i*) $18i+20$

(*b*) $-32i$ (*d*) $\sqrt{2}i$ (*f*) $24\sqrt{2}i$ (*h*) $-2-2\sqrt{2}i$ (*j*) $6\sqrt{3}-6\sqrt{3}i$

9.7 (*a*) $5+5i$ and Fig. 9-4 (*c*) $6-4i$ and Fig. 9-6

(*b*) $-2+4i$ and Fig. 9-5 (*d*) $3i$ and Fig. 9-7

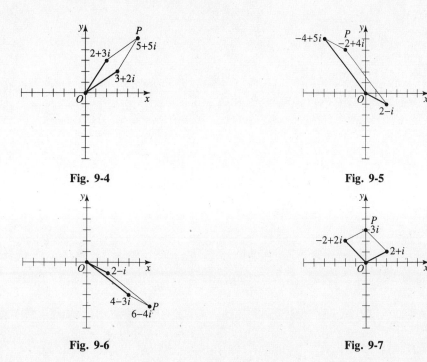

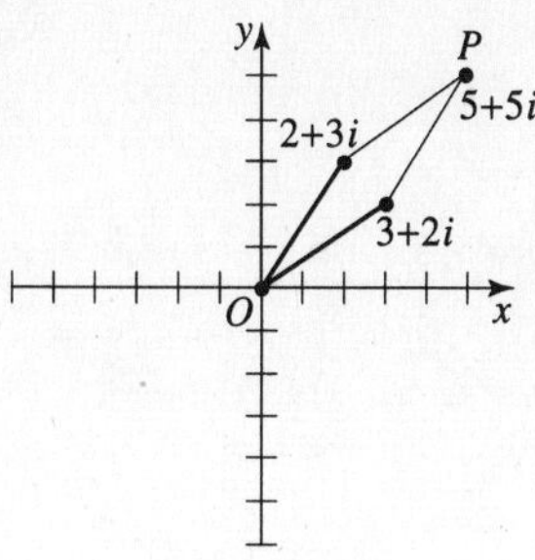

Fig. 9-4

y
−4+5i
P
−2+4i
O
2−i
x

Fig. 9-5

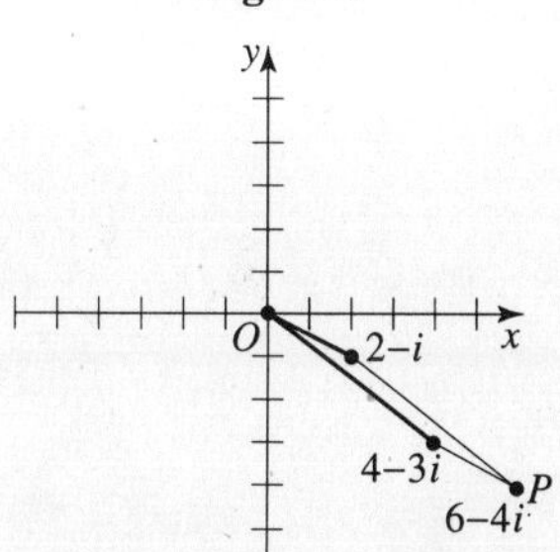

Fig. 9-6

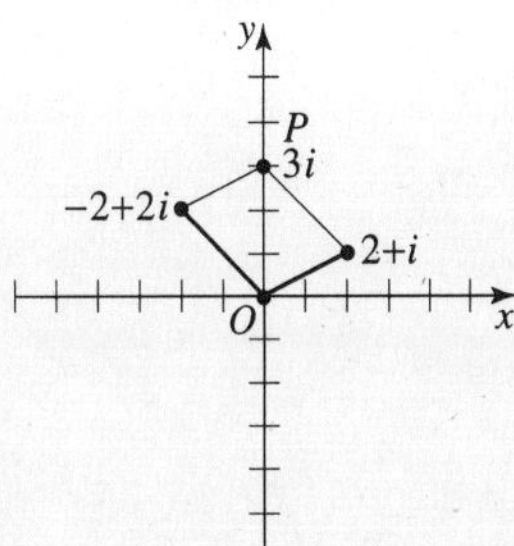

Fig. 9-7

9.8 (*a*) $2-2i$ (*b*) $-5+7i$ (*c*) $1-i$ (*d*) $1+6\sqrt{2}i$ (*e*) -6 (*f*) 2 (*g*) 16 (*h*) $-27/64$ (*i*) $5+10i$ (*j*) 5 (*k*) 25 (*l*) $16-11i$ (*m*) $-7-24i$ (*n*) $4+12i$ (*o*) $2+2i$ (*p*) $-46+9i$ (*q*) -4 (*r*) $-38+41i$

9.9 (*a*) $-\frac{7}{25}-\frac{26}{25}i$ (*b*) $-\frac{1}{4}-\frac{1}{4}i$ (*c*) $\frac{1}{5}+\frac{2}{5}\sqrt{6}i$ (*d*) $-1-\frac{3}{2}\sqrt{2}i$ (*e*) $\frac{32}{85}-\frac{26}{85}i$ (*f*) $\frac{11}{5}-\frac{2}{5}i$ (*g*) 0 (*h*) i (*i*) $3+4i$

Chapter 10

Equations in General

10.1 EQUATIONS

An equation is a statement of equality between two expressions called members.

An equation which is true for only certain values of the variables (sometimes called unknowns) involved is called a *conditional equation* or simply an equation.

An equation which is true for all permissible values of the variables (or unknowns) involved is called an *identity*. By permissible values are meant the values for which the members are defined.

EXAMPLE 10.1. $x + 5 = 8$ is true only for $x = 3$; it is a conditional equation.

EXAMPLE 10.2. $x^2 - y^2 = (x - y)(x + y)$ is true for all values of x and y; it is an identity.

EXAMPLE 10.3.

$$\frac{1}{x-2} + \frac{1}{x-3} = \frac{2x-5}{(x-2)(x-3)}$$

is true for all values except for the nonpermissible values $x = 2$, $x = 3$; these excluded values lead to division by zero which is not allowed. Since the equation is true for all permissible values of x, it is an identity.

The symbol $\equiv$ is sometimes used for identities instead of $=$.

The solutions of a conditional equation are those values of the unknowns which make both members equal. These solutions are said to satisfy the equation. If only one unknown is involved the solutions are also called *roots*. To solve an equation means to find all of the solutions.

Thus $x = 2$ is a solution or root of $2x + 3 = 7$, since if we substitute $x = 2$ into the equation we obtain $2(2) + 3 = 7$ and both members are equal, i.e., the equation is satisfied. Similarly, three (of the many) solutions of $2x + y = 4$ are: $x = 0$, $y = 4$; $x = 1$, $y = 2$; $x = 5$, $y = -6$.

10.2 OPERATIONS USED IN TRANSFORMING EQUATIONS

A. If equals are added to equals, the results are equal.

Thus if $x - y = z$, we may add y to both members and obtain $x = y + z$.

B. If equals are subtracted from equals, the results are equal.

Thus if $x + 2 = 5$, we may subtract 2 from both members to obtain $x = 3$.

Note. Because of rules *A* and *B* we may transpose a term from one member of an equation to the other member merely by changing the sign of the term. Thus if $3x + 2y - 5 = x - 3y + 2$, then $3x - x + 2y + 3y = 5 + 2$ or $2x + 5y = 7$.

C. If equals are multiplied by equals, the results are equal.

Thus if both members of $\frac{1}{4}y = 2x^2$ are multiplied by 4 the result is $y = 8x^2$.

Similarly, if both members of $\frac{9}{5}C = F - 32$ are multiplied by $\frac{5}{9}$ the result is $C = \frac{5}{9}(F - 32)$.

D. If equals are divided by equals, the results are equal provided there is no division by zero.

Thus if $-4x = -12$, we may divide both members by -4 to obtain $x = 3$.

Similarly, if $E = IR$ we may divide both sides by $R \neq 0$ to obtain $I = E/R$.

E. The same powers of equals are equal.

Thus if $T = 2\pi\sqrt{l/g}$, then $T^2 = (2\pi\sqrt{l/g})^2 = 4\pi^2 l/g$.

F. The same roots of equals are equal. Thus:

$$\text{if } r^3 = \frac{3V}{4\pi}, \qquad \text{then} \qquad r = \sqrt[3]{\frac{3V}{4\pi}}.$$

G. Reciprocals of equals are equal provided the reciprocal of zero does not occur.

Thus if $1/x = 1/3$, then $x = 3$. Similarly,

$$\text{if } \frac{1}{R} = \frac{R_1 + R_2}{R_1 R_2} \qquad \text{then} \qquad R = \frac{R_1 R_2}{R_1 + R_2}.$$

Operations *A–F* are sometimes called axioms of equality.

10.3 EQUIVALENT EQUATIONS

Equivalent equations are equations having the same solutions.

Thus $x - 2 = 0$ and $2x = 4$ have the same solution $x = 2$ and so are equivalent. However, $x - 2 = 0$ and $x^2 - 2x = 0$ are not equivalent, since $x^2 - 2x = 0$ has the additional solution $x = 0$.

The above operations used in transforming equations may not all yield equations equivalent to the original equations. The use of such operations may yield derived equations with either more or fewer solutions than the original equation.

If the operations yield an equation with more solutions than the original, the extra solutions are called *extraneous* and the derived equation is said to be *redundant* with respect to the original equation. If the operations yield an equation with fewer solutions than the original, the derived equation is said to be *defective* with respect to the original equation.

Operations *A* and *B* always yield equivalent equations. Operations *C* and *E* may give rise to redundant equations and extraneous solutions. Operations *D* and *F* may give rise to defective equations.

10.4 FORMULAS

A formula is an equation which expresses a general fact, rule, or principle.

For example, in geometry the formula $A = \pi r^2$ gives the area A of a circle in terms of its radius r.

In physics the formula $s = \frac{1}{2}gt^2$, where g is approximately 32.2 ft/s^2, gives the relation between the distance s, in feet, which an object will fall freely from rest during a time t, in seconds.

To solve a formula for one of the variables involved is to perform the same operations on both members of the formula until the desired variable appears on one side of the equation but not on the other side.

Thus if $F = ma$, we may divide by m to obtain $a = F/m$ and the formula is solved for a in terms of the other variables F and m. To check the work, substitute $a = F/m$ into the original equation to obtain $F = m(F/m)$, an identity.

10.5 POLYNOMIAL EQUATIONS

A monomial in a number of unknowns $x, y, z, \ldots$ has the form $ax^p y^q z^r \cdots$ where the exponents $p, q, r, \ldots$ are either positive integers or zero and the coefficient a is independent of the unknowns. The sum of the exponents $p + q + r + \cdots$ is called the *degree* of the term in the unknowns $x, y, z, \ldots$

EXAMPLES 10.1. $3x^2z^3$, $\frac{1}{2}x^4$, 6 are monomials.
$3x^2z^3$ is of degree 2 in x, 3 in z, and 5 in x and z.
$\frac{1}{2}x^4$ is of fourth degree. 6 is of degree zero.
$4y/x = 4yx^{-1}$ is not integral in x; $3x\sqrt{yz^3}$ is not rational in y.

When reference is made to degree without specifying the unknowns considered, the degree in all unknowns is implied.

A polynomial in various unknowns consists of terms each of which is rational and integral. The degree of such a polynomial is defined as the degree of the terms of highest degree.

EXAMPLE 10.2. $3x^3y^4z + xy^2z^5 - 8x + 3$ is a polynomial of degree 3 in x, 4 in y, 5 in z, 7 in x and y, 7 in y and z, 6 in x and z, and 8 in x, y, and z.

A polynomial equation is a statement of equality between two polynomials. The degree of such an equation is the degree of the term of highest degree present in the equation.

EXAMPLE 10.3. $xyz^2 + 3xz = 2x^3y + 3z^2$ is of degree 3 in x, 1 in y, 2 in z, 4 in x and y, 3 in y and z, 3 in x and z, and 4 in x, y, and z.

It should be understood that like terms in the equation have been combined. Thus $4x^3y + x^2z - xy^2 = 4x^3y + z$ should be written $x^2z - xy^2 = z$.

An equation is called *linear* if it is of degree 1 and *quadratic* if it is of degree 2. Similarly the words *cubic*, *quartic* and *quintic* refer to equations of degree 3, 4, and 5 respectively.

EXAMPLES 10.4. $2x + 3y = 7z$ is a linear equation in x, y, and z.
$x^2 - 4xy + 5y^2 = 10$ is a quadratic equation in x and y.
$x^3 + 3x^2 - 4x - 6 = 0$ is a cubic equation in x.

A polynomial equation of degree n in the unknown x may be written

$$a_0x^n + a_1x^{n-1} + a_2x^{n-2} + \cdots + a_{n-1}x + a_n = 0 \qquad a_0 \neq 0$$

where $a_0, a_1, \ldots, a_n$ are given constants and n is a positive integer.
As special cases we see that

$a_0x + a_1 = 0$ or $ax + b = 0$ is of degree 1 (linear equation),
$a_0x^2 + a_1x + a_2 = 0$ or $ax^2 + bx + c = 0$ is of degree 2 (quadratic equation),
$a_0x^3 + a_1x^2 + a_2x + a_s = 0$ is of degree 3 (cubic equation),
$a_0x^4 + a_1x^3 + a_2x^2 + a_3x + a_4 = 0$ is of degree 4 (quartic equation).

Solved Problems

10.1 Which of the following are conditional equations and which are identities?

(*a*) $3x - (x + 4) = 2(x - 2)$, $2x - 4 = 2x - 4$; identity.
(*b*) $(x - 1)(x + 1) = (x - 1)^2$, $x^2 - 1 = x^2 - 2x + 1$; conditional equation.
(*c*) $(y - 3)^2 + 3(2y - 3) = y(y + 1) - y$, $y^2 - 6y + 9 + 6y - 9 = y^2 + y - y$, $y^2 = y^2$; identity.
(*d*) $x + 3y - 5 = 2(x + 2y) + 3$, $x + 3y - 5 = 2x + 4y + 3$; conditional equation.

10.2 Check each of the following equations for the indicated solution or solutions.

(*a*) $\frac{x}{2}+\frac{x}{3}=10;\ x=12.$ $\quad \frac{12}{2}+\frac{12}{3}=10,\ 6+4=10$, and $x=12$ is a solution.

(*b*) $\frac{x^2+6x}{x+2}=3x-2;\ x=2,\ x=-1.$ $\quad \frac{2^2+6(2)}{2+2}=3(2)-2,\ \frac{16}{4}=4$, and $x=2$ is a solution.

$\frac{(-1)^2+6(-1)}{-1+2}=3(-1)-2,\ \frac{-5}{1}=-5$, and $x=-1$ is a solution.

(*c*) $x^2-xy+y^2=19;\ x=-2,\ y=3;\ x=4,\ y=2+\sqrt{7};\ x=2,\ y=-1.$

$x=-2,\ y=3$: $(-2)^2-(-2)3+3^2=19,\ 19=19$, and $x=-2,\ y=3$ is a solution.

$x=4,\ y=2+\sqrt{7}$: $4^2-4(2+\sqrt{7})+(2+\sqrt{7})^2=19,\ 16-8-4\sqrt{7}+(4+4\sqrt{7}+7)=19,\ 19=19$, and $x=4,\ y=2+\sqrt{7}$ is a solution.

$x=2,\ y=-1$: $2^2-2(-1)+(-1)^2=19,\ 7=19$, and $x=2,\ y=-1$ is not a solution.

10.3 Use the axioms of equality to solve each equation.

(*a*) $2(x+3)=3(x-1),\ 2x+6=3x-3.$
Transposing terms: $2x-3x=-6-3,\ -x=-9$. Multiplying by -1: $x=9$.
Check: $2(9+3)=3(9-1),\ 24=24.$

(*b*) $\frac{x}{3}+\frac{x}{6}=1.$

Multiplying by 6: $2x+x=6,\ 3x=6$. Dividing by 3: $x=2$.
Check: $2/3+2/6=1,\ 1=1.$

(*c*) $3y-2(y-1)=4(y+2),\ 3y-2y+2=4y+8,\ y+2=4y+8.$
Transposing: $y-4y=8-2,\ -3y=6$. Dividing by -3: $y=6/(-3)=-2$.
Check: $3(-2)-2(-2-1)=4(-2+2),\ 0=0.$

(*d*) $\frac{2x-3}{x-1}=\frac{4x-5}{x-1}.$ $\quad$ Multiplying by $x-1$, $2x-3=4x-5$ or $x=1$.

Check: Substituting $x=1$ into the given equation, we find $-1/0=-1/0$. This is meaningless, since division by zero is an excluded operation, and the given equation has no solution.
Note that

$$\text{(i)}\ \frac{2x-3}{x-1}=\frac{4x-5}{x-1} \qquad \text{and} \qquad \text{(ii)}\ 2x-3=4x-5$$

are not equivalent equations. When (i) is multiplied by $x-1$ an *extraneous solution* $x=1$ is introduced, and equation (ii) is *redundant* with respect to equation (i).

(*e*) $x(x-3)=2(x-3).$ $\quad$ Dividing each member by $x-3$ gives a solution $x=2$.
Now $x-3=0$ or $x=3$ is also a solution of the given equation which was lost in the division. The required roots are $x=2$ and $x=3$.
The equation $x=2$ is *defective* with respect to the given equation.

(*f*) $\sqrt{x+2}=-1.$ $\quad$ Squaring both sides, $x+2=1$ or $x=-1$.
Check: Substituting $x=-1$ into the given equation, $\sqrt{1}=-1$ or $1=-1$ which is false.
Thus $x=-1$ is an *extraneous solution*. The given equation has no solution.

(*g*) $\sqrt{2x-4}=6.$ $\quad$ Squaring both sides, $2x-4=36$ or $x=20$.
Check: If $x=20$, $\sqrt{2(20)-4}=6$ or $\sqrt{36}=6$ which is true.
Hence $x=20$ is a solution. In this case no extraneous root was introduced.

10.4 In each of the following formulas, solve for the indicated letter.

(*a*) $E=IR$, for R. $\quad$ Dividing both sides by $I\neq 0$, we have $R=E/I$.

(*b*) $s=v_0t+\frac{1}{2}at^2$, for a.
Transposing, $\frac{1}{2}at^2=s-v_0t$. Multiplying by 2, $at^2=2(s-v_0t)$.

Dividing by $t^2 \neq 0$,

$$a = \frac{2(s - v_0 t)}{t^2}.$$

(*c*) $\frac{1}{f} = \frac{1}{p} + \frac{1}{q}$, for p. Transposing,

$$\frac{1}{p} = \frac{1}{f} - \frac{1}{q} = \frac{q - f}{fq}.$$

Taking reciprocals,

$$p = \frac{fq}{q - f} \qquad \text{(assuming } q \neq f\text{)}.$$

(*d*) $T = 2\pi\sqrt{l/g}$, for g. Squaring both sides,

$$T^2 = \frac{4\pi^2 l}{g}.$$

Multiplying by g, $gT^2 = 4\pi^2 l$. Dividing by T^2, $g = 4\pi^2 l/T^2$.

10.5 In each of the following formulas, find the value of the indicated letter, given the values of the other letters.

(*a*) $F = \frac{9}{5}C + 32$, $F = 68$; find C. $68 = \frac{9}{5}C + 32$, $36 = \frac{9}{5}C$, $C = \frac{5}{9}(36) = 20$.

Another method: $\frac{9}{5}C = F - 32$, $C = \frac{5}{9}(F - 32) = \frac{5}{9}(68 - 32) = \frac{5}{9}(36) = 20$.

(*b*) $\frac{1}{R} = \frac{1}{R_1} + \frac{1}{R_2}$, $R = 6$, $R_1 = 15$; find R_2. $\frac{1}{6} = \frac{1}{15} + \frac{1}{R_2}$, $\frac{1}{R_2} = \frac{1}{6} - \frac{1}{15} = \frac{5-2}{30} = \frac{1}{10}$, $R_2 = 10$.

Another method:

$$\frac{1}{R_2} = \frac{1}{R} - \frac{1}{R_1} = \frac{R_1 - R}{RR_1}, \qquad R_2 = \frac{RR_1}{R_1 - R} = \frac{6(15)}{15 - 6} = 10.$$

(*c*) $V = \frac{4}{3}\pi r^3$, $V = 288\pi$; find r. $288\pi = \frac{4}{3}\pi r^3$, $r^3 = \frac{288\pi}{4\pi/3} = 216$, $r = 6$.

Another method:

$$3V = 4\pi r^3, \quad r^3 = \frac{3V}{4\pi}, \quad r = \sqrt[3]{\frac{3V}{4\pi}} = \sqrt[3]{\frac{3(288\pi)}{4\pi}} = \sqrt[3]{216} = 6.$$

10.6 Determine the degree of each of the following equations in each of the indicated unknowns.

(*a*) $2x^2 + xy - 3 = 0$: x; y; x and y.
Degree 2 in x, 1 in y, 2 in x and y.

(*b*) $3xy^2 - 4y^2z + 5x - 3y = x^4 + 2$: x; z; y and z; x, y, and z.
Degree 4 in x, 1 in z, 3 in y and z, 4 in x, y, and z.

(*c*) $x^2 = \frac{3}{y + z}$: x; x and z; x, y, and z.

As it stands it is not a polynomial equation. However, it can be transformed into one by multiplying by $y + z$ to obtain $x^2(y + z) = 3$ or $x^2y + x^2z = 3$. The derived equation is a polynomial equation of degree 2 in x, 3 in x and z, and 3 in x, y, and z.

(*d*) $\sqrt{x + 3} = x + y$: y; x and y.

As given it is not a polynomial equation, but it can be transformed into one by squaring both sides. Thus we obtain $x + 3 = x^2 + 2xy + y^2$, which is of degree 2 in y and 2 in x and y.

It should be mentioned, however, that the equations are not equivalent, since $x^2 + 2xy + y^2 = x + 3$ includes both $\sqrt{x + 3} = x + y$ and $-\sqrt{x + 3} = x + y$.

10.7 Find all values of x for which (*a*) $x^2 = 81$, (*b*) $(x-1)^2 = 4$.

SOLUTION

(*a*) There is nothing to indicate whether x is a positive or negative number, so we must assume either is possible. Taking the square root of both sides of the given equation, we obtain $\sqrt{x^2} = \sqrt{81} = 9$. Now $\sqrt{x^2}$ represents a positive number (or zero) if x is real. Hence we have $\sqrt{x^2} = x$ if x is positive while $\sqrt{x^2} = -x$ if x is negative. Thus when writing $\sqrt{x^2}$ we must consider that it is either x (if $x > 0$) or $-x$ (if $x < 0$). Therefore the equation $\sqrt{x^2} = 9$ may be written as either $x = 9$ or $-x = 9$ (i.e. $x = -9$). The two solutions may be written $x = \pm 9$.

(*b*) $(x-1)^2 = 4$, $\pm(x-1) = 2$ or $(x-1) = \pm 2$, and the two roots are $x = 3$ and $x = -1$.

10.8 Explain the fallacy in the following sequence of steps.

(*a*)	Let $x = y$:	$x = y$
(*b*)	Multiply both sides by x:	$x^2 = xy$
(*c*)	Subtract y^2 from both sides:	$x^2 - y^2 = xy - y^2$
(*d*)	Write as:	$(x-y)(x+y) = y(x-y)$
(*e*)	Divide by $x - y$:	$x + y = y$
(*f*)	Replace x by its equal, y:	$y + y = y$
(*g*)	Hence:	$2y = y$
(*h*)	Divide by y:	$2 = 1$.

SOLUTION

There is nothing wrong in steps (*a*), (*b*), (*c*), (*d*).

However, in (*e*) we divide by $x - y$, which from the original assumption is zero. Since division by zero is not defined, everything we do from (*e*) on is to be looked upon with disfavor.

10.9 Show that $\sqrt{2}$ is an irrational number, i.e., it cannot be the quotient of two integers.

SOLUTION

Assume that $\sqrt{2} = p/q$ where p and q are integers having no common factor except ± 1, i.e., p/q is in lowest terms. Squaring, we have $p^2/q^2 = 2$ or $p^2 = 2q^2$. Since $2q^2$ is an even number, p^2 is even and hence p is even (if p were odd, p^2 would be odd); then $p = 2k$, where k is an integer. Thus $p^2 = 2q^2$ becomes $(2k)^2 = 2q^2$ or $q^2 = 2k^2$; hence q^2 is even and q is even. But if p and q are both even they would have a common factor 2, thus contradicting the assumption that they have no common factor except ± 1. Hence $\sqrt{2}$ is irrational.

Supplementary Problems

10.10 State which of the following are conditional equations and which are identities.

(*a*) $2x + 3 - (2 - x) = 4x - 1$

(*b*) $(2y-1)^2 + (2y+1)^2 = (2y)^2 + 6$

(*c*) $2\{x + 4 - 3(2x-1)\} = 3(4 - 3x) + 2 - x$

(*d*) $(x+2y)(x-2y) - (x-2y)^2 + 4y(2y - x) = 0$

(*e*) $\dfrac{9x^2 - 4y^2}{3x - 2y} = 2x + 3y$

(*f*) $(x-3)(x^2 + 3x + 9) = x^3 - 27$

(*g*) $\dfrac{x^2}{4} + \dfrac{x^2}{12} = x^2$

(*h*) $(x^2 - y^2)^2 + (2xy)^2 = (x^2 + y^2)^2$

10.11 Check each of the following equations for the indicated solution or solutions.

(a) $\dfrac{y^2-4}{y-2}=2y-1;\ y=3$

(b) $x^2-3x=4;\ -1,\ -4$

(c) $\sqrt{3x-2}-\sqrt{x+2}=4$; 34, 2

(d) $x^3-6x^2+11x-6=0;\ 1, 2, 3$

(e) $\dfrac{1}{x}+\dfrac{1}{2x}=\dfrac{1}{x-1};\quad x=3$

(f) $y^3+y^2-5y-5=0;\ \pm\sqrt{5},\ -1$

(g) $x^2-2y=3y^2;\ x=4,\ y=2;\ x=1,\ y=-1$

(h) $(x+y)^2+(x-y)^2=2(x^2+y^2)$; any values of x,y

10.12 Use the axioms of equality to solve each equation. Check the solutions obtained.

(a) $5(x-4)=2(x+1)-7$

(b) $\dfrac{2y}{3}-\dfrac{y}{6}=2$

(c) $\dfrac{1}{y}=8-\dfrac{3}{y}$

(d) $\dfrac{x+1}{x-1}=\dfrac{x-1}{x-2}$

(e) $\dfrac{3x-2}{x-2}=\dfrac{x+2}{x-2}$

(f) $\sqrt{3x-2}=4$

(g) $\sqrt{2x+1}+5=0$

(h) $\sqrt[3]{2x-3}+1=0$

(i) $(y+1)^2=16$

(j) $(2x+1)^2+(2x-1)^2=34$

10.13 In each of the following formulas, solve for the indicated letter.

(a) $\dfrac{P_1V_1}{T_1}=\dfrac{P_2V_2}{T_2};\ T_2$

(b) $t=\sqrt{\dfrac{2s}{g}};\ s$

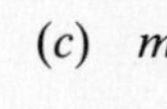

(c) $m=\dfrac{1}{2}\sqrt{2a^2+2b^2-c^2};\ c$

(d) $v^2=v_0^2+2as;\ a$

(e) $T=2\pi\sqrt{\dfrac{m}{k}};\ k$

(f) $S=\dfrac{n}{2}[2a+(n-1)d];\ d$

10.14 In each formula find the value of the indicated letter given the values of the other letters.

(a) $v=v_0+at$; find a if $v=20$, $v_0=30$, $t=5$.

(b) $S=\dfrac{n}{2}(a+d)$; find d if $S=570$, $n=20$, $a=40$.

(c) $\dfrac{1}{f}=\dfrac{1}{p}+\dfrac{1}{q}$; find q if $f=30$, $p=10$.

(d) $Fs=\frac{1}{2}mv^2$; find v if $F=100$, $s=5$, $m=2.5$.

(e) $f=\dfrac{1}{2\pi\sqrt{LC}}$; find C to four decimal places if $f=1000$, $L=4\cdot10^{-6}$.

10.15 Determine the degree of each equation in each of the indicated unknowns.

(a) $x^3-3x+2=0$: x

(b) $x^2+xy+3y^4=6$: x; y; x and y

(c) $2xy^3-3x^2y^2+4xy=2x^3$: x; y; x and y

(d) $xy+yz+xz+z^2x=y^4$: x; y; z; x and z; y and z; x, y, and z

10.16 Classify each equation according as it is (or can be transformed into) an equation which is linear, quadratic, cubic, quartic or quintic in all of the unknowns present.

(*a*) $2x^4 + 3x^3 - x - 5 = 0$
(*b*) $x - 2y = 4$
(*c*) $2x^2 + 3xy + y^2 = 10$
(*d*) $x^2y^3 - 2xyz = 4 + y^5$
(*e*) $\sqrt{x^2 + y^2 - 1} = x + y$
(*f*) $\dfrac{2x + y}{x - 3y} = 4$
(*g*) $3y^2 - 4y + 2 = 2(y - 3)^2$
(*h*) $(z + 1)^2(z - 2) = 0$

10.17 Is the equation $\sqrt{(x + 4)^2} = x + 4$ an identity? Explain.

10.18 Prove that $\sqrt{3}$ is irrational.

ANSWERS TO SUPPLEMENTARY PROBLEMS

10.10 (*a*) Conditional equation (*b*) Conditional equation (*c*) Identity (*d*) Identity (*e*) Conditional equation (*f*) Identity (*g*) Conditional equation (*h*) Identity

10.11 (*a*) $y = 3$ is a solution.
(*b*) $x = -1$ is a solution, $x = -4$ is not.
(*c*) $x = 34$ is a solution, $x = 2$ is not.
(*d*) $x = 1, 2, 3$ are all solutions.
(*e*) $x = 3$ is a solution.
(*f*) $y = \pm\sqrt{5}, -1$ are all solutions.
(*g*) $x = 4, y = 2$; $x = 1, y = -1$ are solutions.
(*h*) The equation is an identity; hence any values of x and y are solutions.

10.12 (*a*) $x = 5$ (*b*) $y = 4$ (*c*) $y = 1/2$ (*d*) $x = 3$ (*e*) no solution (*f*) $x = 6$ (*g*) no solution (*h*) $x = 1$ (*i*) $y = 3, -5$ (*j*) $x = \pm 2$

10.13 (*a*) $T_2 = \dfrac{P_2V_2T_1}{P_1V_1}$ (*b*) $s = \frac{1}{2}gt^2$ (*c*) $c = \pm\sqrt{2a^2 + 2b^2 - 4m^2}$ (*d*) $a = \dfrac{v^2 - v_0^2}{2s}$ (*e*) $k = \dfrac{4\pi^2 m}{T^2}$ (*f*) $d = \dfrac{2S - 2an}{n(n - 1)}$

10.14 (*a*) $a = -2$ (*b*) $d = 17$ (*c*) $q = -15$ (*d*) $v = \pm 20$ (*e*) $C = 0.0063$

10.15 (*a*) 3 (*b*) 2, 4, 4 (*c*) 3, 3, 4 (*d*) 1, 4, 2, 3, 4, 4

10.16 (*a*) quartic (*b*) linear (*c*) quadratic (*d*) quintic (*e*) quadratic (*f*) linear (*g*) quadratic (*h*) cubic

10.17 $\sqrt{(x + 4)^2} = x + 4$ only if $x + 4 \geq 0$; $\sqrt{(x + 4)^2} = -(x + 4)$ if $x + 4 \leq 0$.
The given equation is not an identity.

Chapter 11

Ratio, Proportion, and Variation

11.1 RATIO

The ratio of two numbers a and b, written $a{:}b$, is the fraction a/b provided $b \neq 0$.

Thus $a{:}b = a/b$, $b \neq 0$. If $a = b \neq 0$, the ratio is 1:1 or $1/1 = 1$.

EXAMPLES 11.1.

(1) The ratio of 4 to 6 = $4{:}6 = \dfrac{4}{6} = \dfrac{2}{3}$.

(2) $\dfrac{2}{3}:\dfrac{4}{5} = \dfrac{2/3}{4/5} = \dfrac{5}{6}$ (3) $5x : \dfrac{3y}{4} = \dfrac{5x}{3y/4} = \dfrac{20x}{3y}$

11.2 PROPORTION

A proportion is an equality of two ratios. Thus $a:b = c:d$, or $a/b = c/d$, is a proportion in which a and d are called the *extremes* and b and c the *means*, while d is called the *fourth proportional* to a, b, and c.

In the proportion $a:b = b:c$, c is called the *third proportional* to a and b, and b is called a *mean proportional* between a and c.

Proportions are equations and can be transformed using procedures for equations. Some of the transformed equations are used frequently and are called the laws of proportion. If $a/b = c/d$, then

(1) $ad = bc$ (3) $\dfrac{a}{c} = \dfrac{b}{d}$ (5) $\dfrac{a-b}{b} = \dfrac{c-d}{d}$

(2) $\dfrac{b}{a} = \dfrac{d}{c}$ (4) $\dfrac{a+b}{b} = \dfrac{c+d}{d}$ (6) $\dfrac{a+b}{a-b} = \dfrac{c+d}{c-d}$.

11.3 VARIATION

In reading scientific material, it is common to find such statements as "The pressure of an enclosed gas varies directly as the temperature." This and similar statements have precise mathematical meanings and they represent a specific type of function called variation functions. The three general types of variation functions are direct, inverse, and joint.

(1) If x varies *directly* as y, then $x = ky$ or $x/y = k$, where k is called the constant of proportionality or the constant of variation.

(2) If x varies *directly* as y^2, then $x = ky^2$.

(3) If x varies *inversely* as y, then $x = k/y$.

(4) If x varies *inversely* as y^2, then $x = k/y^2$.

(5) If x varies *jointly* as y and z, then $x = kyz$.

(6) If x varies *directly* as y^2 and *inversely* as z, then $x = ky^2/z$.

The constant k may be determined if one set of values of the variables is known.

11.4 UNIT PRICE

When shopping we find that many items are sold in different sizes. To compare the prices, we must compute the price per unit of measure for each size of the item.

EXAMPLES 11.2. What is the unit price for each item?

(*a*) 3 ounce jar of olives costing 87¢

$$\frac{x¢}{87¢} = \frac{1\text{ oz}}{3\text{ oz}} \qquad x = \frac{87}{3} = 29 \qquad 29¢\text{ per oz}$$

(*b*) 12 ounce box of cereal costing $1.32

$$\frac{x¢}{132¢} = \frac{1\text{ oz}}{12\text{ oz}} \qquad x = \frac{132}{12} = 11 \qquad 11¢\text{ per oz}$$

EXAMPLES 11.3. What is the unit price for each item to the nearest tenth of a cent?

(*a*) 6.5 ounce can of tuna costing $1.09

$$\frac{x¢}{109¢} = \frac{1\text{ oz}}{6.5\text{ oz}} \qquad x = \frac{109}{6.5} = 16.8 \qquad 16.8¢\text{ per oz}$$

(*b*) 14 ounce can of salmon costing $1.95

$$\frac{x¢}{195¢} = \frac{1\text{ oz}}{14\text{ oz}} \qquad x = \frac{195}{14} = 13.9 \qquad 13.9¢\text{ per oz}$$

11.5 BEST BUY

To determine the best buy, we compare the unit price for each size of the item and the size with the lowest unit price is the best buy. In doing this, we make two assumptions – a larger size will not result in any waste and the buyer can afford the total price for each of the sizes of the item. The unit price is usually rounded to the nearest tenth of a cent when finding the best buy.

EXAMPLE 11.4. Which is the best buy on a bottle of vegetable oil when 1 gallon costs $5.99, 1 pint costs 89¢, and 24 ounces costs $1.29?

$$\frac{a¢}{599¢} = \frac{1\text{ oz}}{128\text{ oz}} \qquad a = \frac{599}{128} = 4.7 \qquad 4.7¢\text{ per oz}$$

$$\frac{b¢}{89¢} = \frac{1\text{ oz}}{16\text{ oz}} \qquad b = \frac{89}{16} = 5.6 \qquad 5.6¢\text{ per oz}$$

$$\frac{c¢}{129¢} = \frac{1\text{ oz}}{24\text{ oz}} \qquad c = \frac{129}{24} = 5.4 \qquad 5.4¢\text{ per oz}$$

The best buy is one gallon of vegetable oil for $5.99.

Solved Problems

Ratio and Proportion

11.1 Express each of the following ratios as a simplified fraction.

(*a*) $96:128 = \frac{96}{128} = \frac{3}{4}$ (*b*) $\frac{2}{3}:\frac{3}{4} = \frac{2/3}{3/4} = \frac{8}{9}$ (*c*) $xy^2:x^2y = \frac{xy^2}{x^2y} = \frac{y}{x}$

(*d*) $(xy^2 - x^2y):(x-y)^2 = \dfrac{xy^2 - x^2y}{(x-y)^2} = \dfrac{xy(y-x)}{(y-x)^2} = \dfrac{xy}{y-x}$

11.2 Find the ratio of each of the following quantities.

(*a*) 6 pounds to 12 ounces.
It is customary to express the quantities in the same units.
Then the ratio of 96 ounces to 12 ounces is $96:12 = 8:1$.

(*b*) 3 quarts to 2 gallons.
The required ratio is 3 quarts to 8 quarts or $3:8$.

(*c*) 3 square yards to 6 square feet.
Since 1 square yard = 9 square feet, the required ratio is $27\,\text{ft}^2:6\,\text{ft}^2 = 9:2$.

11.3 In each of the following proportions determine the value of x.

(*a*) $(3-x):(x+1) = 2:1$, $\quad \dfrac{3-x}{x+1} = \dfrac{2}{1}$ and $x = \dfrac{1}{3}$.

(*b*) $(x+3):10 = (3x-2):8$, $\quad \dfrac{x+3}{10} = \dfrac{3x-2}{8}$ and $x = 2$.

(*c*) $(x-1):(x+1) = (2x-4):(x+4)$, $\quad \dfrac{x-1}{x+1} = \dfrac{2x-4}{x+4}$, $\quad x^2 - 5x = 0$,

$x(x-5) = 0$ and $x = 0.5$.

11.4 Find the fourth proportional to each of the following sets of numbers. In each case let x be the fourth proportional.

(*a*) 2, 3, 6. Here $2:3 = 6:x$, $\quad \dfrac{2}{3} = \dfrac{6}{x}$ and $x = 9$.

(*b*) 4, −5, 10. Here $4:-5 = 10:x$ and $x = -\dfrac{25}{2}$.

(*c*) a^2, ab, 2. Here $a^2:ab = 2:x$, $\quad a^2x = 2ab$ and $x = \dfrac{2b}{a}$.

11.5 Find the third proportional to each of the following pairs of numbers. In each case let x be the third proportional.

(*a*) 2, 3. Here $2:3 = 3:x$ and $x = 9/2$.

(*b*) $-2, \dfrac{8}{3}$. Here $-2:\dfrac{8}{3} = \dfrac{8}{3}:x$ and $x = -\dfrac{32}{9}$.

11.6 Find the mean proportional between 2 and 8.

SOLUTION

Let x be the required mean proportional. Then $2:x = x:8$, $x^2 = 16$ and $x = \pm 4$.

11.7 A line segment 30 inches long is divided into two parts whose lengths have the ratio 2:3. Find the lengths of the parts.

SOLUTION

Let the required lengths be x and $30 - x$. Then

$$\frac{x}{30-x} = \frac{2}{3} \quad \text{and} \quad x = 12\text{ in.}, \quad 30 - x = 18\text{ in.}$$

11.8 Two brothers are respectively 5 and 8 years old. In how many years (x) will the ratio of their ages be 3:4?

SOLUTION

In x years their respective ages will be $5+x$ and $8+x$.
Then $(5+x):(8+x) = 3:4$, $4(5+x) = 3(8+x)$, and $x = 4$.

11.9 Divide 253 into four parts proportional to 2, 5, 7, 9.

SOLUTION

Let the four parts be $2k$, $5k$, $7k$, $9k$.
Then $2k+5k+7k+9k = 253$ and $k = 11$. Thus the four parts are 22, 55, 77, 99.

11.10 If $x:y:z = 2:-5:4$ and $x - 3y + z = 63$, find x,y,z.

SOLUTION

Let $x = 2k$, $y = -5k$, $z = 4k$.
Substitute these values in $x - 3y + z = 63$ and obtain $2k - 3(-5k) + 4k = 63$ or $k = 3$.
Hence $x = 2k = 6$, $y = -5k = -15$, $z = 4k = 12$.

Variation

11.11 For each of the following statements write an equation, employing k as the constant of proportionality.

(*a*) The circumference C of a circle varies as its diameter d. *Ans.* $C = kd$

(*b*) The period of vibration T of a simple pendulum at a given place is proportional to the square root of its length l. *Ans.* $T = k\sqrt{l}$

(*c*) The rate of emission of radiant energy E per unit area of a perfect radiator is proportional to the fourth power of its absolute temperature T. *Ans.* $E = kT^4$

(*d*) The heat H in calories developed in a conductor of resistance R ohms when using a current of I amperes, varies jointly as the square of the current, the resistance of the conductor, and the time t during which the conductor draws the current. *Ans.* $H = kI^2Rt$

(*e*) The intensity I of a sound wave varies jointly as the square of its frequency n, the square of its amplitude r, the speed of sound v, and the density d of the undisturbed medium. *Ans.* $I = kn^2r^2vd$

(*f*) The force of attraction F between two masses m_1 and m_2 varies directly as the product of the masses and inversely as the square of the distance r between them. *Ans.* $F = km_1m_2/r^2$

(*g*) At constant temperature, the volume V of a given mass of an ideal gas varies inversely as the pressure p to which it is subjected. *Ans.* $V = k/p$

(*h*) An unbalanced force F acting on a body produces in it an acceleration a which is directly proportional to the force and inversely proportional to the mass m of the body. *Ans.* $a = kF/m$

11.12 The kinetic energy E of a body is proportional to its weight W and to the square of its velocity v. An 8 lb body moving at 4 ft/sec has 2 ft-lb of kinetic energy. Find the kinetic energy of a 3 ton (6000 lb) truck speeding at 60 mi/hr (88 ft/sec).

SOLUTION

To find k: $E = kWv^2$ or $k = \dfrac{E}{Wv^2} = \dfrac{2}{8(4^2)} = \dfrac{1}{64}$.

Thus the kinetic energy of the truck is $E = \dfrac{Wv^2}{64} = \dfrac{6000(88)^2}{64} = 726\,000$ ft-lb.

11.13 The pressure p of a given mass of ideal gas varies inversely as the volume V and directly as the absolute temperature T. To what pressure must 100 cubic feet of helium at 1 atmosphere pressure and 253° temperature be subjected to be compressed to 50 cubic feet when the temperature is 313°?

SOLUTION

To find k: $p = k\dfrac{T}{V}$ or $k = \dfrac{pV}{T} = \dfrac{1(100)}{253} = \dfrac{100}{253}$.

Thus the required pressure is $p = \dfrac{100}{253}\dfrac{T}{V} = \dfrac{100}{253}\left(\dfrac{313}{50}\right) = 2.47$ atmospheres.

Another method: let the subscripts 1 and 2 refer to the initial and final conditions of the gas, respectively.

Then $k = \dfrac{p_1V_1}{T_1} = \dfrac{p_2V_2}{T_2}$, $\dfrac{p_1V_1}{T_1} = \dfrac{p_2V_2}{T_2}$, $\dfrac{1(100)}{253} = \dfrac{p_2(50)}{313}$ and $p_2 = 2.47$ atm.

11.14 If 8 men take 12 days to assemble 16 machines, how many days will it take 15 men to assemble 50 machines?

SOLUTION

The number of days (x) varies directly as the number of machiens (y) and inversely as the number of men (z).

Then $x = \dfrac{ky}{z}$ where $k = \dfrac{xz}{y} = \dfrac{12(8)}{16} = 6$.

Hence the required number of days is $x = \dfrac{6y}{z} = \dfrac{6(50)}{15} = 20$ days.

Unit Price and Best Buy

11.15 What is the unit price for 12 oranges costing 99¢?

SOLUTION

$\dfrac{x¢}{99¢} = \dfrac{1 \text{ orange}}{12 \text{ oranges}}$ $x = \dfrac{99}{12} = 8.25$ 8.25¢ per orange

11.16 What is the unit price for trash bags when 20 bags cost \$2.50?

SOLUTION

$\dfrac{x¢}{250¢} = \dfrac{1 \text{ bag}}{20 \text{ bags}}$ $x = \dfrac{250}{20} = 12.5$ 12.5¢ per bag

11.17 Which is the best buy when 7 cans of soup cost \$2.25 and 3 cans of soup cost 95¢?

SOLUTION

$\dfrac{a¢}{225¢} = \dfrac{1 \text{ can}}{7 \text{ cans}}$ $a = \dfrac{225}{7} = 32.1$ 32.1¢ per can

$\dfrac{b¢}{95¢} = \dfrac{1 \text{ can}}{3 \text{ cans}}$ $b = \dfrac{95}{3} = 31.7$ 31.7¢ per can

The best buy is 3 cans of soup costing 95¢.

11.18 Which is the best buy when a 3 ounce package of cream cheese costs 43¢ and an 8 ounce package of cream cheese costs 87¢?

SOLUTION

$$\frac{a¢}{43¢} = \frac{1\text{ oz}}{3\text{ oz}} \qquad a = \frac{43}{3} = 14.3 \qquad 14.3¢\text{ per oz}$$

$$\frac{b¢}{87¢} = \frac{1\text{ oz}}{8\text{ oz}} \qquad b = \frac{87}{8} = 10.9 \qquad 10.9¢\text{ per oz}$$

The best buy is the 8 ounce package of cream cheese costing 87¢.

Supplementary Problems

11.19 Express each ratio as a simplified fraction.

(*a*) 40:64 (*b*) 4/5:8/3 (*c*) $x^2y^3:3xy^4$ (*d*) $(a^2b+ab^2):(a^2b^{3+}+a^3b^2)$

11.20 Find the ratio of the following quantities.

(*a*) 20 yards to 40 feet (*c*) 2 square feet to 96 square inches
(*b*) 8 pints to 5 quarts (*d*) 6 gallons to 30 pints

11.21 In each proportion determine the value of x.

(*a*) $(x+3):(x-2) = 3:2$ (*c*) $(x+1):4 = (x+6):2x$
(*b*) $(x+4):1 = (2-x):2$ (*d*) $(2x+1):(x+1) = 5x:(x+4)$

11.22 Find the fourth proportional to each set of numbers.

(*a*) 3, 4, 12 (*c*) a, b, c
(*b*) −2, 5, 6 (*d*) $m+2$, $m-2$, 3

11.23 Find the third proportional to each pair of numbers.

(*a*) 3, 5 (*b*) −2, 4 (*c*) a, b (*d*) ab, $\sqrt{ab}$

11.24 Find the mean proportional between each pair of numbers.

(*a*) 3, 27 (*b*) −4, −8 (*c*) $3\sqrt{2}$ and $6\sqrt{2}$ (*d*) $m+2$ and $m+1$

11.25 If $(x+y):(x-y) = 5:2$, find $x:y$.

11.26 Two numbers have the ratio 3:4. If 4 is added to each of the numbers the resulting ratio is 4:5. Find the numbers.

11.27 A line segment of length 120 inches is divided into three parts whose lengths are proportional to 3,4,5. Find the lengths of the parts.

11.28 If $x:y:z = 4:-3:2$ and $2x+4y-3z = 20$, find x, y, z.

11.29 (*a*) If x varies directly as y and if $x = 8$ when $y = 5$, find y when $x = 20$.
(*b*) If x varies directly as y^2 and if $x = 4$ when $y = 3$, find x when $y = 6$.
(*c*) If x varies directly as y and if $x = 8$ when $y = 3$, find y when $x = 2$.

11.30 The distance covered by an object falling freely from rest varies directly as the square of the time of falling. If an object falls 144 ft in 3 sec, how far will it fall in 10 sec?

11.31 The force of wind on a sail varies jointly as the area of the sail and the square of the wind velocity. On a square foot of sail the force is 1 lb when the wind velocity is 15 mi/hr. Find the force of a 45 mi/hr wind on a sail of area 20 square yards.

11.32 If 2 men can plow 6 acres of land in 4 hours, how many men are needed to plow 18 acres in 8 hours?

11.33 What is the unit price to the nearest tenth of a cent for each item?

(*a*) 1.36 liter can of fruit punch costing \$1.09

(*b*) 283 gram jar of jelly costing 79¢

(*c*) 10.4 ounce jar of face cream costing \$3.73

(*d*) 1 dozen cans of peas costing \$4.20

(*e*) 25 pounds of grass seed costing \$27.75

(*f*) 3 doughnuts costing 49¢

11.34 Which is the best buy?

(*a*) 100 aspirin tablets for \$1.75 or 200 aspirin tablets for \$2.69

(*b*) a 6 ounce jar of peanut butter for 85¢ or a 12 ounce jar of peanut butter for \$1.59

(*c*) a 14 ounce bottle of mouthwash for \$1.15 or a 20 ounce bottle of mouthwash for \$1.69

(*d*) a 9 ounce jar of mustard for 35¢ or a 24 ounce jar of mustard for 89¢

(*e*) a 454 gram box of crackers for \$1.05 or a 340 gram box of crackers for 93¢

(*f*) a 0.94 liter bottle of fabric softener for 99¢ or a 2.76 liter bottle of fabric softener for \$2.65

ANSWERS TO SUPPLEMENTARY PROBLEMS

11.19 (*a*) 5/8 (*b*) 3/10 (*c*) $x/3y$ (*d*) $1/ab$

11.20 (*a*) 3:2 (*b*) 4:5 (*c*) 3:1 (*d*) 8:5

11.21 (*a*) 12 (*b*) -2 (*c*) 4, -3 (*d*) 2, $-2/3$

11.22 (*a*) 16 (*b*) -15 (*c*) bc/a (*d*) $3(m-2)/(m+2)$

11.23 (*a*) 25/3 (*b*) -8 (*c*) b^2/a (*d*) 1

11.24 (*a*) ± 9 (*b*) $\pm 4\sqrt{2}$ (*c*) ± 6 (*d*) $\pm\sqrt{m^2+3m+2}$

11.25 7/3

11.26 12, 16

11.27 30, 40, 50 in.

11.28 -8, 6, -4

11.29 (*a*) $12\frac{1}{2}$ (*b*) 16 (*c*) 12

11.30 1600 ft

11.31 1620 lb

11.32 3 men

11.33 (*a*) 80.1¢ per liter (*c*) 35.9¢ per ounce (*e*) 111¢ per pound
(*b*) 0.3¢ per gram (*d*) 35¢ per can (*f*) 16.3¢ per doughnut

11.34 (*a*) 200 aspirins for $2.69 (*c*) 14 ounce bottle for $1.15 (*e*) 454 gram box for $1.05
(*b*) 12 ounce jar for $1.59 (*d*) 24 ounce jar for 89¢ (*f*) 2.76 liter bottle for $2.65

Chapter 12

Functions and Graphs

12.1 VARIABLES

A variable is a symbol which may assume any one of a set of values during a discussion. A *constant* is a symbol which stands for only one particular value during the discussion.

Letters at the end of the alphabet, such as x, y, z, u, v, and w, are usually employed to represent variables, and letters at the beginning of the alphabet, such as a, b, and c, are used as constants.

12.2 RELATIONS

A relation is a set of ordered pairs. The relation may be specified by an equation, a rule, or a table. The set of the first components of the ordered pairs is called the domain of the relation. The set of the second components is called the range of the relation. In this chapter, we shall consider only relations that have sets of real numbers for their domain and range.

EXAMPLE 12.1 What is the domain and range of the relation $\{(1,3), (2,6), (3,9), (4,12)\}$?

$$\text{domain} = \{1,2,3,4\} \qquad \text{range} = \{3,6,9,12\}$$

12.3 FUNCTIONS

A function is a relation such that each element in the domain is paired with exactly one element in the range.

EXAMPLES 12.2. Which relations are functions?

(*a*) $\{(1,2), (2,3), (3,4), (4,5)\}$
function – each first element is paired with exactly one second element

(*b*) $\{(1,2), (1,3), (2,8), (3,9)\}$
not a function – 1 is paired with 2 and with 3

(*c*) $\{(1,3), (2,3), (4,3), (9,3)\}$
function – each first element is paired with exactly one second element

Often functions and relations are stated as equations. When the domain is not stated, we determine the largest subset of the real numbers for which the equation is defined, and that is the domain. Once the domain has been determined, we determine the range by finding the value of the equation for each value of the domain. The variable associated with the domain is called an independent variable and the variable associated with the range is called the dependent variable. In equations with the variables x and y, we generally assume that x is the independent variable and that y is the dependent variable.

EXAMPLE 12.3. What is the domain and range of $y = x^2 + 2$?

The domain is the set of all real numbers since the square of each real number is a real number and that a real number plus 2 is still a real number. Domain = {all real numbers}

The range is the set of all real numbers greater than or equal to 2 since the square of a real number is at least zero and when you add 2 to each value we have the real numbers that are at least 2. Range = {all real numbers ≥ 2}

EXAMPLE 12.4. What is the domain and range of $y = 1/(x - 3)$?

The equation is not defined when $x = 3$, so the domain is the set of all real numbers not equal to 3. Domain = {real numbers $\neq$ 3}

A fraction can be zero only when the numerator can be zero. Since the numerator of this fraction is always 1, the fraction can never equal zero. Thus, the range is the set of all real numbers not equal to 0. Range = {real numbers $\neq$ 0}

12.4 FUNCTION NOTATION

The notation $y = f(x)$, read "y equals f of x," is used to designate that y is a function of x. With this notation $f(a)$ represents the value of the dependent variable y when $x = a$ (provided that there is a value).

Thus $y = x^2 - 5x + 2$ may be written $f(x) = x^2 - 5x + 2$. Then $f(2)$, i.e., the value of $f(x)$ or y when $x = 2$, is $f(2) = 2^2 - 5(2) + 2 = -4$. Similarly, $f(-1) = (-1)^2 - 5(-1) + 2 = 8$.

Any letter may be used in the function notation; thus $g(x)$, $h(x)$, $F(x)$, etc., may represent functions of x.

12.5 RECTANGULAR COORDINATE SYSTEM

A rectangular coordinate system is used to give a picture of the relationship between two variables.

Consider two mutually perpendicular lines $X'X$ and $Y'Y$ intersecting in the point O, as shown in Fig. 12-1.

The line $X'X$, called the x axis, is usually horizontal.

The line $Y'Y$, called the y axis, is usually vertical.

The point O is called the origin.

Using a convenient unit of length, lay off points on the x axis at successive units to the right and left of the origin O, labeling those points to the right $1, 2, 3, 4, \ldots$ and those to the left $-1, -2, -3, -4, \ldots$. Here we have arbitrarily chosen OX to be the positive direction; this is customary but not necessary.

Do the same on the y axis, choosing OY as the positive direction. It is customary (but not necessary) to use the same unit of length on both axes.

The x and y axes divide the plane into 4 parts known as *quadrants*, which are labeled I, II, III, IV as in Fig. 12-1.

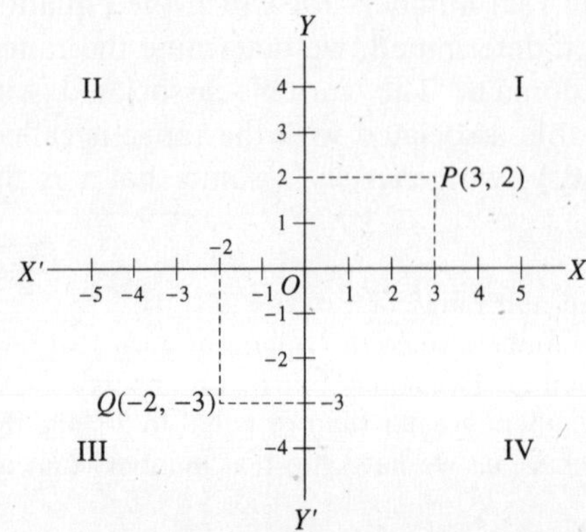

Fig. 12-1

Given a point P in this xy plane, drop perpendiculars from P to the x and y axes. The values of x and y at the points where these perpendiculars meet the x and y axes determine respectively the x *coordinate* (or abscissa) of the point and the y *coordinate* (or ordinate) of the point P. These coordinates are indicated by the symbol (x, y).

Conversely, given the coordinates of a point, we may locate or plot the point in the xy plane.

For example, the point P in Fig. 12-1 has coordinates $(3, 2)$; the point having coordinates $(-2, -3)$ is Q.

The graph of a function $y = f(x)$ is the set of all points (x, y) satisfied by the equation $y = f(x)$.

12.6 FUNCTION OF TWO VARIABLES

The variable z is said to be a function of the variables x and y if there exists a relation such that to each pair of values of x and y there corresponds one or more values of z. Here x and y are independent variables and z is the dependent variable.

The function notation used in this case is $z = f(x, y)$: read "z equals f of x and y." Then $f(a,b)$ denotes the value of z when $x = a$ and $y = b$, provided the function is defined for these values.

Thus if $f(x, y) = x^3 + xy^2 - 2x$, then $f(2, 3) = 2^3 + 2 \cdot 3^2 - 2 \cdot 3 = 20$.

In like manner we may define functions of more than two independent variables.

12.7 SYMMETRY

When the left half of a graph is a mirror image of the right half, we say the graph is symmetric with respect to the y axis (see Fig. 12-2). This symmetry occurs because for any x value, both x and $-x$ result in the same y value, that is $f(x) = f(-x)$. The equation may or may not be a function for y in terms of x.

Some graphs have a bottom half that is the mirror image of the top half, and we say these graphs are symmetric with respect to the x axis. Symmetry with respect to the x axis results when for each y, both y and $-y$ result in the same x value (see Fig. 12-3). In these cases, you do not have a function for y in terms of x.

If substituting for $-x$ for x and $-y$ for y in an equation yields an equivalent equation, we say the graph is symmetric with respect to the origin (see Fig. 12-4). These equations represent relations that are not always functions.

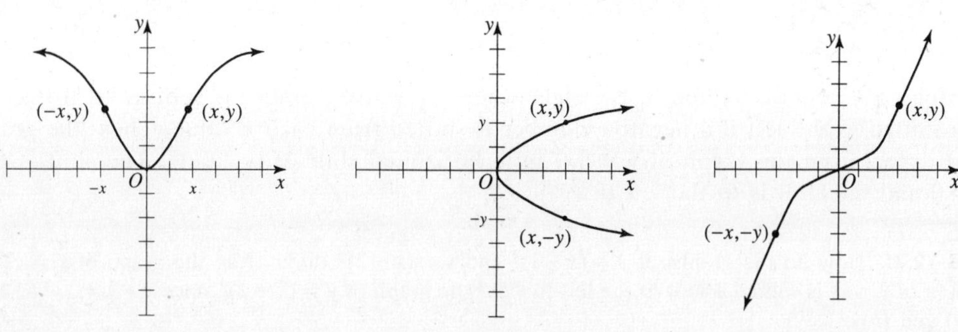

Fig. 12-2 Fig. 12-3 Fig. 12-4

Symmetry can be used to make sketching the graphs of relations and functions easier. Once the type of symmetry, if any, and the shape of half of the graph have been determined, the other half of the graph can be drawn by using this symmetry. Most graphs are not symmetric with respect to the y axis, x axis, or the origin. However, many frequently used graphs do display one of these symmetries and using that symmetry in graphing the relation simplifies the graphing process.

EXAMPLE 12.5. Test the relation $y = 1/x$ for symmetry.

Substituting $-x$ for x, we get $y = -1/x$, so the graph is not symmetric with respect to the y axis.

Substituting $-y$ for y, we get $-y = 1/x$, so the graph is not symmetric with respect to the x axis.

Substituting $-x$ for x and $-y$ for y, we get $-y = -1/x$ which is equivalent to $y = 1/x$, so the graph is symmetric with respect to the origin.

12.8 SHIFTS

The graph of $y = f(x)$ is shifted upward by adding a positive constant to each y value in the graph. It is shifted downward by adding a negative constant to each y value in the graph of $y = f(x)$. Thus, the graph of $y = f(x) + b$ differs from the graph of $y = f(x)$ by a vertical shift of $|b|$ units. The shift is up if $b > 0$ and the shift is down if $b < 0$.

EXAMPLES 12.6. How do the graphs of $y = x^2 + 2$ and $y = x^2 - 3$ differ from the graph of $y = x^2$?

The graph of $y = x^2$ is shifted up 2 units to yield the graph of $y = x^2 + 2$ (see Figs. 12-5(*a*) and (*b*)).

The graph of $y = x^2$ is shifted 3 units down to yield the graph of $y = x^2 - 3$ (see Figs. 12-5(*a*) and (*c*)).

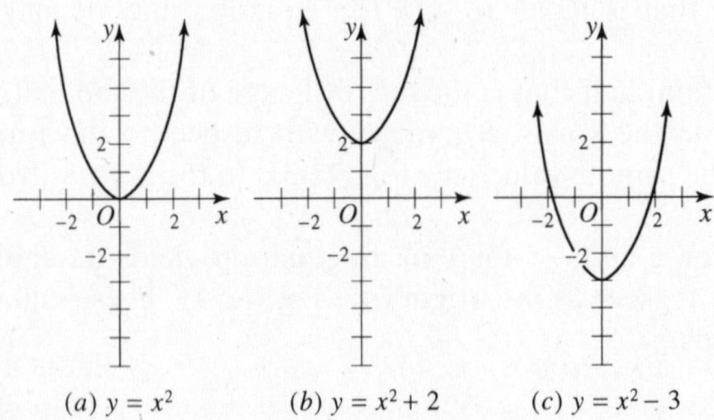

Fig. 12-5

The graph of $y = f(x)$ is shifted to the right when a positive number is subtracted from each x value. It is shifted to the left if a negative number is shifted from each x value. Thus, the graph of $y = f(x - a)$ differs from the graph of $y = f(x)$ by a horizontal shift of $|a|$ units. The shift is to the right if $a > 0$ and the shift is to the left if $a < 0$.

EXAMPLES 12.7. How do the graphs of $y = (x + 1)^2$ and $y = (x - 2)^2$ differ from the graph of $y = x^2$?

The graph of $y = x^2$ is shifted 1 unit to the left to yield the graph of $y = (x + 1)^2$ since $x + 1 = x - (-1)$ (see Figs. 12-6(*a*) and (*b*)).

The graph of $y = x^2$ is shifted 2 units to the right to yield the graph of $y = (x - 2)^2$ (see Figs. 12-6(*a*) and (*c*)).

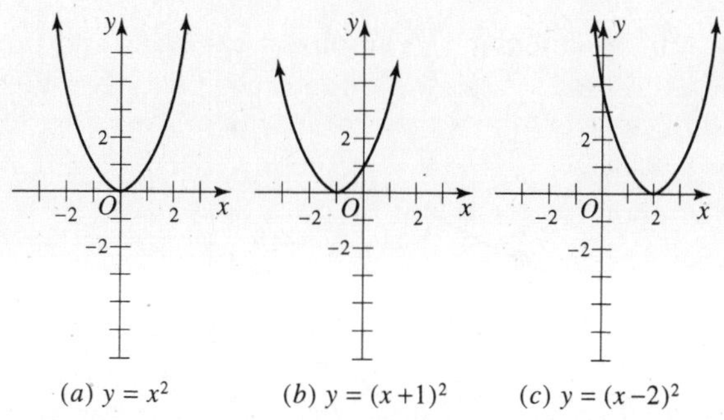

(a) $y = x^2$ (b) $y = (x+1)^2$ (c) $y = (x-2)^2$

Fig. 12-6

12.9 SCALING

If each y value is multiplied by a positive number greater than 1, the rate of change in y is increased from the rate of change in y values for $y = f(x)$. However, if each y value is multiplied by a positive number between 0 and 1, the rate of change in y values is decreased from the rate of change in y values for $y = f(x)$. Thus, the graph of $y = cf(x)$, where c is a positive number, differs from the graph of $y = f(x)$ by the rate of increase in y. If $c > 1$ the rate of change in y is increased and if $0 < c < 1$ the rate of change in y is decreased.

The graph of $y = f(x)$ is reflected across the x axis when each y value is multiplied by a negative number. So the graph of $y = cf(x)$, where $c < 0$, is the reflection of $y = |c|f(x)$ across the x axis.

EXAMPLES 12.8. How do the graphs of $y = -|x|$, $y = 3|x|$, and $y = 1/2|x|$ differ from the graph of $y = |x|$?

The graph of $y = |x|$ is reflected across the x axis to yield $y = -|x|$ (see Figs. 12-7(a) and (b)).

The graph of $y = |x|$ has the y value multiplied by 3 for each x value to yield the graph of $y = 3|x|$ (see Figs. 12-7(a) and (c)).

The graph of $y = |x|$ has the y value multiplied by 1/2 for each x value to yield the graph of $y = 1/2|x|$ (see Figs. 12-7(a) and (d)).

12.10 USING A GRAPHING CALCULATOR

In discussing graphing calculators the information given will be general, but a Texas Instruments TI-81 graphing calculator was used to verify the general procedures. Most graphing calculators operate in a somewhat similar manner to one another, but you need to use the instruction manual for your calculator to see how to do these operations for that particular make and model of graphing calculator.

A graphing calculator allows you to graph functions easily. The key to graphing is setting the graphing window appropriately. To do this, you need to use the domain of the function to set the maximum and minimum x values and the range to set the maximum and minimum y values. When the domain or range is a large interval, it may be necessary to use the scale for x or y to make the graph smaller, increasing the size of the units along either or both axes. Occasionally, it may be necessary to view the graph in parts of its domain or range to see how the graph actually looks.

To compare the graphs of $y = x^2$, $y = x^2 + 2$, and $y = x^2 - 3$ on a graphing calculator, you enter each function in the $y =$ menu. Let $y_1 = x^2$, $y_2 = x^2 + 2$, and $y_3 = x^2 - 3$. Turn off the functions y_2 and y_3 and set the graphing window at the standard setting. When you press the graph key, you will see a graph as shown in Fig. 12-5(a). Turn off the function y_1 and turn on the function y_2, then press the graph key. The graph displayed will be Fig. 12-5(b). Now turn off function y_2 and turn on function y_3. Press the graph key and you will see Fig. 12-5(c). When you turn on functions y_1, y_2, and y_3 and

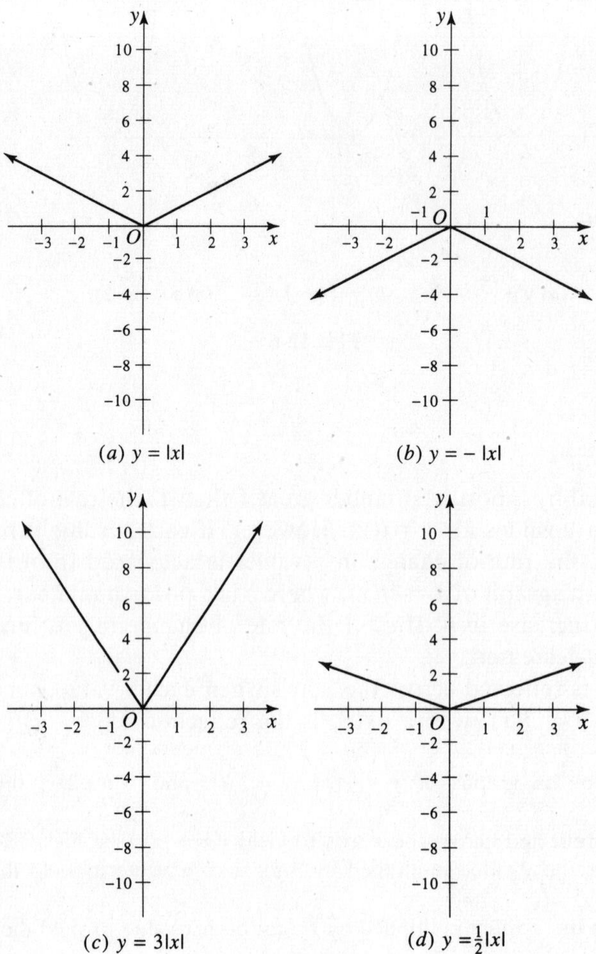

Fig. 12-7

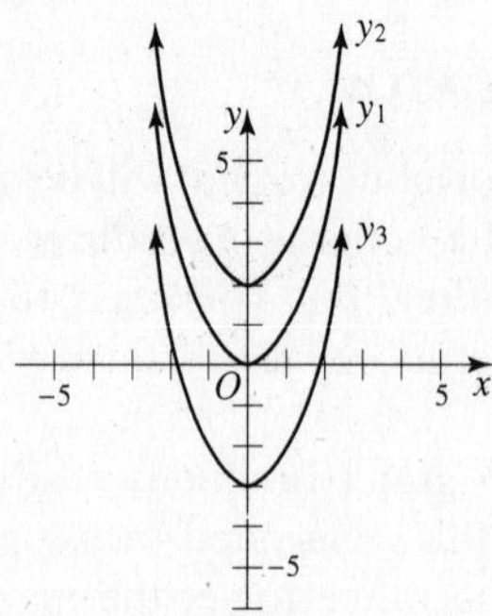

Fig. 12-8

press the graph key and you will see all three functions graphed on the same set of axes (see Fig. 12-8). The graph of $y_2 = x^2 + 2$ is 2 units above the graph of $y_1 = x^2$, while the graph of $y_3 = x^2 - 3$ is 3 units below the graph of y_1.

In a similar fashion, you can compare the graph of $y_1 = f(x)$ and $y_2 = f(x) + b$ for any function $f(x)$. Notice that when $b > 0$, the graph of y_2 is b units above the graph of y_1. When $b < 0$, the graph of y_2 is $|b|$ units below the graph of y_1.

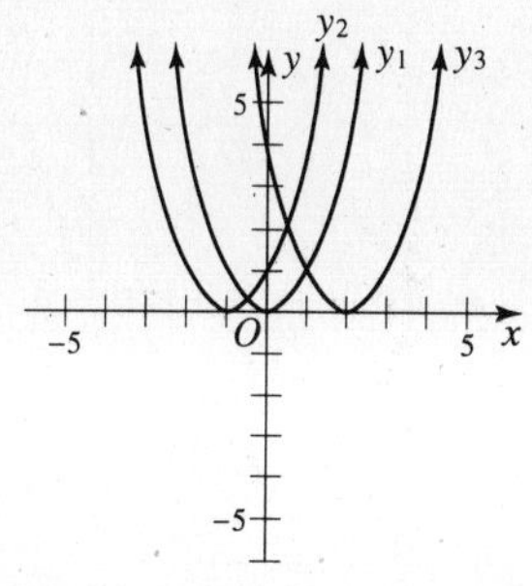

Fig. 12-9

Consider the graphs of $y = x^2$, $y = (x+1)^2$, and $y = (x-2)^2$. To compare these graphs using a calculator, we need to set $y_1 = x^2$, $y_2 = (x+1)^2$, and $y_3 = (x-2)^2$. Using the standard window and graphing all three functions at once, we see that $y_2 = (x+1)^2$ is 1 unit to the left of the graph of $y_1 = x^2$. Also, the graph of $y_3 = (x-2)^2$ is 2 units to the right of the graph of y_1 (see Fig. 12-9).

In general, to compare the graphs of $y_1 = f(x)$ and $y_2 = f(x-a)$ for all functions $f(x)$, we note that the graph of $y_2 = f(x-a)$ is a units to the right of $y_1 = f(x)$ when $a > 0$. When $a < 0$, the graph of y_2 is $|a|$ units to the left of y_1.

EXAMPLE 12.9. Graph $x^2 + y^2 = 9$.

To graph $x^2 + y^2 = 9$ on a calculator, we first solve the equation for y and get $y = \pm\sqrt{9 - x^2}$. We let $y_1 = +\sqrt{9 - x^2}$ and $y_2 = -\sqrt{9 - x^2}$ and graph them on the same set of axes. If we use the standard window, we get a distorted view of this graph because the scale on the y axis is not equal to the scale on the x axis. By multiplying by a factor of 0.67 (for the TI-81), we can adjust the y interval and get a more accurate view of the graph. Thus using the domain $[-10, 10]$ and a range of $[-6.7, 6.7]$, we get the graph of a circle (see Fig. 12-10).

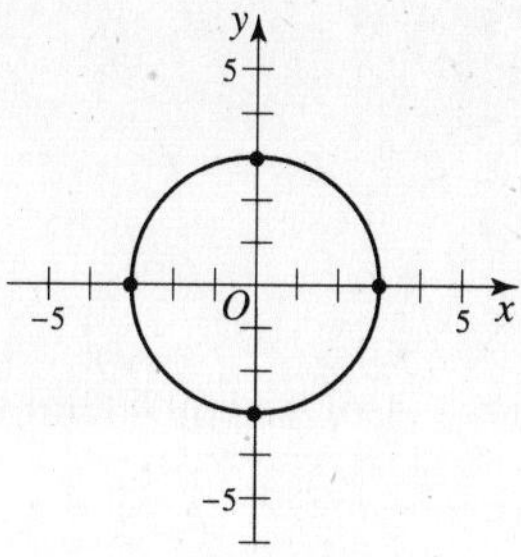

Fig. 12-10

Solved Problems

12.1 Express the area A of a square as a function of its (a) side x, (b) perimeter P, and (c) diagonal D (see Fig. 12-11).

SOLUTION

(a) $A = x^2$

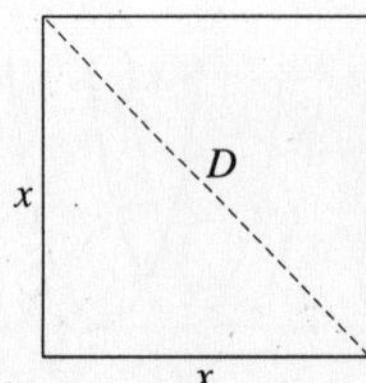

Fig. 12-11

(*b*) $P = 4x$ or $x = \dfrac{P}{4}$. Then $A = x^2 = \left(\dfrac{P}{4}\right)^2$ or $A = \dfrac{P^2}{16}$.

(*c*) $D = \sqrt{x^2 + x^2} = x\sqrt{2}$ or $x = \dfrac{D}{\sqrt{2}}$. Then $A = x^2 = \left(\dfrac{D}{\sqrt{2}}\right)^2$ or $A = \dfrac{D^2}{2}$.

12.2 Express the (*a*) area A, (*b*) perimeter P, and (*c*) diagonal D of a rectangle as a function of its sides x and y. Refer to Fig. 12-12.

SOLUTION

(*a*) $A = xy$, (*b*) $P = 2x + 2y$, (*c*) $D = \sqrt{x^2 + y^2}$

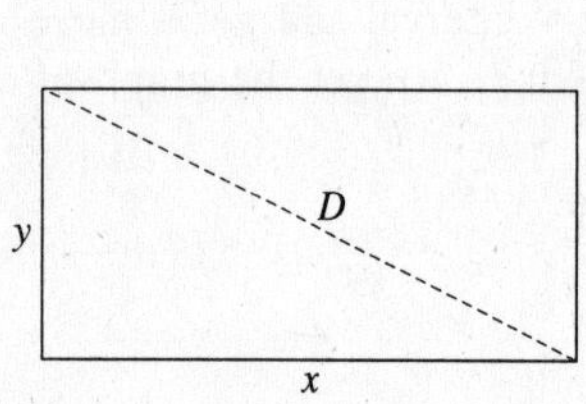

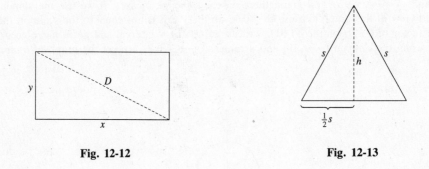

Fig. 12-12

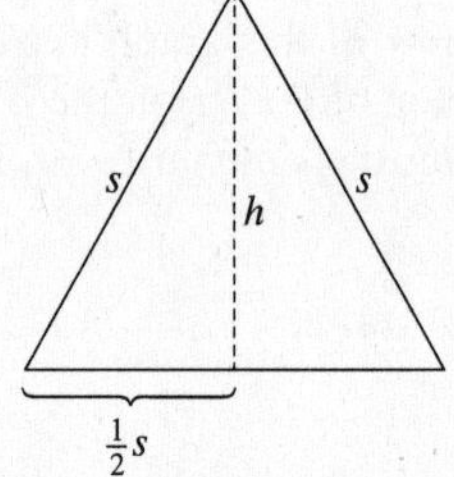

Fig. 12-13

12.3 Express the (*a*) altitude h and (*b*) area A of an equilateral triangle as a function of its side s. Refer to Fig. 12-13.

SOLUTION

(*a*) $h = \sqrt{s^2 - \left(\dfrac{1}{2}s\right)^2} = \sqrt{\dfrac{3}{4}s^2} = \dfrac{s\sqrt{3}}{2}$ (*b*) $A = \dfrac{1}{2}hs = \dfrac{1}{2}\left(\dfrac{s\sqrt{3}}{2}\right)s = \dfrac{s^2\sqrt{3}}{4}$

12.4 The surface area S and volume V of a sphere of radius r are given by $S = 4\pi r^2$ and $V = \frac{4}{3}\pi r^3$. Express (*a*) r as a function of S and also as a function of V, (*b*) V as a function of S, and (*c*) S as a function of V.

SOLUTION

(*a*) From $S = 4\pi r^2$ obtain

$$r = \sqrt{\frac{S}{4\pi}} = \frac{1}{2}\sqrt{\frac{S}{\pi}}.$$

From $V = \frac{4}{3}\pi r^3$ obtain

$$r = \sqrt[3]{\frac{3V}{4\pi}}.$$

(*b*) Set

$$r = \frac{1}{2}\sqrt{\frac{S}{\pi}}$$

in $V = \frac{4}{3}\pi r^3$ and obtain

$$V = \frac{4}{3}\pi\left(\frac{1}{2}\sqrt{\frac{S}{\pi}}\right)^3 = \frac{S}{6}\sqrt{\frac{S}{\pi}}.$$

(*c*) Set

$$r = \sqrt[3]{\frac{3V}{4\pi}}$$

in $S = 4\pi r^2$ and obtain

$$S = 4\pi\sqrt[3]{\left(\frac{3V}{4\pi}\right)^2} = 4\pi\sqrt[3]{\frac{9V^2}{16\pi^2}\cdot\frac{4\pi}{4\pi}} = \sqrt[3]{36\pi V^2}.$$

12.5 Given $y = 3x^2 - 4x + 1$, find the values of y corresponding to $x = -2, -1, 0, 1, 2$.

SOLUTION

For $x = -2$, $y = 3(-2)^2 - 4(-2) + 1 = 21$; for $x = -1$, $y = 3(-1)^2 - 4(-1) + 1 = 8$; for $x = 0$, $y = 3(0)^2 - 4(0) + 1 = 1$; for $x = 1$, $y = 3(1)^2 - 4(1) + 1 = 0$; for $x = 2$, $y = 3(2)^2 - 4(2) + 1 = 5$. These values of x and y are conveniently listed in the following table.

x	-2	-1	0	1	2
y	21	8	1	0	5

12.6 Extend the table of values in Problem 12.5 by finding the values of y corresponding to $x = -3/2$, $-1/2$, $1/2$, $3/2$.

SOLUTION

For $x = -3/2$, $y = 3(-3/2)^2 - 4(-3/2) + 1 = 13\frac{3}{4}$; etc. The following table of values summarizes the results.

x	-2	$-\frac{3}{2}$	-1	$-\frac{1}{2}$	0	$\frac{1}{2}$	1	$\frac{3}{2}$	2
y	21	$13\frac{3}{4}$	8	$3\frac{3}{4}$	1	$-\frac{1}{4}$	0	$1\frac{3}{4}$	5

12.7 State the domain and range for each relation.

(*a*) $y = 3 - x^2$ (*b*) $y = x^3 + 1$ (*c*) $y = \sqrt{x + 2}$ (*d*) $y = \sqrt[3]{x}$

SOLUTION

(*a*) Domain = {all real numbers} Since every real number can be squared, $3 - x^2$ is defined for all real numbers.

Range = {all real numbers ≤ 3} Since x^2 is non-negative for all real numbers, $3 - x^2$ does not exceed 3.

(*b*) Domain = {all real numbers} Since every real number can be cubed, $x^3 + 1$ is defined for all real numbers.

Range = {all real numbers} Since x^3 yields all real numbers, $x^3 + 1$ also yields all real numbers.

(*c*) Domain = {all real numbers ≥ -2} Since the square root yields real numbers only for non-negative real numbers, x must be at least -2.

Range = {all real numbers ≥ 0} Since we want the principal square root, the values will be non-negative numbers.

(*d*) Domain = {all real numbers} Since the cube root yields a real number for all real numbers, x can be any real number.

Range = {all real numbers} Since any real number can be the cube root of a real number, we get all real numbers.

12.8 In which of these equations is y a function of x?

(*a*) $y = 3x^3$ (*c*) $xy = 1$ (*e*) $y = \sqrt{4x}$

(*b*) $y^2 = x$ (*d*) $y = 2x + 5$ (*f*) $y^3 = 8x$

SOLUTION

(*a*) Function — For each value of x, $3x^3$ yields exactly one value.

(*b*) Not a function — For $x = 4$, y can be 2 or -2.

(*c*) Function — $y = 1/x$. For every non-zero real number $1/x$ yields exactly one value.

(*d*) Function — For each value of x, $2x + 5$ yields exactly one value.

(*e*) Function — For each value of $x \geq 0$, $\sqrt{4x}$ yields the principal square root.

(*f*) Function — For each real number, $8x$ is a real number and every real number has exactly one real cube root.

12.9 If $f(x) = x^3 - 5x - 2$, find $f(-2)$, $f(-3/2)$, $f(-1)$, $f(0)$, $f(1)$, $f(2)$.

SOLUTION

$f(-2) = (-2)^3 - 5(-2) - 2 = 0$ $\qquad f(0) = 0^3 - 5(0) - 2 = -2$

$f(-3/2) = (-3/2)^3 - 5(-3/2) - 2 = 17/8$ $\qquad f(1) = 1^3 - 5(1) - 2 = -6$

$f(-1) = (-1)^3 - 5(-1) - 2 = 2$ $\qquad f(2) = 2^3 - 5(2) - 2 = -4$

We may arrange these values in a table.

x	-2	$-3/2$	-1	0	1	2
$f(x)$	0	17/8	2	-2	-6	-4

12.10 If $F(t) = \dfrac{t^3 + 2t}{t - 1}$, find $F(-2)$, $F(x)$, $F(-x)$.

SOLUTION

$$F(-2) = \frac{(-2)^3 + 2(-2)}{-2 - 1} = \frac{-8 - 4}{-3} = 4$$

$$F(x) = \frac{x^3 + 2x}{x - 1}$$

$$F(-x) = \frac{(-x)^3 + 2(-x)}{-x - 1} = \frac{-x^3 - 2x}{-x - 1} = \frac{x^3 + 2x}{x + 1}$$

12.11 Given $R(x) = (3x - 1)/(4x + 2)$, find

(*a*) $R\left(\dfrac{x - 1}{x + 2}\right)$, (*b*) $\dfrac{R(x + h) - R(x)}{h}$, (*c*) $R[R(x)]$.

SOLUTION

(a) $$R\left(\frac{x-1}{x+2}\right)=\frac{3\left(\dfrac{x-1}{x+2}\right)-1}{4\left(\dfrac{x-1}{x+2}\right)+2}=\frac{\left(\dfrac{2x-5}{x+2}\right)}{\left(\dfrac{6x}{x+2}\right)}=\frac{2x-5}{6x}$$

(b) $$\frac{R(x+h)-R(x)}{h}=\frac{1}{h}\{R(x+h)-R(x)\}=\frac{1}{h}\left(\frac{3(x+h)-1}{4(x+h)+2}-\frac{3x-1}{4x+2}\right)$$
$$=\frac{1}{h}\left(\frac{[3(x+h)-1][4x+2]-[3x-1][4(x+h)+2]}{[4(x+h)+2][4x+2]}\right)=\frac{5}{2(2x+2h+1)(2x+1)}$$

(c) $$R[R(x)]=R\left(\frac{3x-1}{4x+2}\right)=\frac{3\left(\dfrac{3x-1}{4x+2}\right)-1}{4\left(\dfrac{3x-1}{4x+2}\right)+2}=\frac{5x-5}{20x}=\frac{x-1}{4x}$$

12.12 If $F(x,y)=x^3-3xy+y^2$, find

(a) $F(2,3)$, (b) $F(-3,0)$, (c) $\dfrac{F(x,y+k)-F(x,y)}{k}$.

SOLUTION

(a) $F(2,3)=2^3-3(2)(3)+3^2=-1$

(b) $F(-3,0)=(-3)^3-3(-3)(0)+0^2=-27$

(c) $$\frac{F(x,y+k)-F(x,y)}{k}=\frac{x^3-3x(y+k)+(y+k)^2-[x^3-3xy+y^2]}{k}=-3x+2y+k$$

12.13 Plot the following points on a rectangular coordinate system: $(2,1)$, $(4,3)$, $(-2,4)$, $(-4,2)$, $(-4,-2)$, $(-5/2,-9/2)$, $(4,-3)$, $(2,-\sqrt{2})$.

SOLUTION

See Fig. 12-14.

12.14 Given $y=2x-1$, obtain the values of y corresponding to $x=-3,-2,-1,0,1,2,3$ and plot the points (x,y) thus obtained.

SOLUTION

The following table lists the values of y corresponding to the given values of x.

x	-3	-2	-1	0	1	2	3
y	-7	-5	-3	-1	1	3	5

The points $(-3,-7)$, $(-2,-5)$, $(-1,-3)$, $(0,-1)$, $(1,1)$, $(2,3)$, $(3,5)$ are plotted, as shown in Fig. 12-15.

Note that all points satisfying $y=2x-1$ lie on a straight line. In general the graph of $y=ax+b$, where a and b are constants, is a straight line; hence $y=ax+b$ or $f(x)=ax+b$ is called a *linear function*. Since two points determine a straight line, only two points need be plotted and the line drawn connecting them.

12.15 Obtain the graph of the function defined by $y=x^2-2x-8$ or $f(x)=x^2-2x-8$.

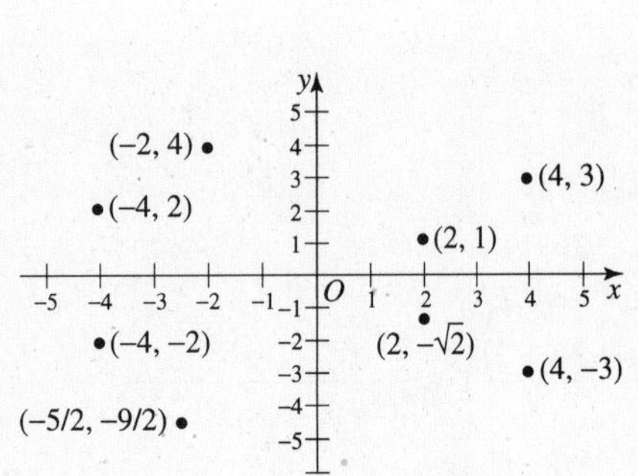

Fig. 12-14

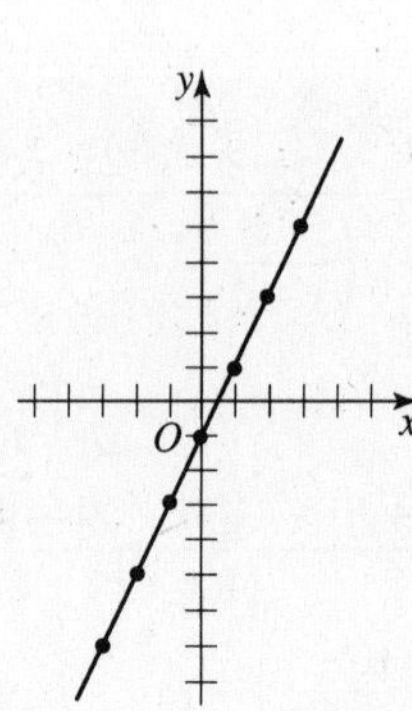

Fig. 12-15

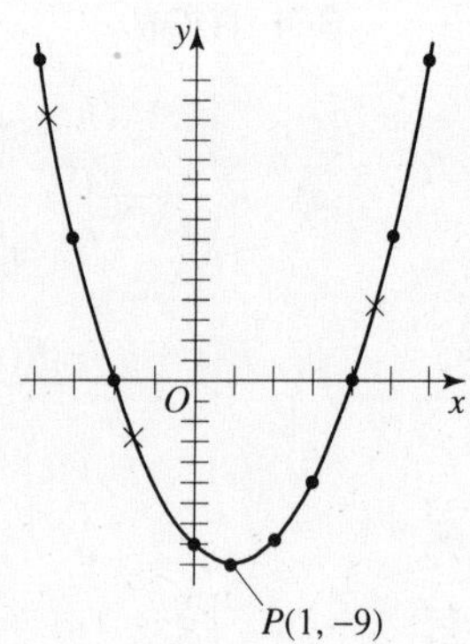

Fig. 12-16

SOLUTION

The following table gives the values of y or $f(x)$ for various values of x.

x	−4	−3	−2	−1	0	1	2	3	4	5	6
y or $f(x)$	16	7	0	−5	−8	−9	−8	−5	0	7	16

Thus the following points lie on the graph: $(-4, 16)$, $(-3, 7)$, $(-2, 0)$, $(-1, -5)$, etc.

In plotting these points it is convenient to use different scales on the x and y axes, as shown in Fig. 12-16. The points marked × were added to those already obtained in order to get a more accurate picture.

The curve thus obtained is called a *parabola*. The lowest point P, called a minimum point, is the *vertex* of the parabola.

12.16 Graph the function defined by $y = 3 - 2x - x^2$.

SOLUTION

x	−5	−4	−3	−2	−1	0	1	2	3	4
y	−12	−5	0	3	4	3	0	−5	−12	−21

The curve obtained is a parabola, as shown in Fig. 12-17. The point $Q(-1, 4)$, the vertex of the parabola, is a maximum point. In general, $y = ax^2 + bx + c$ represents a parabola whose vertex is either a maximum or minimum point depending on whether a is − or +, respectively. The function $f(x) = ax^2 + bx + c$ is sometimes called a quadratic function.

12.17 Obtain the graph of $y = x^3 + 2x^2 - 7x - 3$.

SOLUTION

x	−4	−3	−2	−1	0	1	2	3
y	−7	9	11	5	−3	−7	−1	21

The graph is shown in Fig. 12-18. Points marked × are not listed in the table; they were added in order to improve the accuracy of the graph.

Point A is called a *relative maximum point*; it is not the highest point on the entire curve, but points on either side are lower. Point B is called a *relative minimum point*. The calculus enables us to determine such relative maximum and minimum points.

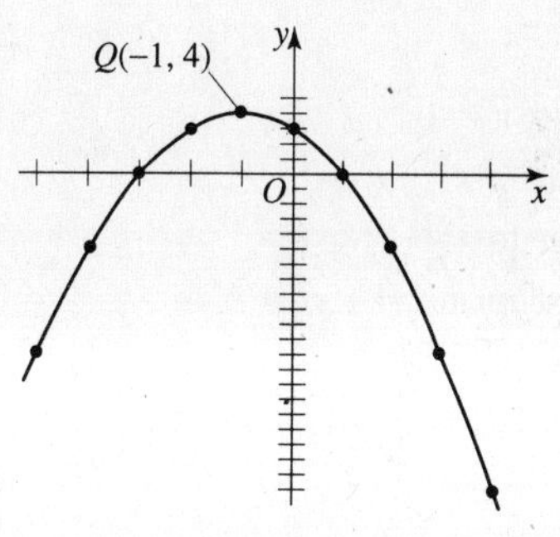

Fig. 12-17

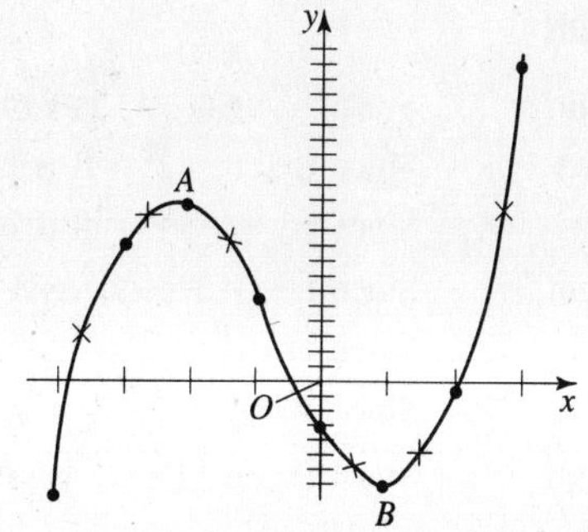

Fig. 12-18

12.18 Obtain the graph of $x^2 + y^2 = 36$.

SOLUTION

We may write $y^2 = 36 - x^2$ or $y = \pm\sqrt{36 - x^2}$. Note that x must have a value between -6 and $+6$ if y is to be a real number.

x	-6	-5	-4	-3	-2	-1	0	1	2	3	4	5	6
y	0	$\pm\sqrt{11}$	$\pm\sqrt{20}$	$\pm\sqrt{27}$	$\pm\sqrt{32}$	$\pm\sqrt{35}$	± 6	$\pm\sqrt{35}$	$\pm\sqrt{32}$	$\pm\sqrt{27}$	$\pm\sqrt{20}$	$\pm\sqrt{11}$	0

The points to be plotted are $(-6, 0)$, $(-5, \sqrt{11})$, $(-5, -\sqrt{11})$, $(-4, \sqrt{20})$, $(-4, -\sqrt{20})$, etc.

Figure 12-19 shows the graph, a circle of radius 6.

In general, the graph of $x^2 + y^2 = a^2$ is a circle with center at the origin and radius a.

It should be noted that if the units had not been taken the same on the x and y axes, the graph would not have looked like a circle.

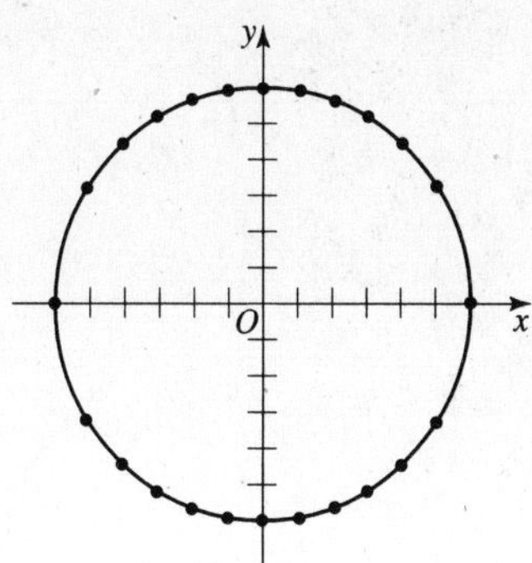

Fig. 12-19

12.19 Determine whether the graph is symmetric with respect to the y axis, x axis, or the origin.

(*a*) $y = 4x$ (*c*) $xy^2 = 1$ (*e*) $y = x^3$

(*b*) $x^2 + y^2 = 8$ (*d*) $x = y^2 + 1$ (*f*) $y = \sqrt{x}$

SOLUTION

(*a*)	Origin	Since $-y = 4(-x)$ is equivalent to $y = 4x$.
(*b*)	y axis	Since $(-x)^2 + y^2 = 8$ is equivalent to $x^2 + y^2 = 8$.
	x axis	Since $x^2 + (-y)^2 = 8$ is equivalent to $x^2 + y^2 = 8$.
	Origin	Since $(-x)^2 + (-y)^2 = 8$ is equivalent to $x^2 + y^2 = 8$.
(*c*)	x axis	Since $x(-y)^2 = 1$ is equivalent to $xy^2 = 1$.
(*d*)	x axis	Since $x = (-y)^2 + 1$ is equivalent to $x = y^2 + 1$.
(*e*)	Origin	Since $(-y) = (-x)^3$ is equivalent to $y = x^3$.
(*f*)	None	

12.20 Use the graph of $y = x^3$ to graph $y = x^3 + 1$.

SOLUTION

The graph of $y = x^3$ is shown in Fig. 12-20. The graph of $y = x^3 + 1$ is the graph of $y = x^3$ 1 unit up and is shown in Fig. 12-21.

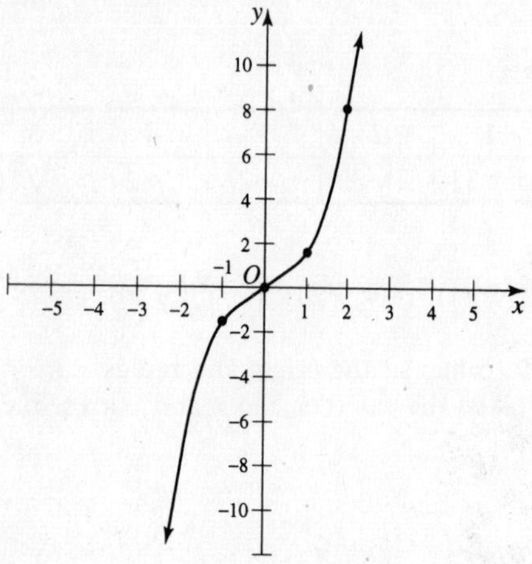

Fig. 12-20

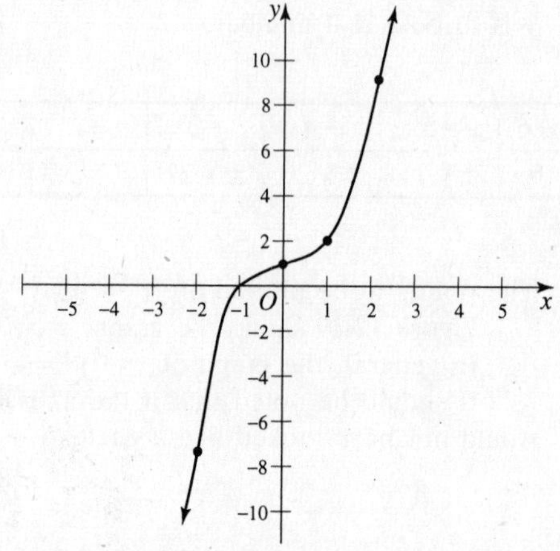

Fig. 12-21

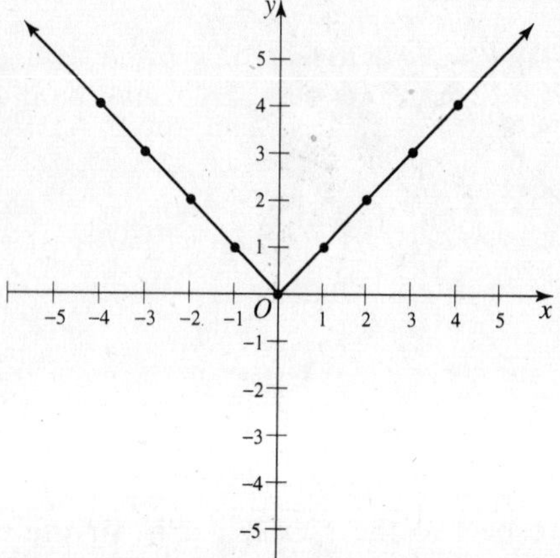

Fig. 12-22

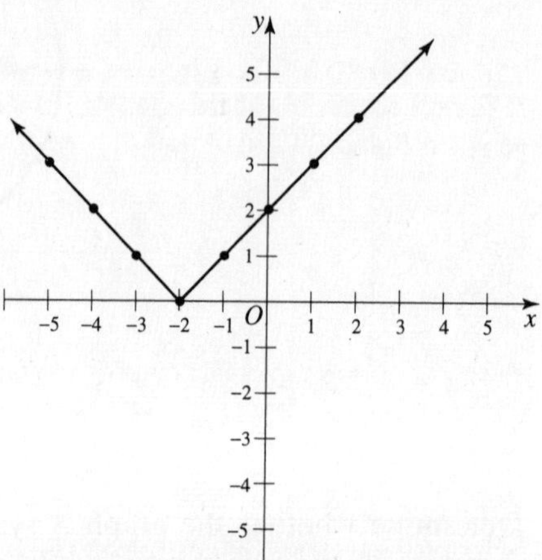

Fig. 12-23

12.21 Use the graph of $y = |x|$ to graph $y = |x + 2|$.

SOLUTION

The graph of $y = |x|$ is shown in Fig. 12-22. The graph of $y = |x + 2|$ is the graph of $y = |x|$ shifted 2 units to the left, since $|x + 2| = |x - (-2)|$ and is shown in Fig. 12-23.

12.22 Use the graph of $y = x^2$ to graph $y = -x^2$.

SOLUTION

The graph of $y = x^2$ is shown in Fig. 12-24. The graph of $y = -x^2$ is the graph of $y = x^2$ reflected across the x axis and is shown in Fig. 12-25.

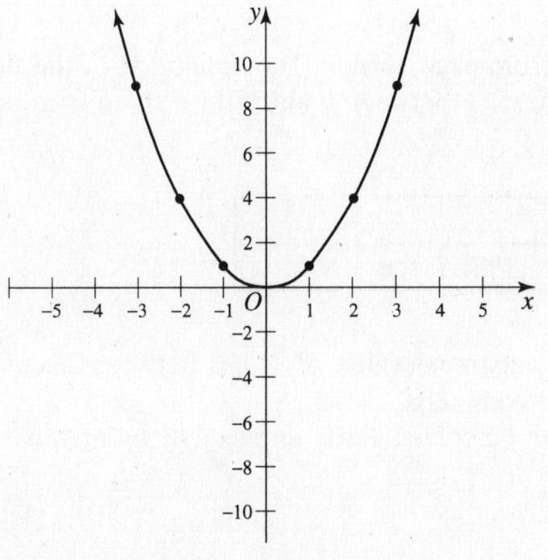

Fig. 12-24

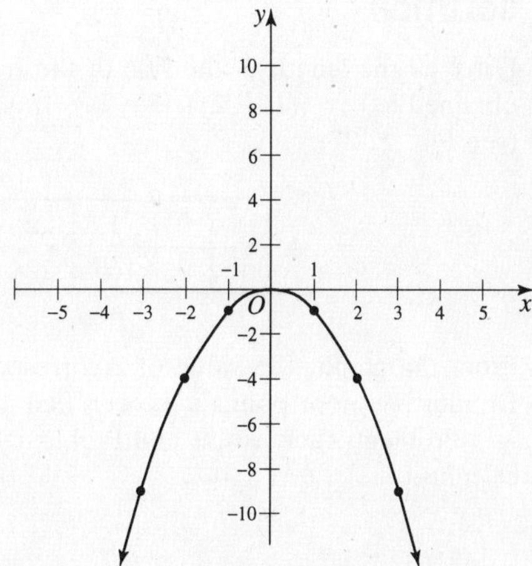

Fig. 12-25

12.23 A man has 40 ft of wire fencing with which to form a rectangular garden. The fencing is to be used only on three sides of the garden, his house providing the fourth side. Determine the maximum area which can be enclosed.

SOLUTION

Let x = length of each of two of the fenced sides of the rectangle; then $40 - 2x$ = length of the third fenced side.

The area A of the garden is $A = x(40 - 2x) = 40x - 2x^2$. We wish to find the maximum value of A. A table of values and the graph of A plotted against x are shown. It is clear that x must have a value between 0 and 20 ft if A is to be positive.

x	0	5	8	10	12	15	20
A	0	150	192	200	192	150	0

From the graph in Fig. 12-26 the maximum point P has coordinates (10, 200) so that the dimensions of the garden are 10 ft by 20 ft and the area is 200 ft^2.

12.24 A rectangular piece of tin has dimensions 12 in. by 18 in. It is desired to make an open box from this material by cutting out equal squares from the corners and then bending up the sides. What are the dimensions of the squares cut out if the volume of the box is to be as large as possible?

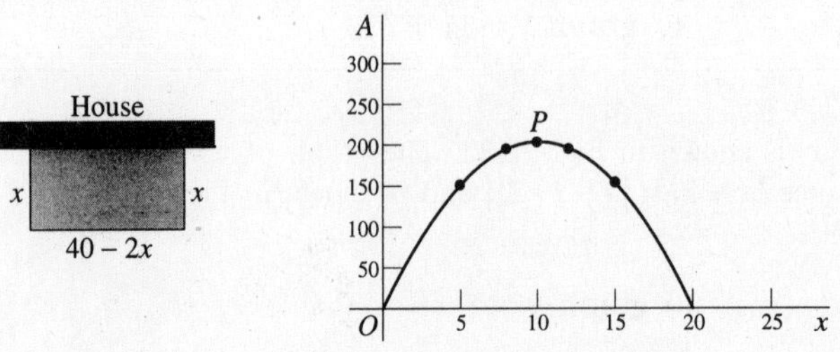

Fig. 12-26

SOLUTION

Let x be the length of the side of the square cut out from each corner. The volume V of the box thus obtained is $V = x(12 - 2x)(18 - 2x)$. It is clear that x must be between 0 and 6 in. if there is to be a box (see Fig. 12-27).

x	0	1	2	$2\frac{1}{2}$	3	$3\frac{1}{2}$	4	5	6
V	0	160	224	$227\frac{1}{2}$	216	$192\frac{1}{2}$	160	80	0

From the graph, the value of x corresponding to the maximum value of V lies between 2 and 2.5 in. By plotting more points it is seen that $x = 2.4$ in. approximately.

Problems such as this and Problem 12.23 may often be solved easily and exactly by methods of the calculus.

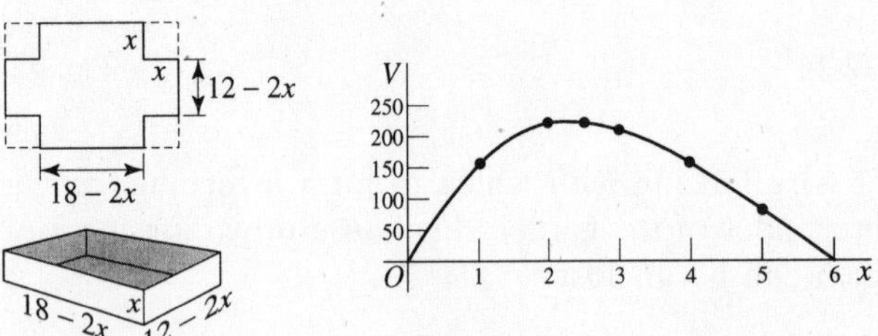

Fig. 12-27

12.25 A cylindrical can is to have a volume of 200 cubic inches. Find the dimensions of the can which is made of the least amount of material.

SOLUTION

Let x be the radius and y the height of the cylinder.

The area of the top or bottom of the can is πx^2 and the lateral area is $2\pi xy$; then the total area $S = 2\pi x^2 + 2\pi xy$.

The volume of the cylinder is $\pi x^2 y$, so that $\pi x^2 y = 200$ and $y = 200/\pi x^2$. Then

$$S = 2\pi x^2 + 2\pi x\left(\frac{200}{\pi x^2}\right) \quad \text{or} \quad S = 2\pi x^2 + \frac{400}{x}.$$

A table of values and the graph of S plotted against x (Fig. 12-28) are shown. We take $\pi = 3.14$ approximately.

x	1	2	3	3.2	3.5	4	4.5	5	6	7	8
S	406	225	190	189	191	200	216	237	293	365	452

From the graph in Fig. 12-28, minimum $S = 189\ \text{in}^2$ occurs when $x = 3.2$ in. approximately; and from $y = 200/\pi x^2$ we have $y = 6.2$ in. approximately.

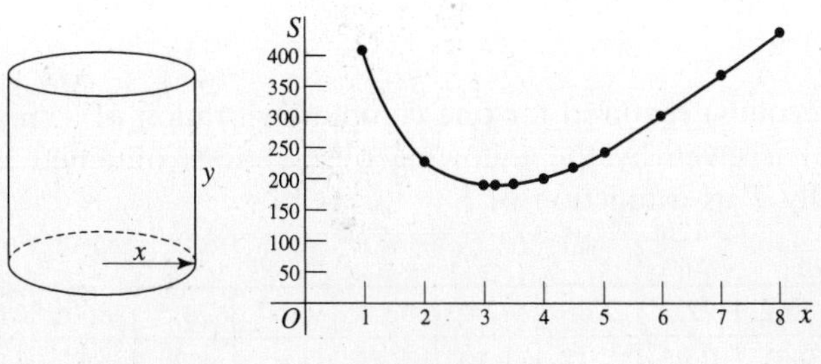

Fig. 12-28

12.26 Find approximately the values of x for which $x^3 + 2x^2 - 7x - 3 = 0$.

SOLUTION

Consider $y = x^3 + 2x^2 - 7x - 3$. We must find values of x for which $y = 0$.

From the graph of $y = x^3 + 2x^2 - 7x - 3$, which is shown in Fig. 12-18, it is clear that there are three real values of x for which $y = 0$ (the values of x where the curve intersects the x axis). These values are $x = -3.7$, $x = -0.4$, and $x = 2.1$ approximately. More exact values may be obtained by advanced techniques.

12.27 The following table shows the population of the United States (in millions) for the years 1840, 1850, . . ., 1950. Graph these data.

Year	1840	1850	1860	1870	1880	1890	1900	1910	1920	1930	1940	1950
Population (millions)	17.1	23.2	31.4	39.8	50.2	62.9	76.0	92.0	105.7	122.8	131.7	150.7

SOLUTION

See Fig. 12-29.

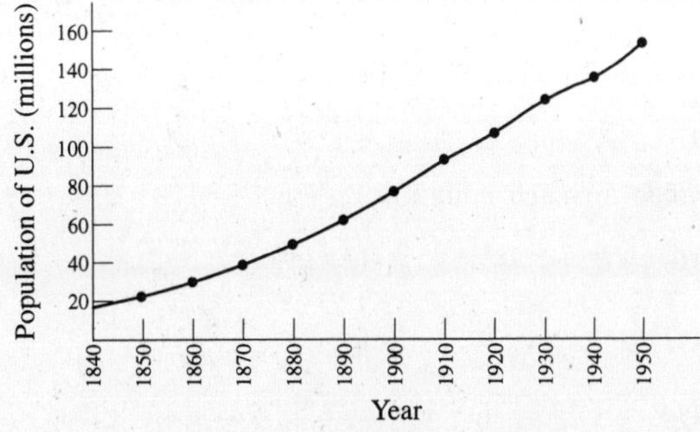

Fig. 12-29

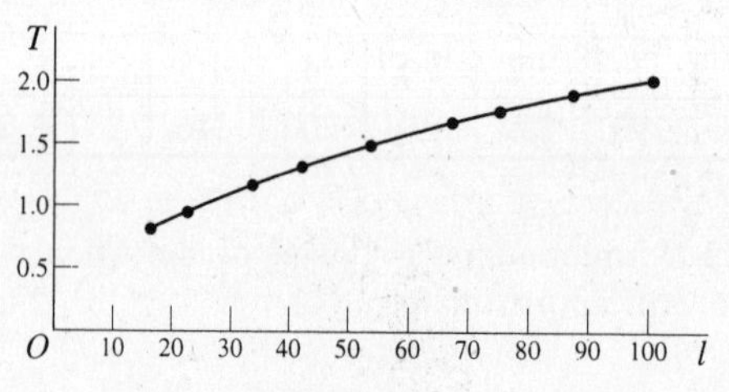

Fig. 12-30

12.28 The time T (in seconds) required for one complete vibration of a simple pendulum of length l (in centimeters) is given by the following observations obtained in a physics laboratory. Exhibit graphically T as a function of l.

l	16.2	22.2	33.8	42.0	53.4	66.7	74.5	86.6	100.0
T	0.81	0.95	1.17	1.30	1.47	1.65	1.74	1.87	2.01

SOLUTION

The observation points are connected by a smooth curve (Fig. 12-30) as is usually done in science and engineering.

Supplementary Problems

12.29 A rectangle has sides of lengths x and $2x$. Express the area A of the rectangle as a function of its (*a*) side x, (*b*) perimeter P, and (*c*) diagonal D.

12.30 Express the area S of a circle in terms of its (*a*) radius r, (*b*) diameter d, and (*c*) perimeter P.

12.31 Express the area A of an isosceles triangle as a function of x and y, where x is the length of the two equal sides and y is the length of the third side.

12.32 The side of a cube has length x. Express (*a*) x as a function of the volume V of the cube, (*b*) the surface area S of the cube as a function of x, and (*c*) the volume V as a function of the surface S.

12.33 Given $y = 5 + 3x - 2x^2$, find the values of y corresponding to $x = -3, -2, -1, 0, 1, 2, 3$.

12.34 Extend the table of values in Problem 12.33 by finding the values of y which correspond to $x = -5/2$, $-3/2$, $-1/2$, $1/2$, $3/2$, $5/2$.

12.35 State the domain and range for each equation.

(*a*) $y = -2x + 3$

(*b*) $y = x^2 - 5$

(*c*) $y = x^3 - 4$

(*d*) $y = 5 - 2x^2$

(*e*) $y = \dfrac{2}{x+6}$

(*f*) $y = \sqrt{x-5}$

(*g*) $y = \sqrt[3]{1-2x}$

(*h*) $y = \dfrac{x}{x+1}$

(*i*) $y = \dfrac{4}{x}$

12.36 For which of these relations is y a function of x?

(*a*) $y = x^3 + 2$ (*d*) $y = x^2 - 5$ (*g*) $x = |y|$

(*b*) $x = y^3 + 2$ (*e*) $x^2 + y^2 = 5$ (*h*) $y = \sqrt[3]{x+1}$

(*c*) $x = y^2 + 4$ (*f*) $y = \pm\sqrt{x-7}$ (*i*) $y = \sqrt{5+x}$

12.37 If $f(x) = 2x^2 + 6x - 1$, find $f(-3)$, $f(-2)$, $f(0)$, $f(1/2)$, $f(3)$.

12.38 If $F(u) = \dfrac{u^2 - 2u}{1+u}$, find (*a*) $F(1)$, (*b*) $F(2)$, (*c*) $F(x)$, (*d*) $F(-x)$.

12.39 If $G(x) = \dfrac{x-1}{x+1}$, find

(*a*) $G\left(\dfrac{x}{x+1}\right)$, (*b*) $\dfrac{G(x+h) - G(x)}{h}$, (*c*) $G(x^2 + 1)$.

12.40 If $F(x, y) = 2x^2 + 4xy - y^2$, find (*a*) $F(1, 2)$, (*b*) $F(-2, -3)$, (*c*) $F(x+1, y-1)$.

12.41 Plot the following points on a rectangular coordinate system:

(*a*) $(1, 3)$, (*b*) $(-2, 1)$, (*c*) $(-1/2, -2)$, (*d*) $(-3, 2/3)$, (*e*) $(-\sqrt{3}, 3)$.

12.42 If $y = 3x + 2$, (*a*) obtain the values of y corresponding to $x = -2, -1, 0, 1, 2$ and (*b*) plot the points (x, y) thus obtained.

12.43 Determine whether the graph of each relation is symmetric with respect to the y axis, x axis, or origin.

(*a*) $y = 2x^4 + 3$ (*d*) $y = 3$ (*g*) $y^2 = x + 2$

(*b*) $y = (x-3)^3$ (*e*) $y = -5x^3$ (*h*) $y = 3x - 1$

(*c*) $y = -\sqrt{9-x}$ (*f*) $y = 7x^2 + 4$ (*i*) $y = 5x$

12.44 State how the graph of the first equation relates to the graph of the second equation.

(*a*) $y = -x^4$ and $y = x^4$ (*f*) $y = |x| + 1$ and $y = |x|$

(*b*) $y = 3x$ and $y = x$ (*g*) $y = |x+5|$ and $y = |x|$

(*c*) $y = x^2 + 10$ and $y = x^2$ (*h*) $y = -x^3$ and $y = x^3$

(*d*) $y = (x-1)^3$ and $y = x^3$ (*i*) $y = x^2/6$ and $y = x^2$

(*e*) $y = x^2 - 7$ and $y = x^2$ (*j*) $y = (x+8)^2$ and $y = x^2$

12.45 Graph the functions (*a*) $f(x) = 1 - 2x$, (*b*) $f(x) = x^2 - 4x + 3$, (*c*) $f(x) = 4 - 3x - x^2$.

12.46 Graph $y = x^3 - 6x^2 + 11x - 6$.

12.47 Graph (*a*) $x^2 + y^2 = 16$, (*b*) $x^2 + 4y^2 = 16$.

12.48 There is available 120 ft of wire fencing with which to enclose two equal rectangular gardens A and B, as shown in Fig. 12-31. If no wire fencing is used along the sides formed by the house, determine the maximum combined area of the gardens.

12.49 Find the area of the largest rectangle which can be inscribed in a right triangle whose legs are 6 and 8 inches respectively (see Fig. 12-32).

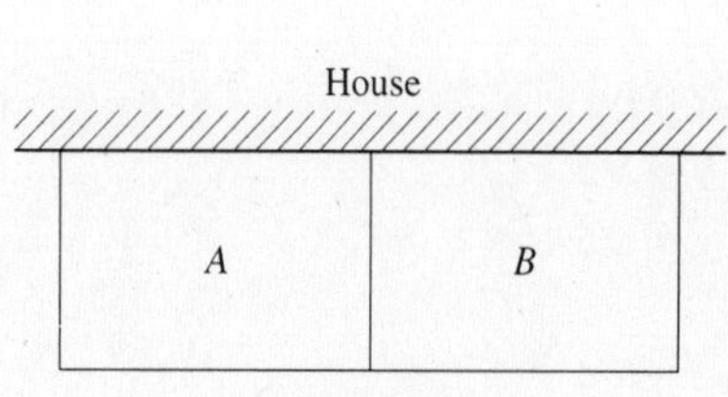

Fig. 12-31

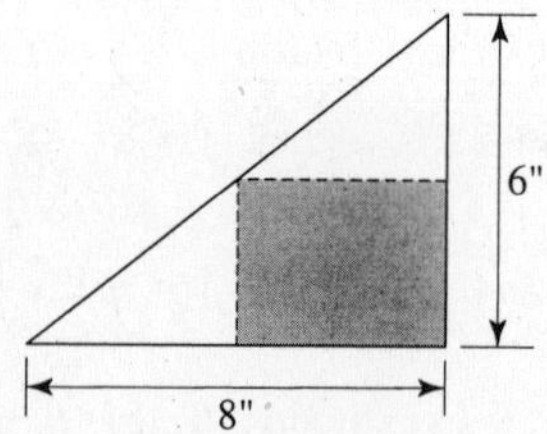

Fig. 12-32

12.50 Obtain the relative maximum and minimum values of the function $f(x) = 2x^3 - 15x^2 + 36x - 23$.

12.51 From the graph of $y = x^3 - 7x + 6$ obtain the roots of the equation $x^3 - 7x + 6 = 0$.

12.52 Show that the equation $x^3 - x^2 + 2x - 3 = 0$ has only one real root.

12.53 Show that $x^4 - x^2 + 1 = 0$ cannot have any real roots.

12.54 The percentage of workers in the USA employed in agriculture during the years 1860, 1870, . . ., 1950 is given in the following table. Graph the data.

Year	1860	1870	1880	1890	1900	1910	1920	1930	1940	1950
% of all workers in agriculture	58.9	53.0	49.4	42.6	37.5	31.0	27.0	21.4	18.0	12.8

12.55 The total time required to bring an automobile to a stop after perceiving danger is composed of the *reaction time* (time between recognition of danger and application of brakes) plus *braking time* (time for stopping after application of brakes). The following table gives the stopping distances d (feet) of an automobile traveling at speeds v (miles per hour) at the instant danger is sighted. Graph d against v.

Speed v (mi/hr)	20	30	40	50	60	70
Stopping distance d (ft)	54	90	138	206	292	396

12.56 The time t taken for an object to fall freely from rest through various heights h is given in the following table.

Time t (sec)	1	2	3	4	5	6
Height h (ft)	16	64	144	256	400	576

(*a*) Graph h against t.

(*b*) How long would it take an object to fall freely from rest through 48 ft? 300 ft?

(*c*) Through what distance can an object fall freely from rest in 3.6 seconds?

ANSWERS TO SUPPLEMENTARY PROBLEMS

12.29 $A = 2x^2, \quad A = \dfrac{P^2}{18}, \quad A = \dfrac{2D^2}{5}$

12.30 $S = \pi r^2, \quad S = \dfrac{\pi d^2}{4}, \quad S = \dfrac{P^2}{4\pi}$

12.31 $A = \frac{y}{2}\sqrt{x^2 - y^2/4} = \frac{y}{4}\sqrt{4x^2 - y^2}$

12.32 $x = \sqrt[3]{V}, \quad S = 6x^2, \quad V = \sqrt{\frac{S^3}{216}} = \frac{S}{36}\sqrt{6S}$

12.33

x	−3	−2	−1	0	1	2	3
y	−22	−9	0	5	6	3	−4

12.34

x	−5/2	−3/2	−1/2	1/2	3/2	5/2
y	−15	−4	3	6	5	0

12.35 (*a*) Domain = {all real numbers}; range = {all real numbers}
(*b*) Domain = {all real numbers}; range = {all real numbers ≥ -5}
(*c*) Domain = {all real numbers}; range = {all real numbers}
(*d*) Domain = {all real numbers}; range = {all real numbers ≤ 5}
(*e*) Domain = {all real numbers $\neq -6$}; range = {all real numbers $\neq 0$}
(*f*) Domain = {all real numbers ≥ 5}; range = {all real numbers ≥ 0}
(*g*) Domain = {all real numbers}; range = {all real numbers}
(*h*) Domain = {all real numbers $\neq -1$}; range = {all real numbers $\neq 1$}
(*i*) Domain = {all real numbers $\neq 0$}; range = {all real numbers $\neq 0$}

12.36 (*a*) Function (*b*) Function (*c*) Not a function (*d*) Function (*e*) Not a function (*f*) Not a function (*g*) Not a function (*h*) Function (*i*) Function

12.37 $f(-3) = -1$, $f(-2) = -5$, $f(0) = -1$, $f(1/2) = 5/2$, $f(3) = 35$

12.38 (*a*) $-1/2$, (*b*) 0, (*c*) $\frac{x^2 - 2x}{1 + x}$, (*d*) $\frac{x^2 + 2x}{1 - x}$

12.39 (*a*) $\frac{-1}{2x + 1}$, (*b*) $\frac{2}{(x + 1)(x + h + 1)}$, (*c*) $\frac{x^2}{x^2 + 2}$

12.40 (*a*) 6, (*b*) 23, (*c*) $2x^2 + 4xy - y^2 + 6y - 3$

12.41 See Fig. 12-33.

12.42 (*a*) −4, −1, 2, 5, 8 (*b*) See Fig. 12-34.

12.43 (*a*) y axis (*b*) Origin (*c*) None (*d*) y axis (*e*) Origin (*f*) y axis (*g*) x axis (*h*) None (*i*) Origin

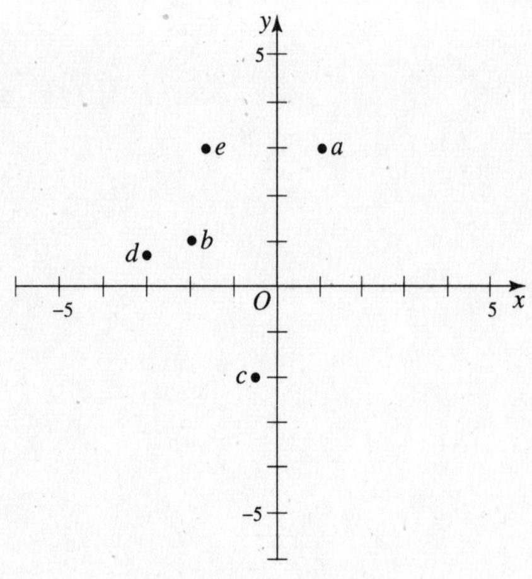

Fig. 12-33

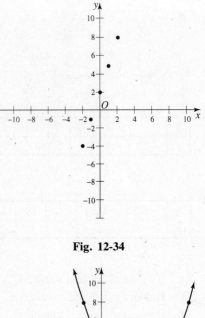

Fig. 12-34

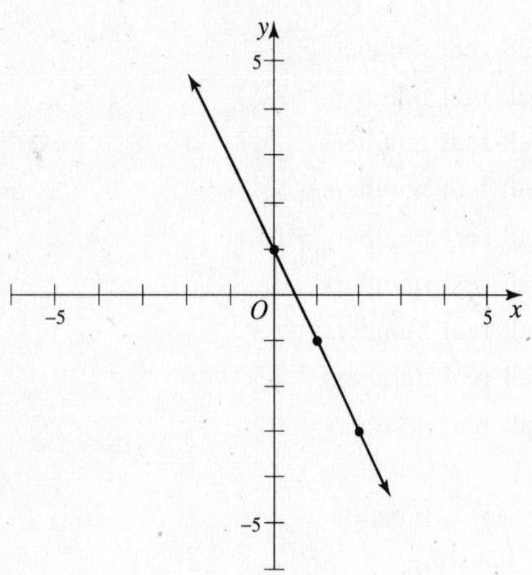

Fig. 12-35

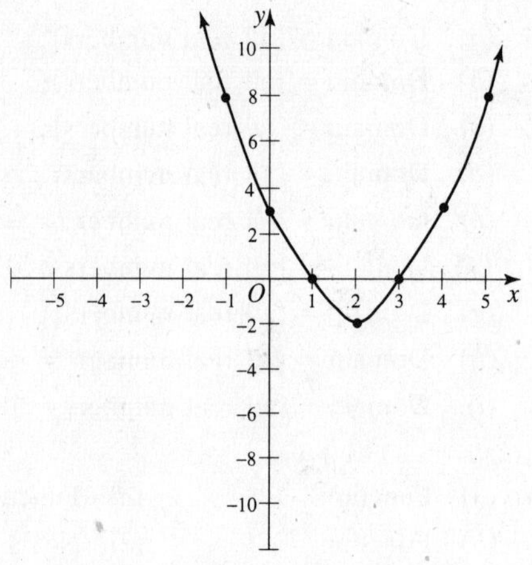

Fig. 12-36

12.44 (*a*) Reflected across x axis
(*b*) y increases 3 times as fast
(*c*) Shifted 10 units up
(*d*) Shifted 1 unit right
(*e*) Shifted 7 units down
(*f*) Shifted 1 unit up
(*g*) Shifted 5 units left
(*h*) Reflected across the x axis
(*i*) y increases 1/6 as fast
(*j*) Shifted 8 units left

12.45 (*a*) See Fig. 12-35. (*b*) See Fig. 12-36. (*c*) See Fig. 12-37.

12.46 See Fig. 12-38.

12.47 (*a*) See Fig. 12-39. (*b*) See Fig. 12-40.

12.48 1200 ft^2

12.49 12 in^2

12.50 Maximum value of $f(x)$ is 5 (at $x = 2$); minimum value of $f(x)$ is 4 (at $x = 3$).

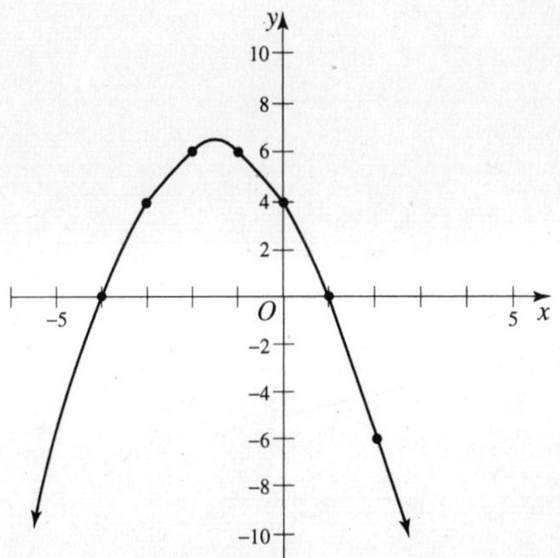

Fig. 12-37

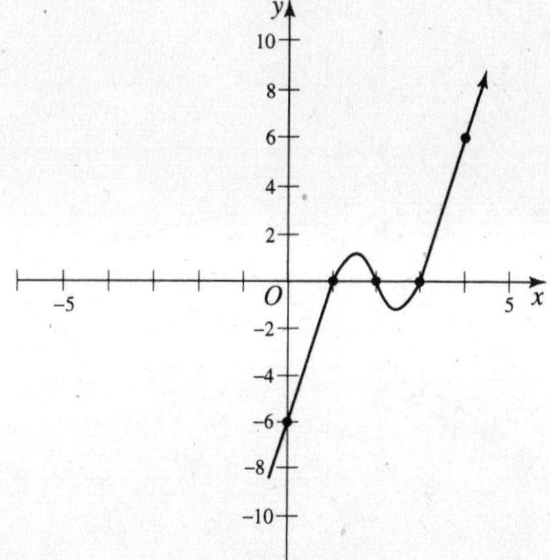

Fig. 12-38

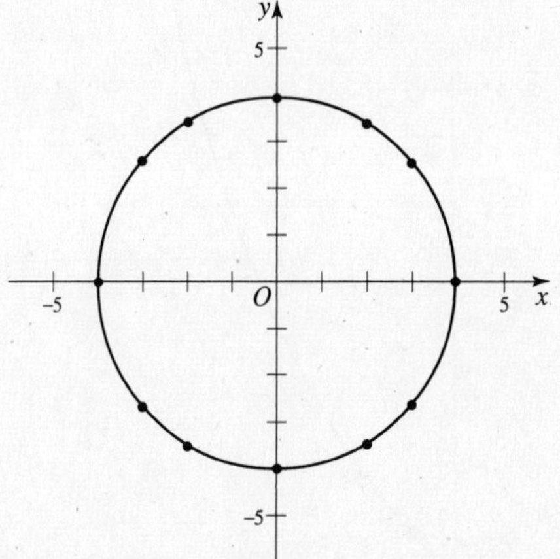

Fig. 12-39

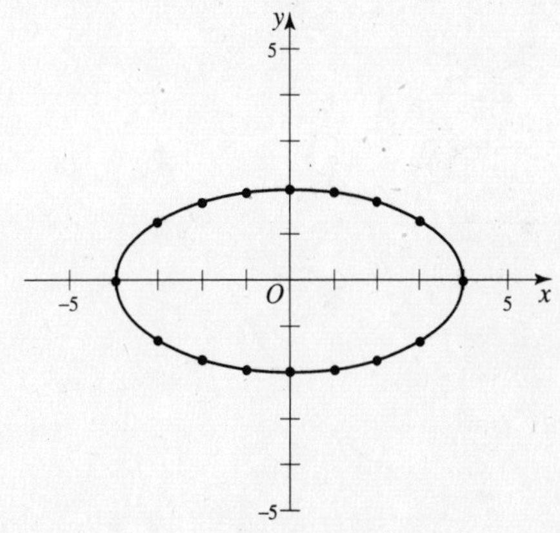

Fig. 12-40

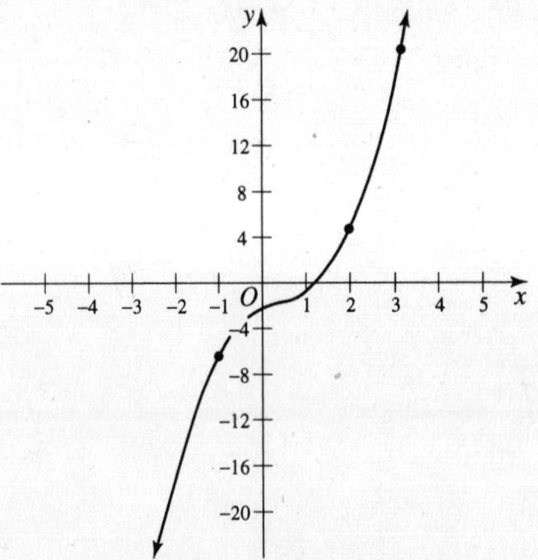

Fig. 12-41

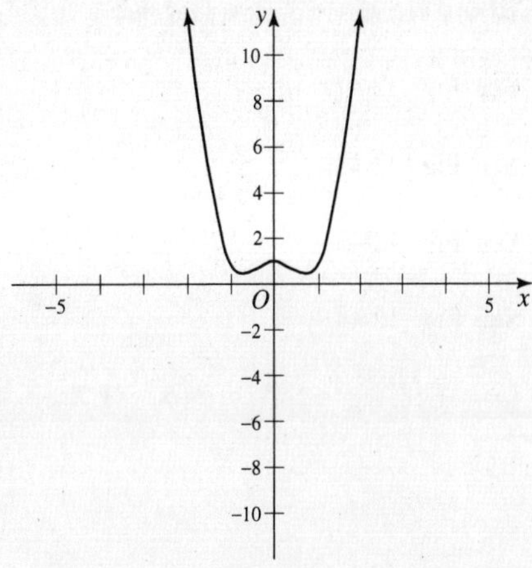

Fig. 12-42

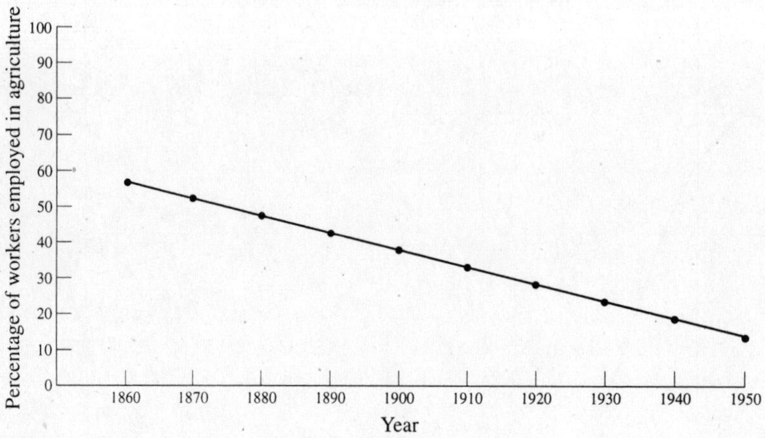

Fig. 12-43

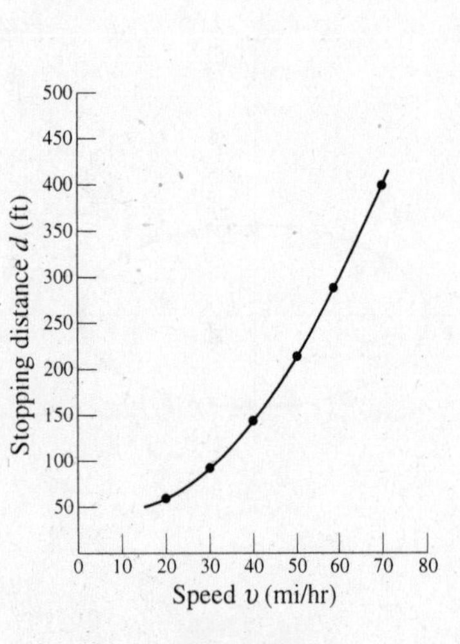

Fig. 12-44

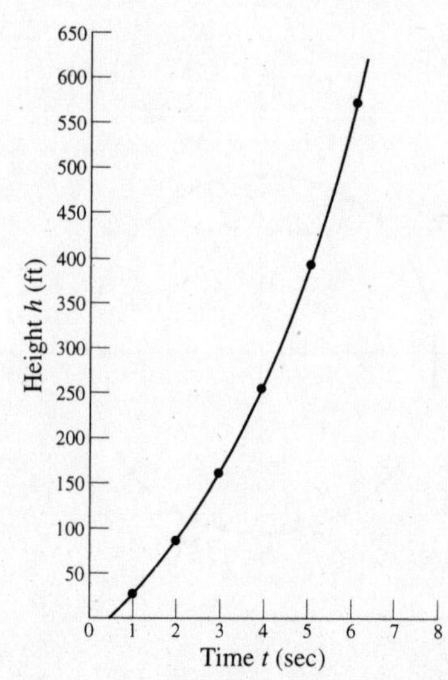

Fig. 12-45

12.51 Roots are $x = -3$, $x = 1$, $x = 2$.

12.52 See Fig. 12-41.

12.53 See Fig. 12-42.

12.54 See Fig. 12-43.

12.55 See Fig. 12-44.

12.56 (*a*) See Fig. 12-45; (*b*) 1.7 sec, 4.3 sec; (*c*) 207 ft

Chapter 13

Linear Equations in One Variable

13.1 LINEAR EQUATIONS

A linear equation in one variable has the form $ax + b = 0$, where $a \neq 0$ and b are constants. The solution of this equation is given by $x = -b/a$.

When a linear equation is not in the form $ax + b = 0$, we simplify the equation by multiplying each term by the LCD for all fractions in the equation, removing any parentheses, or combining like terms. In some equations we do more than one of the procedures.

EXAMPLE 13.1. Solve the equation $x + 8 - 2(x + 1) = 3x - 6$ for x.

$x + 8 - 2(x + 1) = 3x - 6$	First we remove the parentheses.
$x + 8 - 2x - 2 = 3x - 6$	We now combine like terms.
$-x + 6 = 3x - 6$	Now get the variable terms on one side of the equation by
$-x + 6 - 3x = 3x - 6 - 3x$	subtracting $3x$ from each side of the equation.
$-4x + 6 = -6$	Now we subtract 6 from each side of the equation to get the
$-4x + 6 - 6 = -6 - 6$	variable term on one side of the equation by itself.
$-4x = -12$	Finally we divide each side of the equation by the coefficient of the variable, which is -4.
$\frac{-4x}{-4} = \frac{-12}{-4}$	
$x = 3$	Now we check this solution in the original equation.

Check:

$3 + 8 - 2(3 + 1) \mathrel{?} 3(3) - 6$	The question mark indicates that we don't know for sure that
$11 - 2(4) \mathrel{?} 9 - 6$	the two quantities are equal.
$11 - 8 \mathrel{?} 3$	
$3 = 3$	The solution checks.

13.2 LITERAL EQUATIONS

Most literal equations we encounter are formulas. Frequently, we want to use the formula to determine a value other than the standard one. To do this, we consider all variables except the one we are interested in as constants and solve the equation for the desired variable.

EXAMPLE 13.2. Solve $p = 2(l + w)$ for l.

$p = 2(l + w)$	Remove the parentheses first.
$p = 2l + 2w$	Subtract $2w$ from each side of the equation.
$p - 2w = 2l$	Divide by 2, the coefficient of l.
$(p - 2w)/2 = l$	Rewrite the equation.
$l = (p - 2w)/2$	Now we have a formula that can be used to determine l.

13.3 WORD PROBLEMS

In solving a word problem, the first step is to decide what is to be found. The next step is to translate the conditions stated in the problem into an equation or to state a formula that expresses the conditions of the problem. The solution of the equation is the next step.

EXAMPLE 13.3. If the perimeter of a rectangle is 68 meters and the length is 14 meters more than the width, what are the dimensions of the rectangle?

Let $w =$ the number of meters in the width and $w + 14 =$ the number of meters in the length.

$$\begin{aligned} 2[(w+14)+w] &= 68 \\ 2w+28+2w &= 68 \\ 4w+28 &= 68 \\ 4w &= 40 \\ w &= 10 \\ w+14 &= 24 \end{aligned}$$

The rectangle is 24 meters long by 10 meters wide.

EXAMPLE 13.4. The sum of two numbers is -4 and their difference is 6. What are the numbers?

Let $n =$ the smaller number and $n + 6 =$ the larger number.

$$\begin{aligned} n+(n+6) &= -4 \\ n+n+6 &= -4 \\ 2n+6 &= -4 \\ 2n &= -10 \\ n &= -5 \\ n+6 &= 1 \end{aligned}$$

The two numbers are -5 and 1.

EXAMPLE 13.5. If one pump can fill a pool in 16 hours and if two pumps can fill the pool in 6 hours, how fast can the second pump fill the pool?

Let $h =$ the numbers of hours for the second pump to fill the pool.

$$\begin{aligned} \frac{1}{h}+\frac{1}{16} &= \frac{1}{6} \\ 48h\left(\frac{1}{h}+\frac{1}{16}\right) &= 48h\left(\frac{1}{6}\right) \\ 48+3h &= 8h \\ 48 &= 5h \\ 9.6 &= h \end{aligned}$$

The second pump takes 9.6 hours (or 9 hours and 36 minutes) to fill the pool.

EXAMPLE 13.6. How many liters of pure alcohol must be added to 15 liters of a 60% alcohol solution to obtain an 80% alcohol solution?

Let $n =$ the number of liters of pure alcohol to be added.

$$n + 0.60(15) = 0.80(n+15)$$ (The sum of the amount of alcohol in each quantity is equal to the amount of alcohol in the mixture.)

$$\begin{aligned} n+9 &= 0.8n+12 \\ 0.2n &= 3 \\ n &= 15 \end{aligned}$$

Fifteen liters of pure alcohol must be added.

Solved Problems

13.1 Solve each of the following equations.

(*a*) $x + 1 = 5$, $x = 5 - 1$, $x = 4$.
Check: Put $x = 4$ in the original equation and obtain $4 + 1 \,?\, 5$, $5 = 5$.

(*b*) $3x - 7 = 14$, $3x = 14 + 7$, $3x = 21$, $x = 7$.
Check: $3(7) - 7 \,?\, 14$, $14 = 14$.

(*c*) $3x + 2 = 6x - 4$, $3x - 6x = -4 - 2$, $-3x = -6$, $x = 2$.

(*d*) $x + 3(x - 2) = 2x - 4$, $x + 3x - 6 = 2x - 4$, $4x - 2x = 6 - 4$, $2x = 2$, $x = 1$.

(*e*) $3x - 2 = 7 - 2x$, $3x + 2x = 7 + 2$, $5x = 9$, $x = 9/5$.

(*f*) $2(t + 3) = 5(t - 1) - 7(t - 3)$, $2t + 6 = 5t - 5 - 7t + 21$, $4t = 10$, $t = 10/4 = 5/2$.

(*g*) $3x + 4(x - 2) = x - 5 + 3(2x - 1)$, $3x + 4x - 8 = x - 5 + 6x - 3$, $7x - 8 = 7x - 8$.
This is an identity and is true for all values of x.

(*h*) $\dfrac{x-3}{2} = \dfrac{2x+4}{5}$, $5(x - 3) = 2(2x + 4)$, $5x - 15 = 4x + 8$, $x = 23$.

(*i*) $3 + 2[y - (2y + 2)] = 2[y + (3y - 1)]$, $3 + 2[y - 2y - 2] = 2[y + 3y - 1]$,
$3 + 2y - 4y - 4 = 2y + 6y - 2$, $-2y - 1 = 8y - 2$, $-10y = -1$, $y = 1/10$.

(*j*) $(s + 3)^2 = (s - 2)^2 - 5$, $s^2 + 6s + 9 = s^2 - 4s + 4 - 5$, $6s + 4s = -9 - 1$, $s = -1$.

(*k*) $\dfrac{x-2}{x+2} = \dfrac{x-4}{x+4}$, $(x - 2)(x + 4) = (x - 4)(x + 2)$, $x^2 + 2x - 8 = x^2 - 2x - 8$, $4x = 0$, $x = 0$.

Check: $$\frac{0-2}{0+2} \,?\, \frac{0-4}{0+4}, \qquad -1 = -1.$$

(*l*) $\dfrac{3x+1}{x+2} = \dfrac{3x-2}{x+1}$, $(x + 1)(3x + 1) = (x + 2)(3x - 2)$, $3x^2 + 4x + 1 = 3x^2 + 4x - 4$ or $1 = -4$.
There is no value of x which satisfies this equation.

(*m*) $\dfrac{5}{x} + \dfrac{5}{2x} = 6$. Multiplying by $2x$, $5(2) + 5 = 12x$, $12x = 15$, $x = 5/4$.

(*n*) $\dfrac{x+3}{2x} + \dfrac{5}{x-1} = \dfrac{1}{2}$. Multiplying by $2x(x - 1)$, the LCD of the fractions,

$$(x + 3)(x - 1) + 5(2x) = x(x - 1), \qquad x^2 + 2x - 3 + 10x = x^2 - x, \qquad 13x = 3, \qquad x = 3/13.$$

(*o*) $\dfrac{2}{x-3} - \dfrac{4}{x+3} = \dfrac{16}{x^2-9}$. Multiplying by $(x - 3)(x + 3)$ or $x^2 - 9$,

$$2(x + 3) - 4(x - 3) = 16, \qquad 2x + 6 - 4x + 12 = 16, \qquad -2x = -2, \qquad x = 1.$$

(*p*) $\dfrac{1}{y} - \dfrac{1}{y+3} = \dfrac{1}{y+2} - \dfrac{1}{y+5}$, $\quad \dfrac{(y+3)-y}{y(y+3)} = \dfrac{(y+5)-(y+2)}{(y+2)(y+5)}$, $\quad \dfrac{3}{y(y+3)} = \dfrac{3}{(y+2)(y+5)}$,

$$(y + 2)(y + 5) = y(y + 3), \qquad y^2 + 7y + 10 = y^2 + 3y, \qquad 4y = -10,\ y = -5/2.$$

(*q*) $\dfrac{3}{x^2-4x} - \dfrac{2}{2x^2-5x-12} = \dfrac{9}{2x^2+3x}$ or $\dfrac{3}{x(x-4)} - \dfrac{2}{(2x+3)(x-4)} = \dfrac{9}{x(2x+3)}$.
Multiplying by $x(x - 4)(2x + 3)$, the LCD of the fractions,

$$3(2x + 3) - 2x = 9(x - 4), \qquad 6x + 9 - 2x = 9x - 36, \qquad 45 = 5x, \qquad x = 9.$$

13.2 Solve for x.

(*a*) $2x - 4p = 3x + 2p$, $2x - 3x = 2p + 4p$, $-x = 6p$, $x = -6p$.

(*b*) $ax + a = bx + b$, $ax - bx = b - a$, $x(a - b) = b - a$, $x = \dfrac{b-a}{a-b} = -1$ provided $a \neq b$.

If $a = b$ the equation is an identity and is true for all values of x.

(*c*) $2cx + 4d = 3ax - 4b$, $2cx - 3ax = -4b - 4d$, $x = \dfrac{-4b-4d}{2c-3a} = \dfrac{4b+4d}{3a-2c}$ provided $3a \neq 2c$.

If $3a = 2c$ there is no solution unless $d = -b$, in which case the original equation is an identity.

(*d*) $\dfrac{3x+a}{b} = \dfrac{4x+b}{a}$, $3ax + a^2 = 4bx + b^2$, $3ax - 4bx = b^2 - a^2$, $x = \dfrac{b^2-a^2}{3a-4b}$ (provided $3a \neq 4b$).

13.3 Express each statement in terms of algebraic symbols.

(*a*) One more than twice a certain number.
Let x = the number. Then $2x$ = twice the number, and one more than twice the number $= 2x + 1$.

(*b*) Three less than five times a certain number.
Let x = the number. Then three less than five times the number $= 5x - 3$.

(*c*) Each of two numbers whose sum is 100.
If x = one of the numbers, then $100 - x$ = the other number.

(*d*) Three consecutive integers (for example, 5, 6, 7).
If x is the smallest integer, then $(x + 1)$ and $(x + 2)$ are the other two integers.

(*e*) Each of two numbers whose difference is 10.
Let x = the smaller number; then $(x + 10)$ = the larger number.

(*f*) The amount by which 100 exceeds three times a given number.
Let x = given number. Then the excess of 100 over $3x$ is $(100 - 3x)$.

(*g*) Any odd integer.
Let x = any integer. Then $2x$ is always an even integer, and $(2x + 1)$ is an odd integer.

(*h*) Four consecutive odd integers (for example, 1, 3, 5, 7; 17, 19, 21, 23).
The difference between two consecutive odd integers is 2.
Let $2x + 1$ = smallest odd integer. Then the required numbers are $2x + 1$, $2x + 3$, $2x + 5$, $2x + 7$.

(*i*) The number of cents in x dollars.
Since 1 dollar = 100 cents, x dollars = $100x$ cents.

(*j*) John is twice as old as Mary, and Mary is three times as old as Bill. Express each of their ages in terms of a single unknown.
Let x = Bill's age. Then Mary's age is $3x$ and John's age is $2(3x) = 6x$.
Another method. Let y = John's age. Then Mary's age $= \frac{1}{2}y$ and Bill's age $= \frac{1}{3}(\frac{1}{2}y) = \frac{1}{6}y$.

(*k*) The three angles A, B, C of a triangle if angle A has 10° more than twice the number of degrees in angle C.
Let $C = x°$; then $A = (2x + 10)°$. Since $A + B + C = 180°$, $B = 180° - (A + C) = (170 - 3x)°$.

(*l*) The time it takes a boat traveling at a speed of 20 mi/hr to cover a distance of x miles.
Distance = speed × time. Then

$$\text{time} = \frac{\text{distance}}{\text{speed}} = \frac{x \text{ mi}}{20 \text{ mi/hr}} = \frac{x}{20}\text{ hr.}$$

(*m*) The perimeter and area of a rectangle if one side is 4 ft longer than twice the other side.
Let x ft = length of shorter side; then $(2x + 4)$ft = length of longer side. The perimeter $= 2(x) + 2(2x + 4) = (6x + 8)$ft, and the area $= x(2x + 4)\text{ft}^2$.

(*n*) The fraction whose numerator is 3 less than 4 times its denominator.
Let x = denominator; then numerator $= 4x - 3$. The fraction is $(4x - 3)/x$.

(*o*) The number of quarts of alcohol contained in a tank holding x gallons of a mixture which is 40% alcohol by volume.
In x gallons of mixture are $0.40x$ gallons of alcohol or $4(0.40x) = 1.6x$ quarts of alcohol.

13.4 The sum of two numbers is 21, and one number is twice the other. Find the numbers.

SOLUTION

Let x and $2x$ be the two numbers. Then $x + 2x = 21$ or $x = 7$, and the required numbers are $x = 7$ and $2x = 14$.

Check. $7 + 14 = 21$ and $14 = 2(7)$, as required.

13.5 Ten less than four times a certain number is 14. Determine the number.

SOLUTION

Let x = required number. Then $4x - 10 = 14$, $4x = 24$, and $x = 6$.

Check. Ten less than four times 6 is $4(6) - 10 = 14$, as required.

13.6 The sum of three consecutive integers is 24. Find the integers.

SOLUTION

Let the three consecutive integers be x, $x + 1$, $x + 2$. Then $x + (x + 1) + (x + 2) = 24$ or $x = 7$, and the required integers are 7, 8, 9.

13.7 The sum of two numbers is 37. If the larger is divided by the smaller, the quotient is 3 and the remainder is 5. Find the numbers.

SOLUTION

Let x = smaller number, $37 - x$ = larger number.

Then $$\frac{\text{larger number}}{\text{smaller number}} = 3 + \frac{5}{\text{smaller number}} \qquad \text{or} \qquad \frac{37 - x}{x} = 3 + \frac{5}{x}.$$

Solving, $37 - x = 3x + 5$, $4x = 32$, $x = 8$. The required numbers are 8, 29.

13.8 A man is 41 years old and his son is 9. In how many years will the father be three times as old as the son?

SOLUTION

Let x = required number of years.

$$\text{Father's age in } x \text{ years} = 3(\text{son's age in } x \text{ years})$$
$$41 + x = 3(9 + x) \qquad \text{and} \qquad x = 7 \text{ years.}$$

13.9 Ten years ago Jane was four times as old as Bianca. Now she is only twice as old as Bianca. Find their present ages.

SOLUTION

Let x = Bianca's present age; then $2x$ = Jane's present age.

$$\text{Jane's age ten years ago} = 4(\text{Bianca's age ten years ago})$$
$$2x - 10 = 4(x - 10) \qquad \text{and} \qquad x = 15 \text{ years.}$$

Hence Bianca's present age is $x = 15$ years and Jane's present age is $2x = 30$ years.

Check. Ten years ago Bianca was 5 and Jane 20, i.e. Jane was four times as old as Bianca.

13.10 Robert has 50 coins, all in nickels and dimes, amounting to \$3.50. How many nickels does he have?

SOLUTION

Let x = number of nickels; then $50 - x$ = number of dimes.

$$\begin{array}{ccccl} \text{Amount in nickels} & + & \text{amount in dimes} & = & 350¢ \\ 5x¢ & + & 10(50-x)¢ & = & 350¢ \quad \text{from which } x = 30 \text{ nickels.} \end{array}$$

13.11 In a purse are nickels, dimes, and quarters amounting to \$1.85. There are twice as many dimes as quarters, and the number of nickels is two less than twice the numbers of dimes. Determine the number of coins of each kind.

SOLUTION

Let x = number of quarters; then $2x$ = no. of dimes, and $2(2x) - 2 = 4x - 2$ = no. of nickels.

$$\begin{array}{ccccccl} \text{Amount in quarters} & + & \text{amount in dimes} & + & \text{amount in nickels} & = & 185¢ \\ 25(x)¢ & + & 10(2x)¢ & + & 5(4x-2)¢ & = & 185¢ \text{ from which } x = 3. \end{array}$$

Hence there are $x = 3$ quarters, $2x = 6$ dimes, and $4x - 2 = 10$ nickels.
Check. 3 quarters = 75¢, 6 dimes = 60¢. 10 nickels = 50¢, and their sum = \$1.85.

13.12 The tens digit of a certain two-digit number exceeds the units digit by 4 and is 1 less than twice the units digit. Find the two-digit number.

SOLUTION

Let x = units digit; then $x + 4$ = tens digit.
Since the tens digit = 2(units digit) − 1, we have $x + 4 = 2(x) - 1$ or $x = 5$.
Thus $x = 5$, $x + 4 = 9$, and the required number is 95.

13.13 The sum of the digits of a two-digit number is 12. If the digits are reversed, the new number is 4/7 times the original number. Determine the original number.

SOLUTION

Let x = units digit; $12 - x$ = tens digit.
Original number = $10(12 - x) + x$; reversing digits, the new number = $10x + (12 - x)$. Then

$$\text{new number} = \frac{4}{7}\text{ (original number)} \qquad \text{or} \qquad 10x + (12 - x) = \frac{4}{7}[10(12 - x) + x].$$

Solving, $x = 4$, $12 - x = 8$, and the original number is 84.

13.14 A man has \$4000 invested, part at 5% and the remainder at 3% simple interest. The total income per year from these investments is \$168. How much does he have invested at each rate?

SOLUTION

Let x = amount invested at 5%; $\$4000 - x$ = amount at 3%.

$$\begin{array}{ccccl} \text{Interest from 5\% investment} & + & \text{interest from 3\% investment} & = & \$168 \\ 0.05x & + & 0.03(4000-x) & = & 168. \end{array}$$

Solving, $x = \$2400$ at 5%, $\$4000 - x = \$1600$ at 3%.

13.15 What amount should an employee receive as bonus so that she would net \$500 after deducting 30% for taxes?

SOLUTION

Let x = required amount.

Then required amount − taxes = \$500

or $x - 0.30x = \$500$ and $x = \$714.29$.

13.16 At what price should a merchant mark a sofa that costs \$120 in order that it may be offered at a discount of 20% on the marked price and still make a profit of 25% on the selling price?

SOLUTION

Let x = marked price; then sale price $= x - 0.20x = 0.80x$.

Since profit = 25% of sale price, cost = 75% of sale price. Then

$$\text{cost} = 0.75(\text{sale price})$$
$$\$120 = 0.75(0.8x), \qquad \$120 = 0.6x \qquad \text{and} \qquad x = \$200.$$

13.17 When each side of a given square is increased by 4 feet the area is increased by 64 square feet. Determine the dimensions of the original square.

SOLUTION

Let x = side of given square; $x + 4$ = side of new square.

$$\text{New area} = \text{old area} + 64$$
$$(x+4)^2 = x^2 + 64 \qquad \text{from which } x = 6 \text{ ft.}$$

13.18 One leg of a right triangle is 20 inches and the hypotenuse is 10 inches longer than the other leg. Find the lengths of the unknown sides.

SOLUTION

Let x = length of unknown leg; $x + 10$ = length of hypotenuse.

$$\text{Square of hypotenuse} = \text{sum of squares of legs}$$
$$(x+10)^2 = x^2 + (20)^2 \qquad \text{from which } x = 15 \text{ in.}$$

The required sides are $x = 15$ in. and $x + 10 = 25$ in.

13.19 Temperature Fahrenheit $= \frac{9}{5}$(temperature Centigrade) + 32. At what temperature have the Fahrenheit and Centigrade readings the same value?

SOLUTION

Let x = required temperature = temperature Fahrenheit = temperature Centigrade.

Then $x = \frac{9}{5}x + 32$ or $x = -40°$. Thus $-40\,°\text{F} = -40\,°\text{C}$.

13.20 A mixture of 40 lb of candy worth 60¢ a pound is to be made up by taking some worth 45¢/lb and some worth 85¢/lb. How many pounds of each should be taken?

SOLUTION

Let x = weight of 45¢ candy; $40 - x$ = weight of 85¢ candy.

Value of 45¢/lb candy + value of 85¢/lb candy = value of mixture

or $x(45¢) \quad + \quad (40 - x)(85¢) \quad = \quad 40(60¢)$.

Solving, $x = 25$ lb of 45¢/lb candy; $40 - x = 15$ lb of 85¢/lb candy.

13.21 A tank contains 20 gallons of a mixture of alcohol and water which is 40% alcohol by volume. How much of the mixture should be removed and replaced by an equal volume of water so that the resulting solution will be 25% alcohol by volume?

SOLUTION

Let x = volume of 40% solution to be removed.

Volume of alcohol in final solution = volume of alcohol in 20 gal of 25% solution

or $0.40(20 - x) = 0.25(20)$ from which $x = 7.5$ gallons.

13.22 What weight of water must be evaporated from 40 lb of a 20% salt solution to produce a 50% solution? All percentages are by weight.

SOLUTION

Let x = weight of water to be evaporated.

Weight of salt in 20% solution = weight of salt in 50% solution

or $0.20(40\text{ lb}) = 0.50(40\text{ lb} - x)$ from which $x = 24$ lb.

13.23 How many quarts of a 60% alcohol solution must be added to 40 quarts of a 20% alcohol solution to obtain a mixture which is 30% alcohol? All percentages are by volume.

SOLUTION

Let x = number of quarts of 60% alcohol to be added.

Alcohol in 60% solution + alcohol in 20% solution = alcohol in 30% solution

or $0.60x + 0.20(40) = 0.30(x + 40)$ and $x = 13\frac{1}{3}$ qt.

13.24 Two unblended manganese (Mn) ores contain 40% and 25% of manganese respectively. How many tons of each must be mixed to give 100 tons of blended ore containing 35% of manganese? All percentages are by weight.

SOLUTION

Let x = weight of 40% ore required; $100 - x$ = weight of 25% ore required.

Mn from 40% ore + Mn from 25% ore = total Mn in 100 tons of mixture

$0.40x + 0.25(100 - x) = 0.35(100)$

from which $x = 67$ tons of 40% ore and $100 - x = 33$ tons of 25% ore.

13.25 Two cars A and B having average speeds of 30 and 40 mi/hr respectively are 280 miles apart. They start moving toward each other at 3:00 P.M. At what time and where will they meet?

SOLUTION

Let t = time in hours each car travels before they meet. Distance = speed × time.

Distance traveled by A + distance traveled by B = 280 miles

$30t + 40t = 280$ from which $t = 4$ hr.

They meet at 7:00 P.M. at a distance $30t = 120$ mi from initial position of A or at a distance $40t = 160$ mi from initial position of B.

13.26 A and B start from a given point and travel on a straight road at average speeds of 30 and 50 mi/hr respectively. If B starts 3 hr after A, find (a) the time and (b) the distance they travel before meeting.

SOLUTION

Let t and $(t-3)$ be the number of hours A and B respectively travel before meeting.

(*a*) Distance in miles = average speed in mi/hr × time in hours. When they meet,

distance covered by A = distance covered by B

or $30t = 50(t-3)$ from which $t = 7\frac{1}{2}$ hr.

Hence A travels $t = 7\frac{1}{2}$ hr and B travels $(t-3) = 4\frac{1}{2}$ hr.

(*b*) Distance $= 30t = 30(7\frac{1}{2}) = 225$ mi, or distance $= 50(t-3) = 50(4\frac{1}{2}) = 225$ mi.

13.27 A and B can run around a circular mile track in 6 and 10 minutes respectively. If they start at the same instant from the same place, in how many minutes will they pass each other if they run around the track (*a*) in the same direction, (*b*) in opposite directions?

SOLUTION

Let t = required time in minutes.

(*a*) They will pass each other when A covers 1 mile more than B. The speeds A and B are 1/6 and 1/10 mi/min respectively. Then, since distance = speed × time:

Distance by A − distance by B = 1 mile

$$\frac{1}{6}t - \frac{1}{10}t = 1 \quad \text{and } t = 15 \text{ minutes.}$$

(*b*) Distance by A + distance by B = 1 mile

$$\frac{1}{6}t + \frac{1}{10}t = 1 \quad \text{and } t = 15/4 \text{ minutes.}$$

13.28 A boat, propelled to move at 25 mi/hr in still water, travels 4.2 mi against the river current in the same time that it can travel 5.8 mi with the current. Find the speed of the current.

SOLUTION

Let v = speed of current. Then, since time = distance/speed,

time against the current = time in direction of the current

or
$$\frac{4.2 \text{ mi}}{(25-v)\text{ mi/hr}} = \frac{5.8 \text{ mi}}{(25+v)\text{ mi/hr}} \quad \text{and } v = 4 \text{ mi/hr.}$$

13.29 A can do a job in 3 days, and B can do the same job in 6 days. How long will it take them if they work together?

SOLUTION

Let n = number of days it will take them working together.

In 1 day A does 1/3 of the job and B does 1/6 of the job, thus together completing $1/n$ of the job (in 1 day). Then

$$\frac{1}{3} + \frac{1}{6} = \frac{1}{n} \quad \text{from which } n = 2 \text{ days.}$$

Another method. In n days A and B together complete

$$n\left(\frac{1}{3} + \frac{1}{6}\right) = 1 \text{ complete job.} \qquad \text{Solving, } n = 2 \text{ days.}$$

13.30 A tank can be filled by three pipes separately in 20, 30, and 60 minutes respectively. In how many minutes can it be filled by the three pipes acting together?

SOLUTION

Let t = time required, in minutes.

In 1 minute three pipes together fill $(\frac{1}{20}+\frac{1}{30}+\frac{1}{60})$ of the tank. Then in t minutes they together fill

$$t\left(\frac{1}{20}+\frac{1}{30}+\frac{1}{60}\right)=1 \text{ complete tank.} \qquad \text{Solving, } t=10 \text{ minutes.}$$

13.31 A and B working together can complete a job in 6 days. A works twice as fast as B. How many days would it take each of them, working alone, to complete the job?

SOLUTION

Let n, $2n$ = number of days required by A and B respectively, working alone, to do the job.

In 1 day A can do $1/n$ of job and B can do $1/2n$ of job. Then in 6 days they can do

$$6\left(\frac{1}{n}+\frac{1}{2n}\right)=1 \text{ complete job. Solving, } n=9 \text{ days, } 2n=18 \text{ days.}$$

13.32 A's rate of doing work is three times that of B. On a given day A and B work together for 4 hours; then B is called away and A finishes the rest of the job in 2 hours. How long would it take B to do the complete job alone?

SOLUTION

Let t, $3t$ = time in hours required by A and B respectively, working alone, to do the job.

In 1 hour A does $1/t$ of job and B does $1/3t$ of job. Then

$$4\left(\frac{1}{t}+\frac{1}{3t}\right)+2\left(\frac{1}{t}\right)=1 \text{ complete job.} \qquad \text{Solving, } 3t=22 \text{ hours.}$$

13.33 A man is paid \$18 for each day he works and forfeits \$3 for each day he is idle. If at the end of 40 days he nets \$531, how many days was he idle?

SOLUTION

Let x = number of days idle; $40-x$ = number of days worked.

$$\text{Amount earned} - \text{amount forfeited} = \$531$$

or $$\$18(40-x) - 3x = \$531 \quad \text{and } x=9 \text{ days idle.}$$

Supplementary Problems

13.34 Solve each of the following equations.

(*a*) $3x-2=7$

(*b*) $y+3(y-4)=4$

(*c*) $4x-3=5-2x$

(*d*) $x-3-2(6-2x)=2(2x-5)$

(*e*) $\dfrac{2t-9}{3}=\dfrac{3t+4}{2}$

(*f*) $\dfrac{2x+3}{2x-4}=\dfrac{x-1}{x+1}$

(*g*) $(x-3)^2+(x+1)^2=(x-2)^2+(x+3)^2$

(*h*) $(2x+1)^2=(x-1)^2+3x(x+2)$

(*i*) $\dfrac{3}{z}-\dfrac{4}{5z}=\dfrac{1}{10}$

(*j*) $\dfrac{2x+1}{x}+\dfrac{x-4}{x+1}=3$

(*k*) $\dfrac{5}{y-1}-\dfrac{5}{y+1}=\dfrac{2}{y-2}-\dfrac{2}{y+3}$

(*l*) $\dfrac{7}{x^2-4}+\dfrac{2}{x^2-3x+2}=\dfrac{4}{x^2+x-2}$

13.35 Solve for the indicated letter.

(a) $2(x-p) = 3(6p-x):x$

(b) $2by - 2a = ay - 4b:y$

(c) $\dfrac{2x-a}{b} = \dfrac{2x-b}{a}:x$

(d) $\dfrac{x-a}{x-b} = \dfrac{x-c}{x-d}:x$

(e) $\dfrac{1}{ay} + \dfrac{1}{by} = \dfrac{1}{c}:y$

13.36 Express each of the following statements in terms of algebraic symbols.

(a) Two more than five times a certain number.

(b) Six less than twice a certain number.

(c) Each of two numbers whose difference is 25.

(d) The squares of three consecutive integers.

(e) The amount by which five times a certain number exceeds 40.

(f) The square of any odd integer.

(g) The excess of the square of a number over twice the number.

(h) The number of pints in x gallons.

(i) The difference between the squares of two consecutive even integers.

(j) Bob is six years older than Jane who is half as old as Jack. Express each of their ages in terms of a single unknown.

(k) The three angles A, B, C of a triangle ABC if angle A exceeds twice angle B by 20°.

(l) The perimeter and area of a rectangle if one side is 3 ft shorter than three times the other side.

(m) The fraction whose denominator is 4 more than twice the square of the numerator.

(n) The amount of salt in a tank holding x quarts of water if the concentration is 2 lb of salt per gallon.

13.37 (a) One half of a certain number is 10 more than one sixth of the number. Find the number.

(b) The difference between two numbers is 20 and their sum is 48. Find the numbers.

(c) Find two consecutive even integers such that twice the smaller exceeds the larger by 18.

(d) The sum of two numbers is 36. If the larger is divided by the smaller, the quotient is 2 and the remainder is 3. Find the numbers.

(e) Find two consecutive positive odd integers such that the difference of their squares is 64.

(f) The first of three numbers exceeds twice the second number by 4, while the third number is twice the first. If the sum of the three numbers is 54, find the numbers.

13.38 (a) A father is 24 years older than his son. In 8 years he will be twice as old as his son. Determine their present ages.

(b) Mary is fifteen years older than her sister Jane. Six years ago Mary was six times as old as Jane. Find their present ages.

(c) Larry is now twice as old as Bill. Five years ago Larry was three times as old as Bill. Find their present ages.

13.39 (a) In a purse is \$3.05 in nickels and dimes, 19 more nickels than dimes. How many coins are there of each kind?

(b) Richard has twice as many dimes as quarters, amounting to \$6.75 in all. How many coins does he have?

(c) Admission tickets to a theater were 60¢ for adults and 25¢ for children. Receipts for the day showed that 280 persons attended and \$140 was collected. How many children attended that day?

13.40 (a) The tens digit of a certain two-digit number exceeds the units digit by 3. The sum of the digits is 1/7 of the number. Find the number.

(*b*) The sum of the digits of a certain two-digit number is 10. If the digits are reversed, a new number is formed which is one less than twice the original number. Find the original number.

(*c*) The tens digit of a certain two-digit number is 1/3 of the units digit. When the digits are reversed, the new number exceeds twice the original number by 2 more than the sum of the digits. Find the original number.

13.41 (*a*) Goods cost a merchant \$72. At what price should he mark them so that he may sell them at a discount of 10% from his marked price and still make a profit of 20% on the selling price?

(*b*) A woman is paid \$20 for each day she works and forfeits \$5 for each day she is idle. At the end of 25 days she nets \$450. How many days did she work?

(*c*) A labor report states that in a certain factory a total of 400 men and women are employed. The average daily wage is \$16 for a man and \$12 for a woman. If the labor cost is \$5720 per day, how many women are employed?

(*d*) A woman has \$450 invested, part at 2% and the remainder at 3% simple interest. How much is invested at each rate if the total annual income from these investments is \$11?

(*e*) A man has \$2000 invested at 7% and \$5000 at 4% simple interest. What additional sum must he invest at 6% to give him an overall return of 5%?

13.42 (*a*) The perimeter of a rectangle is 110 ft. Find the dimensions if the length is 5 ft less than twice the width.

(*b*) The length of a rectangular floor is 8 ft greater than its width. If each dimension is increased by 2 ft, the area is increased by 60 ft^2. Find the dimensions of the floor.

(*c*) The area of a square exceeds the area of a rectangle by 3 in^2. The width of the rectangle is 3 in. shorter and the length 4 in. longer than the side of the square. Find the side of the square.

(*d*) A piece of wire 40 in. long is bent into the form of a right triangle, one of whose legs is 15 in. long. Determine the lengths of the other two sides.

(*e*) The length of a rectangular swimming pool is twice its width. The pool is surrounded by a cement walk 4 ft wide. If the area of the walk is 748 ft^2, determine the dimensions of the pool.

13.43 (*a*) An excellent solution for cleaning grease stains from cloth or leather consists of the following: carbon tetrachloride 80% (by volume), ligroin 16%, and amyl alcohol 4%. How many pints of each should be taken to make up 75 pints of solution?

(*b*) Lubricating oil worth 28 cents/quart is to be mixed with oil worth 33 cents/quart to make up 45 quarts of a mixture to sell at 30 ¢/qt. What volume of each grade should be taken?

(*c*) What weight of water must be added to 50 lb of a 36% sulfuric acid solution to yield a 20% solution? All percentages are by weight.

(*d*) How many quarts of pure alcohol must be added to 10 quarts of a 15% alcohol solution to obtain a mixture which is 25% alcohol? All percentages are by volume.

(*e*) There is available 60 gallons of a 50% solution of glycerin and water. What volume of water must be added to the solution to reduce the glycerin concentration to 12%? All percentages are by volume.

(*f*) The radiator of a jeep has a capacity of 4 gallons. It is filled with an anti-freeze solution of water and glycol which analyzes 10% glycol. What volume of the mixture must be drawn off and replaced with glycol to obtain a 25% glycol solution? All percentages are by volume.

(*g*) One thousand quarts of milk testing 4% butterfat are to be reduced to 3%. How many quarts of cream testing 23% butterfat must be separated from the milk to produce the required result? All percentages are by volume.

(*h*) There are available 10 tons of coal containing 2.5% sulfur, and also supplies of coal containing 0.80% and 1.10% sulfur respectively. How many tons of each of the latter should be mixed with the original 10 tons to give 20 tons containing 1.7% sulfur?

(*i*) A clay contains 45% silica and 10% water. Determine the percentage of silica in the clay on a dry (water-free) basis. All percentages are by weight.

(*j*) A nugget of gold and quartz weighs 100 grams. Gold weighs 19.3 g/cm^3 (grams per cubic centimeter), quartz weighs 2.6 g/cm^3, and the nugget weighs 6.4 g/cm^3. Find the weight of gold in the nugget.

Hint: Let x = weight of gold in nugget; $(100 - x)$ = weight of quartz in nugget.
Volume of nugget = volume of gold in nugget + volume of quartz in nugget.

(*k*) A cold cream sample weighing 8.41 grams lost 5.83 grams of moisture on heating to 110 °C. The residue on extracting with water and drying lost 1.27 grams of water-soluble glycerin. The balance was oil. Calculate the percentage composition of this cream.

(*l*) A coal contains 2.4% water. After drying, the moisture-free residue contains 71.0% carbon. Determine the percentage of carbon on the "wet basis." All percentages are by weight.

13.44 (*a*) Two motorists start toward each other at 4:30 P.M. from towns 255 miles apart. If their respective average speeds are 40 and 45 mi/hr, at what time will they meet?

(*b*) Two planes start from Chicago at the same time and fly in opposite directions, one averaging a speed of 40 mi/hr greater than the other. If they are 2000 miles apart after 5 hours, find their average speeds.

(*c*) At what rate must motorist A travel to overtake motorist B who is traveling at a rate 20 mi/hr slower if A starts two hours after B and wishes to overtake B in 4 hours?

(*d*) A motorist starts from city A at 2:00 P.M. and travels to city B at an average speed of 30 mi/hr. After resting at B for one hour, she returns over the same route at an average speed of 40 mi/hr and arrives at A that evening at 6:30 P.M. Determine the distance between A and B.

(*e*) Tom traveled a distance of 265 miles. He drove at 40 mi/hr during the first part of the trip and at 35 mi/hr during the remaining part. If he made the trip in 7 hours, how long did he travel at 40 mi/hr?

(*f*) A boat can move at 8 mi/hr in still water. If it can travel 20 miles downstream in the same time it can travel 12 miles upstream, determine the rate of the stream.

(*g*) The speed of a plane is 120 mi/hr in a calm. With the wind it can cover a certain distance in 4 hours, but against the wind it can cover only 3/5 of that distance in the same time. Find the velocity of the wind.

(*h*) An army of soldiers is marching down a road at 5 mi/hr. A messenger on horseback rides from the front to the rear and returns immediately, the total time taken being 10 minutes. Assuming that the messenger rides at the rate of 10 mi/hr, determine the distance from the front to the rear.

(*i*) A train moving at r mi/hr can cover a given distance in h hours. By how many mi/hr must its speed be increased in order to cover the same distance in one hour less time?

13.45 (*a*) A farmer can plow a certain field three times as fast as his son. Working together, it would take them 6 hours to plow the field. How long would it take each to do it alone?

(*b*) A painter can do a given job in 6 hours. Her helper can do the same job in 10 hours. The painter begins the work and after two hours is joined by the helper. In how many hours will they complete the job?

(*c*) One group of workers can do a job in 8 days. After this group has worked 3 days, another group joins it and together they complete the job in 3 more days. In what time could the second group have done the job alone?

(*d*) A tank can be filled by two pipes separately in 10 and 15 minutes respectively. When a third pipe is used simultaneously with the first two pipes, the tank can be filled in 4 minutes. How long would it take the third pipe alone to fill the tank?

ANSWERS TO SUPPLEMENTARY PROBLEMS

13.34	(*a*) $x = 3$	(*d*) $x = 5$	(*g*) $x = -1/2$	(*j*) $x = 1/4$
	(*b*) $y = 4$	(*e*) $t = -6$	(*h*) all values of x (identity)	(*k*) $y = 5$
	(*c*) $x = 4/3$	(*f*) $x = 1/11$	(*i*) $z = 22$	(*l*) $x = -1$

13.35 (a) $x = 4p$ (b) $y = -2$ if $a \neq 2b$ (c) $x = \dfrac{a+b}{2}$ if $a \neq b$ (d) $x = \dfrac{bc - ad}{b + c - a - d}$ (e) $y = \dfrac{ac + bc}{ab}$

13.36 (a) $5x + 2$
(b) $2x - 6$
(c) $x + 25$, x
(d) x^2, $(x+1)^2$, $(x+2)^2$
(e) $5x - 40$
(f) $(2x+1)^2$ where x = integer
(g) $x^2 - 2x$
(h) $8x$
(i) $(2x+2)^2 - (2x)^2$, x = integer
(j) Jane's age x, Bob's age $x + 6$, Jack's age $2x$
(k) $B = x°$, $A = (2x + 20)°$, $C = (160 - 3x)°$
(l) One side is x, adjacent side is $3x - 3$. Perimeter = $8x - 6$, Area = $3x^2 - 3x$
(m) $\dfrac{x}{2x^2 + 4}$
(n) $\dfrac{x}{2}$ lb salt

13.37 (a) 30 (b) 34, 14 (c) 20, 22 (d) 25, 11 (e) 15, 17 (f) 16, 6, 32

13.38 (a) Father 40, son 16 (b) Mary 24, Jane 9 (c) Larry 20, Bill 10

13.39 (a) 14 dimes, 33 nickels (b) 15 quarters, 30 dimes (c) 200 adults, 80 children

13.40 (a) 63 (b) 37 (c) 26

13.41 (a) \$100 (b) 23 days (c) 170 women (d) \$200 at 3%, \$250 at 2% (e) \$1000

13.42 (a) width 20 ft, length 35 ft
(b) width 10 ft, length 18 ft
(c) 9 in.
(d) other leg 8 ft, hypotenuse 17 ft
(e) 30 ft by 60 ft

13.43 (a) 60, 12, 3 pints
(b) 18 qt of 33¢, 27 qt of 28¢
(c) 40 lb
(d) 4/3 qt
(e) 190 gal
(f) 2/3 gal
(g) 50 qt
(h) 6.7 tons of 0.80%, 3.3 tons of 1.10%
(i) 50% silica
(j) 69 grams gold
(k) 69.3% moisture, 15.1% glycerin, 15.6% oil
(l) 69.3% carbon

13.44 (a) 7:30 P.M. (b) 180, 220 mi/hr (c) 60 mi/hr (d) 60 mi (e) 4 hr (f) 2 mi/hr (g) 30 mi/hr (h) 5/8 mi (i) $\dfrac{r}{h-1}$

13.45 (a) Father 8 hr, son 24 hr (b) $2\frac{1}{2}$ hr (c) 12 days (d) 12 minutes

Chapter 14

Equations of Lines

14.1 SLOPE OF A LINE

The equation $ax + by = c$, where not both a and b are 0 and a, b, and c are real numbers is the standard (or general) form of the equation of a line.

The slope of a line is the ratio of the change in y compared to the change in x.

$$\text{slope} = \frac{\text{change in } y}{\text{change in } x}$$

If (x_1, y_1) and (x_2, y_2) are two points on a line and m is the slope of the line, then

$$m = \frac{y_2 - y_1}{x_2 - x_1} \quad \text{when } x_2 \neq x_1.$$

EXAMPLE 14.1. What is the slope of the line through the points $(5, -8)$ and $(6, 2)$?

$$m = \frac{y_2 - y_1}{x_2 - x_1} = \frac{2 - (-8)}{5 - 6} = \frac{10}{-1} = -10$$

The slope of the line through the two points $(5, -8)$ and $(6, 2)$ is -10.

EXAMPLE 14.2. What is the slope of the line $3x - 4y = 12$?

First we need to find two points that satisfy the equation of the line $3x - 4y = 12$. If $x = 0$, then $3(0) - 4y = 12$ and $y = -3$. Thus, one point is $(0, -3)$. If $x = -4$, then $3(-4) - 4y = 12$ and $y = -6$. So, $(-4, -6)$ is another point on the line.

$$m = \frac{y_2 - y_1}{x_2 - x_1} = \frac{-3 - (-6)}{0 - (-4)} = \frac{3}{4}$$

The slope of the line $3x - 4y = 12$ is 3/4.

In Example 14.1, the slope of the line is negative. This means that, as we view the graph of the line from left to right, as x increases y decreases (see Fig. 14-1). In Example 14.2, the slope is positive, which means that as x increases so does y (see Fig. 14-2).

A horizontal line $y = k$, where k is a constant, has zero slope. Since all of the y values are the same, $y_2 - y_1 = 0$.

A vertical line $x = k$, where k is a constant, does not have a slope, that is, the slope is not defined. Since all of the x values are the same, $x_2 - x_1 = 0$ and division by zero is not defined.

14.2 PARALLEL AND PERPENDICULAR LINES

Two non-vertical lines are parallel if and only if the slopes of the lines are equal.

EXAMPLE 14.3. Show that the figure $PQRS$ with vertices $P(0, -2)$, $Q(-2, 3)$, $R(3, 5)$, and $S(5, 0)$ is a parallelogram.

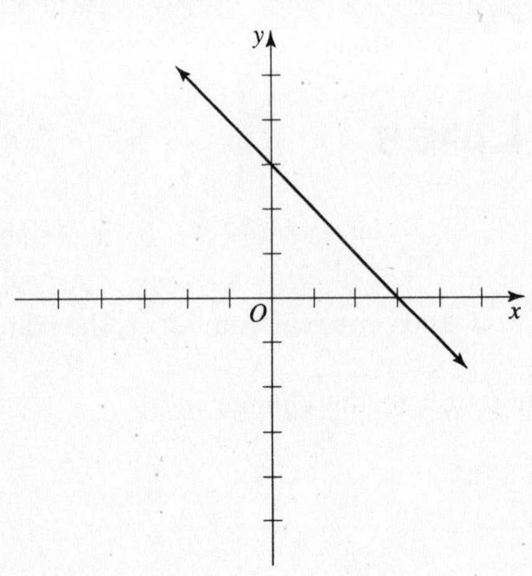

Fig. 14-1

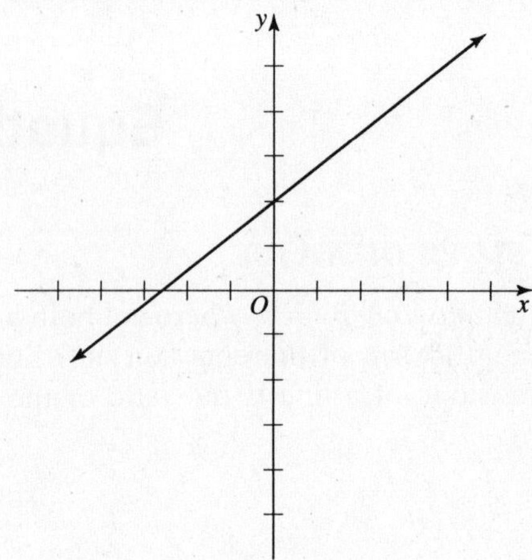

Fig. 14-2

Quadrilateral $PQRS$ is a parallelogram if $\overline{PQ}$ and $\overline{RS}$ are parallel and $\overline{PS}$ and $\overline{QR}$ are parallel.

$$\text{slope }(\overline{PQ}) = \frac{3-(-2)}{-2-0} = \frac{5}{-2} = -\frac{5}{2} \quad \text{and slope }(\overline{RS}) = \frac{0-5}{5-3} = -\frac{5}{2}$$

$$\text{slope }(\overline{PS}) = \frac{0-(-2)}{5-0} = \frac{2}{5} \quad \text{and slope }(\overline{QR}) = \frac{5-3}{3-(-2)} = \frac{2}{5}$$

Since $\overline{PQ}$ and $\overline{RS}$ have the same slope, they are parallel, and since $\overline{PS}$ and $\overline{QR}$ have the same slope, they are parallel. Thus, the opposite sides of $PQRS$ are parallel, so $PQRS$ is a parallelogram.

EXAMPLE 14.4. Show that the points $A(0,4)$, $B(2,3)$, and $C(4,2)$ are collinear, that is, lie on the same line.

The points A, B, and C are collinear if the slopes of the lines through any two pairings of the points are the same.

$$\text{slope }(\overline{AB}) = \frac{3-4}{2-0} = -\frac{1}{2} \quad \text{and slope }(\overline{BC}) = \frac{2-3}{4-2} = -\frac{1}{2}$$

The lines AB and BC have the same slope and share a common point, B, so the lines are the same line. Thus, the points A, B, and C are collinear.

Two non-vertical lines are perpendicular if and only if the product of their slopes is -1. The slope of each line is said to be the negative reciprocal of the slope of the other line.

EXAMPLE 14.5. Show that the line through the points $A(3,3)$ and $B(6,-3)$ is perpendicular to the line through the points $C(4,2)$ and $D(8,4)$.

$$\text{slope }(\overline{AB}) = \frac{-3-3}{6-3} = \frac{-6}{3} = -2 \quad \text{and slope }(\overline{CD}) = \frac{4-2}{8-4} = \frac{2}{4} = \frac{1}{2}$$

Since $(-2)(1/2) = -1$, the lines AB and CD are perpendicular.

14.3 SLOPE–INTERCEPT FORM OF EQUATION OF A LINE

If a line has slope m and y intercept $(0,b)$, then for any point (x,y), where $x \neq 0$, on the line we have

$$m = \frac{y-b}{x-0} \quad \text{and } y = mx + b.$$

The slope–intercept form of the equation of a line with slope m and y intercept b is $y = mx + b$.

EXAMPLE 14.6. Find the slope and y intercept of the line $3x + 2y = 12$.

We solve the equation $3x + 2y = 12$ for y to get $y = -\frac{3}{2}x + 6$. The slope of the line is $-\frac{3}{2}$ and the y intercept is 6.

EXAMPLE 14.7. Find the equation of the line with slope -4 and y intercept 6.

The slope of the line is -4, so $m = -4$ and the y intercept is 6, so $b = 6$. Substituting into $y = mx + b$, we get $y = -4x + 6$ for the equation of the line.

14.4 SLOPE–POINT FORM OF EQUATION OF A LINE

If a line has slope m and goes through a point (x_1, y_1), then for any other point (x, y) on the line, we have $m = (y - y_1)/(x - x_1)$ and $y - y_1 = m(x - x_1)$.

The slope–point form of the equation of a line is $y - y_1 = m(x - x_1)$.

EXAMPLE 14.8. Write the equation of the line passing through the point $(1, -2)$ and having slope $-2/3$.

Since $(x_1, y_1) = (1, -2)$ and $m = -2/3$, we substitute into $y - y_1 = m(x - x_1)$ to get $y + 2 = -2/3(x - 1)$. Simplifying we get $3(y + 2) = -2(x - 1)$, and finally $2x + 3y = -4$.

The equation of the line through $(1, -2)$ with slope $-2/3$ is $2x + 3y = -4$.

14.5 TWO-POINT FORM OF EQUATION OF A LINE

If a line goes through the points (x_1, y_1) and (x_2, y_2), it has slope $m = (y_2 - y_1)/(x_2 - x_1)$ if $x_2 \neq x_1$. Substituting into the equation $y - y_1 = m(x - x_1)$, we get

$$y - y_1 = \frac{y_2 - y_1}{x_2 - x_1}(x - x_1).$$

The two-point form of the equation of a line is

$$y - y_1 = \frac{y_2 - y_1}{x_2 - x_1}(x - x_1) \qquad \text{if } x_2 \neq x_1.$$

If $x_2 = x_1$, we get the vertical line $x = x_1$. If $y_2 = y_1$, we get the horizontal line $y = y_1$.

EXAMPLE 14.9. Write the equation of the line passing through $(3, 6)$ and $(-4, 4)$.

Let $(x_1, y_1) = (3, 6)$ and $(x_2, y_2) = (-4, 4)$ and substitute into

$$\begin{aligned} y - y_1 &= \frac{y_2 - y_1}{x_2 - x_1}(x - x_1). \\ y - 6 &= \frac{4 - 6}{-4 - 3}(x - 3) \\ -7(y - 6) &= -2(x - 3) \\ -7y + 42 &= -2x + 6 \\ 2x - 7y &= -36 \end{aligned}$$

The equation of the line through the points $(3, 6)$ and $(-4, 4)$ is $2x - 7y = -36$.

14.6 INTERCEPT FORM OF EQUATION OF A LINE

If a line has x intercept a and y intercept b, it goes through the points $(a, 0)$ and $(0, b)$. The equation of the line is

$$y - b = \frac{0 - b}{a - 0}(x - 0) \qquad \text{if } a \neq 0,$$

which simplifies to $bx + ay = ab$. If both a and b are non-zero we get $x/a + y/b = 1$.

If a line has x intercept a and y intercept b and both a and b are non-zero, the equation of the line is

$$\frac{x}{a} + \frac{y}{b} = 1.$$

EXAMPLE 14.10. Find the intercepts of the line $4x - 3y = 12$.
We divide the equation $4x - 3y = 12$ by 12 to get

$$\frac{x}{3} + \frac{y}{-4} = 1.$$

The x intercept is 3 and the y intercept is -4 for the line $4x - 3y = 12$.

EXAMPLE 14.11. Write the equation of the line that has an x intercept of 2 and a y intercept of 5.
We have $a = 2$ and $b = 5$ for the equation

$$\frac{x}{a} + \frac{y}{b} = 1.$$

Substituting we get $$\frac{x}{2} + \frac{y}{5} = 1.$$

Simplifying we get $$5x + 2y = 10.$$

The line with x intercept 2 and y intercept 5 is $5x + 2y = 10$.

Solved Problems

14.1 What is the slope of the line through each pair of points?

(*a*) $(4, 1)$ and $(7, 6)$ (*b*) $(3, 9)$ and $(7, 4)$ (*c*) $(-4, 1)$ and $(-4, 3)$ (*d*) $(-3, 2)$ and $(2, 2)$

SOLUTION

(*a*) $m = \frac{6 - 1}{7 - 4} = \frac{5}{3}$ The slope of the line is 5/3.

(*b*) $m = \frac{4 - 9}{7 - 3} = -\frac{5}{4}$ The slope of the line is $-5/4$.

(*c*) $m = \frac{3 - 1}{-4 - (-4)} = \frac{2}{0}$ Slope is not defined for this line.

(*d*) $m = \frac{2 - 2}{2 - (-3)} = \frac{0}{5} = 0$ The slope of the line is 0.

14.2 Determine whether the line containing the points A and B is parallel, perpendicular, or neither to the line containing the points C and D.

(a) $A(2,4)$, $B(3,8)$, $C(5,1)$, and $D(4,-3)$

(b) $A(2,-3)$, $B(-4,5)$, $C(0,-1)$, and $D(-4,-4)$

(c) $A(1,9)$, $B(4,0)$, $C(0,6)$ and $D(5,3)$

(d) $A(8,-1)$, $B(2,3)$, $C(5,1)$, and $D(2,-1)$

SOLUTION

(a) slope $(\overline{AB}) = \dfrac{8-4}{3-2} = \dfrac{4}{1} = 4;$ slope $(\overline{CD}) = \dfrac{-3-1}{4-5} = \dfrac{-4}{-1} = 4$

Since the slopes are equal, the lines AB and CD are parallel.

(b) slope $(\overline{AB}) = \dfrac{5-(-3)}{-4-2} = \dfrac{8}{-6} = -\dfrac{4}{3};$ slope $(\overline{CD}) = \dfrac{-4-(-1)}{-4-0} = \dfrac{-3}{-4} = \dfrac{3}{4}$

Since $(-4/3)(3/4) = -1$, the lines AB and CD are perpendicular.

(c) slope $(\overline{AB}) = \dfrac{0-9}{4-1} = \dfrac{-9}{3} = -3;$ slope $(\overline{CD}) = \dfrac{3-6}{5-0} = -\dfrac{3}{5}$

Since the slopes are not equal and do not have a product of -1, the lines AB and CD are neither parallel nor perpendicular.

(d) slope $(\overline{AB}) = \dfrac{3-(-1)}{2-8} = \dfrac{4}{-6} = -\dfrac{2}{3};$ slope $(\overline{CD}) = \dfrac{-1-1}{2-5} = \dfrac{-2}{-3} = \dfrac{2}{3}$

Since the slopes are not equal and do not have a product of -1, the lines AB and CD are neither parallel nor perpendicular.

14.3 Determine whether the given three points are collinear or not.

(a) $(0,3)$, $(1,1)$, and $(2,-1)$ (b) $(1,5)$, $(-2,-1)$, and $(-3,-4)$

SOLUTION

(a) $m_1 = \dfrac{1-3}{1-0} = \dfrac{-2}{1} = -2$ and $m_2 = \dfrac{-1-1}{2-1} = \dfrac{-2}{1} = -2$

Since the line between $(0,3)$ and $(1,1)$ and the line between $(1,1)$ and $(2,-1)$ have the same slope, the points $(0,3)$, $(1,1)$, and $(2,-1)$ are collinear.

(b) $m_1 = \dfrac{-1-5}{-2-1} = \dfrac{-6}{-3} = 2$ and $m_2 = \dfrac{-4-(-1)}{-3-(-2)} = \dfrac{-3}{-1} = 3$

Since the slope of the line between $(1,5)$ and $(-2,-1)$ and the slope of the line between $(-2,-1)$ and $(-3,-4)$ are different, the points $(1,5)$, $(-2,-1)$, and $(-3,-4)$ are not collinear.

14.4 Write the equation of the line with slope m and y intercept b.

(a) $m = 2/3$, $b = 6$ (b) $m = -3$, $b = -4$ (c) $m = 0$, $b = 8$ (d) $m = 3$, $b = 0$

SOLUTION

(a) $y = mx + b = -2/3x + 6$ $\quad 2x + 3y = 18$

(b) $y = mx + b = -3x - 4$ $\quad 3x + y = -4$

(c) $y = mx + b = 0x + 8$ $\quad y = 8$

(d) $y = mx + b = 3x + 0$ $\quad 3x - y = 0$

14.5 Write the equation of the line that contains point P and has slope m.

(*a*) $P(2,5), m=4$ (*b*) $P(1,4), m=0$ (*c*) $P(-1,-6), m=1/4$ (*d*) $P(2,-3)$, $m=-3/7$

SOLUTION

We use the formula $y - y_1 = m(x - x_1)$.

(*a*) $y-5=4(x-4)$ $\quad 4x-y=3$

(*b*) $y=4=0(x-1)$ $\quad y=4$

(*c*) $y-(-3)=1/4(x-(-1))$ $\quad x-4y=23$

(*d*) $y-(-3)=-3/7(x-2)$ $\quad 3x+7y=-15$

14.6 Write the equation of the line passing through points P and Q.

(*a*) $P(1,-4)$, $Q(2,3)$ (*c*) $P(-1,4)$, $Q(3,4)$ (*e*) $P(7,1)$, $Q(8,3)$

(*b*) $P(6,-1)$, $Q(0,2)$ (*d*) $P(1,5)$, $Q(-2,3)$ (*f*) $P(4,-1)$, $Q(4,3)$

SOLUTION

(*a*) $y-3=\dfrac{3-(-4)}{2-1}(x-2)$ $\quad y-3=7(x-2)$ $\quad 7x-y=11$

(*b*) $y-2=\dfrac{2-(-1)}{0-6}$ $\quad y-2=-1/2x$ $\quad x+2y=4$

(*c*) $y-4=\dfrac{4-4}{3-(-1)}(x-3)$ $\quad y-2=0(x-3)$ $\quad y=2$

(*d*) $y-3=\dfrac{3-5}{-2-1}(x-(-2))$ $\quad y-3=2/3(x+2)$ $\quad 2x-3y=-13$

(*e*) $y-3=\dfrac{3-1}{8-7}(x-8)$ $\quad y-3=2(x-8)$ $\quad 2x-y=13$

(*f*) Since P and Q have the same x value, slope is not defined. However, the line through P and Q has to have 4 for its x coordinate in all points. Thus, the line is $x=4$.

14.7 Write the equation of the line that has x intercept -3 and y intercept 4.

SOLUTION

$$\frac{x}{a}+\frac{y}{b}=1 \quad \text{so} \quad \frac{x}{-3}+\frac{y}{4}=1 \quad 4x-3y=-12$$

14.8 Write the equation of the line through $(-5,6)$ that is parallel to the line $3x-4y=5$.

SOLUTION

We write the equation $3x-4y=5$ in slope–intercept form to identify its slope, $y=3/4x-5/4$. Since the form is $y=mx+b$, $m=3/4$. Parallel lines have the same slope, so the line we want has a slope of 3/4.

Now that we have the slope and a point the line goes through, so we can write the equation using the point–slope form: $y-y_1=m(x-x_1)$. Substituting we get $y-6=3/4(x+5)$. Simplifying we get $4y-24=3x+15$ and finally $3x-4y=-39$. The equation we want is $3x-4y=-39$.

14.9 Write the equation of the line through $(4,6)$ that is perpendicular to the line $2x-y=8$.

SOLUTION

In the slope–intercept form the given line is $y = 2x - 8$. The slope of the line is 2, so the slope of a perpendicular line is the negative reciprocal of 2, which is $-1/2$. We want to write the equation of the line with slope of $-1/2$ and goes through $(4, 6)$. Thus, $y - 6 = -1/2(x - 4)$, so $2y - 12 = -x + 4$, and finally $x + 2y = 16$. The line we want is $x + 2y = 16$.

Supplementary Problems

14.10 What is the slope of the line through each pair of points?

(*a*) $(-1, 2)$, $(4, -3)$ (*c*) $(5, 4)$, $(5, -2)$ (*e*) $(-1, 5)$, $(-2, 3)$
(*b*) $(3, 4)$, $(-4, -3)$ (*d*) $(-5, 3)$, $(2, 3)$ (*f*) $(7, 3)$, $(8, -3)$

14.11 Determine whether the line containing the points P and Q is parallel, perpendicular, or neither to the line containing the points R and S.

(*a*) $P(4, 2)$, $Q(8, 3)$, $R(-2, 8)$, and $S(1, -4)$
(*b*) $P(0, -5)$, $Q(15, 0)$, $R(1, 2)$, and $S(0, 5)$
(*c*) $P(-7, 8)$, $Q(8, -7)$, $R(-8, 10)$, and $S(6, -4)$
(*d*) $P(8, -2)$, $Q(2, 8)$, $R(-2, -8)$, and $S(-8, -2)$

14.12 Determine a constant real number k such that the lines AB and CD are (1) parallel and (2) perpendicular.

(*a*) $A(2, 1)$, $B(6, 3)$, $C(4, k)$, and $D(3, 1)$
(*b*) $A(1, k)$, $B(2, 3)$, $C(1, 7)$, and $D(3, 6)$
(*c*) $A(9, 4)$, $B(k, 10)$, $C(11, -2)$, and $D(-2, 4)$
(*d*) $A(1, 2)$, $B(4, 0)$, $C(k, 2)$, and $D(1, -3)$

14.13 Determine whether the given three points are collinear or not.

(*a*) $(-3, 1)$, $(-11, -1)$, and $(-15, -2)$ (*b*) $(1, 1)$, $(4, 2)$, and $(2, 3)$

14.14 Write the equation of the line with slope m and y intercept b.

(*a*) $m = -3$, $b = 4$ (*c*) $m = 2/3$, $b = -2$ (*e*) $m = -1/2$, $b = 3$
(*b*) $m = 0$, $b = -3$ (*d*) $m = 4$, $b = 0$ (*f*) $m = -5/6$, $b = 1/6$

14.15 Write the equation of the line that goes through point P and has slope m.

(*a*) $P(-5, 2)$, $m = -1$ (*c*) $P(4, -1)$, $m = 2/3$ (*e*) $P(2, 6)$, $m = -5$
(*b*) $P(-4, -3)$, $m = 4$ (*d*) $P(0, 4)$, $m = -4/3$ (*f*) $P(-1, 6)$, $m = 0$

14.16 Write the equation of the line through the points P and Q.

(*a*) $P(1, 2)$, $Q(2, 4)$ (*d*) $P(10, 2)$, $Q(5, 2)$ (*g*) $P(-1, 3)$, $Q(0, 6)$
(*b*) $P(1.6, 3)$, $Q(0.3, 1.4)$ (*e*) $P(3, 6)$, $Q(-3, 8)$ (*h*) $P(0, 0)$, $Q(-3, 6)$
(*c*) $P(0.7, 3)$, $Q(0.7, -3)$ (*f*) $P(-4, 2)$, $Q(2, 4)$

14.17 Write the equation of the line that has x intercept a and y intercept b.

(a) $a = -2$, $b = -2$ (b) $a = 6$, $b = -3$ (c) $a = -1/2$, $b = 4$ (d) $a = 6$, $b = 1/3$

14.18 Write the equation of the line through point P and parallel to the line l.

(a) $P(2, -4)$, line l: $y = 4x - 6$ (c) $P(-1, -1)$, line l: $4x + 5y = 5$
(b) $P(1, 0)$, line l: $y = 3x + 1$ (d) $P(3, 5)$, line l: $3x - 2y = 18$

14.19 Write the equation of the line through the point P and perpendicular to the line l.

(a) $P(2, -1)$, line l: $x = 4y$ (c) $P(1, 1)$, line l: $3x - 2y = 4$
(b) $P(0, 6)$, line l: $2x + 3y = 5$ (d) $P(1, -2)$, line l: $4x + y = 7$

14.20 Determine if the triangle with vertices A, B, and C is a right triangle.

(a) $A(4, 0)$, $B(7, -7)$, and $C(2, -5)$ (c) $A(2, 1)$, $B(3, -1)$, and $C(1, -2)$
(b) $A(5, 8)$, $B(-2, 1)$, and $C(2, -3)$ (d) $A(-6, 3)$, $B(3, -5)$, and $C(-1, 5)$

14.21 Show by using slopes that the diagonals $\overline{PR}$ and $\overline{QS}$ of quadrilateral $PQRS$ are perpendicular.

(a) $P(0, 0)$, $Q(5, 0)$, $R(8, 4)$, and $S(3, 4)$ (b) $P(-3, 0)$, $Q(6, -3)$, $R(7, 5)$, and $S(3, 3)$

14.22 Show that the points P, Q, R, and S are the vertices of a parallelogram $PQRS$.

(a) $P(5, 0)$, $Q(8, 2)$, $R(6, 5)$, and $S(3, 3)$ (b) $P(-9, 0)$, $Q(-10, -6)$, $R(4, 8)$, and $S(5, 14)$

14.23 Write the equation of the line through $(7, 3)$ that is parallel to the x axis.

14.24 Write the equation of the horizontal line through the point $(-2, -3)$.

14.25 Write the equation of the vertical line through the point $(2, 4)$.

14.26 Write the equation of the line through $(5, 8)$ that is perpendicular to the x axis.

ANSWERS TO SUPPLEMENTARY PROBLEMS

14.10 (a) -1 (b) 1 (c) not defined (d) 0 (e) 2 (f) -6

14.11 (a) perpendicular, slopes 1/4 and -4 (c) parallel, slopes -1 and -1
(b) perpendicular, slopes 1/3 and -3 (d) neither, slopes $-5/3$ and -1

14.12 (a) (1) 3/2 (2) -1 (c) (1) -4 (2) 153/13
(b) (1) 7/2 (2) 1 (d) (1) $-13/2$ (2) $-3/2$

14.13 (a) yes: $m = 1/4$ (b) no: slopes are not equal

14.14 (a) $y = -3x + 4$ (c) $2x - 3y = 6$ (e) $x + 2y = 6$
(b) $y = -3$ (d) $y = 4x$ (f) $5x + 6y = 1$

14.15 (a) $x + y = -3$ (c) $2x - 3y = 11$ (e) $5x + y = 16$
(b) $4x - y = -13$ (d) $4x + 3y = 12$ (f) $y = 6$

14.16 (a) $y = 2x$ (c) $10x = 7$ (e) $x + 3y = 21$ (g) $3x - y = -6$
(b) $80x - 65y = -67$ (d) $y = 2$ (f) $x - 3y = -10$ (h) $y = -2x$

14.17 (*a*) $x + y = -2$ (*b*) $x - 2y = 6$ (*c*) $8x - y = -4$ (*d*) $x + 18y = 6$

14.18 (*a*) $y = 4x - 12$ (*b*) $y = 3x - 3$ (*c*) $4x + 5y = -9$ (*d*) $3x - 2y = -1$

14.19 (*a*) $4x + y = 7$ (*b*) $3x - 2y = -12$ (*c*) $2x + 3y = 5$ (*d*) $x - 4y = 9$

14.20 (*a*) yes $\overline{AC} \perp \overline{BC}$ (*b*) yes $\overline{AB} \perp \overline{BC}$ (*c*) yes $\overline{AB} \perp \overline{BC}$ (*d*) yes $\overline{AC} \perp \overline{BC}$

14.21 (*a*) yes: slopes are 1/2 and -2 (*b*) yes: slopes are 1/2 and -2

14.22 (*a*) yes: $\overline{PQ} \parallel \overline{RS}$ and $\overline{QR} \parallel \overline{SP}$ (*b*) yes: $\overline{PQ} \parallel \overline{RS}$ and $\overline{QR} \parallel \overline{SP}$

14.23 $y = 3$

14.24 $y = -3$

14.25 $x = 2$

14.26 $x = 5$

Chapter 15

Simultaneous Linear Equations

15.1 SYSTEMS OF TWO LINEAR EQUATIONS

A linear equation in two variables x and y is of the form $ax + by = c$ where a, b, c are constants and a, b are not both zero. If we consider two such equations

$$a_1x + b_1y = c_1$$
$$a_2x + b_2y = c_2$$

we say that we have two simultaneous linear equations in two unknowns or a system of two linear equations in two unknowns. A pair of values for x and y, (x, y), which satisfies both equations is called a *simultaneous solution* of the given equations.

Thus the simultaneous solution of $x + y = 7$ and $x - y = 3$ is $(5, 2)$.

Three methods of solving such systems of linear equations are illustrated here.

A. Solution by addition or subtraction. If necessary, multiply the given equations by such numbers as will make the coefficients of one unknown in the resulting equations numerically equal. If the signs of the equal coefficients are unlike, add the resulting equations; if like, subtract them. Consider

$$(1)\ 2x - y = 4$$
$$(2)\ x + 2y = -3.$$

To eliminate y, multiply (1) by 2 and add to (2) to obtain

$$\begin{aligned} 2 \times (1)\text{:}\ & 4x - 2y = 8 \\ (2)\text{:}\ & \underline{\ x + 2y = -3\ } \\ \text{Addition:}\ & 5x \qquad = 5 \qquad \text{or } x = 1. \end{aligned}$$

Substitute $x = 1$ in (1) and obtain $2 - y = 4$ or $y = -2$. Thus the simultaneous solution of (1) and (2) is $(1, 2)$.

Check: Put $x = 1$, $y = -2$ in (2) and obtain $1 + 2(-2) ?-3, -3, = -3$.

B. Solution by substitution. Find the value of one unknown in either of the given equations and substitute this value in the other equation.

For example, consider the system (1), (2) above. From (1) obtain $y = 2x - 4$ and substitute this value into (2) to get $x + 2(2x - 4) = -3$ which reduces to $x = 1$. Then put $x = 1$ into either (1) or (2) and obtain $y = -2$. The solution is $(1, -2)$.

C. Graphical solution. Graph both equations, obtaining two straight lines. The simultaneous solution is given by the coordinates (x, y) of the point of intersection of these lines. Fig. 15-1 shows that the simultaneous solution of (1) $2x - y = 4$ and (2) $x + 2y = -3$ is $x = 1$, $y = -2$, also written $(1, -2)$.

If the lines are parallel, the equations are *inconsistent* and have no simultaneous solution. For example, (3) $x + y = 2$ and (4) $2x + 2y = 8$ are inconsistent, as indicated in Fig. 15-2. Note that if equation (3) is multiplied by 2 we obtain $2x + 2y = 4$, which is obviously inconsistent with (4).

Dependent equations are represented by the same line. Thus every point on the line represents a solution and, since there are an infinite number of points, there are an infinite number of simultaneous solutions. For example, (5) $x + y = 1$ and (6) $4x + 4y = 4$ are dependent equations as indicated in Fig. 15-3. Note that if (5) is multiplied by 4 the result is (6).

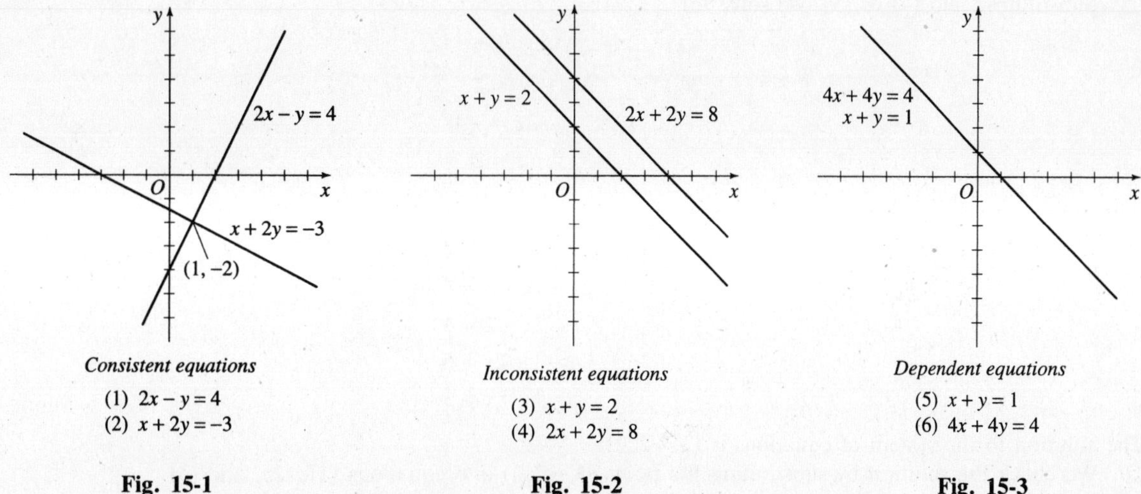

Consistent equations
(1) $2x - y = 4$
(2) $x + 2y = -3$

Fig. 15-1

Inconsistent equations
(3) $x + y = 2$
(4) $2x + 2y = 8$

Fig. 15-2

Dependent equations
(5) $x + y = 1$
(6) $4x + 4y = 4$

Fig. 15-3

15.2 SYSTEMS OF THREE LINEAR EQUATIONS

A system of three linear equations in three variables is solved by eliminating one unknown from any two of the equations and then eliminating the same unknown from any other pair of equations.

Linear equations in three variables represent planes and can result in two or more parallel planes, which are thus inconsistent and have no solution. The three planes can coincide or all three can intersect in a common line and be dependent. The three planes can intersect in a single point, like the ceiling and two walls forming a corner in a room, and are consistent.

Linear equations in three variables x, y, and z are of the form $ax + by + cz = d$, where a, b, c, and d are real numbers and not all three of a, b, and c are zero. If we consider three such equations

$$a_1x + b_1y + c_1z = d_1$$
$$a_2x + b_2y + c_2z = d_2$$
$$a_3x + b_3y + c_3z = d_3$$

and find a value (x, y, z) that satisfies all three equations, we say we have a simultaneous solution to the system of equations.

EXAMPLE 15.1. Solve the system of equations $2x + 5y + 4z = 4$, $x + 4y + 3z = 1$, and $x - 3y - 2z = 5$.

$$(1)\quad 2x + 5y + 4z = 4$$
$$(2)\quad x + 4y + 3z = 1$$
$$(3)\quad x - 3y - 2z = 5$$

First we will eliminate x from (1) and (2) and from (2) and (3).

$$\begin{array}{rr} & 2x + 5y + 4z = 4 \\ & -2x - 8y - 6z = -2 \\ \hline (4) & -3y - 2z = 2 \end{array} \qquad \begin{array}{rr} & x + 4y + 3z = 1 \\ & -x + 3y + 2z = -5 \\ \hline (5) & 7y + 5z = -4 \end{array}$$

Now we eliminate z from equations (4) and (5).

$$\begin{array}{rr} & -15y - 10z = 10 \\ & 14y + 10z = -8 \\ \hline (6) & -y \qquad = 2 \end{array}$$

We solve (6) and get $y = -2$.

Substituting into (4) or (5), we solve for z.

$$\begin{aligned} (4)\quad -3(-2) - 2z &= 2 \\ +6 - 2z &= 2 \\ -2z &= -4 \\ z &= 2 \end{aligned}$$

Substituting into (1), (2), or (3), we solve for x.

$$\begin{aligned} (1)\quad 2x + 5(-2) + 4(2) &= 4 \\ 2x - 10 + 8 &= 4 \\ 2x - 2 &= 4 \\ 2x &= 6 \\ x &= 3 \end{aligned}$$

The solution to the system of equations is $(3, -2, 2)$.

We check the solution by substituting the point $(3, -2, 2)$ into equations (1), (2), and (3).

(1) $2(3) + 5(-2) + 4(2)\ ?\ 4$	(2) $3 + 4(-2) + 3(2)\ ?\ 1$	(3) $3 - 3(-2) - 2(2)\ ?\ 5$
$6 - 10 + 8\ ?\ 4$	$3 - 8 + 6\ ?\ 1$	$3 + 6 - 4\ ?\ 5$
$4 = 4$	$1 = 1$	$5 = 5$

Thus $(3, -2, 2)$ checks in each of the given equations and is the answer to the problem.

Solved Problems

Solve the following systems.

15.1 (1) $2x - y = 4$

(2) $x + y = 5$

SOLUTION

Add (1) and (2) and obtain $3x = 9$, $x = 3$.

Now put $x = 3$ in (1) or (2) and get $y = 2$. The solution is $x = 3$, $y = 2$ or $(3, 2)$.

Another method. From (1) obtain $y = 2x - 4$ and substitute this value into equation (2) to get $x + 2x - 4 = 5$, $3x = 9$, $x = 3$. Now put $x = 3$ into (1) or (2) and obtain $y = 2$.

Check: $2x - y = 2(3) - 2 = 4$ and $x + y = 3 + 2 = 5$.

Graphical solution. The graph of a linear equation is a straight line. Since a straight line is determined by two points, we need plot only two points for each equation. However, to insure accuracy we shall plot three points for each line.

For $2x - y = 4$:

x	-1	0	1
y	-6	-4	-2

For $x + y = 5$:

x	-1	0	1
y	6	5	4

The simultaneous solution is the point of intersection $(3, 2)$ of the lines (see Fig. 15-4).

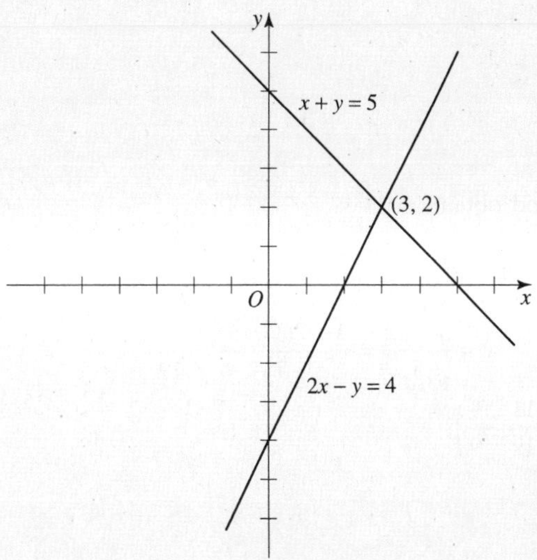

Fig. 15-4

15.2 (1) $5x + 2y = 3$
(2) $2x + 3y = -1$

SOLUTION

To eliminate y, multiply (1) by 3 and (2) by 2 and subtract the results.

$$\begin{array}{rl} 3 \times (1): & 15x + 6y = 9 \\ 2 \times (2): & 4x + 6y = -2 \\ \hline \text{Subtraction:} & 11x \qquad\; = 11 \qquad \text{or } x = 1. \end{array}$$

Now put $x = 1$ in (1) or (2) and obtain $y = -1$. The simultaneous solution is $(1, -1)$.

15.3 (1) $2x + 3y = 3$
(2) $6y - 6x = 1$

SOLUTION

Rearranging (2),

$$\begin{array}{ll} (1) & 2x + 3y = 3 \\ (2) & -6x + 6y = 1 \end{array}$$

To eliminate x, multiply (1) by 3 and add the result with (2) to get

$$\begin{array}{rl} 3 \times (1): & 6x + 9y = 9 \\ (2): & -6x + 6y = 1 \\ \hline & 15y = 10 \qquad \text{or } y = 2/3. \end{array}$$

Now put $y = 2/3$ into (1) or (2) and obtain $x = 1/2$. The solution is $(1/2, 2/3)$.

15.4 (1) $5y = 3 - 2x$
(2) $3x = 2y + 1$

SOLUTION

Solve by substitution.

From (1), $$y = \frac{3-2x}{5}.$$

Put this value into (2) and obtain

$$3x = 2\left(\frac{3-2x}{5}\right) + 1 \quad \text{or} \quad x = \frac{11}{19}.$$

Then $$y = \frac{3-2x}{5} = \frac{3-2(11/19)}{5} = \frac{7}{19}$$

and the solution is $\left(\frac{11}{19}, \frac{7}{19}\right)$.

15.5 (1) $\dfrac{x-2}{3} + \dfrac{y+1}{6} = 2$

(2) $\dfrac{x+3}{4} - \dfrac{2y-1}{2} = 1$

SOLUTION

To eliminate fractions, multiply (1) by 6 and (2) by 4 and simplify to obtain

(1_1) $2x + y = 15$

(2_1) $x - 4y = -1$

Solving, we find

$$x = \frac{59}{9}, \quad y = \frac{17}{9} \quad \text{and the solution is} \quad \left(\frac{59}{9}, \frac{17}{9}\right).$$

15.6 (1) $x - 3y = 2a$

(2) $2x + y = 5a$

SOLUTION

To eliminate x, multiply (1) by 2 and subtract (2); then $y = a/7$.

To eliminate y, multiply (2) by 3 and add with (1); then $x = 17a/7$.

The solution is $\left(\frac{17a}{7}, \frac{a}{7}\right)$.

15.7 (1) $3u + 2v = 7r + s$

(2) $2u - v = 3s$

Solve for u and v in terms of r and s.

SOLUTION

To eliminate v, multiply (2) by 2 and add with (1); then $7u = 7r + 7s$ or $u = r + s$.

To eliminate u, multiply (1) by 2, (2) by -3, and add the results; then $v = 2r - s$.

The solution is $(r + s, 2r - s)$.

15.8 (1) $ax + by = 2a^2 - 3b^2$

(2) $x + 2y = 2a - 6b$

SOLUTION

Multiply (2) by a and subtract from (1); then $by - 2ay = 6ab - 3b^2$, $y(b - 2a) = 3b(2a - b)$,

and $y = \dfrac{3b(2a-b)}{(b-2a)} = \dfrac{-3b(b-2a)}{b-2a} = -3b$ provided $b - 2a \neq 0$.

Similarly, we obtain $x = 2a$ provided $b - 2a \neq 0$.
Check: (1) $a(2a) + b(-3b) = 2a^2 - 3b^2$, (2) $2a + 2(-3b) = 2a - 6b$.
Note. If $b - 2a = 0$, or $b = 2a$, the given equations become

$$(1_1) \quad ax + 2ay = -10a^2$$
$$(2_1) \quad x + 2y = -10a$$

which are dependent since (1_1) may be obtained by multiplying (2_1) by a. Thus if $b = 2a$ the system possesses an infinite number of solutions, i.e. any values of x and y which satisfy $x + 2y = -10a$.

15.9 The sum of two numbers is 28 and their difference is 12. Find the numbers.

SOLUTION

Let x and y be the two numbers. Then (1) $x + y = 28$ and (2) $x - y = 12$.
Add (1) and (2) to obtain $2x = 40$, $x = 20$. Subtract (2) from (1) to obtain $2y = 16$, $y = 8$.
Note. Of course this problem may also be solved easily by using one unknown. Let the numbers by n and $28 - n$. Then $n - (28 - n) = 12$ or $n = 20$, and $28 - n = 8$.

15.10 If the numerator of a certain fraction is increased by 2 and the denominator is increased by 1, the resulting fraction equals 1/2. If, however, the numerator is increased by 1 and the denominator decreased by 2, the resulting fraction equals 3/5. Find the fraction.

SOLUTION

Let x = numerator, y = denominator, and x/y = required fraction. Then

(1) $\dfrac{x+2}{y+1} = \dfrac{1}{2}$ or $2x - y = -3$, and (2) $\dfrac{x+1}{y-2} = \dfrac{3}{5}$ or $5x - 3y = -11$.

Solve (1) and (2) simultaneously and obtain $x = 2$, $y = 7$. The required fraction is 2/7.

15.11 Two years ago a man was six times as old as his daughter. In 18 years he will be twice as old as his daughter. Determine their present ages.

SOLUTION

Let x = father's present age in years, y = daughter's present age in years.

Equation for condition 2 years ago: (1) $(x - 2) = 6(y - 2)$.
Equation for condition 18 years hence: (2) $(x + 18) = 2(y + 18)$.

Solve (1) and (2) simultaneously and obtain $x = 32$, $y = 7$.

15.12 Find the two-digit number satisfying the following two conditions. (1) Four times the units digit is six less than twice the tens digit. (2) The number is nine less than three times the number obtained by reversing the digits.

SOLUTION

Let t = tens digit, u = units digit.

The required number = $10t + u$; reversing digits, the new number = $10u + t$. Then

$$(1)\quad 4u = 2t - 6 \qquad \text{and} \qquad (2)\quad 10t + u = 3(10u + t) - 9.$$

Solving (1) and (2) simultaneously, $t = 7$, $u = 2$, and the required number is 72.

15.13 Five tables and eight chairs cost \$115; three tables and five chairs cost \$70. Determine the cost of each table and of each chair.

SOLUTION

Let x = cost of a table, y = cost of a chair. Then

$$(1)\quad 5x + 8y = \$115 \qquad \text{and} \qquad (2)\quad 3x + 5y = \$70.$$

Solve (1) and (2) simultaneously and obtain $x = \$15$, $y = \$5$.

15.14 A merchant sold his entire stock of shirts and ties for \$1000, the shirts being priced at 3 for \$10 and the ties at \$2 each. If he had sold only 1/2 of the shirts and 2/3 of the ties he would have collected \$600. How many of each kind did he sell?

SOLUTION

Let s = number of shirts sold, t = number of ties sold. Then

$$(1)\quad \frac{10}{3}s + 2t = 1000 \qquad \text{and} \qquad (2)\quad \frac{10}{3}\left(\frac{1}{2}s\right) + 2\left(\frac{2}{3}t\right) = 600.$$

Solving (1) and (2) simultaneously, $s = 120$, $t = 300$.

15.15 An investor has \$1100 income from bonds bearing 4% and 5%. If the amounts at 4% and 5% were interchanged she would earn \$50 more per year. Find the total sum invested.

SOLUTION

Let x = amount invested at 4%, y = amount at 5%. Then

$$(1)\quad 0.04x + 0.05y = 1100 \qquad \text{and} \qquad (2)\quad 0.05x + 0.04y = 1150.$$

Solving (1) and (2) simultaneously, $x = \$15\,000$, $y = \$10\,000$, and their sum is \$25 000.

15.16 Tank A contains a mixture of 10 gallons water and 5 gallons pure alcohol. Tank B has 12 gallons water and 3 gallons alcohol. How many gallons should be taken from each tank and combined in order to obtain an 8 gallon solution containing 25% alcohol by volume?

SOLUTION

In 8 gal of required mixture are $0.25(8) = 2$ gal alcohol.

Let x, y = volumes taken from tanks A, B respectively; then (1) $x + y = 8$.

$$\text{Fraction of alcohol in tank } A = \frac{5}{10+5} = \frac{1}{3}, \text{ in tank } B = \frac{3}{12+3} = \frac{1}{5}.$$

Thus in x gal of A are $x/3$ gal alcohol, and in y gal of B are $y/5$ gal alcohol; then (2) $x/3 + y/5 = 2$.

Solving (1) and (2) simultaneously, $x = 3$ gal, $y = 5$ gal.

Another method, using only one unknown. Let x = volume taken from tank A, $8 - x$ = volume taken from tank B.

Then $\frac{1}{3}x + \frac{1}{5}(8 - x) = 2$ from which $x = 3$ gal, $8 - x = 5$ gal.

15.17 A given alloy contains 20% copper and 5% tin. How many pounds of copper and of tin must be melted with 100 lb of the given alloy to produce another alloy analyzing 30% copper and 10% tin? All percentages are by weight.

SOLUTION

Let x, y = number of pounds of copper and tin to be added, respectively.

In 100 lb of given alloy are 20 lb copper and 5 lb tin. Then, in the new alloy,

$$\text{fraction of copper} = \frac{\text{pounds copper}}{\text{pounds alloy}} \quad \text{or} \quad (1)\quad 0.30 = \frac{20+x}{100+x+y}$$

$$\text{fraction of tin} = \frac{\text{pounds tin}}{\text{pounds alloy}} \quad \text{or} \quad (2)\quad 0.10 = \frac{5+y}{100+x+y}.$$

The simultaneous solution of (1) and (2) is $x = 17.5$ lb copper, $y = 7.5$ lb tin.

15.18 Determine the rate of a woman's rowing in still water and the rate of the river current, if it takes her 2 hours to row 9 miles with the current and 6 hours to return against the current.

SOLUTION

Let x = rate of rowing in still water, y = rate of current.

With current: $2\text{ hr} \times (x+y)\text{ mi/hr} = 9\text{ mi}$ or (1) $2x + 2y = 9$.
Against current: $6\text{ hr} \times (x-y)\text{ mi/hr} = 9\text{ mi}$ or (2) $6x - 6y = 9$.

Solving (1) and (2) simultaneously, $x = 3$ mi/hr, $y = 3/2$ mi/hr.

15.19 Two particles move at different but constant speeds along a circle of circumference 276 ft. Starting at the same instant and from the same place, when they move in opposite directions they pass each other every 6 sec and when they move in the same direction they pass each other every 23 sec. Determine their rates.

SOLUTION

Let x, y = their respective rates in ft/sec.

Opposite directions: $6\text{ sec} \times (x+y)\text{ ft/sec} = 276$ or (1) $6x + 6y = 276$.
Same direction: $23\text{ sec} \times (x-y)\text{ ft/sec} = 276$ or (2) $23x - 23y = 276$.

Solving (1) and (2) simultaneously, $x = 29$ ft/sec, $y = 17$ ft/sec.

15.20 Fahrenheit temperature = m(Centigrade temperature) + n, or $F = mC + n$, where m and n are constants. At one atmosphere pressure, the boiling point of water is 212 °F or 100 °C and the freezing point of water is 32 °F or 0 °C. (*a*) Find m and n. (*b*) What Fahrenheit temperature corresponds to −273 °C, the lowest temperature obtainable?

SOLUTION

(*a*) (1) $212 = m(100) + n$ and (2) $32 = m(0) + n$. Solving, $m = 9/5$, $n = 32$.

(*b*) $F = \frac{9}{5}C + 32 = \frac{9}{5}(-273) + 32 = -491.4 + 32 = -459$ °F, to the nearest degree.

Solve the following systems.

15.21 (1) $2x - y + z = 3$

(2) $x + 3y - 2z = 11$

(3) $3x - 2y + 4z = 1$

SOLUTION

To eliminate y between (1) and (2) multiply (1) by 3 and add with (2) to obtain

$$(1_1) \quad 7x + z = 20.$$

To eliminate y between (2) and (3) multiply (2) by 2, (3) by 3, and add the results to get

$$(2_1) \quad 11x + 8z = 25.$$

Solving (1_1) and (2_1) simultaneously, we find $x = 3$, $z = -1$. Substituting these values into any of the given equations, we find $y = 2$.

Thus the solution is $(3, 2, -1)$.

15.22 (1) $\dfrac{x}{3} + \dfrac{y}{2} - \dfrac{z}{4} = 2$, (2) $\dfrac{x}{4} + \dfrac{y}{3} - \dfrac{z}{2} = \dfrac{1}{6}$, (3) $\dfrac{x}{2} - \dfrac{y}{4} + \dfrac{z}{3} = \dfrac{23}{6}$.

SOLUTION

To remove fractions, multiply the equations by 12 and obtain the system

$$(1_1) \quad 4x + 6y - 3z = 24$$
$$(2_1) \quad 3x + 4y - 6z = 2$$
$$(3_1) \quad 6x - 3y + 4z = 46.$$

To eliminate x between (1_1) and (2_1) multiply (1_1) by 3, (2_1) by -4, and add the results to obtain

$$(1_2) \quad 2y + 15z = 64.$$

To eliminate x between (2_1) and (3_1) multiply (2_1) by 2 and subtract (3_1) to obtain

$$(2_2) \quad 11y - 16z = -42.$$

The simultaneous solution of (1_2) and (2_2) is $y = 2$, $z = 4$. Substituting these values of y and z into any of the given equations, we find $x = 6$.

Thus the simultaneous solution of the three given equations is $(6, 2, 4)$.

15.23 (1) $\dfrac{1}{x} - \dfrac{2}{y} - \dfrac{2}{z} = 0$, (2) $\dfrac{2}{x} + \dfrac{3}{y} + \dfrac{1}{z} = 1$, (3) $\dfrac{3}{x} - \dfrac{1}{y} - \dfrac{3}{z} = 3$.

SOLUTION

Let $$\frac{1}{x} = u, \quad \frac{1}{y} = v, \quad \frac{1}{z} = w$$

so that the given equations may be written

$$(1_1) \quad u - 2v - 2w = 0$$
$$(2_1) \quad 2u + 3v + w = 1$$
$$(3_1) \quad 3u - v - 3w = 3$$

from which we find $u = -2$, $v = 3$, $w = -4$.

Thus $\dfrac{1}{x} = -2$ or $x = -1/2$, $\dfrac{1}{y} = 3$ or $y = 1/3$, $\dfrac{1}{z} = -4$ or $z = -1/4$.

The solution is $(-1/2, 1/3, -1/4)$.

Check: (1) $\dfrac{1}{-1/2} - \dfrac{2}{1/3} - \dfrac{2}{-1/4} = 0$, (2) $\dfrac{2}{-1/2} + \dfrac{3}{1/3} + \dfrac{1}{-1/4} = 1$, (3) $\dfrac{3}{-1/2} - \dfrac{1}{1/3} - \dfrac{3}{-1/4} = 3$.

15.24 (1) $3x + y - z = 4$, (2) $x + y + 4z = 3$, (3) $9x + 5y + 10z = 8$.

SOLUTION

Subtracting (2) from (1), we obtain (1_1) $2x - 5z = 1$.

Multiplying (2) by 5 and subtracting (3), we obtain (2_1) $-4x + 10z = 7$.

Now (1_1) and (2_1) are inconsistent since (1_1) multiplied by -2 gives $-4x + 10z = -2$, thus contradicting (2_1). This indicates that the original system is inconsistent and hence has no simultaneous solution.

15.25 A and B working together can do a given job in 4 days, B and C together can do the job in 3 days, and A and C together can do it in 2.4 days. In how many days can each do the job working alone?

SOLUTION

Let a, b, c = number of days required by each working alone to do the job, respectively. Then $1/a$, $1/b$, $1/c$ = fraction of complete job done by each in 1 day, respectively. Thus

$$(1)\quad \frac{1}{a}+\frac{1}{b}=\frac{1}{4}, \qquad (2)\quad \frac{1}{b}+\frac{1}{c}=\frac{1}{3}, \qquad (3)\quad \frac{1}{a}+\frac{1}{c}=\frac{1}{2.4}.$$

Solving (1), (2), (3) simultaneously, we find $a = 6$, $b = 12$, $c = 4$ days.

Supplementary Problems

15.26 Solve each of the following pairs of simultaneous equations by the methods indicated.

(a) $\begin{cases} 2x - 3y = 7 \\ 3x + y = 5 \end{cases}$ Solve (1) by addition or subtraction, (2) by substitution.

(b) $\begin{cases} 3x - y = -6 \\ 2x + 3y = 7 \end{cases}$ Solve (1) graphically, (2) by addition or subtraction.

(c) $\begin{cases} 4x + 2y = 5 \\ 5x - 3y = -2 \end{cases}$ Solve (1) graphically, (2) by addition or subtraction, (3) by substitution.

15.27 Solve each of the following pairs of simultaneous equations by any method.

(a) $\begin{cases} 2x - 5y = 10 \\ 4x + 3y = 7 \end{cases}$

(b) $\begin{cases} 2y - x = 1 \\ 2x + y = 8 \end{cases}$

(c) $\begin{cases} \dfrac{2x}{3}+\dfrac{y}{5}=6 \\ \dfrac{x}{6}-\dfrac{y}{2}=-4 \end{cases}$

(d) $\begin{cases} \dfrac{2x-1}{3}+\dfrac{y+2}{4}=4 \\ \dfrac{x+3}{2}-\dfrac{x-y}{3}=3 \end{cases}$

(e) $\begin{cases} 2x - 3y = 9t \\ 4x - y = 8t \end{cases}$

(f) $\begin{cases} 2x + y + 1 = 0 \\ 3x - 2y + 5 = 0 \end{cases}$

(g) $\begin{cases} 2u - v = -5s \\ 3u + 2v = 7r - 4s \end{cases}$ Find u and v in terms of r and s.

(h) $\begin{cases} 5/x - 3/y = 1 \\ 2/x + 1/y = 7 \end{cases}$

(i) $\begin{cases} ax - by = a^2 + b^2 \\ 2bx - ay = 2b^2 + 3ab - a^2 \end{cases}$ Find x and y in terms of a and b.

15.28 Indicate which of the following systems are (1) consistent, (2) dependent, (3) inconsistent.

(*a*) $\begin{cases} x + 3y = 4 \\ 2x - y = 1 \end{cases}$ (*c*) $\begin{cases} 3x = 2y + 3 \\ x - 2y/3 = 1 \end{cases}$ (*e*) $\begin{cases} 2x - y = 1 \\ 2y - x = 1 \end{cases}$

(*b*) $\begin{cases} 2x - y = 5 \\ 2y = 7 + 4x \end{cases}$ (*d*) $\begin{cases} (x + 3)/4 = (2y - 1)/6 \\ 3x - 4y = 2 \end{cases}$ (*f*) $\begin{cases} (x + 2)/4 - (y - 2)/12 = 5/4 \\ y = 3x - 7 \end{cases}$

15.29 (*a*) When the first of two numbers is added to twice the second the result is 21, but when the second number is added to twice the first the result is 18. Find the two numbers.

(*b*) If the numerator and denominator of a certain fraction are both increased by 3, the resulting fraction equals 2/3. If, however, the numerator and denominator are both decreased by 2, the resulting fraction equals 1/2. Determine the fraction.

(*c*) Twice the sum of two numbers exceeds three times their difference by 8, while half the sum is one more than the difference. What are the numbers?

(*d*) If three times the larger of two numbers is divided by the smaller, the quotient is 6 and the remainder is 6. If five times the smaller is divided by the larger, the quotient is 2 and the remainder is 3. Find the numbers.

15.30 (*a*) Six years ago Bob was four times as old as Mary. In four years he will be twice as old as Mary. How old are they now?

(*b*) *A* is eleven times as old as *B*. In a certain number of years *A* will be five times as old as *B*, and five years after that she will be three times as old as *B*. How old are they now?

15.31 (*a*) Three times the tens digit of a certain two-digit number is two more than four times the units digit. The difference between the given number and the number obtained by reversing the digits is two less than twice the sum of the digits. Find the number.

(*b*) When a certain two-digit number is divided by the number obtained by reversing the digits, the quotient is 2 and the remainder is 7. If the number is divided by the sum of its digits, the quotient is 7 and the remainder 6. Find the number.

15.32 (*a*) Two pounds of coffee and 3 lb of butter cost \$4.20. A month later the price of coffee advanced 10% and that of butter 20%, making the total cost of a similar order \$4.86. Determine the original cost of a pound of each.

(*b*) If 3 gallons of Grade A oil are mixed with 7 gal of Grade B oil the resulting mixture is worth 43 ¢/gal. However, if 3 gal of Grade A oil are mixed with 2 gal of Grade B oil the resulting mixture is worth 46 ¢/gal. Find the price per gallon of each grade.

(*c*) An investor has \$116 annual income from bonds bearing 3% and 5% interest. Then he buys 25% more of the 3% bonds and 40% more of the 5% bonds, thereby increasing his annual income by \$41. Find his initial investment in each type of bond.

15.33 (*a*) Tank A contains 32 gallons of solution which is 25% alcohol by volume. Tank B has 50 gal of solution which is 40% alcohol by volume. What volume should be taken from each tank and combined in order to make up 40 gal of solution containing 30% alcohol by volume?

(*b*) Tank A holds 40 gal of a salt solution containing 80 lb of dissolved salt. Tank B has 120 gal of solution containing 60 lb of dissolved salt. What volume should be taken from each tank and combined in order to make up 30 gal of solution having a salt concentration of 1.5 lb/gal?

(*c*) A given alloy contains 10% zinc and 20% copper. How many pounds of zinc and of copper must be melted with 1000 lb of the given alloy to produce another alloy analyzing 20% zinc and 24% copper? All percentages are by weight.

(*d*) An alloy weighing 600 lb is composed of 100 lb copper and 50 lb tin. Another alloy weighing 1000 lb is composed of 300 lb copper and 150 lb tin. What weights of copper and tin must be melted with the two given alloys to produce a third alloy analyzing 32% copper and 28% tin. All percentages are by weight.

15.34 (*a*) Determine the speed of a motorboat in still water and the speed of the river current, if it takes 3 hr to travel a distance of 45 mi upstream and 2 hr to travel 50 mi downstream.

(*b*) When two cars race around a circular mile track starting from the same place and at the same instant, they pass each other every 18 seconds when traveling in opposite directions and every 90 seconds when traveling in the same direction. Find their speeds in mi/hr.

(*c*) A passenger on the front of train A observes that she passes the complete length of train B in 33 seconds when traveling in the same direction as B and in 3 seconds when traveling in the opposite direction. If B is 330 ft long, find the speeds of the two trains.

15.35 Solve each of the following systems of equations.

(*a*) $\begin{cases} 2x - y + 2z = -8 \\ x + 2y - 3z = 9 \\ 3x - y - 4z = 3 \end{cases}$

(*b*) $\begin{cases} x = y - 2z \\ 2y = x + 3z + 1 \\ z = 2y - 2x - 3 \end{cases}$

(*c*) $\begin{cases} \dfrac{x}{3} + \dfrac{y}{2} - z = 7 \\ \dfrac{x}{4} - \dfrac{3y}{2} + \dfrac{z}{2} = -6 \\ \dfrac{x}{6} - \dfrac{y}{4} - \dfrac{z}{3} = 1 \end{cases}$

(*d*) $\begin{cases} \dfrac{1}{x} + \dfrac{1}{y} + \dfrac{1}{z} = 5 \\ \dfrac{2}{x} - \dfrac{3}{y} - \dfrac{4}{z} = -11 \\ \dfrac{3}{x} + \dfrac{2}{y} - \dfrac{1}{z} = -6 \end{cases}$

15.36 Indicate which of the following systems are (1) consistent, (2) dependent, (3) inconsistent.

(*a*) $\begin{cases} x + y - z = 2 \\ x - 3y + 2z = 1 \\ 3x - 5y + 3z = 4 \end{cases}$ (*b*) $\begin{cases} 2x - y + z = 1 \\ x + 2y - 3z = -2 \\ 3x - 4y + 5z = 1 \end{cases}$ (*c*) $\begin{cases} x + y + 2z = 3 \\ 3x - y + z = 1 \\ 2x + 3y - 4z = 8 \end{cases}$

15.37 The first of three numbers exceeds the third by one-half of the second. The sum of the second and third numbers is one more than the first. If the second is subtracted from the sum of the first and third numbers the result is 5. Determine the numbers.

15.38 When a certain three-digit number is divided by the number with digits reversed, the quotient is 2 and the remainder 25. The tens digit is one less than twice the sum of the hundreds digit and units digit. If the units digit is subtracted from the tens digit, the result is twice the hundreds digit. Find the number.

ANSWERS TO SUPPLEMENTARY PROBLEMS

15.26 (*a*) $x = 2, y = -1$ (*b*) $x = -1, y = 3$ (*c*) $x = 1/2, y = 3/2$

15.27 (*a*) $x = 5/2, y = -1$ (*d*) $x = 5, y = 2$ (*g*) $u = r - 2s, v = 2r + s$
(*b*) $x = 3, y = 2$ (*e*) $x = 3t/2, y = -2t$ (*h*) $x = 1/2, y = 1/3$
(*c*) $x = 6, y = 10$ (*f*) $x = -1, y = 1$ (*i*) $x = a + b, y = a - b$ if $a^2 \neq 2b^2$

15.28 (*a*) Consistent, (*c*) Dependent, (*e*) Consistent,
(*b*) Inconsistent, (*d*) Inconsistent, (*f*) Dependent.

15.29 (*a*) 5, 8 (*b*) 7/12 (*c*) 7, 3 (*d*) 16, 7

15.30 (*a*) Mary 11 yr, Bob 26 yr (*b*) *A* is 22 yr, *B* is 2 yr

15.31 (*a*) 64 (*b*) 83

15.32 (*a*) Coffee 90 ¢/lb, butter 80 ¢/lb (*b*) Grade A 50 ¢/gal, Grade B 40 ¢/gal
(*c*) \$1200 at 3%, \$1600 at 5%

15.33 (*a*) 26 2/3 gal from A, 13 1/3 gal from B (*b*) 20 gal from A, 10 gal from B (*c*) 150 lb zinc, 100 lb copper (*d*) 400 lb copper, 500 lb tin

15.34 (*a*) Boat 20 mi/hr, river 5 mi/hr (*b*) 120 mi/hr, 80 mi/hr (*c*) 60 ft/sec, 50 ft/sec

15.35 (*a*) $x = -1,\ y = 2,\ z = -2$ (*b*) $x = 0,\ y = 2,\ z = 1$ (*c*) $x = 6,\ y = 4,\ z = -3$ (*d*) $x = 1/2,\ y = -1/3,\ z = 1/6$

15.36 (*a*) Dependent (*b*) Inconsistent (*c*) Consistent

15.37 4, 2, 3

15.38 371

Chapter 16

Quadratic Equations in One Variable

16.1 QUADRATIC EQUATIONS

A quadratic equation in the variable x has the form $ax^2 + bx + c = 0$ where a, b, and c are constants and $a \neq 0$.

Thus $x^2 - 6x + 5 = 0$, $2x^2 + x - 6 = 0$, $x^2 + 3x = 0$, and $3x^2 - 5 = 0$ are quadratic equations in one variable. The last two equations may be divided by 2 and 4 respectively to obtain $x^2 + \frac{1}{2}x - 3 = 0$ and $x^2 - \frac{5}{3} = 0$ where the coefficient of x^2 in each case is 1.

An incomplete quadratic equation is one which either $b = 0$ or $c = 0$, e.g., $4x^2 - 5 = 0$, $7x^2 - 2x = 0$, and $3x^2 = 0$.

To solve a quadratic equation $ax^2 + bx + c = 0$ is to find values of x which satisfy the equation. These values of x are called *zeros* or *roots* of the equation.

For example, $x^2 - 5x + 6 = 0$ is satisfied by $x = 2$ and $x = 3$. Then $x = 2$ and $x = 3$ are zeros or roots of the equation.

16.2 METHODS OF SOLVING QUADRATIC EQUATIONS

A. Solution by square root

EXAMPLES 16.1. Solve each quadratic equation for x.

(*a*) $x^2 - 4 = 0$ (*b*) $2x^2 - 21 = 0$ (*c*) $x^2 + 9 = 0$

(*a*) $x^2 - 4 = 0$. Then $x^2 = 4$, $x = \pm 2$, and the roots are $x = 2, -2$.

(*b*) $2x^2 - 21 = 0$. Then $x^2 = 21/2$ and the roots are $x = \pm\sqrt{21/2} = \pm\frac{1}{2}\sqrt{42}$.

(*c*) $x^2 + 9 = 0$. Then $x^2 = -9$ and the roots are $x = \pm\sqrt{-9} = \pm 3i$.

B. Solution by factoring

EXAMPLES 16.2. Solve each quadratic equation for x.

(*a*) $7x^2 - 5x = 0$ (*b*) $x^2 - 5x + 6 = 0$ (*c*) $3x^2 + 2x - 5 = 0$ (*d*) $x^2 - 4x + 4 = 0$

(*a*) $7x^2 - 5x = 0$ may be written as $x(7x - 5) = 0$. Since the product of the two factors is zero, we set each factor equal to 0 and solve the resulting linear equations. $x = 0$ or $7x - 5 = 0$. So $x = 0$ and $x = 5/7$ are the roots of the equation.

(*b*) $x^2 - 5x + 6 = 0$ may be written as $(x - 3)(x - 2) = 0$. Since the product is equal to 0, we set each factor equal to 0 and solve the resulting linear equations. $x - 3 = 0$ or $x - 2 = 0$. So $x = 3$ and $x = 2$ are the roots of the equation.

(*c*) $3x^2 + 2x - 5 = 0$ may be written as $(3x + 5)(x - 1) = 0$. Thus, $3x + 5 = 0$ or $x - 1 = 0$ and the roots of the equation are $x = -5/3$ and $x = 1$.

(*d*) $x^2 - 4x + 4 = 0$ may be written as $(x - 2)(x - 2) = 0$. Thus, $x - 2 = 0$ and the equation has a double root $x = 2$.

C. Solution by completing the square

EXAMPLE 16.3. Solve $x^2 - 6x - 2 = 0$.

Write the unknowns on one side and the constant term on the other; then

$$x^2 - 6x = 2.$$

Add 9 to both sides, thus making the left-hand side a perfect square; then

$$x^2 - 6x + 9 = 2 + 9 \qquad \text{or} \qquad (x-3)^2 = 11.$$

Hence $x - 3 = \pm\sqrt{11}$ and the required roots are $x = 3 \pm \sqrt{11}$.

Note. In the method of completing the square (1) the coefficient of the x^2 term must be 1 and (2) the number added to both sides is the square of half the coefficient of x.

EXAMPLE 16.4. Solve $3x^2 - 5x + 1 = 0$.

Dividing by 3, $$x^2 - \frac{5x}{3} = -\frac{1}{3}.$$

Adding $[\frac{1}{2}(-\frac{5}{3})]^2 = \frac{25}{36}$ to both sides,

$$x^2 - \frac{5}{3}x + \frac{25}{36} = -\frac{1}{3} + \frac{25}{36} = \frac{13}{36}, \qquad \left(x - \frac{5}{6}\right)^2 = \frac{13}{36},$$

$$x - \frac{5}{6} = \pm\frac{\sqrt{13}}{6} \qquad \text{and} \qquad x = \frac{5}{6} \pm \frac{\sqrt{13}}{6}.$$

D. Solution by quadratic formula

The solutions of the quadratic equation $ax^2 + bx + c = 0$ are given by the formula

$$x = \frac{-b \pm \sqrt{b^2 - 4ac}}{2a}$$

where $b^2 - 4ac$ is called the *discriminant* of the quadratic equation.

For a derivation of the quadratic formula see Problem 16.5.

EXAMPLE 16.5. Solve $3x^2 - 5x + 1 = 0$. Here $a = 3$, $b = -5$, $c = 1$ so that

$$x = \frac{-(-5) \pm \sqrt{(-5)^2 - 4(3)(1)}}{2(3)} = \frac{5 \pm \sqrt{13}}{6} \qquad \text{as in Example 16.4.}$$

EXAMPLE 16.6. Solve $4x^2 - 6x + 3 = 0$.

Here $a = 4$, $b = -6$, and $c = 3$.

$$x = \frac{-(-6) \pm \sqrt{(-6)^2 - 4(4)(3)}}{2(4)} = \frac{6 \pm \sqrt{-12}}{8} = \frac{6 \pm 2i\sqrt{3}}{8} = \frac{2(3 \pm i\sqrt{3})}{8}$$

$$x = \frac{3 \pm i\sqrt{3}}{4}$$

E. Graphical solution

The real roots or zeros of $ax^2 + bx + c = 0$ are the values of x corresponding to $y = 0$ on the graph of the parabola $y = ax^2 + bx + c$. Thus the solutions are the abscissas of the points where the parabola intersects the x axis. If the graph does not intersect the x axis the roots are imaginary.

16.3 SUM AND PRODUCT OF THE ROOTS

The sum S and the product P of the roots of the quadratic equation $ax^2 + bx + c = 0$ are given by $S = -b/a$ and $P = c/a$.

Thus in $2x^2 + 7x - 6 = 0$ we have $a = 2$, $b = 7$, $c = -6$ so that $S = -7/2$ and $P = -6/2 = -3$.

It follows that a quadratic equation whose roots are r_1, r_2 is given by $x^2 - Sx + P = 0$ where $S = r_1 + r_2$ and $P = r_1 r_2$. Thus the quadratic equation whose roots are $x = 2$ and $x = -5$ is $x^2 - (2-5)x + 2(-5) = 0$ or $x^2 + 3x - 10 = 0$.

16.4 NATURE OF THE ROOTS

The nature of the roots of the quadratic equation $ax^2 + bx + c = 0$ is determined by the discriminant $b^2 - 4ac$. When the roots involve the imaginary unit i, we say the roots are imaginary.

Assuming a, b, c are *real numbers* then

(1) if $b^2 - 4ac > 0$, the roots are *real* and *unequal*,

(2) if $b^2 - 4ac = 0$, the roots are *real* and *equal*,

(3) if $b^2 - 4ac < 0$, the roots are *imaginary*.

Assuming a, b, c are *rational numbers* then

(1) if $b^2 - 4ac$ is a perfect square $\neq 0$, the roots are real, rational, and unequal,

(2) if $b^2 - 4ac = 0$, the roots are real, rational, and equal,

(3) if $b^2 - 4ac > 0$ but not a perfect square, the roots are real, irrational, and unequal,

(4) if $b^2 - 4ac < 0$, the roots are imaginary.

Thus $2x^2 + 7x - 6 = 0$, with discriminant $b^2 - 4ac = 7^2 - 4(2)(-6) = 97$, has roots which are real, irrational and unequal.

16.5 RADICAL EQUATIONS

A radical equation is an equation having one or more unknowns under a radical.

Thus $\sqrt{x+3} - \sqrt{x} = 1$ and $\sqrt[3]{y} = \sqrt{y-4}$ are radical equations.

To solve a radical equation, isolate one of the radical terms on one side of the equation and transpose all other terms to the other side. If both members of the equation are then raised to a power equal to the index of the isolated radical, the radical will be removed. This process is continued until radicals are no longer present.

EXAMPLE 16.7. Solve $\sqrt{x+3} - \sqrt{x} = 1$.

Transposing, $\sqrt{x+3} = \sqrt{x} + 1$.

Squaring, $x + 3 = x + 2\sqrt{x} + 1$ or $\sqrt{x} = 1$.

Finally, squaring both sides of $\sqrt{x} = 1$ gives $x = 1$.

Check. $\sqrt{1+3} - \sqrt{1} \mathrel{?} 1$, $2 - 1 = 1$.

It is very important to check the values obtained, as this method often introduces extraneous roots which are to be rejected.

16.6 QUADRATIC-TYPE EQUATIONS

An equation of quadratic type has the form $az^{2n} + bz^n + c = 0$ where $a \neq 0$, b, c, and $n \neq 0$ are constants, and where z depends on x. Upon letting $z^n = u$ this equation becomes $au^2 + bu + c = 0$, which may be solved for u. These values of u may be used to obtain z from which it may be possible to obtain x.

EXAMPLE 16.8. Solve $x^4 - 3x^2 - 10 = 0$.

Let $u = x^2$ and substituting	$u^2 - 3u - 10 = 0$
Factoring	$(u-5)(u-2) = 0$
Solving for u	$u = 5$ or $u = -2$
Substituting $x^2 = u$	$x^2 = 5$ or $x^2 = -2$
Solving for x	$x = \pm\sqrt{5}$ or $x = \pm i\sqrt{2}$

EXAMPLE 16.9. Solve $(2x-1)^2 + 7(2x-1) + 12 = 0$.

Let $u = 2x - 1$ and substituting	$u^2 + 7u + 12 = 0$
Factoring	$(u+4)(u+3) = 0$
Solving for u	$u = -4$ or $u = -3$
Substituting $2x - 1 = u$	$2x - 1 = -4$ or $2x - 1 = -3$
Solving for x	$x = -3/2$ or $x = -1$

Solved Problems

16.1 Solve.

(*a*) $x^2 - 16 = 0$. Then $x^2 = 16$, $x = \pm 4$.

(*b*) $4t^2 - 9 = 0$. Then $4t^2 = 9$, $t^2 = 9/4$, $t = \pm 3/2$.

(*c*) $3 - x^2 = 2x^2 + 1$. Then $3x^2 = 2$, $x^2 = 2/3$, $x = \pm\sqrt{2/3} = \pm\frac{1}{3}\sqrt{6}$.

(*d*) $4x^2 + 9 = 0$. Then $x^2 = -9/4$, $x = \pm\sqrt{-9/4} = \pm\frac{3}{2}i$.

(*e*) $\dfrac{2x^2-1}{x-3} = x + 3 + \dfrac{17}{x-3}$. Then $2x^2 - 1 = (x+3)(x-3) + 17$, $2x^2 - 1 = x^2 - 9 + 17$, $x^2 = 9$, and $x = \pm 3$.

Check. If $x = 3$ is substituted into the original equation, we have division by zero which is not allowed. Hence $x = 3$ is not a solution.

If $x = -3$, $\dfrac{2(-3)^2 - 1}{-3-3} \,?\, -3 + 3 + \dfrac{17}{-3-3}$ or $\dfrac{17}{6} = \dfrac{17}{6}$ and $x = -3$ is a solution.

16.2 Solve by factoring.

(*a*) $x^2 + 5x - 6 = 0$, $(x+6)(x-1) = 0$, $x = -6, 1$.

(*b*) $t^2 = 4t$, $t^2 - 4t = 0$, $t(t-4) = 0$, $t = 0, 4$.

(*c*) $x^2 + 3x = 28$, $x^2 + 3x - 28 = 0$, $(x+7)(x-4) = 0$, $x = -7, 4$.

(*d*) $5x - 2x^2 = 2$, $2x^2 - 5x + 2 = 0$, $(2x-1)(x-2) = 0$, $x = 1/2, 2$.

(*e*) $\dfrac{1}{t-1} + \dfrac{1}{t-4} = \dfrac{5}{4}$. Multiplying by $4(t-1)(t-4)$,

$4(t-4) + 4(t-1) = 5(t-1)(t-4)$, $5t^2 - 33t + 40 = 0$, $(t-5)(5t-8) = 0$, $t = 5, 8/5$.

(*f*) $\dfrac{y}{2p} = \dfrac{3p}{6y - 5p}$, $6y^2 - 5py - 6p^2 = 0$, $(3y + 2p)(2y - 3p) = 0$, $y = -2p/3, 3p/2$.

16.3 What term must be added to each of the following expressions in order to make it a perfect square trinomial?

(*a*) $x^2 - 2x$. Add $[\frac{1}{2}(\text{coefficient of } x)]^2 = [\frac{1}{2}(-2)]^2 = 1$. *Check*: $x^2 - 2x + 1 = (x-1)^2$.

(*b*) $x^2 + 4x$. Add $[\frac{1}{2}(\text{coefficient of } x)]^2 = [\frac{1}{2}(4)]^2 = 4$. *Check*: $x^2 + 4x + 4 = (x + 2)^2$.

(*c*) $u^2 + \frac{5}{4}u$. Add $\left[\frac{1}{2}\left(\frac{5}{4}\right)\right]^2 = \frac{25}{64}$. *Check*: $u^2 + \frac{5}{4}u + \frac{25}{64} = \left(u + \frac{5}{8}\right)^2$.

(*d*) $x^4 + px^2$. Add $[\frac{1}{2}(p)]^2 = p^2/4$. *Check*: $x^4 + px^2 + p^2/4 = (x^2 + p/2)^2$.

16.4 Solve by completing the square.

(*a*) $x^2 - 6x + 8 = 0$. Then $x^2 - 6x = -8$, $x^2 - 6x + 9 = -8 + 9$, $(x - 3)^2 = 1$.
Hence $x - 3 = \pm 1$, $x = 3 \pm 1$, and the roots are $x = 4$ and $x = 2$.
Check: For $x = 4$, $4^2 - 6(4) + 8 \,?\, 0$, $0 = 0$. For $x = 2$, $2^2 - 6(2) + 8 \,?\, 0$, $0 = 0$.

(*b*) $t^2 = 4 - 3t$. Then $t^2 + 3t = 4$, $t^2 + 3t + \left(\frac{3}{2}\right)^2 = 4 + \left(\frac{3}{2}\right)^2$, $\left(t + \frac{3}{2}\right)^2 = \frac{25}{4}$.

Hence $t + \frac{3}{2} = \pm\frac{5}{2}$, $t = -\frac{3}{2} \pm \frac{5}{2}$, and the roots are $t = 1, -4$.

(*c*) $3x^2 + 8x + 5 = 0$. Then $x^2 + \frac{8}{3}x = -\frac{5}{3}$, $x^2 + \frac{8}{3}x + \left(\frac{4}{3}\right)^2 = -\frac{5}{3} + \left(\frac{4}{3}\right)^2$, $\left(x + \frac{4}{3}\right)^2 = \frac{1}{9}$.

Hence $x + \frac{4}{3} = \pm\frac{1}{3}$, $x = -\frac{4}{3} \pm \frac{1}{3}$, and the roots are $x = -1, -5/3$.

(*d*) $x^2 + 4x + 1 = 0$. Then $x^2 + 4x = -1$, $x^2 + 4x + 4 = 3$, $(x + 2)^2 = 3$.
Hence $x + 2 = \pm\sqrt{3}$, and roots are $x = -2 \pm \sqrt{3}$.
Check: For $x = -2 + \sqrt{3}$, $(-2 + \sqrt{3})^2 + 4(-2 + \sqrt{3}) + 1 = (4 - 4\sqrt{3} + 3) - 8 + 4\sqrt{3} + 1 = 0$.
For $x = -2 - \sqrt{3}$, $(-2 - \sqrt{3})^2 + 4(-2 - \sqrt{3}) + 1 = (4 + 4\sqrt{3} + 3) - 8 - 4\sqrt{3} + 1 = 0$.

(*e*) $5x^2 - 6x + 5 = 0$. Then $5x^2 - 6x = -5$, $x^2 - \frac{6x}{5} + \left(\frac{3}{5}\right)^2 = -1 + \left(\frac{3}{5}\right)^2$, $\left(x - \frac{3}{5}\right)^2 = -\frac{16}{25}$.

Hence $x - 3/5 = \pm\sqrt{-16/25}$, and the roots are $x = \frac{3}{5} \pm \frac{4}{5}i$.

16.5 Solve the equation $ax^2 + bx + c = 0$, $a \neq 0$, by the method of completing the square.

SOLUTION

Dividing both sides by a, $x^2 + \frac{b}{a}x + \frac{c}{a} = 0$ or $x^2 + \frac{b}{a}x = -\frac{c}{a}$.

Adding $\left[\frac{1}{2}\left(\frac{b}{a}\right)\right]^2 = \frac{b^2}{4a^2}$ to both sides, $x^2 + \frac{b}{a}x + \frac{b^2}{4a^2} = -\frac{c}{a} + \frac{b^2}{4a^2} = \frac{b^2 - 4ac}{4a^2}$.

Then $\left(x + \frac{b}{2a}\right)^2 = \frac{b^2 - 4ac}{4a^2}$, $x + \frac{b}{2a} = \pm\frac{\sqrt{b^2 - 4ac}}{2a}$, and $x = \frac{-b \pm \sqrt{b^2 - 4ac}}{2a}$.

16.6 Solve by the quadratic formula.

(*a*) $x^2 - 3x + 2 = 0$. Here $a = 1$, $b = -3$, $c = 2$. Then

$$x = \frac{-b \pm \sqrt{b^2 - 4ac}}{2a} = \frac{-(-3) \pm \sqrt{(-3)^2 - 4(1)(2)}}{2(1)} = \frac{3 \pm 1}{2} \quad \text{or} \quad x = 1, 2.$$

(*b*) $4t^2 + 12t + 9 = 0$. Here $a = 4$, $b = 12$, $c = 9$. Then

$$t = \frac{-12 \pm \sqrt{(12)^2 - 4(4)(9)}}{2(4)} = \frac{-12 \pm 0}{8} = -\frac{3}{2} \quad \text{and } t = -\frac{3}{2} \text{ is a double root.}$$

(*c*) $9x^2 + 18x - 17 = 0$. Here $a = 9$, $b = 18$, $c = -17$. Then

$$x = \frac{-18 \pm \sqrt{(18)^2 - 4(9)(-17)}}{2(9)} = \frac{-18 \pm \sqrt{936}}{18} = \frac{-18 \pm 6\sqrt{26}}{18} = \frac{-3 \pm \sqrt{26}}{3}.$$

(*d*) $6u(2-u)=7$. Then $6u^2-12u+7=0$ and

$$u=\frac{-(-12)\pm\sqrt{(-12)^2-4(6)(7)}}{2(6)}=\frac{12\pm\sqrt{-24}}{12}=\frac{12\pm2\sqrt{6}i}{12}=1\pm\frac{\sqrt{6}}{6}i.$$

16.7 Solve graphically: (*a*) $2x^2+3x-5=0$, (*b*) $4x^2-12x+9=0$, (*c*) $4x^2-4x+5=0$.

SOLUTION

(*a*) $y=2x^2+3x-5$

x	-3	-2	-1	0	1	2
y	4	-3	-6	-5	0	9

The graph of $y=2x^2+3x-5$ indicates that when $y=0$, $x=1$, and -2.5.
Thus the roots of $2x^2+3x-5=0$ are $x=1$, -2.5 (see Fig. 16-1(a)).

(*b*) $y=4x^2-12x+9$

x	-1	0	1	2	3	4
y	25	9	1	1	9	25

The graph of $y=4x^2-12x+9$ is tangent to the x axis at $x=1.5$, i.e. when $y=0$, $x=1.5$.
Thus $4x^2-12x+9=0$ has the equal roots $x=1.5$ (see Fig. 16-1(b)).

(*c*) $y=4x^2-4x+5$

x	-2	-1	0	1	2	3
y	29	13	5	5	13	29

The graph of $y=4x^2-4x+5$ does not intersect the x axis, i.e., there is no real value of x for which $y=0$.
Hence the roots of $4x^2-4x+5=0$ are imaginary (see Fig. 16-1(c)).
(By the quadratic formula the roots are $x=\frac{1}{2}\pm i$.)

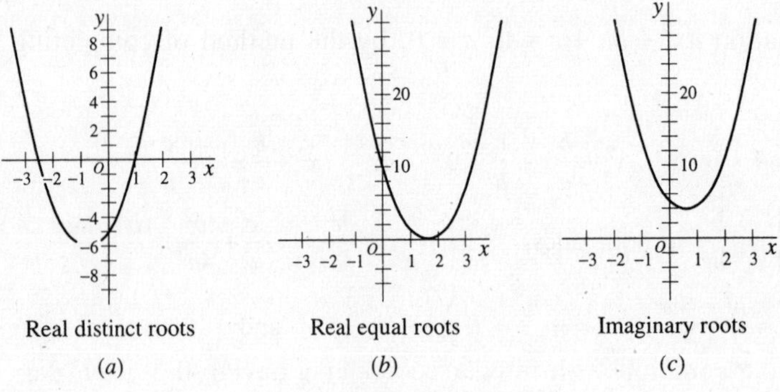

Fig. 16-1

16.8 Prove that the sum S and product P of the roots of the quadratic equation $ax^2+bx+c=0$ are $S=-b/a$ and $P=c/a$.

SOLUTION

By the quadratic formula the roots are

$$\frac{-b+\sqrt{b^2-4ac}}{2a} \quad \text{and} \quad \frac{-b-\sqrt{b^2-4ac}}{2a}.$$

The sum of the roots is

$$S=\frac{-2b}{2a}=-\frac{b}{a}.$$

The product of the roots is

$$P = \left(\frac{-b+\sqrt{b^2-4ac}}{2a}\right)\left(\frac{-b-\sqrt{b^2-4ac}}{2a}\right) = \frac{(-b)^2-(b^2-4ac)}{4a^2} = \frac{c}{a}.$$

16.9 Without solving, find the sum S and product P of the roots.

(*a*) $x^2-7x+6=0$. Here $a=1$, $b=-7$, $c=6$; then $S=-\dfrac{b}{a}=7$, $P=\dfrac{c}{a}=6$.

(*b*) $2x^2+6x-3=0$. Here $a=2$, $b=6$, $c=-3$; then $S=-\dfrac{6}{2}=-3$, $P=\dfrac{-3}{2}$.

(*c*) $x+3x^2+5=0$. Write as $3x^2+x+5=0$. Then $S=-\dfrac{1}{3}$, $P=\dfrac{5}{3}$.

(*d*) $3x^2-5x=0$. Here $a=3$, $b=-5$, $c=0$; then $S=\dfrac{5}{3}$, $P=0$.

(*e*) $2x^2+3=0$. Here $a=2$, $b=0$, $c=3$; then $S=0$, $P=\dfrac{3}{2}$.

(*f*) $mnx^2+(m^2+n^2)x+mn=0$. Then $S=-\dfrac{m^2+n^2}{mn}$, $P=\dfrac{mn}{mn}=1$.

(*g*) $0.3x^2-0.01x+4=0$. Then $S=-\dfrac{-0.01}{0.3}=\dfrac{1}{30}$, $P=\dfrac{4}{0.3}=\dfrac{40}{3}$.

16.10 Find the discriminant b^2-4ac of each of the following equations and thus determine the character of their roots.

(*a*) $x^2-8x+12=0$. $b^2-4ac=(-8)^2-4(1)(12)=16$; the roots are real, rational, unequal.
(*b*) $3y^2+2y-4=0$. $b^2-4ac=52$; the roots are real, irrational, unequal.
(*c*) $2x^2-x+4=0$. $b^2-4ac=-31$; the roots are conjugate imaginaries.
(*d*) $4z^2-12z+9=0$. $b^2-4ac=0$; the roots are real, rational, equal.
(*e*) $2x-4x^2=1$ or $4x^2-2x+1=0$. $b^2-4ac=-12$; the roots are conjugate imaginaries.
(*f*) $\sqrt{2}x^2-4\sqrt{3}x+4\sqrt{2}=0$. Here the coefficients are real but not rational numbers.
$b^2-4ac=16$; the roots are real and unequal.

16.11 Find a quadratic equation with integer coefficients having the given pair of roots. (S = sum of roots, P = product of roots.)

(*a*) $1, 2$
Method 1. $S=1+2=3, P=2$; hence $x^2-3x+2=0$.
Method 2. $(x-1)$ and $(x-2)$ must be factors of the quadratic expression.
Then $(x-1)(x-2)=0$ or $x^2-3x+2=0$.

(*b*) $-3, 2$
Method 1. $S=-1, P=-6$; hence $x^2+x-6=0$.
Method 2. $[x-(-3)]$ and $(x-2)$ are factors of the quadratic expression.
Then $(x+3)(x-2)=0$ or $x^2+x-6=0$.

(*c*) $\dfrac{4}{3}, -\dfrac{3}{5}$. $S=\dfrac{11}{15}, P=-\dfrac{4}{5}$; hence $x^2-\dfrac{11}{15}x-\dfrac{4}{5}=0$ or $15x^2-11x-12=0$.

(*d*) $2+\sqrt{2}, 2-\sqrt{2}$
Method 1. $S=4, P=(2+\sqrt{2})(2-\sqrt{2})=2$; hence $x^2-4x+2=0$.

Method 2. $[x-(2+\sqrt{2})]$ and $[x-(2-\sqrt{2})]$ are factors of the quadratic expression.
Then $[x-(2+\sqrt{2})][x-(2-\sqrt{2})]=[(x-2)-\sqrt{2}][(x-2)+\sqrt{2}]=0$,
$(x-2)^2-2=0$ or $x^2-4x+2=0$.

Method 3. Since $x=2\pm\sqrt{2}$, $x-2=\pm\sqrt{2}$. Squaring, $(x-2)^2=2$ or $x^2-4x+2=0$.

(*e*) $-3+2i, -3-2i$

Method 1. $S=-6, P=(-3+2i)(-3-2i)=13$; hence $x^2+6x+13=0$.

Method 2. $[x-(-3+2i)]$ and $[x-(-3-2i)]$ are factors of the quadratic expression.
Then $[(x+3)-2i][(x+3)+2i]=0$, $(x+3)^2+4=0$ or $x^2+6x+13=0$.

16.12 In each quadratic equation find the value of the constant p subject to the given condition.

(*a*) $2x^2-px+4=0$ has one root equal to -3.
Since $x=-3$ is a root, it must satisfy the given equation.
Then $2(-3)^2-p(-3)+4=0$ and $p=-22/3$.

(*b*) $(p+2)x^2+5x+2p=0$ has the product of its roots equal to 2/3.
The product of the roots is

$$\frac{2p}{p+2};\quad \text{then}\quad \frac{2p}{p+2}=\frac{2}{3}\quad \text{and}\quad p=1.$$

(*c*) $2px^2+px+2x=x^2+7p+1$ has the sum of its roots equal to $-4/3$.
Write the equation in the form $(2p-1)x^2+(p+2)x-(7p+1)=0$.
Then the sum of the roots is

$$-\frac{p+2}{2p-1}=-\frac{4}{3}\quad \text{and}\quad p=2.$$

(*d*) $3x^2+(p+1)+24=0$ has one root equal to twice the other. Let the roots be r, $2r$.
Product of the roots is $r(2r)=8$; then $r^2=4$ and $r=\pm 2$.
Sum of the roots is $3r=-(p+1)/3$. Substitute $r=2$ and $r=-2$ into this equation and obtain $p=-19$ and $p=17$ respectively.

(*e*) $2x^2-12x+p+2=0$ has the difference between its roots equal to 2.
Let the roots be r, s; then (1) $r-s=2$. The sum of the roots is 6; then (2) $r+s=6$. The simultaneous solution of (1) and (2) is $r=4$, $s=2$.
Now put $x=2$ or $x=4$ into the given equation and obtain $p=14$.

16.13 Find the roots of each quadratic equation subject to the given conditions.

(*a*) $(2k+2)x^2+(4-4k)x+k-2=0$ has roots which are reciprocals of each other.
Let r and $1/r$ be the roots, their product being 1.

Product of roots is $\dfrac{k-2}{2k+2}=1$, from which $k=-4$.

Put $k=-4$ into the given equation; then $3x^2-10x+3=0$ and the roots are $1/3, 3$.

(*b*) $kx^2-(1+k)x+3k+2=0$ has the sum of its roots equal to twice the product of its roots.
Sum of roots = 2(product of roots); then

$$\frac{1+k}{k}=2\left(\frac{3k+2}{k}\right)\quad \text{and}\quad k=-\frac{3}{5}.$$

Put $k=-3/5$ into the given equation; then $3x^2+2x-1=0$ and the roots are $-1, 1/3$.

(*c*) $(x+k)^2=2-3k$ has equal roots.
Write the equation as $x^2+2kx+(k^2+3k-2)=0$ where $a=1$, $b=2k$, $c=k^2+3k-2$. The roots are equal if the discriminant $(b^2-4ac)=0$.
Then from $b^2-4ac=(2k)^2-4(1)(k^2+3k-2)=0$ we get $k=2/3$.
Put $k=2/3$ into the given equation and solve to obtain the double root $-2/3$.

16.14 Solve.

(*a*) $\sqrt{2x+1}=3$. Squaring both sides, $2x+1=9$ and $x=4$.
Check. $\sqrt{2(4)+1}\ ?\ 3$, $3=3$.

(*b*) $\sqrt{5+2x}=x+1$. Squaring both sides, $5+2x=x^2+2x+1$, $x^2=4$ and $x=\pm 2$.
Check. For $x=2$, $\sqrt{5+2(2)}\ ?\ 2+1$ or $3=3$.
For $x=-2$, $\sqrt{5+2(-2)}\ ?\ -2+1$ or $\sqrt{1}=-1$ which is not true since $\sqrt{1}=1$.
Thus $x=2$ is the only solution; $x=-2$ is an extraneous root.

(*c*) $\sqrt{3x-5}=x-1$. Squaring, $3x-5=x^2-2x+1$, $x^2-5x+6=0$ and $x=3,2$.
Check. For $x=3$, $\sqrt{3(3)-5}\ ?\ 3-1$ or $2=2$. For $x=2$, $\sqrt{3(2)-5}\ ?\ 2-1$ or $1=1$.
Thus both $x=3$ and $x=2$ are solutions of the given equation.

(*d*) $\sqrt[3]{x^2-x+6}-2=0$. Then $\sqrt[3]{x^2-x+6}=2$, $x^2-x+6=8$, $x^2-x-2=0$ and $x=2,-1$.
Check. For $x=2$, $\sqrt[3]{2^2-2+6}-2\ ?\ 0$ or $2-2=0$.
For $x=-1$, $\sqrt[3]{(-1)^2-(-1)+6}-2\ ?\ 0$ or $2-2=0$.

16.15 Solve.

(*a*) $\sqrt{2x+1}-\sqrt{x}=1$. Rearranging, (1) $\sqrt{2x+1}=\sqrt{x}+1$.
Squaring both sides of (1), $2x+1=x+2\sqrt{x}+1$ or (2) $x=2\sqrt{x}$.
Squaring (2), $x^2=4x$; then $x(x-4)=0$ and $x=0,4$.
Check. For $x=0$, $\sqrt{2(0)+1}-\sqrt{0}\ ?\ 1$, $1=1$. For $x=4$, $\sqrt{2(4)+1}-\sqrt{4}\ ?\ 1$, $1=1$.

(*b*) $\sqrt{4x-1}+\sqrt{2x+3}=1$. Rearranging, (1) $\sqrt{4x-1}=1-\sqrt{2x+3}$.
Squaring (1), $4x-1=1-2\sqrt{2x+3}+2x+3$ or (2) $2\sqrt{2x+3}=5-2x$.
Squaring (2), $4(2x+3)=25-20x+4x^2$, $4x^2-28x+13=0$ and $x=1/2$, $13/2$.
Check. For $x=1/2$, $\sqrt{4(1/2)-1}+\sqrt{2(1/2)+3}\ ?\ 1$ or $3=1$ which is not true.
For $x=13/2$, $\sqrt{4(13/2)-1}+\sqrt{2(13/2)+3}\ ?\ 1$ or $9=1$ which is not true.
Hence $x=1/2$ and $x=13/2$ are extraneous roots; the equation has no solution.

(*c*) $\sqrt{\sqrt{x+16}-\sqrt{x}}=2$. Squaring, $\sqrt{x+16}-\sqrt{x}=4$ or (1) $\sqrt{x+16}=\sqrt{x}+4$.
Squaring (1), $x+16=x+8\sqrt{x}+16$, $8\sqrt{x}=0$, and $x=0$ is a solution.

16.16 Solve.

(*a*) $\sqrt{x^2+6x}=x+\sqrt{2x}$.
Squaring, $x^2+6x=x^2+2x\sqrt{2x}+2x$, $2x\sqrt{2x}=4x$, $x(\sqrt{2x}-2)=0$.
Then $x=0$; and from $\sqrt{2x}-2=0$, $\sqrt{2x}=2$, $2x=4$, $x=2$.
Both $x=0$ and $x=2$ satisfy the given equation.

(*b*) $\sqrt{x}-\dfrac{2}{\sqrt{x}}=1$. Multiply by $\sqrt{x}$ and obtain (1) $x-2=\sqrt{x}$.
Squaring (1), $x^2-4x+4=x$, $x^2-5x+4=0$, $(x-1)(x-4)=0$, and $x=1,4$.
Only $x=4$ satisfies the given equation; $x=1$ is extraneous.

16.17 Solve the equation $x^2-6x-\sqrt{x^2-6x-3}=5$.

SOLUTION

Let $x^2-6x=u$; then $u-\sqrt{u-3}=5$ or (1) $\sqrt{u-3}=u-5$.
Squaring (1), $u-3=u^2-10u+25$, $u^2-11u+28=0$, and $u=7,4$.
Since only $u=7$ satisfies (1), substitute $u=7$ in $x^2-6x=u$ and obtain

$$x^2-6x-7=0, \qquad (x-7)(x+1)=0, \qquad \text{and} \qquad x=7,-1.$$

Both $x=7$ and $x=-1$ satisfy the original equation and are thus solutions.

Note. If we write the given equation as $\sqrt{x^2-6x-3}=x^2-6x-5$ and square both sides, the resulting fourth degree equation would be difficult to solve.

16.18 Solve the equation

$$\frac{4-x}{\sqrt{x^2-8x+32}}=\frac{3}{5}.$$

SOLUTION

Squaring,

$$\frac{16-8x+x^2}{x^2-8x+32}=\frac{9}{25};$$

then $25(16-8x+x^2)=9(x^2-8x+32)$, $x^2-8x+7=0$, and $x=7,1$. The only solution is $x=1$; reject $x=7$, an extraneous solution.

16.19 Solve.

(*a*) $x^4-10x^2+9=0$. Let $x^2=u$; then $u^2-10u+9=0$ and $u=1,9$.
For $u=1$, $x^2=1$ and $x=\pm1$; for $u=9$, $x^2=9$ and $x=\pm3$.
The four solutions are $x=\pm1,\pm3$; each satisfies the given equation.

(*b*) $2x^4+x^2-1=0$. Let $x^2=u$; then $2u^2+u-1=0$ and $u=\frac{1}{2},-1$.
If $u=\frac{1}{2}$, $x^2=\frac{1}{2}$ and $x=\pm\frac{1}{2}\sqrt{2}$; if $u=-1$, $x^2=-1$ and $x=\pm i$.
The four solutions are $x=\pm\frac{1}{2}\sqrt{2}$, $\pm i$.

(*c*) $\sqrt{x}-\sqrt[4]{x}-2=0$. Let $\sqrt[4]{x}=u$; then $u^2-u-2=0$ and $u=2,-1$.
If $u=2$, $\sqrt[4]{x}=2$ and $x=2^4=16$. Since $\sqrt[4]{x}$ is positive, it cannot equal -1.
Hence $x=16$ is the only solution of the given equation.

(*d*) $2\left(x+\frac{1}{x}\right)^2-7\left(x+\frac{1}{x}\right)+5=0$.

Let $x+\frac{1}{x}=u$; then $2u^2-7u+5=0$ and $u=5/2,1$.

For $u=\frac{5}{2}$, $x+\frac{1}{x}=\frac{5}{2}$, $2x^2-5x+2=0$ and $x=2,\frac{1}{2}$.

For $u=1$, $x+\frac{1}{x}=1$, $x^2-x+1=0$ and $x=\frac{1}{2}\pm\frac{1}{2}\sqrt{3}i$.

The four solutions are $x=2,\ \frac{1}{2},\ \frac{1}{2}\pm\frac{1}{2}\sqrt{3}i$.

(*e*) $9(x+2)^{-4}+17(x+2)^{-2}-2=0$. Let $(x+2)^{-2}=u$; then $9u^2+17u-2=0$ and $u=1/9,-2$.
If $(x+2)^{-2}=1/9$, $(x+2)^2=9$, $(x+2)=\pm3$ and $x=1,-5$.
If $(x+2)^{-2}=-2$, $(x+2)^2=-\frac{1}{2}$, $(x+2)=\pm\frac{1}{2}\sqrt{2}i$ and $x=-2\pm\frac{1}{2}\sqrt{2}i$.
The four solutions are $x=1,\ -5,\ -2\pm\frac{1}{2}\sqrt{2}i$.

16.20 Find the values of x which satisfy each of the following equations.

(*a*) $16\left(\frac{x}{x+1}\right)^4-25\left(\frac{x}{x+1}\right)^2+9=0$.

Let $\left(\frac{x}{x+1}\right)^2=u$; then $16u^2-25u+9=0$ and $u=1,\ 9/16$.

If $u=1$, $\left(\frac{x}{x+1}\right)^2=1$ or $\frac{x}{x+1}=\pm1$.

The equation $$\frac{x}{x+1}=1$$

has no solution; the equation $$\frac{x}{x+1}=-1$$

has solution $x=-1/2$.

If $u = 9/16$,

$$\left(\frac{x}{x+1}\right)^2 = \frac{9}{16} \quad \text{or} \quad \frac{x}{x+1} = \pm\frac{3}{4} \text{ so that } x = 3, -3/7.$$

The required solutions are $x = -1/2, -3/7, 3$.

(*b*) $(x^2+3x+2)^2 - 8(x^2+3x) = 4$. Let $x^2+3x = u$; then $(u+2)^2 - 8u = 4$ and $u = 0, 4$.
If $u = 0$, $x^2+3x = 0$ and $x = 0, -3$; if $u = 4$, $x^2+3x = 4$ and $x = -4, 1$.
The solutions are $x = -4, -3, 0, 1$.

16.21 One positive number exceeds three times another positive number by 5. The product of the two numbers is 68. Find the numbers.

SOLUTION

Let x = smaller number; then $3x+5$ = larger number.

Then $x(3x+5) = 68$, $3x^2+5x-68 = 0$, $(3x+17)(x-4) = 0$, and $x = 4, -17/3$.
We exclude $-17/3$ since the problem states that the numbers are positive.
The required numbers are $x = 4$ and $3x+5 = 17$.

16.22 When three times a certain number is added to twice its reciprocal, the result is 5. Find the number.

SOLUTION

Let x = the required number and $1/x$ = its reciprocal.

Then $3x + 2(1/x) = 5$, $3x^2 - 5x + 2 = 0$, $(3x-2)(x-1) = 0$, and $x = 1, 2/3$.
Check. For $x = 1$, $3(1) + 2(1/1) = 5$; for $x = 2/3$, $3(2/3) + 2(3/2) = 5$.

16.23 Determine the dimensions of a rectangle having perimeter 50 feet and area 150 square feet.

SOLUTION

Sum of four sides = 50 ft; hence, sum of two adjacent sides = 25 ft (see Fig. 16-2). Let x and $25-x$ be the lengths of two adjacent sides.

The area is $x(25-x) = 150$; then $x^2 - 25x + 150 = 0$, $(x-10)(x-15) = 0$, and $x = 10, 15$.
Then $25 - x = 15, 10$; and the rectangle has dimensions 10 ft by 15 ft.

16.24 The hypotenuse of a right triangle is 34 inches. Find the lengths of the two legs if one leg is 14 inches longer than the other.

SOLUTION

Let x and $x+14$ be the lengths of the legs (see Fig. 16-3).

Then $x^2 + (x+14)^2 = (34)^2$, $x^2 + 14x - 480 = 0$, $(x+30)(x-16) = 0$, and $x = -30, 16$.
Since $x = -30$ has no physical significance, we have $x = 16$ in. and $x + 14 = 30$ in.

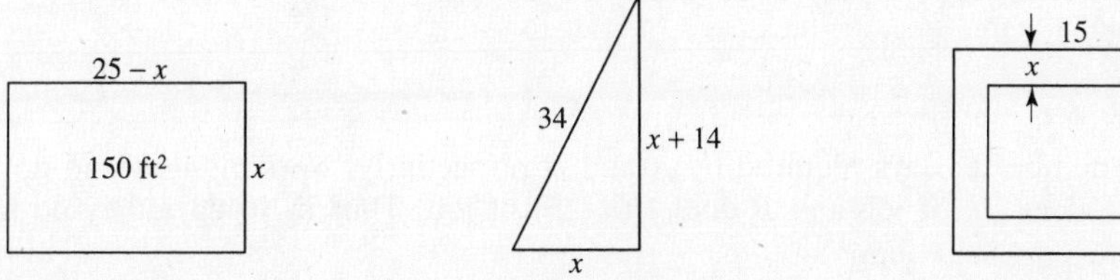

Fig. 16-2 Fig. 16-3 Fig. 16-4

16.25 A picture frame of uniform width has outer dimensions 12 in. by 15 in. Find the width of the frame (*a*) if 88 square inches of picture shows, (*b*) if 100 square inches of picture shows.

SOLUTION

Let x = width of frame; then the dimensions of the picture are $(15-2x)$, $(12-2x)$ (see Fig. 16-4).

(*a*) Area of picture $= (15-2x)(12-2x) = 88$; then $2x^2 - 27x + 46 = 0$, $(x-2)(2x-23) = 0$, and $x = 2, 11\frac{1}{2}$. Clearly, the width cannot be $11\frac{1}{2}$ in. Hence the width of the frame is 2 in.
Check. The area of the picture is $(15-4)(12-4) = 88\text{ in.}^2$, as given.

(*b*) Here $(15-2x)(12-2x) = 100$, $2x^2 - 27x + 40 = 0$ and, by the quadratic formula,

$$x = \frac{-b \pm \sqrt{b^2-4ac}}{2a} = \frac{27 \pm \sqrt{409}}{4} \qquad \text{or} \qquad x = 11.8, 1.7 \text{ (approximately)}.$$

Reject $x = 11.8$ in., which cannot be the width. The required width is 1.7 in.

16.26 A pilot flies a distance of 600 miles. He could fly the same distance in 30 minutes less time by increasing his average speed by 40 mi/hr. Find his actual average speed.

SOLUTION

Let x = actual average speed in mi/hr.

$$\text{Time in hours} = \frac{\text{distance in mi}}{\text{speed in mi/hr}}.$$

Time to fly 600 mi at x mi/hr − time to fly 600 mi at $(x+40)$ mi/hr $= \frac{1}{2}$ hr.

Then
$$\frac{600}{x} - \frac{600}{x+40} = \frac{1}{2}.$$

Solving, the required speed is $x = 200$ mi/hr.

16.27 A retailer bought a number of shirts for \$180 and sold all but 6 at a profit of \$2 per shirt. With the total amount received she could buy 30 more shirts than before. Find the cost per shirt.

SOLUTION

Let x = cost per shirt in dollars; $180/x$ = number of shirts bought.

Then
$$\left(\frac{180}{x} - 6\right)(x+2) = x\left(\frac{180}{x} + 30\right).$$

Solving, $x = \$3$ per shirt.

16.28 *A* and *B* working together can do a job in 10 days. It takes *A* 5 days longer than *B* to do the job when each works alone. How many days would it take each of them, working alone, to do the job?

SOLUTION

Let n, $n-5$ = number of days required by *A* and *B* respectively, working alone, to do the job.
In 1 day *A* does $1/n$ of job and *B* does $1/(n-5)$ of job. Thus in 10 days they do together

$$10\left(\frac{1}{n} + \frac{1}{n-5}\right) = 1 \text{ complete job.}$$

Then $10(2n-5) = n(n-5)$, $n^2 - 25n + 50 = 0$, and

$$n = \frac{25 \pm \sqrt{625 - 200}}{2} = 22.8, 2.2.$$

Rejecting $n = 2.2$, the required solution is $n = 22.8$ days, $n - 5 = 17.8$ days.

16.29 A ball projected vertically upward with initial speed v_0 ft/sec is at time t sec at a distance s ft from the point of projection as given by the formula $s = v_0 t - 16t^2$. If the ball is given an initial upward speed of 128 ft/sec, at what times would it be 100 ft above the point of projection?

SOLUTION

$$s = v_0 t - 16t^2, \quad 100 = 128t - 16t^2, \quad 4t^2 - 32t + 25 = 0, \quad \text{and} \quad t = \frac{32 \pm \sqrt{624}}{8} = 7.12, 0.88.$$

At $t = 0.88$ sec, $s = 100$ ft and the ball is rising; at $t = 7.12$ sec, $s = 100$ ft and the ball is falling. This is seen from the graph of s plotted against t (see Fig. 16-5).

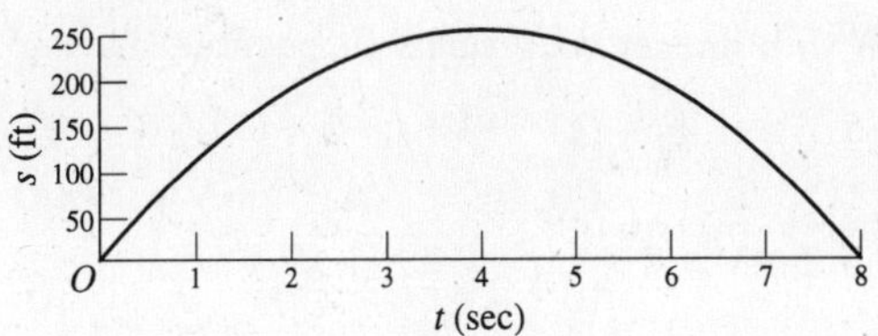

Fig. 16-5

Supplementary Problems

16.30 Solve each equation.

(a) $x^2 - 40 = 9$
(b) $2x^2 - 400 = 0$
(c) $x^2 + 36 = 9 - 2x^2$
(d) $\dfrac{x}{16} = \dfrac{4}{x}$
(e) $\dfrac{y^2}{3} = \dfrac{y^2}{6} + 2$
(f) $\dfrac{1-2x}{3-x} = \dfrac{x-2}{3x-1}$
(g) $\dfrac{1}{2x-1} - \dfrac{1}{2x+1} = \dfrac{1}{4}$
(h) $x - \dfrac{2x}{x+1} = \dfrac{5}{x+1} - 1$

16.31 Solve each equation by factoring.

(a) $x^2 - 7x = -12$
(b) $x^2 + x = 6$
(c) $x^2 = 5x + 24$
(d) $2x^2 + 2 = 5x$
(e) $9x^2 = 9x - 2$
(f) $4x - 5x^2 = -12$
(g) $\dfrac{x}{2a} = \dfrac{4a}{x+2a}$
(h) $\dfrac{1}{4-x} - \dfrac{1}{2+x} = \dfrac{1}{4}$
(i) $\dfrac{2x-1}{x+2} + \dfrac{x+2}{2x-1} = \dfrac{10}{3}$
(j) $\dfrac{2c-3y}{y-c} - \dfrac{y}{2y-c} = \dfrac{2}{3}$

16.32 Solve each equation by completing the square.

(a) $x^2 + 4x - 5 = 0$
(b) $x(x-3) = 4$
(c) $2x^2 = x + 1$
(d) $3x^2 - 2 = 5x$
(e) $4x^2 = 12x - 7$
(f) $6y^2 = 19y - 15$
(g) $2x^2 + 3a^2 = 7ax$
(h) $12x - 9x^2 = 5$

16.33 Solve each equation by the quadratic formula.

(a) $x^2 - 5x = 6$
(b) $x^2 - 6 = x$
(c) $3x^2 - 2x = 8$
(d) $16x^2 - 8x + 1 = 0$
(e) $x(5x - 4) = 2$
(f) $9x^2 + 6x = -4$
(g) $\dfrac{5x^2 - 2p^2}{x} = \dfrac{p}{3}$
(h) $\dfrac{2x+3}{4x-1} = \dfrac{3x-2}{3x+2}$

16.34 Solve each equation graphically.

(*a*) $2x^2+x-3=0$ (*c*) $x^2-2x=2$ (*e*) $6x^2-7x-5=0$

(*b*) $4x^2-8x+4=0$ (*d*) $2x^2+2=3x$ (*f*) $2x^2+8x+3=0$

16.35 Without solving, find the sum S and product P of the roots of each equation.

(*a*) $2x^2+3x+1=0$ (*d*) $2x^2+6x-5=0$ (*g*) $2x^2+5kx+3k^2=0$

(*b*) $x-x^2=2$ (*e*) $3x^2-4=0$ (*h*) $0.2x^2-0.1x+0.03=0$

(*c*) $2x(x+3)=1$ (*f*) $4x^2+3x=0$ (*i*) $\sqrt{2}x^2-\sqrt{3}x+1=0$

16.36 Find the discriminant b^2-4ac and thus determine the character of the roots.

(*a*) $2x^2-7x+4=0$ (*c*) $3x-x^2=4$ (*e*) $2x^2=5+3x$ (*g*) $1+2x+2x^2=0$

(*b*) $3x^2=5x-2$ (*d*) $x(4x+3)=5$ (*f*) $4x\sqrt{3}=4x^2+3$ (*h*) $3x+25/3x=10$

16.37 Find a quadratic equation with integer coefficients (if possible) having the given roots.

(*a*) $2,-3$ (*d*) $-2,-5$ (*g*) $-1+i,-1-i$ (*j*) $\sqrt{3}-\sqrt{2},\sqrt{3}+\sqrt{2}$

(*b*) $-3,0$ (*e*) $-1/3,1/2$ (*h*) $-2-\sqrt{6},-2+\sqrt{6}$ (*k*) $a+bi,a-bi$ a,b integers

(*c*) $8,-4$ (*f*) $2+\sqrt{3},2-\sqrt{3}$ (*i*) $2+\frac{3}{2}i,2-\frac{3}{2}i$ (*l*) $\frac{m+\sqrt{n}}{2},\frac{m-\sqrt{n}}{2}$ m,n integers

16.38 In each quadratic equation, evaluate the constant p subject to the given condition.

(*a*) $px^2-x+5-3p=0$ has one root equal to 2.

(*b*) $(2p+1)x^2+px+p=4(px+2)$ has the sum of its roots equal to the product of its roots.

(*c*) $3x^2+p(x-2)+1=0$ has roots which are reciprocals.

(*d*) $4x^2-8x+2p-1=0$ has one root equal to three times the other.

(*e*) $4x^2-20x+p^2-4=0$ has one root equal to two more than the other.

(*f*) $x^2=5x-3p+3$ has the difference between its roots equal to 11.

16.39 Find the roots of each equation subject to the given condition.

(*a*) $2px^2-4px+5p=3x^2+x-8$ has the product of its roots equal to twice their sum.

(*b*) $x^2-3(x-p)-2=0$ has one root equal to 3 less than twice the other root.

(*c*) $p(x^2+3x-9)=x-x^2$ has one root equal to the negative of the other.

(*d*) $(m+3)x^2+2m(x+1)+3=0$ has one root equal to half the reciprocal of the other.

(*e*) $(2m+1)x^2-4mx=1-3m$ has equal roots.

16.40 Solve each equation.

(*a*) $\sqrt{x^2-x+2}=2$ (*e*) $\sqrt{2x+7}=\sqrt{x}+2$ (*i*) $\sqrt{x^2-\sqrt{2x+1}}=2-x$

(*b*) $\sqrt{2x-2}=x-1$ (*f*) $\sqrt{2x^2-7}-x=3$ (*j*) $\sqrt{2x-10}+\sqrt{x+9}=2$

(*c*) $\sqrt{4x+1}=3-3x$ (*g*) $\sqrt{2+x}-4+\sqrt{10-3x}=0$ (*k*) $\sqrt{2x+8}+\sqrt{2x+5}=\sqrt{8x+25}$

(*d*) $2-\sqrt[3]{x^2+2x}=0$ (*h*) $2\sqrt{x}-\sqrt{4x-3}=\frac{1}{\sqrt{4x-3}}$ (*l*) $\sqrt[3]{2x-1}=\sqrt[6]{x+1}$

16.41 Solve each equation.

(*a*) $x^4 - 13x^2 + 36 = 0$

(*b*) $x^4 - 3x^2 - 10 = 0$

(*c*) $4x^{-4} - 17x^{-2} + 4 = 0$

(*d*) $x^{-4/3} - 5x^{-2/3} + 4 = 0$

(*e*) $(x^2 - 6x)^2 - 2(x^2 - 6x) = 35$

(*f*) $x^2 + x = 7\sqrt{x^2 + x + 2} - 12$

(*g*) $\left(x + \frac{1}{x}\right)^2 - \frac{7}{2}\left(x + \frac{1}{x}\right) = 2$

(*h*) $\sqrt{x+2} - \sqrt[4]{x+2} = 6$

(*i*) $x^3 - 7x^{3/2} - 8 = 0$

(*j*) $\frac{x^2+2}{x} + \frac{8x}{x^2+2} = 6$

16.42 (*a*) The sum of the squares of two numbers is 34, the first number being one less than twice the second number. Determine the numbers.

(*b*) The sum of the squares of three consecutive integers is 110. Find the numbers.

(*c*) The difference between two positive numbers is 3, and the sum of their reciprocals is 1/2. Determine the numbers.

(*d*) A number exceeds twice its square root by 3. Find the number.

16.43 (*a*) The length of a rectangle is three times its width. If the width is diminished by 1 ft and the length increased by 3 ft, the area will be 72 ft^2. Find the dimensions of the original rectangle.

(*b*) A piece of wire 60 in. long is bent into the form of a right triangle having hypotenuse 25 in. Find the other two sides of the triangle.

(*c*) A picture 8 in. by 12 in. is placed in a frame which has uniform width. If the area of the frame equals the area of the picture, find the width of the frame.

(*d*) A rectangular piece of tin 9 in. by 12 in. is to be made into an open box with base area 60 in.^{2} by cutting out equal squares from the four corners and then bending up the edges. Find to the nearest tenth of an inch the length of the side of the square cut from each corner.

16.44 (*a*) The tens digit of a certain two-digit number is twice the units digit. If the number is multiplied by the sum of its digits, the product is 63. Find the number.

(*b*) Find a number consisting of two digits such that the tens digit exceeds the units digit by 3 and the number is 4 less than the sum of the squares of its digits.

16.45 (*a*) Two men start at the same time from the same place and travel along roads that are at right angles to each other. One man travels 4 mi/hr faster than the other, and at the end of 2 hours they are 40 miles apart. Determine their rates of travel.

(*b*) By increasing her average speed by 10 mi/hr a motorist could save 36 minutes in traveling a distance of 120 miles. Find her actual average speed.

(*c*) A woman travels 36 miles down a river and back in 8 hours. If the rate of her boat in still water is 12 mi/hr, what is the rate of the river current?

16.46 (*a*) A merchant purchased a number of coats, each at the same price, for a total of \$720. He sold them at \$40 each, thus realizing a profit equal to his cost of 8 coats. How many did he buy?

(*b*) A grocer bought a number of cans of corn for \$14.40. Later the price increased 2 cents a can and as a result she received 24 fewer cans for the same amount of money. How many cans were in his first purchase and what was the cost per can?

16.47 (*a*) It takes *B* 6 hours longer than *A* to assemble a machine. Together they can do it in 4 hr. How long would it take each working alone to do the job?

(*b*) Pipe *A* can fill a given tank in 4 hr. If pipe *B* works alone, it takes 3 hr longer to fill the tank than if pipes *A* and *B* act together. How long will it take pipe *B* working alone?

16.48 A ball projected vertically upward is distant *s* ft from the point of projection after *t* seconds, where $s = 64t - 16t^2$.

(*a*) At what times will the ball be 40 ft above the ground?

(*b*) Will the ball ever be 80 ft above the ground?

(*c*) What is the maximum height reached?

ANSWERS TO SUPPLEMENTARY PROBLEMS

16.30 (*a*) $x = \pm 7$ (*c*) $x = \pm 3i$ (*e*) $y = \pm 2\sqrt{3}$ (*g*) $x = \pm 3/2$
(*b*) $x = \pm 10\sqrt{2}$ (*d*) $x = \pm 8$ (*f*) $x = \pm 1$ (*h*) $x = \pm 2$

16.31 (*a*) $3, 4$ (*c*) $8, -3$ (*e*) $1/3, 2/3$ (*g*) $2a, -4a$ (*i*) $1, -7$
(*b*) $2, -3$ (*d*) $2, 1/2$ (*f*) $2, -6/5$ (*h*) $2, -8$ (*j*) $2c/5, 4c/5$

16.32 (*a*) $1, -5$ (*c*) $1, -1/2$ (*e*) $\dfrac{3 \pm \sqrt{2}}{2}$ (*f*) $3/2, 5/3$ (*h*) $\dfrac{2}{3} \pm \dfrac{i}{3}$
(*b*) $4, -1$ (*d*) $2, -1/3$ (*g*) $3a, a/2$

16.33 (*a*) $6, -1$ (*c*) $2, -4/3$ (*e*) $\dfrac{2 \pm \sqrt{14}}{5}$ (*f*) $\dfrac{-1 \pm i\sqrt{3}}{3}$ (*g*) $\dfrac{2p}{3}, -\dfrac{3p}{5}$ (*h*) $\dfrac{6 \pm \sqrt{42}}{3}$
(*b*) $3, -2$ (*d*) $1/4, 1/4$

16.34 (*a*) $x = -3/2$ and $x = 1$ (see Fig. 16-6).
(*b*) Double root of $x = 1$ (see Fig. 16-7).
(*c*) Real zeros between -1 and 0 and between 2 and 3 (see Fig. 16-8).
(*d*) No real zeros (see Fig. 16-9).

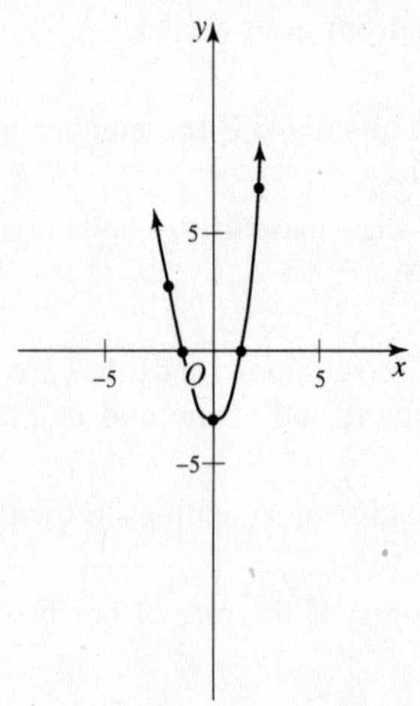

Fig. 16-6

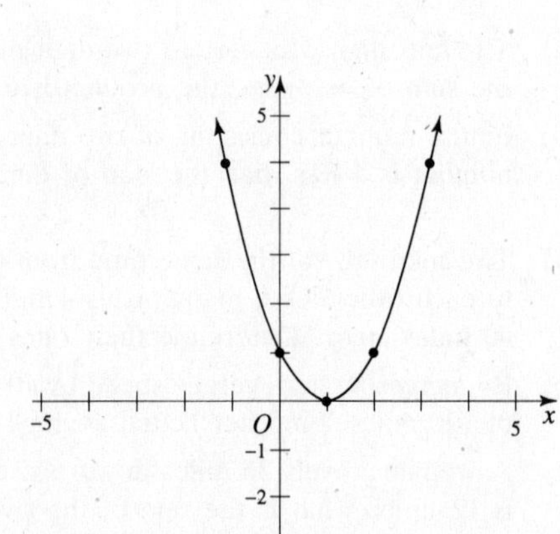

Fig. 16-7

(*e*) Real zeros between -1 and 0 and between 1 and 2 (see Fig. 16-10).
(*f*) Real zeros between -4 and -3 and between -1 and 0 (see Fig. 16-11).

16.35 (*a*) $S = -3/2, P = 1/2$ (*d*) $S = -3, P = -5/2$ (*g*) $S = -5k/2, P = 3k^2/2$
(*b*) $S = 1, P = 2$ (*e*) $S = 0, P = -4/3$ (*h*) $S = 0.5, P = 0.15$
(*c*) $S = -3, P = -1/2$ (*f*) $S = -3/4, P = 0$ (*i*) $S = \frac{1}{2}\sqrt{6}, P = \frac{1}{2}\sqrt{2}$

16.36 (*a*) 17; real, irrational, unequal (*e*) 49; real, rational, unequal
(*b*) 1; real, rational, unequal (*f*) 0; real, equal
(*c*) -7; imaginary (*g*) -4; imaginary
(*d*) 89; real, irrational, unequal (*h*) 0; real, rational, equal

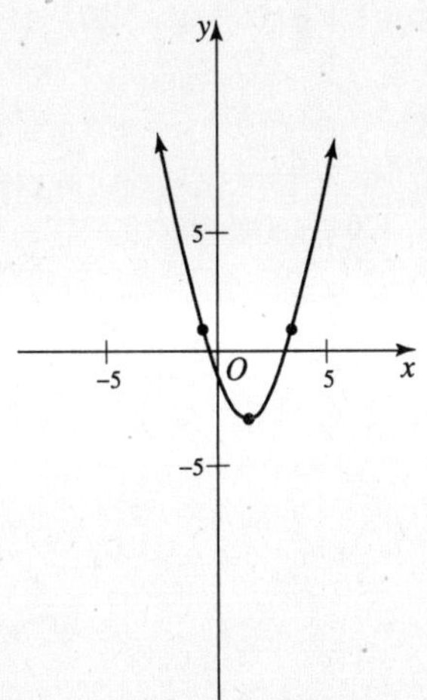

Fig. 16-8

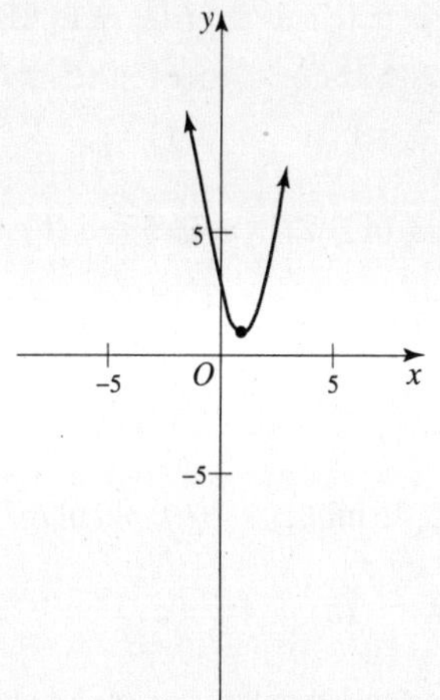

Fig. 16-9

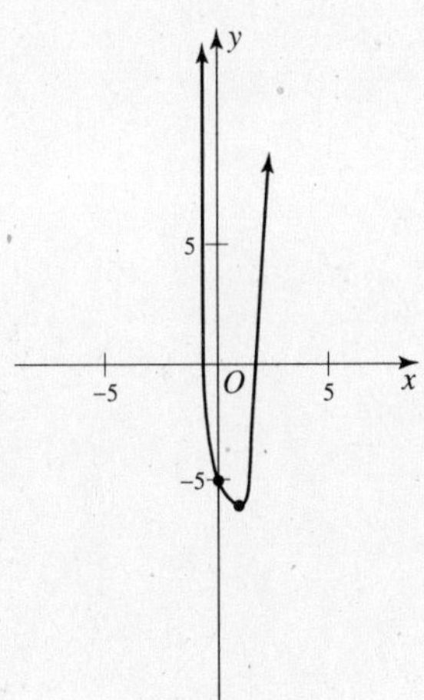

Fig. 16-10

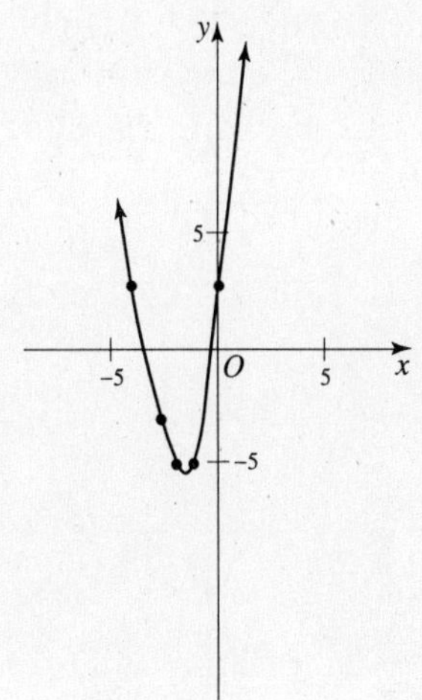

Fig. 16-11

16.37 (*a*) $x^2 + x - 6 = 0$ (*e*) $6x^2 - x - 1 = 0$ (*i*) $4x^2 - 16x + 25 = 0$
(*b*) $x^2 + 3x = 0$ (*f*) $x^2 - 4x + 1 = 0$ (*j*) not possible
(*c*) $x^2 - 4x - 32 = 0$ (*g*) $x^2 + 2x + 2 = 0$ (*k*) $x^2 - 2ax + a^2 + b^2 = 0$
(*d*) $x^2 + 7x + 10 = 0$ (*h*) $x^2 + 4x - 2 = 0$ (*l*) $4x^2 - 4mx + m^2 - n = 0$

16.38 (*a*) $p = -3$ (*b*) $p = -4$ (*c*) $p = -1$ (*d*) $p = 2$ (*e*) $p = \pm 5$ (*f*) $p = -7$

16.39 (*a*) 3, 6 (*b*) 1, 2 (*c*) $\pm 3/2$ (*d*) $1/2 \pm i/2$
(*e*) If $m = -1$, the roots are 2, 2; if $m = 1/2$, the roots are 1/2, 1/2.

16.40 (*a*) 2, −1 (*c*) 4/9 (*e*) 9, 1 (*g*) ±2 (*i*) 3/2 (*k*) −2
(*b*) 1, 3 (*d*) −4, 2 (*f*) 8, −2 (*h*) 1 (*j*) no solution (*l*) 5/4

16.41 (*a*) $\pm 2, \pm 3$ (*d*) $\pm 1, \pm 1/8$ (*g*) $2 \pm \sqrt{3}, -1/4 \pm i\sqrt{15}/4$ (*j*) $1 \pm i, 2 \pm \sqrt{2}$
(*b*) $\pm\sqrt{5}, \pm i\sqrt{2}$ (*e*) $7, 5, \pm 1$ (*h*) 79
(*c*) $\pm 2, \pm 1/2$ (*f*) $1, -2, (-1 \pm \sqrt{93})/2$ (*i*) 4

16.42 (*a*) $5, 3$ or $-27/5, -11/5$ (*b*) $5, 6, 7$ or $-7, -6, -5$ (*c*) $3, 6$ (*d*) 9

16.43 (*a*) 5, 15 ft (*b*) 15, 20 in. (*c*) 2 in. (*d*) 1.3 in.

16.44 (*a*) 21 (*b*) 85

16.45 (*a*) 12, 16 mi/hr (*b*) 40 mi/hr (*c*) 6 mi/hr

16.46 (*a*) 24 (*b*) 144, 10¢

16.47 (*a*) A, 6 hr; B, 12 hr (*b*) 5.3 hr approx.

16.48 (*a*) 0.78 and 3.22 seconds after projection (*b*) No (*c*) 64 ft

Chapter 17

Conic Sections

17.1 GENERAL QUADRATIC EQUATIONS

The general quadratic equation in the two variables x and y has the form

$$ax^2 + bxy + cy^2 + dx + ey + f = 0 \tag{1}$$

where a, b, c, d, e, f are given constants and a, b, c are not all zero.

Thus $3x^2 + 5xy = 2$, $x^2 - xy + y^2 + 2x + 3y = 0$, $y^2 = 4x$, $xy = 4$ are quadratic equations in x and y.

The graph of equation (1), if a, b, c, d, e, f are real, depends on the value of $b^2 - 4ac$.

(1) If $b^2 - 4ac < 0$, the graph is in general an ellipse. However, if $b = 0$ and $a = c$ the graph may be a circle, a point, or non-existent. The point and non-existent situations are called the degenerate cases.

(2) If $b^2 - 4ac = 0$, the graph is a parabola, two parallel or coincident lines, or non-existent. The parallel or coincident lines and non-existent situations are called the degenerate cases.

(3) If $b^2 - 4ac > 0$, the graph is a hyperbola or two intersecting lines. The two intersecting lines situation is called the degenerate case.

These graphs are the intersections of a plane and a right circular cone, and for this reason are called conic sections.

EXAMPLES 17.1. Identify the type of conic section described by each equation.

(*a*) $x^2 + xy = 6$ (*c*) $2x^2 - y^2 = 7$ (*e*) $3x^2 + 3y^2 - 4x + 3y + 10 = 0$

(*b*) $x^2 + 5xy - 4y^2 = 10$ (*d*) $3x^2 + 2y^2 = 14$ (*f*) $y^2 + 4x + 3y + 4 = 0$

(*a*) $a = 1,\ b = 1,\ c = 0 \qquad b^2 - 4ac = 1 - 4 < 0$
So the figure is an ellipse or a degenerate case.

(*b*) $a = 1,\ b = 5,\ c = -4 \qquad b^2 - 4ac = 25 + 16 > 0$
So the figure is a hyperbola or a degenerate case.

(*c*) $a = 2,\ b = 0,\ c = -1 \qquad b^2 - 4ac = 0 + 8 > 0$
So the figure is a hyperbola or a degenerate case.

(*d*) $a = 3,\ b = 0,\ c = 2 \qquad b^2 - 4ac = 0 - 24 < 0$
So the figure is an ellipse or a degenerate case.

(*e*) $a = 3,\ b = 0,\ c = 3 \qquad b^2 - 4ac = 0 - 36 < 0$
So the figure is a circle or a degenerate case since $a = c$ and $b = 0$.

(*f*) $a = 0,\ b = 0,\ c = 1 \qquad b^2 - 4ac = 0 - 0 = 0$
So the figure is a parabola or a degenerate case.

17.2 CONIC SECTIONS

Each conic section is the locus (set) of all points in a plane meeting a given set of conditions. The set of points can be described by an equation. When the locus is positioned at the origin, the figure is called a central conic section. A general equation used to describe a conic section is called the standard equation, which may have more than one form for a conic section. The conic sections are the circle, parabola, ellipse, and hyperbola. We will consider only conic sections in which $b = 0$, thus having the general quadratic equation $Ax^2 + Cy^2 + Dx + Ey + F = 0$. Trigonometry is needed to discuss fully the general quadratic equations in which $b \neq 0$.

17.3 CIRCLES

A circle is the locus of all points in a plane which are at a fixed distance from a fixed point in the plane. The fixed point is the center of the circle and the fixed distance is the radius of the circle.

When the center of the circle is the origin, $(0, 0)$, and the radius is r, the standard form of the equation of a circle is $x^2 + y^2 = r^2$. If the center of the circle is the point (h, k) and the radius is r, the standard form of the equation of a circle is $(x-h)^2 + (y-k)^2 = r^2$. If $r^2 = 0$, we have the degenerate case of a single point which is sometimes called the point circle. If $r^2 < 0$, we have the non-existent degenerate case, which is sometimes called the imaginary circle, since the radius would have to be an imaginary number.

The graph of the circle $(x-2)^2 + (y+3)^2 = 9$ has its center at $(2, -3)$ and a radius of 3 (see Fig. 17-1).

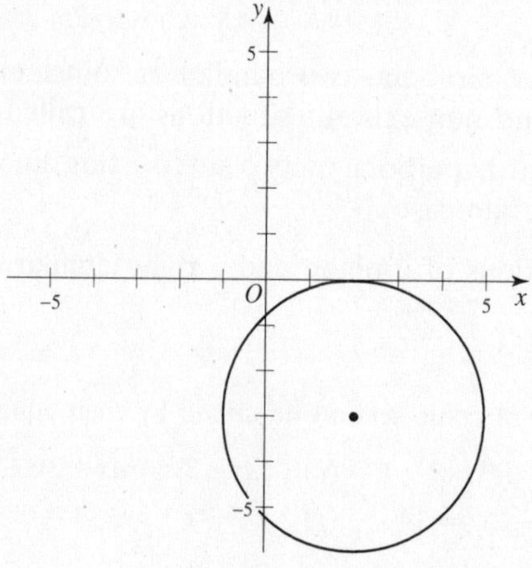

Fig. 17-1

EXAMPLES 17.2. For each circle state the center and radius.

(*a*) $x^2 + y^2 = 5$ (*b*) $x^2 + y^2 = 28$ (*c*) $(x+2)^2 + (y-4)^2 = 81$

(*a*) $C(0, 0)$, $r = \sqrt{5}$

(*b*) $C(0, 0)$, $r = \sqrt{28} = \sqrt{4}\sqrt{7} = 2\sqrt{7}$

(*c*) $(x+2)^2 + (y-4)^2 = 81$ so $(x-(-2))^2 + (y-4)^2 = 9^2$ $C(-2, 4)$, $r = 9$

EXAMPLES 17.3. Write the equation of each circle in standard form.

(*a*) $x^2 + y^2 - 8x + 12y - 48 = 0$ (*b*) $x^2 + y^2 - 4x + 6y + 100 = 0$

(*a*) $x^2 + y^2 - 8x + 12y - 48 = 0$
$(x^2 - 8x) + (y^2 + 12y) = 48$ rearrange terms
$(x^2 - 8x + 16) + (y^2 + 12y + 36) = 48 + 16 + 36$ complete the square for x and y
$(x-4)^2 + (y+6)^2 = 100$ standard form (1)

(*b*) $x^2 + y^2 - 4x + 6y + 100 = 0$
$(x^2 - 4x) + (y^2 + 6y) = -100$ rearrange terms
$(x^2 - 4x + 4) + (y^2 + 6y + 9) = -100 + 4 + 9$ complete the square for x and y
$(x-2)^2 + (y+3) = -87$ standard form (2)

Note: In (1) $r^2 = 100$, so we have a circle, but in (2) $r^2 = -87$ so we have the non-existent case.

EXAMPLE 17.4. Write the equation of the circle going through the points $P(2, -1)$, $Q(-3, 0)$, and $R(1, 4)$.

By substituting the points P, Q, and R into the general form of a circle, $x^2 + y^2 + Dx + Ey + F = 0$, we get a system of three linear equations.

for $P(2, 1-)$	$2^2 + (-1)^2 + 2D - E + F = 0$	then (1)	$2D - E + F = -5$
for $Q(-3, 0)$	$(-3)^2 + 0^2 - 3D + 0E + F = 0$	then (2)	$-3D + F = -9$
for $R(1, 4)$	$1^2 + 4^2 + D + 4E + F = 0$	then (3)	$D + 4E + F = -17$

Eliminating F from (1) and (2) and from (1) and (3), we get

$$(4)\quad 5D - E = 4 \qquad \text{and} \qquad (5)\quad D - 5E = 12$$

Solving (4) and (5) we get $D = 1/3$ and $E = -7/3$, and by substituting D and E in (1) we get $F = -8$.

The equation of the circle is $x^2 + y^2 + 1/3x - 7/3y - 8 = 0$ or $3x^2 + 3y^2 + x - 7y - 24 = 0$.

17.4 PARABOLAS

A parabola is the locus of all points in a plane equidistant from a fixed line, the directrix, and a fixed point, the focus.

Central parabolas have their vertex at the origin, focus on one axis, and directrix parallel to the other axis. We denote the distance from the focus to the vertex by $|p|$. The distance from the directrix to the vertex is also $|p|$. The equations of the central parabolas are (1) and (2) below.

$$(1)\quad y^2 = 4py \qquad \text{and} \qquad (2)\quad x^2 = 4py$$

In (1) the focus is on the x axis and the directrix is parallel to the other axis. If p is positive, the curve opens to the right and if p is negative, the curve opens to the left (see Fig. 17-2). In (2) the

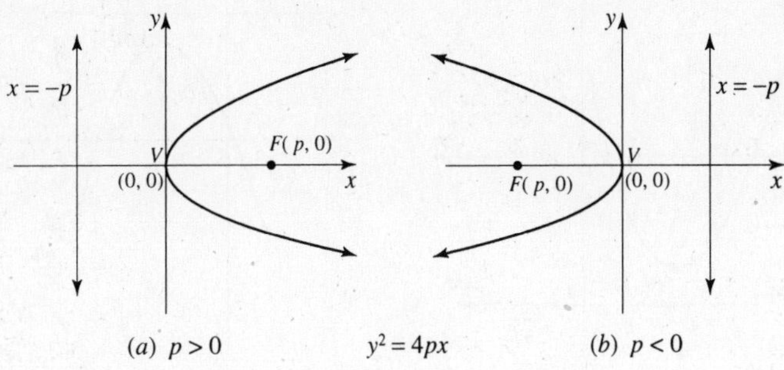

Fig. 17-2

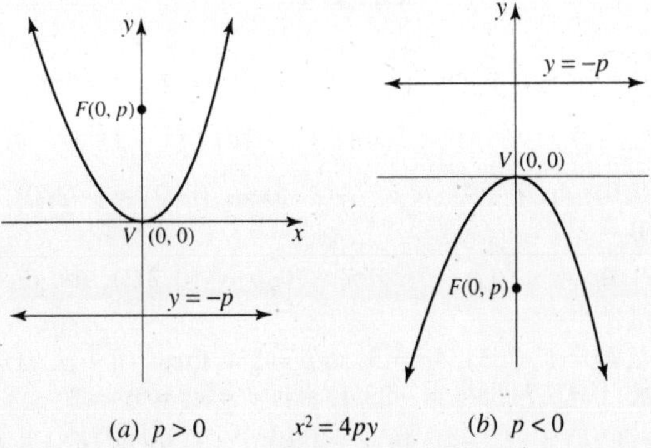

Fig. 17-3

focus is on the y axis and the directrix is parallel to the x axis. If p is positive, the curve opens up and if p is negative, the curve opens down (see Fig. 17-3).

The line through the vertex and the focus is the axis of the parabola and the graph is symmetric with respect to this line.

The parabolas with vertex at the point (h, k) and with the axis and the directrix parallel to the x axis and the y axis have the standard forms listed in (3) and (4) below.

$$(3) \quad (y-k)^2 = 4p(x-h) \qquad \text{and} \qquad (4) \quad (x-h)^2 = 4p(y-k)$$

In (3) the focus is $F(h+p, k)$, the directrix is $x = h - p$, and the axis is $y = k$ (see Fig. 17-4). However, in (4) the focus is $F(h, k+p)$, the directrix is $y = k - p$, and the axis is $x = h$ (see Fig. 17-5).

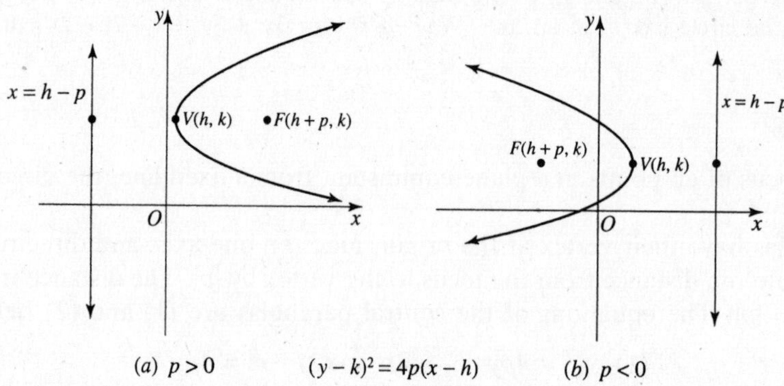

Fig. 17-4

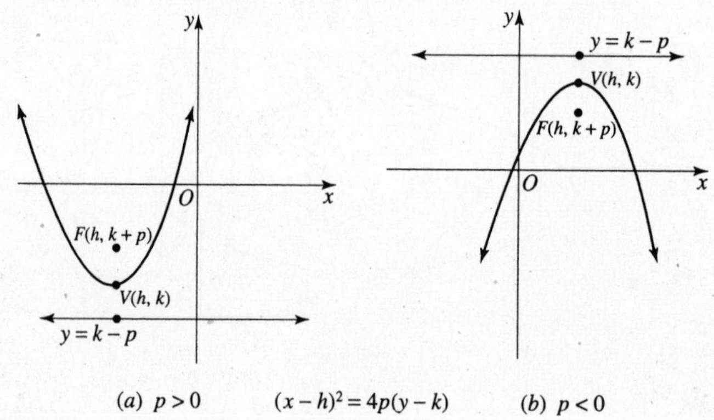

Fig. 17-5

EXAMPLES 17.5. Determine the vertex, focus, directrix, and axis for each parabola.

(a) $y^2 = -8x$ (b) $x^2 = 6y$ (c) $(y-3)^2 = 5(x+7)$ (d) $(x-1)^2 = -4(y+4)$

(a) $y^2 = -8x$: vertex $(h, k) = (0, 0)$, $4p = -8$, so $p = -2$, focus $(p, 0) = (-2, 0)$, and directrix is $x = -p$, so $x = -(-2) = 2$, axis is $y = 0$

(b) $x^2 = 6y$: vertex $(h, k) = (0, 0)$, $4p = 6$, so $p = 3/2$, focus $(0, p) = (0, 3/2)$, and directrix is $y = -p$, so $y = -3/2$, axis is $y = 0$

(c) $(y-3)^2 = 5(x+7)$: vertex $(h, k) = (-7, 3)$, $4p = 5$, so $p = 5/4$, focus $(h+p, k) = (-7 + 5/4, 3) = (-23/4, 3)$, and directrix is $x = h - p$, so $x = -7 - 5/4 = -33/4$, axis $y = k$, so $y = 3$

(d) $(x-1)^2 = -4(y+4)$: vertex $(h, k) = (1, -4)$, $4p = -4$, so $p = 1$, focus $(h, k+p) = (1, -4 + (-1)) = (1, 5)$, and directrix is $y = k - p$, so $y = -4 - (-1) = -3$, axis is $h = k$, so $x = 1$

EXAMPLES 17.6. Write the equation of the parabola with the given characteristics.

(*a*) vertex $(4,6)$ and focus $(4,8)$ (*b*) focus $(3,5)$ and directrix $y = 3$

(*a*) Since the vertex (4, 6) and the focus (4, 8) lie on the line $x = 4$ (see Fig. 17-4), we have a parabola of the form $(y-k)^2 = 4p(x-h)$.

Since the vertex is (4, 6), we have $h = 4$ and $k = 6$.

The focus is $(h, k+p)$, so $k+p = 8$ and $6+p = 8$, so $p = 2$.

The equation of the parabola is $(y-6)^2 = 8(x-4)$.

(*b*) Since the directrix is $y = 3$ (see Fig. 17-5), the parabola has the form $(x-h)^2 = 4p(y-k)$.

The focus $(3, -5)$ is 2 units above the directrix $y = 3$, so $p > 0$. The distance from the focus to the directrix is $2|p|$, so $2p = 2$ and $p = 1$.

The focus is $(h, p+k)$, so $h = 3$ and $k+p = 5$. Since $p = 1$, $k = 4$.

The equation of the parabola is $(x-3)^2 = 4(y-4)$.

EXAMPLES 17.7. Write the equation of each parabola in standard form.

(*a*) $x^2 - 4x - 12y - 32 = 0$ (*b*) $y^2 + 3x - 6y = 0$

(*a*) $x^2 - 4x - 12y - 32 = 0$

$x^2 - 4x = 12y + 32$	reorganize terms
$x^2 - 4x + 4 = 12y + 32 + 4$	complete the square for x
$(x-4)^2 = 12y + 36$	factor right-hand side of equation
$(x-4)^2 = 12(y+3)$	standard form

(*b*) $y^2 + 3x - 6y = 0$

$y^2 - 6y = -3x$	reorganize terms
$y^2 - 6y + 9 = -3x + 9$	complete the square for y
$(y-3)^2 = -3(x-3)$	standard form

17.5 ELLIPSES

An ellipse is the locus of all points in a plane such that the sum of the distances from two fixed points, the foci, to any point on the locus is a constant.

Central ellipses have their center at the origin, vertices and foci lie on one axis, and the covertices lie on the other axis. We will denote the distance from a vertex to the center by a, the distance from a covertex to the center by b, and the distance from a focus to the center by c. For an ellipse, the values a, b, and c are related by $a^2 = b^2 + c^2$ and $a > b$. We call the line segment between the vertices the major axis and the line segment between the covertices the minor axis.

The standard forms for the central ellipses are:

$$(1)\quad \frac{x^2}{a^2} + \frac{y^2}{b^2} = 1 \quad \text{and} \quad (2)\quad \frac{y^2}{a^2} + \frac{x^2}{b^2} = 1$$

The larger denominator is always a^2 for an ellipse. If the numerator for a^2 is x^2, then the major axis lies on the x axis. In (1) the vertices have coordinates $V(a,0)$ and $V'(-a,0)$, the foci have coordinates $F(c,0)$ and $F'(-c,0)$, and the covertices have coordinates $B(0,b)$ and $B'(0,-b)$ (see Fig. 17-6). If the numerator for a^2 is y^2, then the major axis lies on the y axis. In (2) the vertices are at $V(0,a)$ and $V'(0-a)$, the foci are at $F(0,c)$ and $F'(0,-c)$, and the covertices are at $B(b,0)$ and $B'(-b,0)$ (see Fig. 17-7).

If the center of an ellipse is $C(h, k)$ then the standard forms for the ellipses are:

$$(3)\quad \frac{(x-h)^2}{a^2} + \frac{(y-k)^2}{b^2} = 1 \quad \text{and} \quad (4)\quad \frac{(y-k)^2}{a^2} + \frac{(x-h)^2}{b^2} = 1$$

In (3) the major axis is parallel to the x axis and the minor axis is parallel to the y axis. The foci have coordinates $F(h+c,k)$ and $F'(h-c,k)$, the vertices are at $V(h+a,k)$ and $V'(h-a,k)$, and the

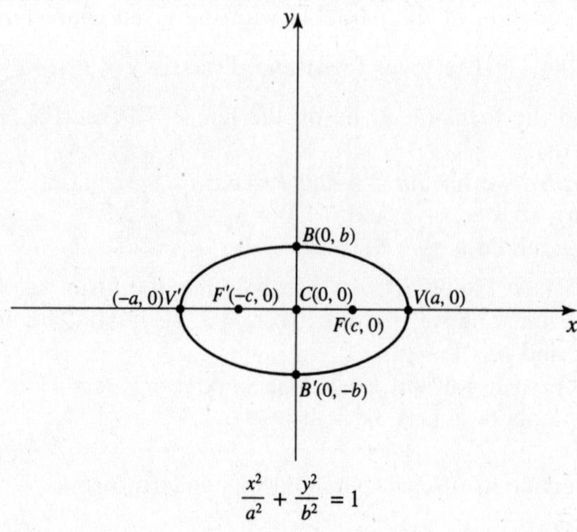

Fig. 17-6

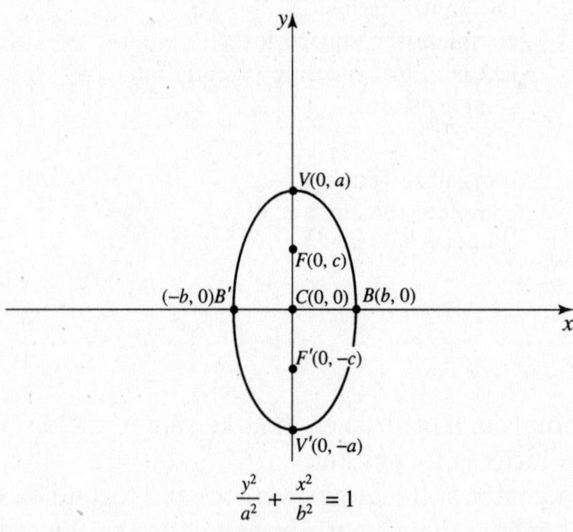

Fig. 17-7

covertices are at $B(h, k+b)$ and $B'(h, k-b)$ (see Fig. 17-8). In (4) the major axis is parallel to the y axis and the minor axis is parallel to the x axis. The foci are at $F(h, k+c)$ and $F'(h, k-c)$, the vertices have coordinates $V(h, k+a)$ and $V'(h, k-a)$, and the covertices are at $B(h+b, k)$ and $B'(h-b, k)$ (see Fig. 17-9).

EXAMPLES 17.8. Determine the center, foci, vertices, and covertices for each ellipse.

(a) $\dfrac{x^2}{25}+\dfrac{y^2}{9}=1$ (c) $\dfrac{(x-3)^2}{225}+\dfrac{(y-4)^2}{289}=1$

(b) $\dfrac{x^2}{3}+\dfrac{y^2}{10}=1$ (d) $\dfrac{(x+1)^2}{100}+\dfrac{(y-2)^2}{64}=1$

(a) $\dfrac{x^2}{25}+\dfrac{y^2}{9}=1$

Since a^2 is the greater denominator, $a^2=25$ and $b^2=9$, so $a=5$ and $b=3$. From $a^2=b^2+c^2$, we get $25=9+c^2$ and $c=4$. The center is at (0, 0). The vertices are at $(a, 0)$ and $(-a, 0)$, so $V(5, 0)$ and $V'(-5, 0)$. The foci are at $(c, 0)$ and $(-c, 0)$, so $F(4, 0)$ and $F(-4, 0)$. The covertices are at $(0, b)$ and $(0, -b)$, so $B(0, 3)$ and $B'(0, -3)$.

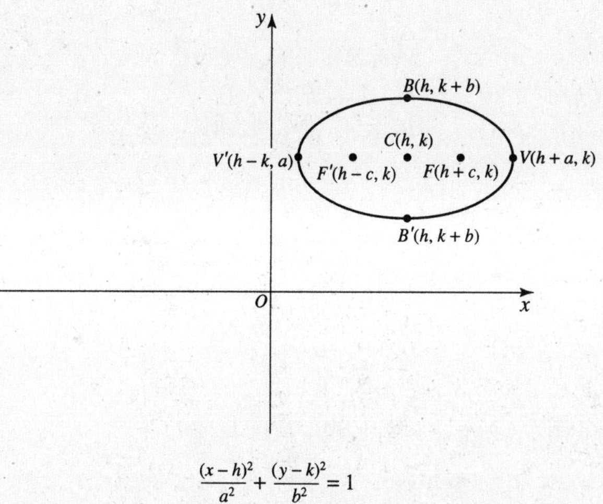

Fig. 17-8

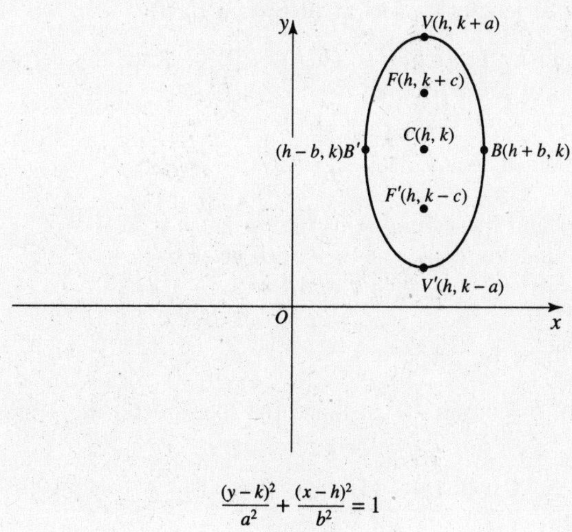

Fig. 17-9

(*b*) $\dfrac{y^2}{10} + \dfrac{x^2}{3} = 1$

$a^2 = 10$ and $b^2 = 3$, so $a = \sqrt{10}$, $b = \sqrt{3}$, and since $a^2 = b^2 + c^2$, $c = \sqrt{7}$.

Since y^2 is over the larger denominator, the vertices and foci are on the y axis. The center is $(0, 0)$.

vertices $(0, a)$ and $(0, -a)$ $\quad V(0, \sqrt{10})$, $V'(0, -\sqrt{10})$
foci $(0, c)$ and $(0, -c)$ $\quad F(0, \sqrt{7})$, $F'(0, -\sqrt{7})$
covertices $(b, 0)$ and $(-b, 0)$ $\quad B(\sqrt{3}, 0)$, $B'(-\sqrt{3}, 0)$

(*c*) $\dfrac{(y-4)^2}{289} + \dfrac{(x-3)^2}{225} = 1$

$a^2 = 289$ and $b^2 = 225$, so $a = 17$ and $b = 15$ and from $a^2 = b^2 + c^2$, $c = 8$. Since $(y-4)^2$ is over a^2, the vertices and foci are on a line parallel to the y axis.

center $(h, k) = (3, 4)$
vertices $(h, k + a)$ and $(h, k - a)$ $V(3, 21)$, $V'(3, -13)$
foci $(h, k + c)$ and $(h, k - c)$ $F(3, 12)$, $F'(3, -4)$
covertices $(h + b, k)$ and $(h - b, k)$ $B(18, 4)$, $B'(-12, 4)$

(*d*) $\dfrac{(x+1)^2}{100} + \dfrac{(y-2)^2}{64} = 1$

$a^2 = 100$, $b^2 = 64$, so $a = 10$ and $b = 8$. From $a^2 = b^2 + c^2$, we get $c = 6$. Since $(x + 1)^2$ is over a^2 the vertices and foci are on a line parallel to the x axis.

center $(h, k) = (-1, 2)$
vertices $(h + a, k)$ and $(h - a, k)$ $V(9, 2)$, $V'(-11, 2)$
foci $(h + c, k)$ and $(h - c, k)$ $F(5, 2)$, $F'(-7, 2)$
covertices $(h, k + b)$ and $(h, k - b)$ $B(-1, 10)$, $B'(-1, -6)$

EXAMPLES 17.9. Write the equation of the ellipse having the given characteristics.

(*a*) central ellipse, foci at $(\pm 4, 0)$, and vertices at $(\pm 5, 0)$
(*b*) center at $(0, 3)$, major axis of length 12, foci at $(0, 6)$ and $(0, 0)$

(*a*) A central ellipse has it center at the origin, so $(h, k) = (0, 0)$. Since the vertices are on the x axis and the center is at (0, 0), the form of the ellipse is

$$\frac{x^2}{a^2} + \frac{y^2}{b^2} = 1$$

From a vertex at $(5, 0)$ and the center at $(0, 0)$, we get $a = 5$.
From a focus at $(4, 0)$ and the center at $(0, 0)$, we get $c = 4$.
Since $a^2 = b^2 + c^2$, $25 = b^2 + 16$, so $b^2 = 9$ and $b = 3$.

The equation of the ellipse is $\dfrac{x^2}{25} + \dfrac{y^2}{9} = 1$

(*b*) Since the center is at $(0, 3)$, $h = 0$ and $k = 3$. Since the foci are on the y axis, the form of the equation of the ellipse is

$$\frac{(y-k)^2}{a^2} + \frac{(x-h)^2}{b^2} = 1$$

The foci are $(h, k + c)$ and $(h, k - c)$, so $(0, 6) = (h, k + c)$ and $3 + c = 6$ and $c = 3$.
The major axis length is 12, so we know $2a = 12$ and $a = 6$.
From $a^2 = b^2 + c^2$, we get $36 = b^2 + 9$ and $b^2 = 27$.

The equation of the ellipse is $\dfrac{(y-3)^2}{36} + \dfrac{x^2}{27} = 1$

EXAMPLE 17.10. Write the equation of the ellipse $18x^2 + 12y^2 - 144x + 48y + 120 = 0$ in standard form.

$18x^2 + 12y^2 - 144x + 48y + 120 = 0$	
$(18x^2 - 144x) + (12y^2 + 48y) = -120$	reorganize terms
$18(x^2 - 8x) + 12(y^2 + 4y) = -120$	factor to get x^2 and y^2
$18(x^2 - 8x + 16) + 12(y^2 + 4y + 4) = -120 + 18(16) + 12(4)$	complete square on x and y
$18(x-4)^2 + 12(y+2)^2 = 216$	simplify
$\dfrac{18(x-4)^2}{216} + \dfrac{12(y+2)^2}{216} = 1$	divide by 216
$\dfrac{(x-4)^2}{12} + \dfrac{(y+2)^2}{18} = 1$	standard form

17.6 HYPERBOLAS

The hyperbola is the locus of all points in a plane such that for any point of the locus the difference of the distances from two fixed points, the foci, is a constant.

Central hyperbolas have their center at the origin and their vertices and foci on one axis, and are symmetric with respect to the other axis. The standard form equations for central hyperbolas are:

$$(1)\quad \frac{x^2}{a^2}-\frac{y^2}{b^2}=1 \qquad \text{and} \qquad (2)\quad \frac{y^2}{a^2}-\frac{x^2}{b^2}=1$$

The distance from the center to a vertex is denoted by a and the distance from the center to a focus is c. For a hyperbola, $c^2=a^2+b^2$ and b is a positive number. The line segment between the vertices is called the transverse axis. The denominator of the positive fraction for the standard form is always a^2.

In (1) the transverse axis $\overline{VV'}$ lies on the x axis, the vertices are $V(a,0)$ and $V'(-a,0)$, and the foci are at $F(c,0)$ and $F'(-c,0)$ (see Fig. 17-10). In (2) the transverse axis $\overline{VV'}$ lies on the y axis, the vertices are at $V(0,a)$ and $V'(0,-a)$, and the foci are at $F(0,c)$ and $F'(0,-c)$ (see Fig. 17-11). When lines are drawn through the points R and C and the points S and C, we have the asymptotes of the hyperbola. The asymptote is a line that the graph of the hyperbola approaches but does not reach.

If the center of the hyperbola is at (h,k) the standard forms are (3) and (4):

$$(3)\quad \frac{(x-h)^2}{a^2}-\frac{(y-k)^2}{b^2}=1 \qquad \text{and} \qquad (4)\quad \frac{(y-k)^2}{a^2}-\frac{(x-h)^2}{b^2}=1$$

In (3) the transverse axis is parallel to the x axis, the vertices have coordinates $V(h+a,k)$ and $V'(h-a,k)$, the foci have coordinates $F(h+c,k)$ and $F'(h-c,k)$, and the points R and S have coordinates $R(h+a,k+b)$ and $S(h+a,k-b)$. The lines through R and C and S and C are the asymptotes of the hyperbola (see Fig. 17-12). In equation (4) the transverse axis is parallel to the y axis, the vertices are at $V(h,k+a)$ and $V'(h,k-a)$, the foci are at $F(h,k+c)$ and $F'(h,k-c)$, and the points R and S have coordinates $R(h+b,k+a)$ and $S(h-b,k+a)$ (see Fig. 17-13).

EXAMPLES 17.11. Find the coordinates of the center, vertices, and foci for each hyperbola.

(*a*) $\dfrac{(x-4)^2}{9}-\dfrac{(y-5)^2}{16}=1$ (*b*) $\dfrac{(y+5)^2}{25}-\dfrac{(x+9)^2}{144}=1$ (*c*) $\dfrac{(x+3)^2}{225}-\dfrac{(y-4)^2}{64}=1$

(*a*) $\dfrac{(x-4)^2}{9}-\dfrac{(y-5)^2}{16}=1$

Since $a^2=9$ and $b^2=16$ we have $a=3$ and $b=4$.

From $c^2=a^2+b^2$, we get $c=5$.

center is $(h,k)=(4,5)$

vertices are $V(h+a,k)$ and $V'(h-a,k)$ — $V(7,5)$ and $V'(1,5)$

foci are $F(h+c,k)$ and $F'(h-c,k)$ — $F(9,5)$ and $F'(-1,5)$

(*b*) $\dfrac{(y+5)^2}{25}-\dfrac{(x+9)^2}{144}=1$

Since $a^2=25$ and $b^2=144$, $a=5$ and $b=12$.

From $c^2=a^2+b^2$, we get $c=13$.

center $C(h,k)=(-5,-9)$

vertices are $V(h,k+a)$ and $V'(h,k-a)$ — $V(-5,-4)$ and $V'(-5,-2)$

foci are $F(h,k+c)$ and $F'(h,k-c)$ — $F(-5,4)$ and $F'(-5,-22)$

(c) $\dfrac{(x+3)^2}{225} - \dfrac{(y-4)^2}{64} = 1$

Since $a^2 = 225$ and $b^2 = 64$, we get $a = 15$ and $b = 8$.

From $c^2 = a^2 + b^2$, we get $c = 17$.

center $C(h, k) = (-3, 4)$

vertices are $V(h+a, k)$ and $V'(h-a, k)$ $\qquad V(12, 4)$ and $V'(-18, 4)$

foci are $F(h+c, k)$ and $F'(h-c, k)$ $\qquad F(14, 4)$ and $F'(-20, 4)$

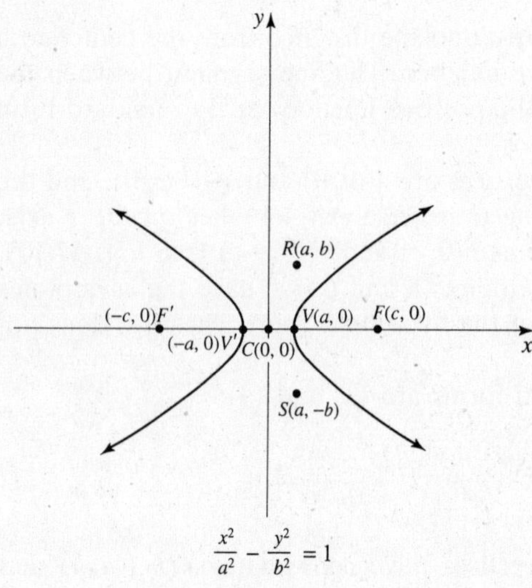

Fig. 17-10

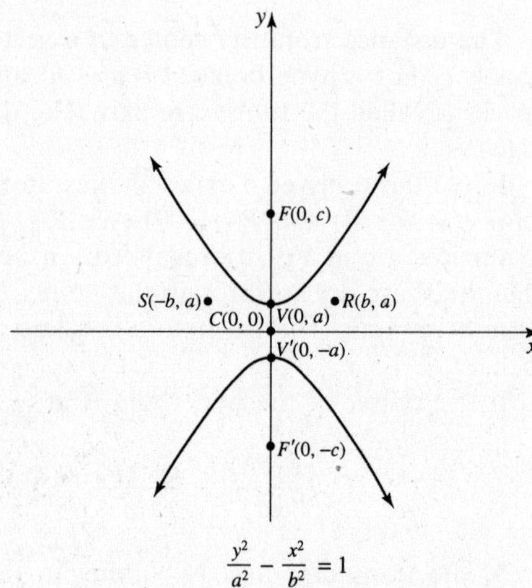

Fig. 17-11

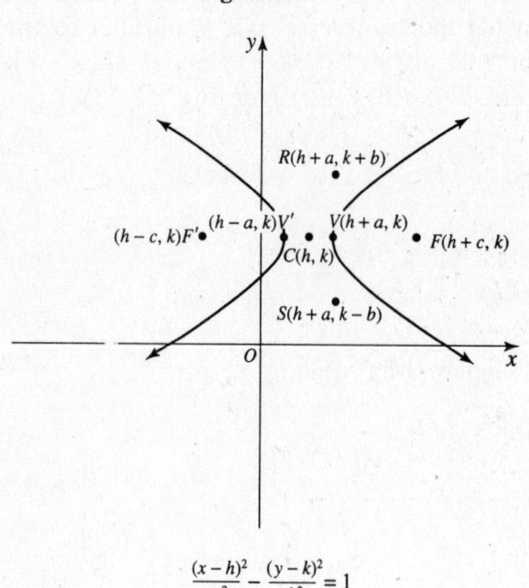

Fig. 17-12

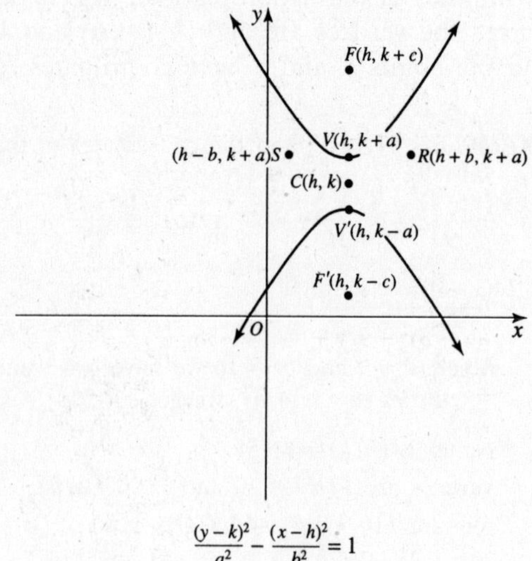

Fig. 17-13

EXAMPLES 17.12. Write the equation of the hyperbola that has the given characteristics.

(a) foci are at $(2, 5)$ and $(-4, 5)$ and transverse axis has length 4

(b) center at $(1, -3)$, a focus is at $(1, 2)$ and a vertex is at $(1, 1)$

(*a*) The foci are on a line parallel to the x axis, so the form is

$$\frac{(x-h)^2}{a^2}-\frac{(y-k)^2}{b^2}=1$$

The center is half-way between the foci, so $c = 3$ and the center is at $C(-1, 5)$.
The transverse axis joins the vertices, so its length is $2a$, so $2a = 4$ and $a = 2$.
Since $c^2 = a^2 + b^2$, $c = 3$ and $a = 2$, so $b^2 = 5$.
The equation of the hyperbola is

$$\frac{(x+1)^2}{4}-\frac{(y-5)^2}{5}=1$$

(*b*) The distance from the vertex $(1, 1)$ to the center $(1, -3)$ is a, so $a = 4$.
The distance from the focus $(1, 2)$ to the center $(1, -3)$ is c, so $c = 5$.
Since $c^2 = a^2 + b^2, a = 4$, and $c = 5, b^2 = 9$.
Since the center, vertex, and focus lie on a line parallel to the y axis, the hyperbola has the form

$$\frac{(y-k)^2}{a^2}-\frac{(x-h)^2}{b^2}=1$$

The center is $(1, -3)$, so $h = 1$, and $k = -3$.
The equation of the hyperbola is

$$\frac{(y+3)^2}{16}-\frac{(x-1)^2}{9}=1$$

EXAMPLES 17.13. Write the equation of each hyperbola in standard form.

(*a*) $25x^2 - 9y^2 - 100x - 72y - 269 = 0$ (*b*) $4x^2 - 9y^2 - 24x - 90y - 153 = 0$

(*a*) $25x^2 - 9y^2 - 100x - 72y - 269 = 0$

$(25x^2 - 100x) + (-9y^2 - 72y) = 269$	rearrange terms
$25(x^2 - 4x) - 9(y^2 + 8y) = 269$	factor to get x^2 and y^2
$25(x^2 - 4x + 4) - 9(y^2 + 8y + 16) = 269 + 25(18) - 9(16)$	complete square for x and y
$25(x-2)^2 - 9(y+4)^2 = 225$	simplify then divide by 225
$\frac{(x-2)^2}{9}-\frac{(y+4)^2}{25}=1$	standard form

(*b*) $4x^2 - 9y^2 - 24x - 90y - 153 = 0$

$(4x^2 - 24x) + (-9y^2 - 90y) = 153$	reorganize terms
$4(x^2 - 6x) - 9(y^2 + 10y) = 153$	factor to get x^2 and y^2
$4(x^2 - 6x + 9) - 9(y^2 + 10y + 25) = 153 + 4(9) - 9(25)$	complete square for x and y
$4(x-3)^2 - 9(y+5)^2 = -36$	simplify then divide by -36
$\frac{(x-3)^2}{-9}-\frac{(y+5)^2}{-4}=1$	simplify signs
$\frac{(y+5)^2}{4}-\frac{(x-3)^2}{9}=1$	standard form

17.7 GRAPHING CONIC SECTIONS WITH A CALCULATOR

Since most conic sections are not functions, an important step is to solve the standard form equation for y. If y is equal to an expression in x that contains a $\pm$ quantity, we need to separate the expression into two parts: $y_1 =$ the expression using the $+$ quantity and $y_2 =$ the expression using the $-$ expression. Otherwise, set $y_1 =$ the expression. Graph either y_1 or y_1 and y_2 simultaneously. The window may need to be adjusted to correct for the distortion caused by unequal scales used on the

x axis and the y axis in many graphing calculators' standard windows. Using the y scale to be 0.67 often corrects for this distortion.

For the circle, ellipse, and hyperbola, it is usually necessary to center the graphing window at the point (h, k) the center of the conic section. However, the parabola is viewed better if the vertex (h, k) is at one end of the viewing window.

Solved Problems

17.1 Draw the graph of each of the following equations:

(*a*) $4x^2 + 9y^2 = 36$, (*b*) $4x^2 - 9y^2 = 36$, (*c*) $4x + 9y^2 = 36$.

SOLUTION

(*a*) $4x^2 + 9y^2 = 36$, $y^2 = \frac{4}{9}(9 - x^2)$, $y = \pm\frac{2}{3}\sqrt{9 - x^2}$.

Note that y is real when $9 - x^2 \geq 0$, i.e., when $-3 \leq x \leq 3$. Hence values of x greater than 3 or less than -3 are excluded.

x	-3	-2	-1	0	1	2	3
y	0	± 1.49	± 1.89	± 2	± 1.89	± 1.49	0

The graph is an ellipse with center at the origin (see Fig. 17-14(*a*)).

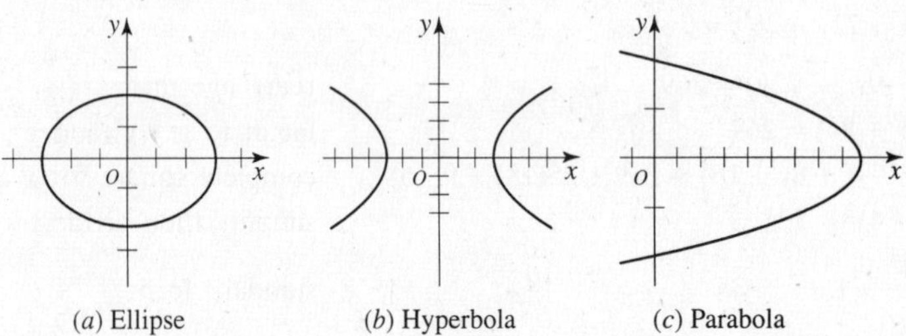

Fig. 17-14

(*b*) $4x^2 - 9y^2 = 36$, $y^2 = \frac{4}{9}(x^2 - 9)$, $y = \pm\frac{2}{3}\sqrt{x^2 - 9}$.

Note that x cannot have a value between -3 and 3 if y is to be real.

x	6	5	4	3	-3	-4	-5	-6
y	± 3.46	± 2.67	± 1.76	0	0	± 1.76	± 2.67	± 3.46

The graph consists of two branches and is called a hyperbola (see Fig. 17-14(*b*)).

(*c*) $4x + 9y^2 = 36$, $y^2 = \frac{4}{9}(9 - x)$, $y = \pm\frac{2}{3}\sqrt{9 - x}$.

Note that if x is greater than 9, y is imaginary.

x	-1	0	1	5	8	9
y	± 2.11	± 2	± 1.89	± 1.33	± 0.67	0

The graph is a parabola (see Fig. 17-14(*c*)).

17.2 Plot the graph of each of the following equations:

(a) $xy = 8$, (b) $2x^2 - 3xy + y^2 + x - 2y - 3 = 0$, (c) $x^2 + y^2 - 4x + 8y + 25 = 0$.

SOLUTION

(a) $xy = 8$, $y = 8/x$. Note that if x is any real number except zero, y is real. The graph is a hyperbola (see Fig. 17-15(a)).

x	4	2	1	$\frac{1}{2}$	$-\frac{1}{2}$	-1	-2	-4
y	2	4	8	16	-16	-8	-4	-2

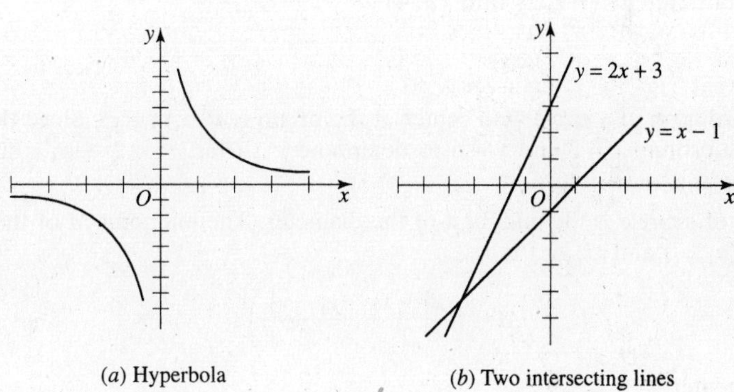

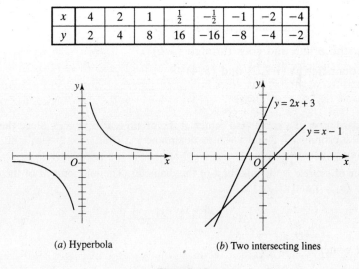

(a) Hyperbola (b) Two intersecting lines

Fig. 17-15

(b) $2x^2 - 3xy + y^2 + x - 2y - 3 = 0$. Write as $y^2 - (3x + 2)y + (2x^2 + x - 3) = 0$ and solve by the quadratic formula to obtain

$$y = \frac{3x + 2 \pm \sqrt{x^2 + 8x + 16}}{2} = \frac{(3x + 2) \pm (x + 4)}{2} \quad \text{or} \quad y = 2x + 3,\ y = x - 1$$

The given equation is equivalent to two linear equations, as can be seen by writing the given equation as $(2x - y + 3)(x - y - 1) = 0$. The graph consists of two intersecting lines (see Fig. 17-15(b)).

(c) Write as $y^2 + 8y + (x^2 - 4x + 25) = 0$; solving,

$$y = \frac{-4 \pm \sqrt{-4(x^2 - 4x + 9)}}{2}.$$

Since $x^2 - 4x + 9 = x^2 - 4x + 4 + 5 = (x - 2)^2 + 5$ is always positive, the quantity under the radical sign is negative. Thus y is imaginary for all real values of x and the graph does not exist.

17.3 For each equation of a circle, write it in standard form and determine the center and radius.

(a) $x^2 + y^2 - 8x + 10y - 4 = 0$ (b) $4x^2 + 4y^2 + 28y + 13 = 0$

SOLUTION

(a) $x^2 + y^2 - 8x + 10y - 4 = 0$

$(x^2 - 8x + 16) + (y^2 + 10y + 25) = 4 + 16 + 25$

$(x - 4)^2 + (y + 5)^2 = 45$ standard form

center: $C(4, -5)$ radius: $r = \sqrt{45} = 3\sqrt{5}$

(*b*) $4x^2 + 4y^2 + 28y + 13 = 0$

$x^2 + y^2 + 7y = -13/4$

$x^2 + (y^2 + 7y + 49/4) = -13/4 + 49/4$

$x^2 + (y + 7/2)^2 = 9$ standard form

center: $C(0, -7/2)$ radius: $r = 3$

17.4 Write the equation of the following circles.

(*a*) center at the origin and goes through $(2, 6)$

(*b*) ends of diameter at $(-7, 2)$ and $(5, 4)$

SOLUTION

(*a*) The standard form of a circle with center at the origin is $x^2 + y^2 = r^2$. Since the circle goes through $(2, 6)$, we substitute $x = 2$ and $y = 6$ to determine r^2. Thus, $r^2 = 2^2 + 6^2 = 40$. The standard form of the circle is $x^2 + y^2 = 40$.

(*b*) The center of a circle is the midpoint of the diameter. The midpoint M of the line segment having endpoints (x_1, y_1) and (x_2, y_2) is

$$M = \left(\frac{x_1 + x_2}{2}, \frac{y_1 + y_2}{2}\right).$$

Thus, the center is

$$C\left(\frac{-7+5}{2}, \frac{2+4}{2}\right) = C\left(\frac{-2}{2}, \frac{6}{2}\right) = C(-1, 3).$$

The radius of a circle is the distance from the center to the endpoint of the diameter. The distance, d, between two points (x_1, y_1) and (x_2, y_2) is $d = \sqrt{(x_2 - x_1)^2 + (y_2 - y_1)^2}$. Thus, the distance from the center $C(-1, 3)$ to $(5, 4)$ is $r = \sqrt{(5 - (-1))^2 + (4 - 3)^2} = \sqrt{6^2 + 1^2} = \sqrt{37}$.

The equation of the circle is $(x + 1)^2 + (y - 3)^2 = 37$.

17.5 Write the equation of the circle passing through three points $(3, 2)$, $(-1, 4)$, and $(2, 3)$.

SOLUTION

The general form of the equation of a circle is $x^2 + y^2 + Dx + Ey + F = 0$, so we must substitute the given points into this equation to get a system of equations in D, E, and F.

For $(3, 2)$	$3^2 + 2^2 + D(3) + E(2) + F = 0$	then (1)	$3D + 2E + F = -13$
For $(-1, 4)$	$(-1)^2 + 4^2 + D(-1) + E(4) + F = 0$	then (2)	$-D + 4E + F = -17$
For $(2, 3)$	$2^2 + 3^2 + D(2) + E(3) + F = 0$	then (3)	$2D + 3E + F = -13$

We eliminate F from (1) and (2) and from (1) and (3) to get

(4) $4D - 2E = 4$ and (5) $D - E = 0$.

We solve the system of (4) and (5) to get $D = 2$ and $E = 2$ and substituting into (1) we get $F = -23$.

The equation of the circle is $x^2 + y^2 + 2x + 2y - 23 = 0$.

17.6 Write the equation of the parabola in standard form and determine the vertex, focus, directrix, and axis.

(*a*) $y^2 - 4x + 10y + 13 = 0$ (*b*) $3x^2 + 18x + 11y + 5 = 0$.

SOLUTION

(*a*) $y^2 - 4x + 10y + 13 = 0$

$y^2 + 10y = 4x - 13$	rearrange terms
$y^2 + 10y + 25 = 4x + 12$	complete the square for y
$(y+5)^2 = 4(x+3)$	standard form
vertex $(h, k) = (-3, -4)$	$4p = 4$ so $p = 1$
focus $(h + p, k) = (-3 + 1, -4) = (-2, -4)$	
directrix: $x = h - p = -4$	axis: $y = k = -4$

(*b*) $3x^2 + 18x + 11y + 5 = 0$

$x^2 + 6x = -11/3y - 5/3$	
$x^2 + 6x + 9 = -11/3y + 22/3$	
$(x+3)^2 = -11/3(y+2)$	standard form
vertex $(h, k) = (-3, -2)$	$4p = -11/3 = -11/12$
focus $(h, k + p) = (-3, -2 + (-11/12)) = (-3, -35/12)$	
directrix $= k - p = -2 - (-11/12) = -13/12$	axis: $x = h = -3$

17.7 Write the equation of the parabola with the given characteristics.

(*a*) vertex at origin and directrix $y = 2$ (*b*) vertex $(-1, -3)$ and focus $(-3, -3)$

SOLUTION

(*a*) Since the vertex is at the origin, we have the form $y^2 = 4px$ or $x^2 = 4py$. However, since the directrix is $y = 2$, the form is $x^2 = 4py$.

The vertex is $(0, 0)$ and the directrix is $y = k - p$. Since $y = 2$ and $k = 0$, we have $p = -2$. The equation of the parabola is $x^2 = -8y$.

(*b*) The vertex is $(-1, -3)$ and the focus is $(-3, -3)$ and since they lie on a line parallel to the x axis, the standard form is $(y - k)^2 = 4p(x - h)$.

From the vertex we get $h = -1$ and $k = -3$, and since the focus is $(h + p, k)$, $h + p = -3$ and $-1 + p = -3$, we get $p = -2$.

Thus, the standard form of the parabola is $(y + 3)^2 = -8(x + 1)$.

17.8 Write the equation of the ellipse in standard form and determine its center, vertices, foci, and covertices.

(*a*) $64x^2 + 81y^2 = 64$ (*b*) $9x^2 + 5y^2 + 36x + 10y - 4 = 0$

SOLUTION

(*a*) $64x^2 + 81y^2 = 64$

$x^2 + \dfrac{81y^2}{64} = 1$	divide by 64
$\dfrac{x^2}{1} + \dfrac{y^2}{\left(\dfrac{64}{81}\right)} = 1$	divide the numerator and denominator by 81
	standard form
center is the origin $(0, 0)$	$a^2 = 1$ and $b^2 = 64/81$, so $a = 1$ and $b = 8/9$

For an ellipse, $a^2 = b^2 + c^2$, so $1 = 64/81 + c^2$ and $c^2 = 17/81$, giving $c = \sqrt{17}/9$.
The vertices are $(a, 0)$ and $(-a, 0)$, so $V(1, 0)$ and $V'(-1, 0)$.
The foci are $(c, 0)$, and $(-c, 0)$, so $F(\sqrt{17}/9,\ 0)$ and $F'(-\sqrt{17}/9, 0)$.
The covertices are $(0, b)$ and $(0, -b)$, so $B(0, 8/9)$ and $B'(0, -8/9)$.

(*b*) $9x^2 + 5y^2 + 36x + 10y - 4 = 0$

$9(x^2 + 4x + 4) + 5(y^2 + 2y + 1) = 4 + 36 + 5$

$9(x + 2)^2 + 5(y + 1)^2 = 45$

$\frac{(x+2)^2}{5} + \frac{(y+1)^2}{9} = 1$ standard form

center $(h, k) = (-2, -1)$ $a^2 = 9$, $b^2 = 5$, so $a = 3$ and $b = \sqrt{5}$

Since $a^2 = b^2 + c^2$, $c^2 = 4$ and $c = 2$.
The vertices are $(h, k + a)$ and $(h, k - a)$, so $V(-2, 2)$ and $V'(-2, 3)$.
The foci are $(h, k + c)$ and $(h, k - c)$, so $F(-2, 1)$ and $F'(-2, -3)$.
The covertices are $(h + b, k)$ and $(h - b, k)$ so $B(-2 + \sqrt{5}, -1)$ and $B'(-2 - \sqrt{5}, -1)$.

17.9 Write the equation of the ellipse that has these characteristics.

(*a*) foci are $(-1, 0)$ and $(-1, 0)$ and length of minor axis is $2\sqrt{2}$.

(*b*) vertices are at $(5, -1)$ and $(-3, -1)$ and $c = 3$.

SOLUTION

(*a*) The midpoint of the line segment between the foci is the center, so the center is $C(0, 0)$ and we have a central ellipse. The standard form is

$$\frac{x^2}{a^2} + \frac{y^2}{b^2} = 1 \quad \text{or} \quad \frac{y^2}{a^2} + \frac{x^2}{b^2} = 1$$

The foci are $(c, 0)$ and $(-c, 0)$ so $(c, 0) = (1, 0)$ and $c = 1$.
The minor axis has length $2\sqrt{2}$, so $2b = 2\sqrt{2}$ and $b = \sqrt{2}$ and $b^2 = 2$.
For the ellipse, $a^2 = b^2 + c^2$ and $a^2 = 1 + 2 = 3$.
Since the foci are on the x axis, the standard form is

$$\frac{x^2}{a^2} + \frac{y^2}{b^2} = 1$$

The equation of the ellipse is

$$\frac{x^2}{3} + \frac{y^2}{2} = 1.$$

(*b*) The midpoint of the line segment between the vertices is the center, so the center is

$$C\left(\frac{5-3}{2}, \frac{-1-1}{2}\right) = (1, -1).$$

We have an ellipse with center at (h, k) where $h = 1$ and $k = -1$.
The standard form of the ellipse is

$$\frac{(x-h)^2}{a^2} + \frac{(y-k)^2}{b^2} = 1 \quad \text{or} \quad \frac{(y-k)^2}{a^2} + \frac{(x-h)^2}{b^2} = 1.$$

The vertices are $(h + a, k)$ and $(h - a, k)$, so $(h + a, k) = (1 + a, -1) = (5, -1)$. Thus, $1 + a = 5$ and $a = 4$.

For the ellipse, $a^2 = b^2 + c^2, c$ is given to be 3, and we found a to be 4. Thus, $a^2 = 4^2 = 16$ and $c^2 = 3^2 = 9$. Therefore, $a^2 = b^2 + c^2$ yields $16 = b^2 + 9$ and $b^2 = 7$.

Since the vertices are on a line parallel to the x axis, the standard form is

$$\frac{(x-h)^2}{a^2} + \frac{(y-k)^2}{b^2} = 1.$$

The equation of the ellipse is

$$\frac{(x-1)^2}{16} + \frac{(y+1)^2}{7} = 1.$$

17.10 For each hyperbola, write the equation in standard form and determine the center, vertices, and foci.

(*a*) $16x^2 - 9y^2 + 144 = 0$ (*b*) $9x^2 - 16y^2 + 90x + 64y + 17 = 0$

SOLUTION

(*a*) $16x^2 - 9y^2 + 144 = 0$

$16x^2 - 9y^2 - 144$

$$\frac{x^2}{-9} - \frac{y^2}{-16} = 1$$

$$\frac{y^2}{16} - \frac{x^2}{9} = 1 \qquad \text{standard form}$$

center $(h, k) = (0, 0)$ $a^2 = 16$ and $b^2 = 9$, so $a = 4$ and $b = 3$

Since $c^2 = a^2 + b^2$ for a hyperbola, $c^2 = 16 + 9 = 25$ and $c = 5$.
The foci are $(0, c)$ and $(0, -c)$, so $F(0, 5)$ and $F'(0, -5)$.
The vertices are $(0, a)$ and $(0, -a)$, so $V(0, 4)$ and $V'(0, -4)$

(*b*) $9x^2 - 16y^2 + 90x + 64y + 17 = 0$

$9(x^2 + 10x + 25) - 16(y^2 - 4y + 4) = -17 + 225 - 64$

$9(x + 5)^2 - 16(y - 2)^2 = 144$

$$\frac{(x + 5)^2}{16} - \frac{(y - 2)^2}{9} = 1 \qquad \text{standard form}$$

center $(h, k) = (-5, 2)$ $a^2 = 16$ and $b^2 = 9$, so $a = 4$ and $b = 3$

Since $c^2 = a^2 + b^2$, $c^2 = 16 + 9 = 25$ and $c = 5$.
The foci are $(h + c, k)$ and $(h - c, k)$, so $F(0, 2)$ and $F'(-10, 2)$.
The vertices are $(h + a, k)$ and $(h - a, k)$, so $V(-1, 2)$ and $V'(-9, 2)$.

17.11 Write the equation of the hyperbola with the given characteristics.

(*a*) vertices are $(0, \pm 2)$ and foci are $(0, \pm 3)$

(*b*) foci $(1, 2)$ and $(-11, 2)$ and the transverse axis has length 4

SOLUTION

(*a*) Since the vertices are $(0, \pm 2)$, the center is at $(0, 0)$, and since they are on a vertical line the standard form is

$$\frac{y^2}{a^2} - \frac{x^2}{b^2} = 1$$

The vertices are at $(0, \pm a)$ so $a = 2$ and the foci are at $(0, \pm 3)$ so $c = 3$.
Since $c^2 = a^2 + b^2$, $9 = 4 + b^2$ so $b^2 = 5$.
The equation of the hyperbola is

$$\frac{y^2}{4} - \frac{x^2}{5} = 1$$

(*b*) Since the foci are $(1, 2)$ and $(-11, 2)$, they are on a line parallel to the x axis, so the form is

$$\frac{(x - h)^2}{a^2} - \frac{(y - k)^2}{b^2} = 1$$

The midpoint of the line segment between the foci $(1, 2)$ and $(-11, 2)$ is the center, so $C(h, k) = (-5, 2)$. The foci are at $(h + c, k)$ and $(h - c, k)$, so $(h + c, k) = (1, 2)$ and $-5 + c = 1$, with

$c = 6$. The transverse axis has length 4 so $2a = 4$ and $a = 2$. From $c^2 = a^2 + b^2$, we get $36 = 4 + b^2$ and $b^2 = 32$.

The equation of the hyperbola is

$$\frac{(x+5)^2}{4} - \frac{(y-2)^2}{32} = 1$$

Supplementary Problems

17.12 Graph each of the following equations.

(*a*) $x^2 + y^2 = 9$
(*b*) $xy = -4$
(*c*) $4x^2 + y^2 = 16$
(*d*) $x^2 - 4y^2 = 36$
(*e*) $y^2 = 4x$
(*f*) $x^2 + 3y^2 - 1 = 0$
(*g*) $x^2 + 3xy + y^2 = 16$
(*h*) $x^2 + 4y = 4$
(*i*) $x^2 + y^2 - 2x + 2y + 2 = 0$
(*j*) $2x^2 - xy - y^2 - 7x - 2y + 3 = 0$

17.13 Write the equation of the circle that has the given characteristics.

(*a*) center (4, 1) and radius 3
(*b*) center (5, −3) and radius 6
(*c*) goes through (0, 0), (−4, 0), and (0, 6)
(*d*) goes through (2, 3), (−1, 7), and (1, 5)

17.14 Write the equation of the circle in standard form and state the center and radius.

(*a*) $x^2 + y^2 + 6x - 12y - 20 = 0$
(*b*) $x^2 + y^2 + 12x - 4y - 5 = 0$
(*c*) $x^2 + y^2 + 7x + 3y - 10 = 0$
(*d*) $2x^2 + 2y^2 - 5x - 9y + 11 = 0$

17.15 Write the equation of the parabola that has the given characteristics.

(*a*) vertex (3, −2) and directrix $x = -5$
(*b*) vertex (3, 5) and focus (3, 10)
(*c*) passes through (5, 10), vertex is at the origin, and axis is the x axis
(*d*) vertex (5, 4) and focus (2, 4)

17.16 Write the equation of the parabola in standard form and determine its vertex, focus, directrix, and axis.

(*a*) $y^2 + 4x - 8y + 28 = 0$
(*b*) $x^2 - 4x + 8y + 36 = 0$
(*c*) $y^2 - 24x + 6y - 15 = 0$
(*d*) $5x^2 + 20x - 9y + 47 = 0$

17.17 Write the equation of the ellipse that has these characteristics.

(*a*) vertices (±4, 0), foci $(\pm 2\sqrt{3}, 0)$
(*b*) covertices (±3, 0), major axis length 10
(*c*) center (−3, 2), vertex (2, 2), $c = 4$
(*d*) vertices (3, 2) and (3, −6), covertices (1, −2) and (5, −2)

17.18 Write the equation of the ellipse in standard form and determine the center, vertices, foci, and covertices.

(*a*) $3x^2 + 4y^2 - 30x - 8y + 67 = 0$
(*b*) $16x^2 + 7y^2 - 32x + 28y - 20 = 0$
(*c*) $9x^2 + 8y^2 + 54x + 80y + 209 = 0$
(*d*) $4x^2 + 5y^2 - 24x - 10y + 17 = 0$

17.19 Write the equations of the hyperbola that has the given characteristics.

(*a*) vertices (±3, 0), foci (±5, 0)

(b) vertices $(0, \pm 8)$, foci $(0, \pm 10)$

(c) foci $(4, -1)$ and $(4, 5)$, transverse axis length is 2

(d) vertices $(-1, -1)$ and $(-1, 5)$, $b = 5$

17.20 Write the equation of the hyperbola in standard form and determine the center, vertices, and foci.

(a) $4x^2 - 5y^2 - 8x - 30y - 21 = 0$

(b) $5x^2 - 4y^2 - 10x - 24y - 51 = 0$

(c) $3x^2 - y^2 - 18x + 10y - 10 = 0$

(d) $4x^2 - y^2 + 8x + 6y + 11 = 0$

ANSWERS TO SUPPLEMENTARY PROBLEMS

17.12 (a) circle, Fig. 17-16

(b) hyperbola, Fig. 17-17

(c) ellipse, Fig. 17-18

(d) hyperbola, Fig. 17-19

(e) parabola, Fig. 17-20

(f) ellipse, Fig. 17-21

(g) hyperbola, Fig. 17-22

(h) parabola, Fig. 17-23

(i) single point, $(1, -1)$

(j) two intersecting lines, Fig. 17-24

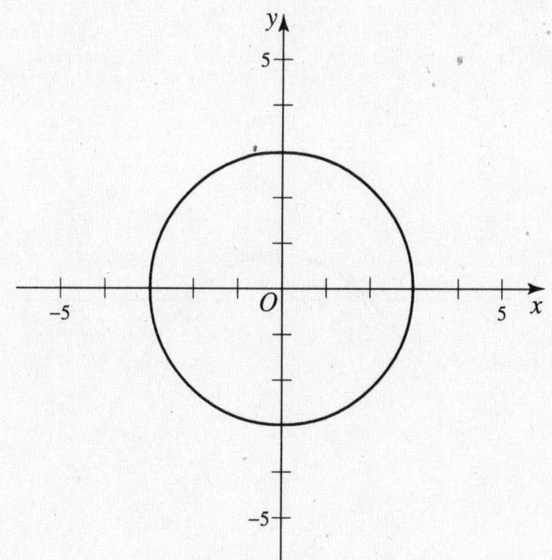

Fig. 17-16

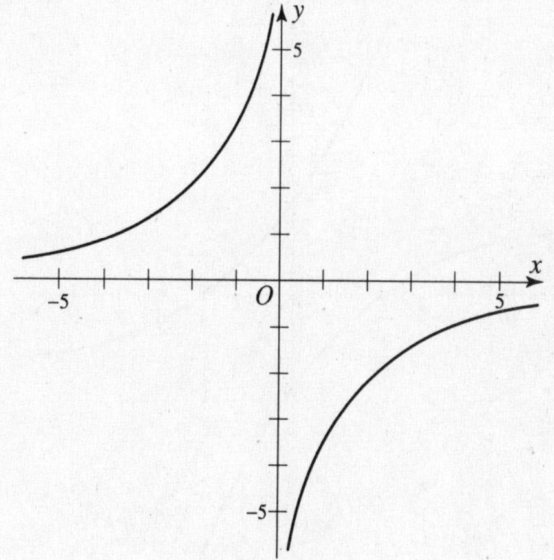

Fig. 17-17

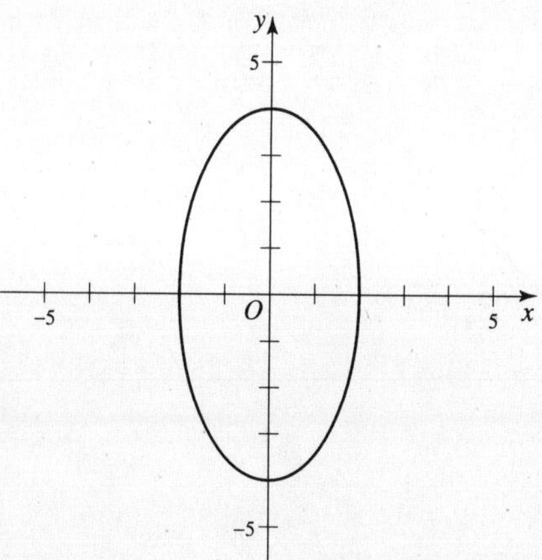

Fig. 17-18

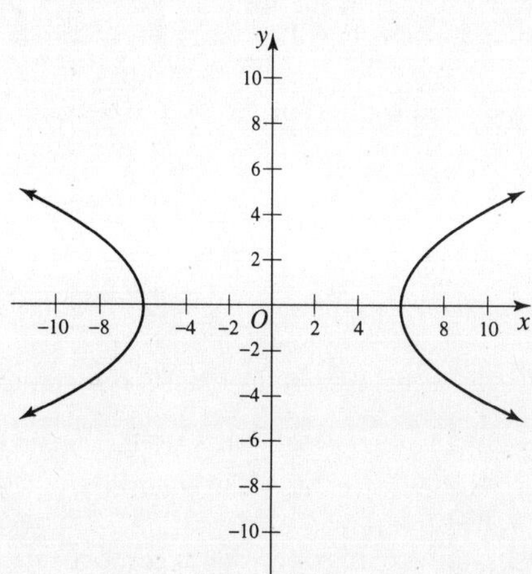

Fig. 17-19

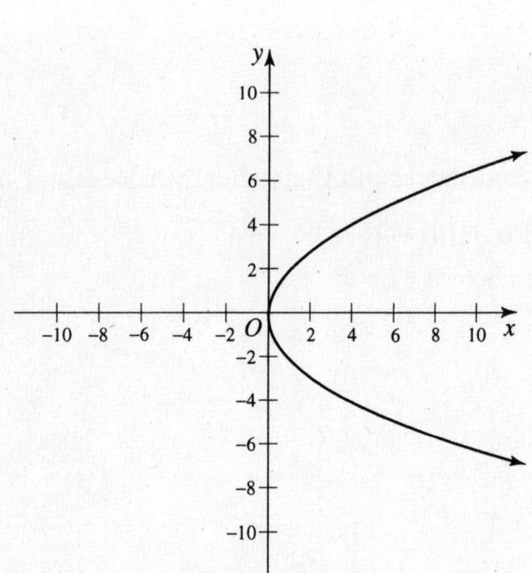

Fig. 17-20

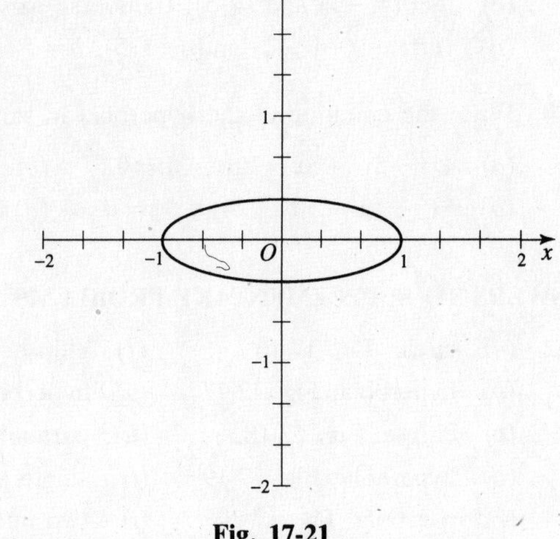

Fig. 17-21

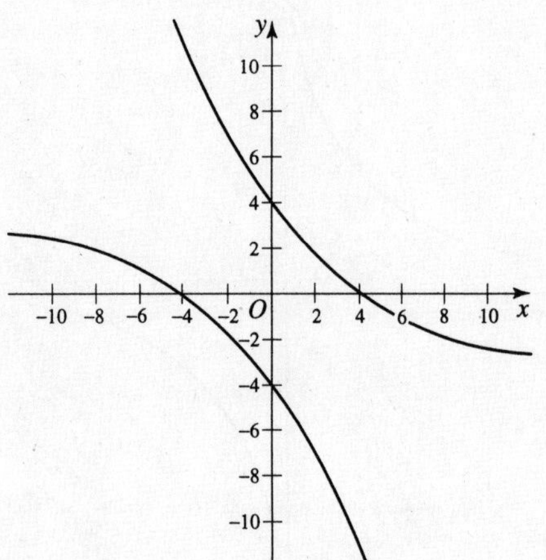

Fig. 17-22

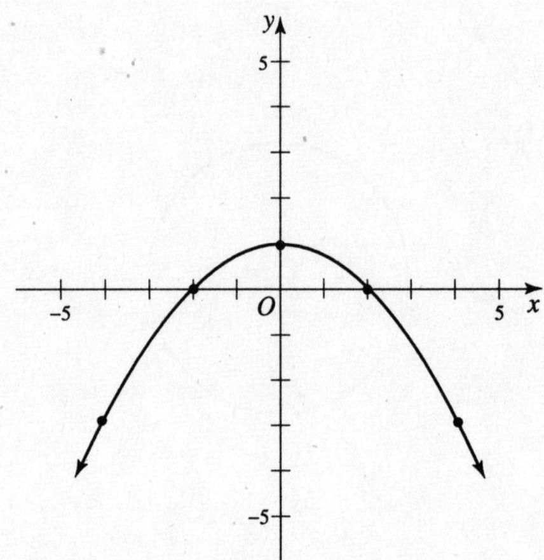

Fig. 17-23

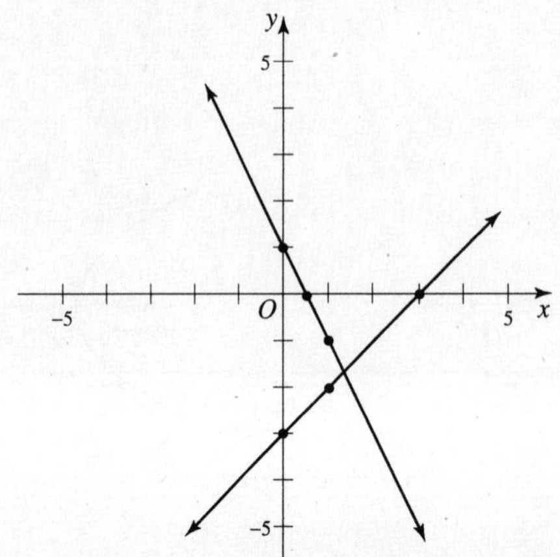

Fig. 17-24

17.13 (*a*) $(x-4)^2+(y-1)^2=9$
(*b*) $(x-5)^2+(y+3)^2=36$ (*c*) $x^2+y^2+4x-6y=0$
(*d*) $x^2+y^2+11y-y-32=0$

17.14 (*a*) $(x+3)^2+(y-6)^2=65$, $C(-3,6)$, $r=\sqrt{65}$
(*b*) $(x+6)^2+(y-2)^2=45$, $C(-6,2)$, $r=3\sqrt{5}$
(*c*) $(x+7/2)^2+(y+3/2)^2=49/2$, $C(-7/2,-3/2)$, $r=7\sqrt{2}/2$
(*d*) $(x-5/4)^2+(y-9/4)^2=9/8$, $C(5/4,9/4)$, $r=3\sqrt{2}/4$

17.15 (*a*) $(y+2)^2=8(x-3)$ (*b*) $(x-3)^2=20(y-5)$ (*c*) $y^2=20x$ (*d*) $(x-5)^2=-12(y-4)$

17.16 (*a*) $(y-4)^2=-4(x+3)$, $V(-3,4)$, $F(-4,4)$, directrix: $x=-2$, axis: $y=4$
(*b*) $(x-2)^2=-8(y+4)$, $V(2,-4)$, $F(2,\ -6)$, directrix: $y=-2$, axis: $x=2$
(*c*) $(y+3)^2=24(x+1)$, $V(-1,-3)$, $F(5,-3)$, directrix: $x=-7$, axis: $y=-3$
(*d*) $(x+2)^2=9(y-3)/5$, $V(-2,3)$, $F(-2,69/20)$, directrix: $y=51/20$, axis: $x=-2$.

17.17 (*a*) $\dfrac{x^2}{16}+\dfrac{y^2}{4}=1$ (*c*) $\dfrac{(x+3)^2}{25}+\dfrac{(y-4)^2}{9}=1$
(*b*) $\dfrac{y^2}{25}+\dfrac{x^2}{9}=1$ (*d*) $\dfrac{(y+2)^2}{16}+\dfrac{(x-3)^2}{4}=1$

17.18 (*a*) $\dfrac{(x-5)^2}{4}+\dfrac{(y-1)^2}{3}=1$, center $(5,1)$, vertices $(7,1)$ and $(3,1)$, foci $(6,1)$ and $(4,1)$, covertices $(5,1+\sqrt{3})$ and $(5,1-\sqrt{3})$

(*b*) $\dfrac{(y+2)^2}{16}+\dfrac{(x-2)^2}{7}=1$, center $(2,-2)$, vertices $(2,2)$ and $(2,-6)$, foci $(2,1)$ and $(2,-5)$, covertices $(2+\sqrt{7},-2)$ and $(2-\sqrt{7},-2)$

(*c*) $\dfrac{(y+5)^2}{9}+\dfrac{(x+3)^2}{8}=1$, center $(-3,-5)$, vertices $(-3,-2)$ and $(-3,-8)$, foci $(-3,-4)$ and $(-3,-6)$, covertices $(-3+2\sqrt{2},-5)$ and $(-3-2\sqrt{2},-5)$

(*d*) $\dfrac{(x-3)^2}{5}+\dfrac{(y-1)^2}{4}=1$, center $(3,1)$, vertices $(3+\sqrt{5},1)$ and $(3-\sqrt{5},1)$, foci $(4,1)$ and $(2,1)$, covertices $(3,3)$ and $(3,-1)$

17.19 (*a*) $\dfrac{x^2}{9}-\dfrac{y^2}{16}=1$ (*c*) $\dfrac{(y-2)^2}{1}-\dfrac{(x-4)^2}{8}=1$
(*b*) $\dfrac{y^2}{64}-\dfrac{x^2}{36}=1$ (*d*) $\dfrac{(y-1)^2}{9}-\dfrac{(x+1)^2}{25}=1$

17.20 (*a*) $\dfrac{(y+3)^2}{4}-\dfrac{(x-1)^2}{5}=1$, center $(1,-3)$, vertices $(1,-1)$ and $(1,-5)$, foci $(1,0)$ and $(1,-8)$

(*b*) $\dfrac{(x-1)^2}{4}-\dfrac{(y-3)^2}{5}=1$, center $(1,3)$, vertices $(-1,3)$ and $(3,3)$, foci $(4,3)$ and $(-2,3)$

(*c*) $\dfrac{(x-3)^2}{4}-\dfrac{(y+5)^2}{12}=1$, center $(3,-5)$, vertices $(5,-5)$ and $(1,-5)$, foci $(7,-5)$ and $(-1,-5)$

(*d*) $\dfrac{(y-3)^2}{16}-\dfrac{(x+1)^2}{4}=1$, center $(-1,3)$, vertices $(-1,7)$ and $(-1,-1)$, foci $(1,3+2\sqrt{5})$ and $(1,3-2\sqrt{5})$

Chapter 18

Systems of Equations Involving Quadratics

18.1 GRAPHICAL SOLUTION

The real simultaneous solutions of two quadratic equations in x and y are the values of x and y corresponding to the points of intersection of the graphs of the two equations. If the graphs do not intersect, the simultaneous solutions are imaginary.

18.2 ALGEBRAIC SOLUTION

A. One linear and one quadratic equation
Solve the linear equation for one of the unknowns and substitute in the quadratic equation.

EXAMPLE 18.1. Solve the system

(1) $x+y=7$
(2) $x^2+y^2=25$

Solving (1) for y, $y=7-x$. Substitute in (2) and obtain $x^2+(7-x)^2=25$, $x^2-7x+12=0$, $(x-3)(x-4)=0$, and $x=3,\ 4$. When $x=3$, $y=7-x=4$; when $x=4$, $y=7-x=3$. Thus the simultaneous solutions are $(3,4)$ and $(4,3)$.

B. Two equations of the form $ax^2+by^2=c$
Use the method of addition or subtraction.

EXAMPLE 18.2. Solve the system

(1) $2x^2-y^2=7$
(2) $3x^2+2y^2=14$

To eliminate y, multiply (1) by 2 and add to (2); then

$$7x^2=28, \qquad x^2=4 \qquad \text{and} \qquad x=\pm 2.$$

Now put $x=2$ or $x=-2$ in (1) and obtain $y=\pm 1$.
The four solutions are:

$$(2,1); \qquad (-2,1); \qquad (2,-1); \qquad (-2,-1)$$

C. Two equations of the form $ax^2+bxy+cy^2=d$

EXAMPLE 18.3. Solve the system

(1) $x^2+xy=6$
(2) $x^2+5xy-4y^2=10$

Method 1.
Eliminate the constant term between both equations. Multiply (1) by 5, (2) by 3, and subtract; then

$$x^2-5xy+6y^2=0,\ (x-2y)(x-3y)=0,\ x=2y \text{ and } x=3y.$$

Now put $x=2y$ in (1) or (2) and obtain $y^2=1$, $y=\pm 1$.
When $y=1$, $x=2y=2$; when $y=-1$, $x=2y=-2$. Thus two solutions are: $x=2$, $y=1$; $x=-2$, $y=-1$.

Then put $x = 3y$ in (1) or (2) and get

$$y^2 = \frac{1}{2}, \qquad y = \pm\frac{\sqrt{2}}{2}.$$

When

$$y = \frac{\sqrt{2}}{2}, \qquad x = 3y = \frac{3\sqrt{2}}{2};$$

when

$$y = -\frac{\sqrt{2}}{2}, \qquad x = -\frac{3\sqrt{2}}{2}.$$

Thus the four solutions are:

$$(2, 1); \qquad (-2, -1); \qquad \left(\frac{3\sqrt{2}}{2}, \frac{\sqrt{2}}{2}\right); \qquad \left(-\frac{3\sqrt{2}}{2}, -\frac{\sqrt{2}}{2}\right)$$

Method 2.
Let $y = mx$ in both equations.

From (1): $x^2 + mx^2 = 6, \qquad x^2 = \dfrac{6}{1+m}.$

From (2): $x^2 + 5mx^2 - 4m^2x^2 = 10, \qquad x^2 = \dfrac{10}{1+5m-4m^2}.$

Then

$$\frac{6}{1+m} = \frac{10}{1+5m-4m^2}$$

from which $m = \frac{1}{2}, \frac{1}{3}$; hence $y = x/2$, $y = x/3$. The solution proceeds as in Method 1.

D. Miscellaneous methods

(1) Some systems of equations may be solved by replacing them by equivalent and simpler systems (see Problems 18.8–18.10).

(2) An equation is called symmetric in x and y if interchange of x and y does not change the equation. Thus $x^2 + y^2 - 3xy + 4x + 4y = 8$ is symmetric in x and y. Systems of symmetric equations may often be solved by the substitutions $x = u + v$, $y = u - v$ (see Problems 18.11–18.12).

Solved Problems

18.1 Solve graphically the following systems:

$$(a)\ \begin{matrix} x^2 + y^2 = 25 \\ x + 2y = 10 \end{matrix}, \qquad (b)\ \begin{matrix} x^2 + 4y^2 = 16 \\ xy = 4 \end{matrix}, \qquad (c)\ \begin{matrix} x^2 + 2y = 9 \\ 2x^2 - 3y^2 = 1 \end{matrix}$$

SOLUTION

See Fig. 18-1.

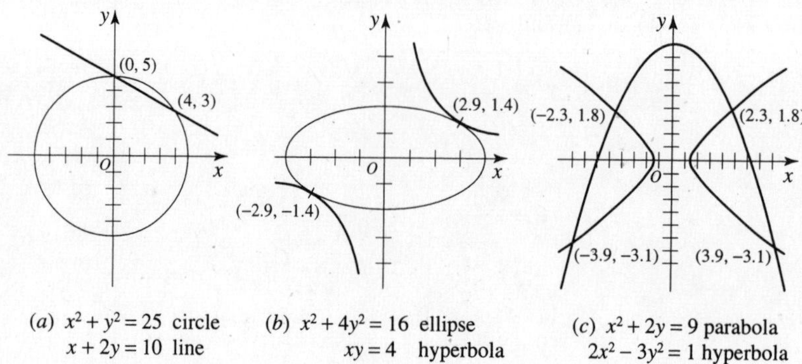

(*a*) $x^2 + y^2 = 25$ circle
$x + 2y = 10$ line

(*b*) $x^2 + 4y^2 = 16$ ellipse
$xy = 4$ hyperbola

(*c*) $x^2 + 2y = 9$ parabola
$2x^2 - 3y^2 = 1$ hyperbola

Fig. 18-1

18.2 Solve the following systems:

(*a*) $\begin{matrix} x + 2y = 4 \\ y^2 - xy = 7 \end{matrix}$, (*b*) $\begin{matrix} 3x - 1 + 2y = 0 \\ 3x^2 - y^2 + 4 = 0 \end{matrix}$

SOLUTION

(*a*) Solving the linear equation for x, $x = 4 - 2y$. Substituting in the quadratic equation,

$$y^2 - y(4 - 2y) = 7,\ 3y^2 - 4y - 7 = 0,\ (y + 1)(3y - 7) = 0 \text{ and } y = -1,\ 7/3.$$

If $y = -1$, $x = 4 - 2y = 6$; if $y = 7/3$, $x = 4 - 2y = -2/3$.
The solutions are $(6, -1)$ and $(-2/3, 7/3)$.

(*b*) Solving the linear equations for y, $y = \frac{1}{2}(1 - 3x)$. Substituting in the quadratic equation,

$$3x^2 - [\tfrac{1}{2}(1 - 3x)]^2 + 4 = 0, \qquad x^2 + 2x + 5 = 0 \qquad \text{and} \qquad x = \frac{-2 \pm \sqrt{2^2 - 4(1)(5)}}{2(1)} = -1 \pm 2i.$$

If $x = -1 + 2i$, $y = \frac{1}{2}(1 - 3x) = \frac{1}{2}[1 - 3(-1 + 2i)] = \frac{1}{2}(4 - 6i) = 2 - 3i$.
If $x = -1 - 2i$, $y = \frac{1}{2}(1 - 3x) = \frac{1}{2}[1 - 3(-1 - 2i)] = \frac{1}{2}(4 + 6i) = 2 + 3i$.
The solutions are $(-1 + 2i, 2 - 3i)$ and $(-1 - 2i, 2 + 3i)$.

18.3 Solve the system: (1) $2x^2 - 3y^2 = 6$, (2) $3x^2 + 2y^2 = 35$.

SOLUTION

To eliminate y, multiply (1) by 2, (2) by 3 and add; then $13x^2 = 117$, $x^2 = 9$, $x = \pm 3$.
Now put $x = 3$ or $x = -3$ in (1) and obtain $y = \pm 2$.
The solutions are: $(3, 2)$; $(-3, 2)$; $(3, -2)$; $(-3, -2)$.

18.4 Solve the system:

(1) $\dfrac{8}{x^2} - \dfrac{3}{y^2} = 5$, (2) $\dfrac{5}{x^2} + \dfrac{2}{y^2} = 38$.

SOLUTION

The equations are quadratic in $\dfrac{1}{x}$ and $\dfrac{1}{y}$. Substituting $u = \dfrac{1}{x}$ and $v = \dfrac{1}{y}$, we obtain

$$8u^2 - 3v^2 = 5 \qquad \text{and} \qquad 5u^2 + 2v^2 = 38.$$

Solving simultaneously, $u^2 = 4$, $v^2 = 9$ or $x^2 = 1/4$, $y^2 = 1/9$; then $x = \pm 1/2$, $y = \pm 1/3$.
The solutions are:

$$\left(\frac{1}{2}, \frac{1}{3}\right); \quad \left(-\frac{1}{2}, \frac{1}{3}\right), \quad \left(\frac{1}{2}, -\frac{1}{3}\right); \quad \left(-\frac{1}{2}, -\frac{1}{3}\right).$$

18.5 Solve the system

$$(1)\quad 5x^2 + 4y^2 = 48$$
$$(2)\quad x^2 + 2xy = 16$$

by eliminating the constant terms.

SOLUTION

Multiply (2) by 3 and subtract from (1) to obtain

$2x^2 - 6xy + 4y^2 = 0, \quad x^2 - 3xy + 2y^2 = 0, \quad (x - y)(x - 2y) = 0 \quad$ and $\quad x = y, \quad x = 2y.$

Substituting $x = y$ in (1) or (2), we have $y^2 = \dfrac{16}{3}$ and $y = \pm\dfrac{4}{3}\sqrt{3}$.

Substituting $x = 2y$ in (1) or (2), we have $y^2 = 2$ and $y = \pm\sqrt{2}$.

The four solutions are:

$$\left(\frac{4\sqrt{3}}{3}, \frac{4\sqrt{3}}{3}\right); \quad \left(-\frac{4\sqrt{3}}{3}, -\frac{4\sqrt{3}}{3}\right); \quad (2\sqrt{2}, \sqrt{2}); \quad (-2\sqrt{2}, -\sqrt{2}).$$

18.6 Solve the system

$$(1)\quad 3x^2 - 4xy = 4$$
$$(2)\quad x^2 - 2y^2 = 2$$

by using the substitution $y = mx$.

SOLUTION

Put $y = mx$ in (1); then $3x^2 - 4mx^2 = 4$ and $x^2 = \dfrac{4}{3 - 4m}$.

Put $y = mx$ in (2); then $x^2 - 2m^2x^2 = 2$ and $x^2 = \dfrac{2}{1 - 2m^2}$.

Thus $\dfrac{4}{3 - 4m} = \dfrac{2}{1 - 2m^2}, \quad 4m^2 - 4m + 1 = 0, \quad (2m - 1)^2 = 0$ and $m = \dfrac{1}{2}, \dfrac{1}{2}$.

Now substitute $y = mx = \frac{1}{2}x$ in (1) or (2) and obtain $x^2 = 4$, $x = \pm 2$.

The solutions are $(2,1)$ and $(-2,-1)$.

18.7 Solve the system: (1) $x^2 + y^2 = 40$, (2) $xy = 12$.

SOLUTION

From (2), $y = 12/x$; substituting in (1), we have

$$x^2 + \frac{144}{x^2} = 40, \quad x^4 - 40x^2 + 144 = 0, \quad (x^2 - 36)(x^2 - 4) = 0 \quad \text{and} \quad x = \pm 6, \quad \pm 2.$$

For $x = \pm 6$, $y = 12/x = \pm 2$; for $x = \pm 2$, $y = \pm 6$.

The four solutions are: $(6, 2)$; $(-6, -2)$; $(2, 6)$; $(-2, -6)$.

Note. Equation (2) indicates that those solutions in which the product xy is negative (e.g. $x = 2$, $y = -6$) are extraneous.

18.8 Solve the system: (1) $x^2 + y^2 + 2x - y = 14$, (2) $x^2 + y^2 + x - 2y = 9$.

SOLUTION

Subtract (2) from (1): $x + y = 5$ or $y = 5 - x$.

Substitute $y = 5 - x$ in (1) or (2): $2x^2 - 7x + 6 = 0$, $(2x - 3)(x - 2) = 0$ and $x = 3/2, 2$.

The solutions are $(\frac{3}{2}, \frac{7}{2})$ and $(2, 3)$.

18.9 Solve the system: (1) $x^3 + y^3 = 35$, (2) $x + y = 5$.

SOLUTION

Dividing (1) by (2),

$$\frac{x^3 + y^3}{x + y} = \frac{35}{5} \quad \text{and} \quad (3)\ x^2 - xy + y^2 = 7.$$

From (2), $y = 5 - x$; substituting in (3), we have

$$x^2 - x(5 - x) + (5 - x)^2 = 7, \quad x^2 - 5x + 6 = 0, \quad (x - 3)(x - 2) = 0 \quad \text{and} \quad x = 3, 2.$$

The solutions are $(3, 2)$ and $(2, 3)$.

18.10 Solve the system: (1) $x^2 + 3xy + 2y^2 = 3$, (2) $x^2 + 5xy + 6y^2 = 15$.

SOLUTION

Dividing (1) by (2),

$$\frac{x^2 + 3xy + 2y^2}{x^2 + 5xy + 6y^2} = \frac{(x + y)(x + 2y)}{(x + 3y)(x + 2y)} = \frac{x + y}{x + 3y} = \frac{1}{5}.$$

From $\dfrac{x + y}{x + 3y} = \dfrac{1}{5}$, $y = -2x$. Substituting $y = -2x$ in (1) or (2), $x^2 = 1$ and $x = \pm 1$.

The solutions are $(1, -2)$ and $(-1, 2)$.

18.11 Solve the system: (1) $x^2 + y^2 + 2x + 2y = 32$, (2) $x + y + 2xy = 22$.

SOLUTION

The equations are symmetric in x and y since interchange of x and y yields the same equation. Substituting $x = u + v$, $y = u - v$ in (1) and (2), we obtain

$$(3)\quad u^2 + v^2 + 2u = 16 \quad \text{and} \quad (4)\quad u^2 - v^2 + u = 11.$$

Adding (3) and (4), we get $2u^2 + 3u - 27 = 0$, $(u - 3)(2u + 9) = 0$ and $u = 3, -9/2$.

When $u = 3$, $v^2 = 1$ and $v = \pm 1$; when $u = -9/2$, $v^2 = 19/4$ and $v = \pm\sqrt{19}/2$. Thus the solutions of (3) and (4) are: $u = 3, v = 1$; $u = 3, v = -1$; $u = -9/2, v = \sqrt{19}/2$; $u = -9/2, v = -\sqrt{19}/2$.

Then, since $x = u + v$, $y = u - v$, the four solutions of (1) and (2) are:

$$(4, 2); \quad (2, 4); \quad \left(\frac{-9 + \sqrt{19}}{2}, \frac{-9 - \sqrt{19}}{2}\right); \quad \left(\frac{-9 - \sqrt{19}}{2}, \frac{-9 + \sqrt{19}}{2}\right).$$

18.12 Solve the system:

$$(1)\quad x^2 + y^2 = 180, \qquad (2)\quad \frac{1}{x} + \frac{1}{y} = \frac{1}{4}.$$

SOLUTION

From (2) obtain (3) $4x + 4y - xy = 0$. Since (1) and (3) are symmetric in x and y, substitute $x = u + v$, $y = u - v$ in (1) and (3) and obtain

$$(4)\quad u^2 + v^2 = 90 \quad \text{and} \quad (5)\quad 8u - u^2 + v^2 = 0.$$

Subtracting (5) from (4), we have $u^2 - 4u - 45 = 0$, $(u - 9)(u + 5) = 0$ and $u = 9, -5$.

When $u = 9$, $v = \pm 3$; when $u = -5$, $v = \pm\sqrt{65}$. Thus the solutions of (4) and (5) are: $u = 9, v = 3$; $u = 9, v = -3$; $u = -5, v = \sqrt{65}$; $u = -5, v = -\sqrt{65}$.

Hence the four solutions of (1) and (2) are:

$$(12, 6); \quad (6, 12); \quad (-5 + \sqrt{65}, -5 - \sqrt{65}); \quad (-5 - \sqrt{65}, -5 + \sqrt{65}).$$

18.13 The sum of two numbers is 25 and their product is 144. What are the numbers?

SOLUTION

Let the numbers be x, y. Then (1) $x + y = 25$ and (2) $xy = 144$.

The simultaneous solutions of (1) and (2) are $x = 9$, $y = 16$ and $x = 16$, $y = 9$. Hence the required numbers are 9, 16.

18.14 The difference of two positive numbers is 3 and the sum of their squares is 65. Find the numbers.

SOLUTION

Let the numbers be p, q. Then (1) $p - q = 3$ and (2) $p^2 + q^2 = 65$.

The simultaneous solutions of (1) and (2) are $p = 7$, $q = 4$ and $p = -4$, $q = -7$. Hence the required (positive) numbers are 7, 4.

18.15 A rectangle has perimeter 60 ft and area 216 ft^2. Find its dimensions.

SOLUTION

Let the rectangle have sides of lengths x, y. Then (1) $2x + 2y = 60$ and (2) $xy = 216$.

Solving (1) and (2) simultaneously, the required sides are 12 and 18 ft.

18.16 The hypotenuse of a right triangle is 41 ft long and the area of the triangle is 180 ft^2. Find the lengths of the two legs.

SOLUTION

Let the legs have lengths x, y. Then (1) $x^2 + y^2 = (41)^2$ and (2) $\frac{1}{2}(xy) = 180$.

Solving (1) and (2) simultaneously, we find the legs have lengths 9 and 40 ft.

Supplementary Problems

18.17 Solve the following systems graphically.

(a) $x^2 + y^2 = 20$, $3x - y = 2$

(b) $x^2 + 4y^2 = 25$, $x^2 - y^2 = 5$

(c) $y^2 = x$, $x^2 + 2y^2 = 24$

(d) $x^2 + 1 = 4y$, $3x - 2y = 2$

18.18 Solve the following systems algebraically.

(a) $2x^2 - y^2 = 14$, $x - y = 1$

(b) $xy + x^2 = 24$, $y - 3x + 4 = 0$

(c) $3xy - 10x = y$, $2 - y + x = 0$

(d) $4x + 5y = 6$, $xy = -2$

(e) $2x^2 - y^2 = 5$, $3x^2 + 4y^2 = 57$

(f) $9/x^2 + 16/y^2 = 5$, $18/x^2 - 12/y^2 = -1$

(g) $x^2 - xy = 12$, $xy - y^2 = 3$

(h) $x^2 + 3xy = 18$, $x^2 - 5y^2 = 4$

(i) $x^2 + 2xy = 16$, $3x^2 - 4xy + 2y^2 = 6$

(j) $x^2 - xy + y^2 = 7$, $x^2 + y^2 = 10$

(k) $x^2 - 3y^2 + 10y = 19$, $x^2 - 3y^2 + 5x = 9$

(l) $x^3 - y^3 = 9$, $x - y = 3$

(m) $x^3 - y^3 = 19$, $x^2y - xy^2 = 6$

(n) $1/x^3 + 1/y^3 = 35$, $1/x^2 - 1/xy + 1/y^2 = 7$

18.19 The square of a certain number exceeds twice the square of another number by 16. Find the numbers if the sum of their squares is 208.

18.20 The diagonal of a rectangle is 85 ft. If the short side is increased by 11 ft and the long side decreased by 7 ft, the length of the diagonal remains the same. Find the dimensions of the original rectangle.

ANSWERS TO SUPPLEMENTARY PROBLEMS

18.17 (*a*) $(2, 4)$, $(-0.8, -4.4)$
(*b*) $(3, 2)$, $(-3, 2)$, $(3, -2)$, $(-3, -2)$
(*c*) $(4, 2)$, $(4, -2)$
(*d*) $(1, 0.5)$, $(5, 6.5)$

18.18 (*a*) $(3, 2)$, $(-5, -6)$
(*b*) $(3, 5)$, $(-2, -10)$
(*c*) $(2, 4)$, $(-1/3, 5/3)$
(*d*) $(-1, 2)$, $(5/2, -4/5)$
(*e*) $(\sqrt{7}, 3)$, $(\sqrt{7}, -3)$, $(-\sqrt{7}, 3)$, $(-\sqrt{7}, -3)$
(*f*) $(3, 2)$, $(3, -2)$, $(-3, 2)$, $(-3, -2)$
(*g*) $(4, 1)$, $(-4, -1)$
(*h*) $(3, 1)$, $(-3, -1)$, $\left(3i\sqrt{5}, \dfrac{-7i\sqrt{5}}{5}\right)$, $\left(-3i\sqrt{5}, \dfrac{7i\sqrt{5}}{5}\right)$
(*i*) $(2, 3)$, $(-2, -3)$
(*j*) $(1, 3)$, $(-1, -3)$, $(3, 1)$, $(-3, -1)$
(*k*) $(-12, -5)$, $(4, 3)$
(*l*) $(1, -2)$, $(2, -1)$
(*m*) $(-2, -3)$, $(3, 2)$
(*n*) $(1/2, 1/3)$, $(1/3, 1/2)$

18.19 $12, 8$; $-12, -8$; $12, -8$; $-12, 8$

18.20 40 ft, 75 ft

Chapter 19

Inequalities

19.1 DEFINITIONS

An *inequality* is a statement that one real quantity or expression is greater or less than another real quantity or expression.

The following indicate the meaning of inequality signs.

(1) $a>b$ means "a is greater than b" (or $a-b$ is a positive number).

(2) $a<b$ means "a is less than b" (or $a-b$ is a negative number).

(3) $a \geq b$ means "a is greater than or equal to b."

(4) $a \leq b$ means "a is less than or equal to b."

(5) $0<a<2$ means "a is greater than zero but less than 2."

(6) $-2 \leq x<2$ means "x is greater than or equal to -2 but less than 2."

An *absolute inequality* is true for all real values of the letters involved. For example, $(a-b)^2>-1$ holds for all real values of a and b, since the square of any real number is positive or zero.

A *conditional inequality* holds only for particular values of the letters involved. Thus $x-5>3$ is true only when x is greater than 8.

The inequalities $a>b$ and $c>d$ have the *same sense*. The inequalities $a>b$ and $x<y$ have *opposite sense*.

19.2 PRINCIPLES OF INEQUALITIES

(1) The sense of an inequality is unchanged if each side is increased or decreased by the same real number. It follows that any term may be transposed from one side of an inequality to the other, provided the sign of the term is changed.

Thus if $a>b$, then $a+c>b+c$, and $a-c>b-c$, and $a-b>0$.

(2) The sense of an inequality is unchanged if each side is multiplied or divided by the same positive number.

Thus if $a>b$ and $k>0$, then

$$ka>kb \quad \text{and} \quad \frac{a}{k}>\frac{b}{k}.$$

(3) The sense of an inequality is reversed if each side is multiplied or divided by the same negative number.

Thus if $a>b$ and $k<0$, then

$$ka<kb \quad \text{and} \quad \frac{a}{k}<\frac{b}{k}.$$

(4) If $a>b$ and a,b,n are positive, then $a^n>b^n$ but $a^{-n}<b^{-n}$.

EXAMPLES 19.1.

$$5>4;\ \text{then } 5^3>4^3 \quad \text{or} \quad 125>64,\ \text{but } 5^{-3}<4^{-3} \quad \text{or} \quad \frac{1}{125}<\frac{1}{64}.$$

$$16>9;\ \text{then } 16^{1/2}>9^{1/2} \quad \text{or} \quad 4>3,\ \text{but } 16^{-1/2}<9^{-1/2} \quad \text{or} \quad \frac{1}{4}<\frac{1}{3}.$$

(5) If $a > b$ and $c > d$, then $(a + c) > (b + d)$.

(6) If $a > b > 0$ and $c > d > 0$, then $ac > bd$.

19.3 ABSOLUTE VALUE INEQUALITIES

The absolute value of a quantity represents the distance that the value of the expression is from zero on a number line. So $|x - a| = b$, where $b > 0$, says that the quantity $x - a$ is b units from 0, $x - a$ is b units to the right of 0, or $x - a$ is b units to the left of 0. When we say $|x - a| > b$, $b > 0$, then $x - a$ is at a distance from 0 that is greater than b. Thus, $x - a > b$ or $x - a < -b$. Similarly, if $|x - a| < b$, $b > 0$, then $x - a$ is at a distance from 0 that is less than b. Hence, $x - a$ is between b units below 0, $-b$, and b units above 0.

EXAMPLES 19.2. Solve each of these inequalities for x.

(*a*) $|x - 3| > 4$ (*b*) $|x + 4| < 7$ (*c*) $|x - 5| < -3$ (*d*) $|x - 5| > -5$

(*a*) $|x - 3| > 4$, then $x - 3 > 4$ or $x - 3 < -4$. Thus, $x > 7$ or $x < -1$. The solution interval is $(-\infty, -1) \cup (7, \infty)$, (where $\cup$ represents the union of the two intervals).

(*b*) $|x + 4| < 7$ then $-7 < x + 4 < 7$. Thus, $-11 < x < 3$. The solution interval is $(-11, 3)$.

(*c*) $|x - 5| < -3$ Since the absolute value of a number is always greater than or equal to zero, there are no values for which the absolute value will be less than -3. Thus, there is no solution and we may write $\varnothing$ for the solution interval.

(*d*) $|x + 3| > -5$ Since the absolute value of a number is always at least zero, it is always greater than -5. Thus the solution is all real numbers, and for the solution interval we write $(-\infty, \infty)$.

19.4 HIGHER DEGREE INEQUALITIES

Solving higher degree inequalities is similar to solving higher degree equations: we must always compare the expression to zero. If $f(x) > 0$, then we are interested in the values of x that will produce a product and/or quotient of factors that is positive, while if $f(x) < 0$, we wish to find the values of x that will produce a product and/or quotient that is negative.

If $f(x)$ is a quadratic expression we have just two factors to consider, and we can do this by examining cases based on the possible signs of the two factors that will produce the desired sign for the expression (see Problems 19.3(*c*) and 19.14). When the number of factors in $f(x)$ increases by one, the number of cases to consider doubles. Thus, for an expression with 2 factors there are 4 cases, with 3 factors there are 8 cases, and with 4 factors there are 16 cases. In each instance half of the cases will produce a positive expression and half a negative one. Thus, the case procedure gets to be a very long one quite quickly. An alternative procedure to the case method is the sign chart.

EXAMPLE 19.3. Solve the inequality $x^2 + 15 < 8x$.

The inequality $x^2 + 8x < 15$ is equivalent to $x^2 - 8x + 15 < 0$ and to $(x - 3)(x - 5) < 0$ and is true when the product of $x - 3$ and $x - 5$ is negative. The critical values of the product are the values that make these factors 0, because they represent where the product may change signs.

The critical values of x, 3 and 5, are placed on a number line and divide it into three intervals. We need to find the sign of the product of $x - 3$ and $x - 5$ on each of these intervals to find the solution (see Fig. 19-1). Vertical lines are drawn through each critical value. A dashed line indicates that the critical value is not in the solution and a solid line indicates that the critical value is in the solution.

The signs above the number line are the signs for the factors and are found by selecting an arbitrary value in the interval as a test value and determining whether each factor is positive or negative for the test value. For the interval to the left of 3, we choose a test value of 1 and substitute it into $x - 3$ and see that the value is -2, so we record a $-$ sign, and for $x - 5$ the value is -4 and again we record a $-$ sign. For the interval between 3 and 5 we choose any value, such as 3.5, and determine that $x - 3$ is positive and $x - 5$ is negative. Finally, for the interval to the right of 5, we choose a value of 12 and see that both $x - 3$ and $x - 5$ are positive. The

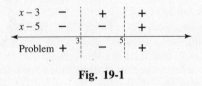

		3		5	
$x-3$	$-$		$+$		$+$
$x-5$	$-$		$-$		$+$
Problem	$+$		$-$		$+$

Fig. 19-1

sign for the problem, written below the line, in each interval is determined by the signs of the factors in that interval. If an even number of factors in a product or quotient are negative, the product or quotient is positive. If an odd number of factors are negative, the product or quotient is negative.

We select the intervals that satisfy our problem $(x-3)(x-5)<0$, so we select the intervals that are negative in the sign chart. In the interval between 3 and 5 the problem is negative (see Fig. 19-1), so the solution is the interval (3, 5). The parentheses mean that the 3 and 5 are not included in the interval, and we know this since the boundary lines are dashed. If they had been in the solution, we would have used a bracket instead of a parenthesis at the end of the interval next to the 3.

The solution for $x^2+15<8x$ is the interval (3, 5).

EXAMPLE 19.4. Solve the inequality

$$\frac{x-3}{x(x+4)} \geq 0.$$

The inequality is compared to 0 and the numerator and denominator are factored, so we can see that the critical values for the problem are the solution of $x=0$, $x-3=0$, and $x+4=0$. Thus, the critical values are $x=0$, $x=3$ and $x=-4$. Since there are three critical values, the number line is divided into four distinct intervals, as shown in Fig. 19-2.

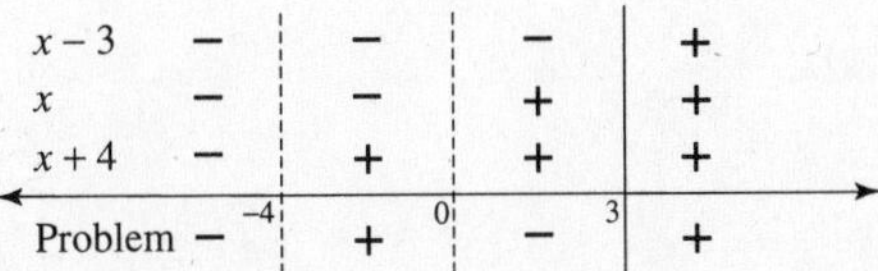

Fig. 19-2

The signs above the line are the signs of each factor in each interval. The sign below is the sign for the problem and it is + when an even number of factors are negative and − when an odd number of factors are negative. Since the problem uses the $\geq$ sign, values that make the numerator zero are solutions, so a solid line is drawn through 3. Since 0 and -4 make the denominator of the fraction 0, they are not solutions and dashed lines were drawn through 0 and -4 (see Fig. 19-2).

Since the problem

$$\frac{x-3}{x(x+4)} \geq 0$$

indicates that a positive or zero value is wanted, we want the regions with a + sign in the sign chart. Thus, the solutions are the intervals, $(-4, 0)$ and $[3, \infty)$, and the solution is written $(-4, 0) \cup [3, \infty)$. The $\cup$ indicates that we want the union of the two intervals. Note that the bracket, [, is used because the critical value 3 is in the solution and a parenthesis,), is always used for the infinite, ∞, side of an interval.

19.5 LINEAR INEQUALITIES IN TWO VARIABLES

The solution of linear inequalities in two variables x and y consists of all points (x, y) that satisfy the inequality. Since a linear equation represents a line, a linear inequality is the points on one side of a line. The points on the line are included when the sign $\geq$ or $\leq$ is used in the statement of the inequality. The solutions of linear inequalities are usually found by graphical methods.

EXAMPLE 19.5. Find the solution for $2x - y \le 3$.

We graph the line related to the inequality $2x - y \le 3$, which is $2x - y = 3$. Since the symbol $\le$ is used, the line is part of the solution and a solid line is used to indicate this (see Fig. 19-3). If the line is not part of the solution, we use a dashed line to indicate that fact. We shade the region on the side of the line where the points are solutions of the inequality. The solution region is determined by selecting a test point that is not on the line. If the test point satisfies the inequality, then all points on that side of the line are in the solution. If the test point does not satisfy the inequality, no points on that side of the line are in the solution. Hence the solution points are on the opposite side of the line from the test point.

The point $P(2, 4)$ is not on the line $2x - y = 3$, so it can be used as a test point. When we substitute $(2, 4)$ into the inequality $2x - y \le 3$, we get $2(2) - 4 \le 3$, which is true, since $0 \le 3$. We shade on the side of the line that contains the test point $(2, 4)$ to indicate the solution region. If we had selected $Q(5, -2)$ and substituted into $2x - y \le 3$, we would have obtained $12 \le 3$, which is false, and would have shaded on the opposite side of the line from Q. This is the same region we found using the test point P.

The solution for $2x - y \le 3$ is shown in Fig. 19-3 and consists of the shaded region and the line.

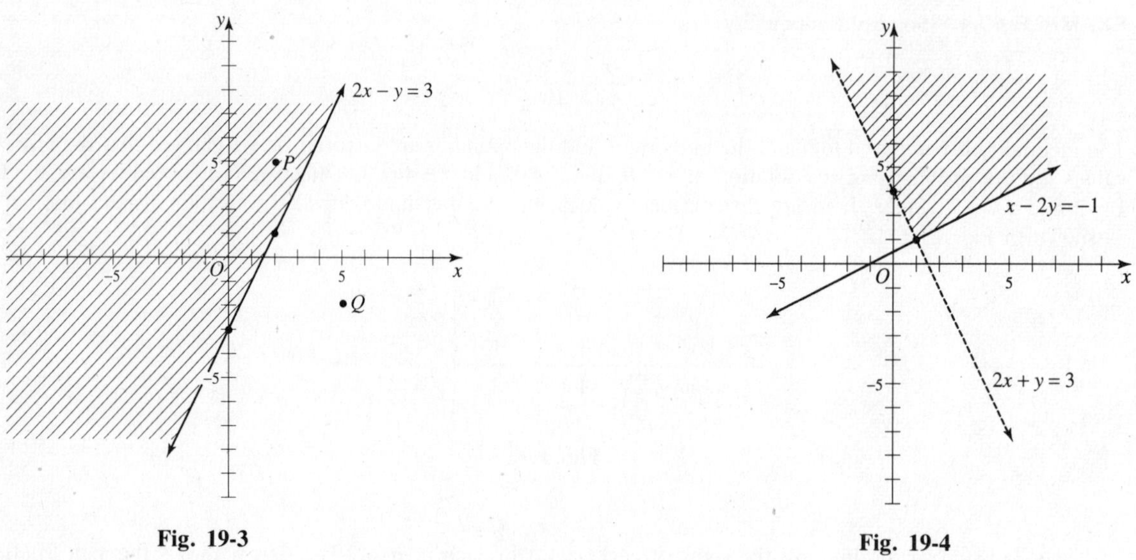

Fig. 19-3

Fig. 19-4

19.6 SYSTEMS OF LINEAR INEQUALITIES

If we have two or more linear inequalities in two variables, we say we have a system of linear inequalities and the solution of the system is the intersection, or common region, of the solution regions for the inequalities.

A system with two inequalities whose related equations intersect always has a solution region. If the related equations are parallel, the system may or may not have a solution. Systems with three or more inequalities may or may not have a solution.

EXAMPLE 19.6. Solve the system of inequalities $2x + y > 3$ and $x - 2y \le -1$.

We graph the related equations $2x + y = 3$ and $x - 2y = -1$ on the same set of axes. The line $2x + y = 3$ is dashed, since it is not included in $2x + y > 3$, but the line $x - 2y = -1$ is solid, since it is included in $x - 2y \le -1$.

Now we select a test such as $(0, 5)$ that is not on either line, determine which side of each line to shade and shade only the common region. Since $2(0) + 5 > 3$ is true, the solution region is to the right and above the line $2x + y = 3$. Since $0 - 2(5) \le -1$ is true, the solution region is to the left and above the line $x - 2y = -1$.

The solution region of $2x + y > 3$ and $x - 2y \le -1$ is the shaded region of Fig. 19-4, which includes the part of the solid line bordering the shaded region.

19.7 LINEAR PROGRAMMING

Many practical problems from business involve a function (objective) that is to be either maximized or minimized subject to a set of conditions (constraints). If the objective is a linear function and the constraints are linear inequalities, the values, if any, that maximize or minimize the objective occur at the corners of the region determined by the constraints.

EXAMPLE 19.7. The Green Company uses three grades of recycled paper, called grades A, B, and C, produced from scrap paper it collects. Companies that produce these grades of recycled paper do so as the result of a single operation, so the proportion of each grade of paper is fixed for each company. The Ecology Company process produces 1 unit of grade A, 2 units of grade B, and 3 units of grade C for each ton of paper processed and charges \$300 for the processing. The Environment Company process produces 1 unit of grade A, 5 units of grade B, and 1 unit of grade C for each ton of paper processed and charges \$500 for processing. The Green Company needs at least 100 units of grade A paper, 260 units of grade B paper, and 180 units of grade C paper. How should the company place its order so that costs are minimized?

If x represents the number of tons of paper to be recycled by the Ecology Company and y represents the number of tons of paper to be processed by the Environment Company, then the objective function is $C(x, y) = 300x + 500y$, and we want to minimize $C(x, y)$.

The constraints stated in terms of x and y are for grade A: $1x + 1y \geq 100$; for grade B: $2x + 5y \geq 260$; and for grade C: $3x + 1y \geq 180$. Since you can not have a company process a negative number of tons of paper, $x \geq 0$ and $y \geq 0$. These last two constraints are called natural or implied constraints, because these conditions are true as a matter of fact and need not be stated in the problem.

We graph the inequalities determined from the constraints (see Fig. 19-5). The vertices of the region are $A(0, 180)$, $B(40, 60)$, $C(80, 20)$, and $D(130, 0)$.

The minimum for $C(x, y)$, if it exists, will occur at point A, B, C, or D, so we evaluate the obective function at these points.

$$C(0, 180) = 300(0) + 500(180) = 0 + 90\,000 = 90\,000$$
$$C(40, 60) = 300(40) + 500(60) = 12\,000 + 30\,000 = 42\,000$$
$$C(80, 20) = 300(80) + 500(20) = 24\,000 + 10\,000 = 34\,000$$
$$C(130, 0) = 300(130) + 500(0) = 39\,000 + 0 = 39\,000$$

The Green Printing Company can minimize the cost of recycled paper to \$34 000 by having the Ecology Company process 80 tons of paper and the Environment Company process 20 tons of paper.

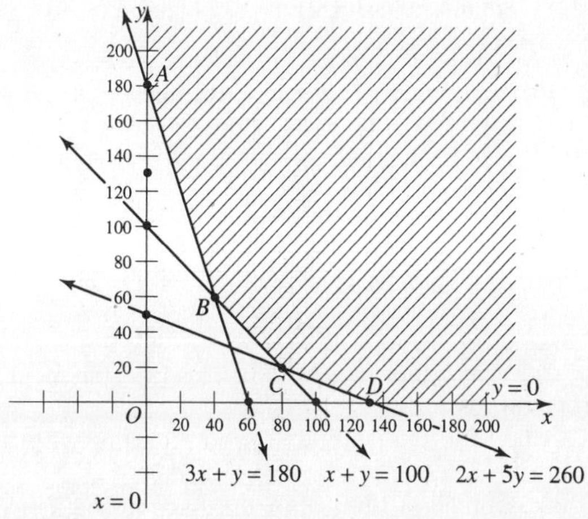

Fig. 19-5

Solved Problems

19.1 If $a > b$ and $c > d$, prove that $a + c > b + d$.

SOLUTION

Since $(a - b)$ and $(c - d)$ are positive, $(a - b) + (c - d)$ is positive.

Hence $(a - b) + (c - d) > 0$, $(a + c) - (b + d) > 0$ and $(a + c) > (b + d)$.

19.2 Find the fallacy.

(*a*) Let $a = 3$, $b = 5$; then $a < b$
(*b*) Multiply by a: $a^2 < ab$
(*c*) Subtract b^2: $a^2 - b^2 < ab - b^2$
(*d*) Factor: $(a + b)(a - b) < b(a - b)$
(*e*) Divide by $a - b$: $a + b < b$
(*f*) Substitute $a = 3$, $b = 5$: $8 < 5$

SOLUTION

There is nothing wrong with steps (*a*), (*b*), (*c*), (*d*). The error is made in step (*e*) where the inequality is divided by $a - b$, a negative number, without reversing the inequality sign.

19.3 Find the values of x for which each of the following inequalities holds.

(*a*) $4x + 5 > 2x + 9$. We have $4x - 2x > 9 - 5$, $2x > 4$ and $x > 2$.

(*b*) $\frac{x}{2} - \frac{1}{3} < \frac{2x}{3} + \frac{1}{2}$. Multiplying by 6, we obtain

$$3x - 2 < 4x + 3,\ 3x - 4x < 2 + 3,\ -x < 5,\ x > -5.$$

(*c*) $x^2 < 16$.

Method 1. $x^2 - 16 < 0$, $(x - 4)(x + 4) < 0$. The product of the factors $(x - 4)$ and $(x + 4)$ is negative. Two cases are possible.

(1) $x - 4 > 0$ and $x + 4 < 0$ simultaneously. Thus $x > 4$ and $x < -4$. This is impossible, as x cannot be both greater than 4 and less than -4 simultaneously.

(2) $x - 4 < 0$ and $x + 4 > 0$ simultaneously. Thus $x < 4$ and $x > -4$. This is possible if and only if $-4 < x < 4$. Hence $-4 < x < 4$.

Method 2. $(x^2)^{1/2} < (16)^{1/2}$. Now $(x^2)^{1/2} = x$ if $x \geq 0$, and $(x^2)^{1/2} = -x$ if $x \leq 0$.
If $x \geq 0$, $(x^2)^{1/2} < (16)^{1/2}$ may be written $x < 4$. Hence $0 \leq x < 4$.
If $x \leq 0$, $(x^2)^{1/2} < (16)^{1/2}$ may be written $-x < 4$ or $x > -4$. Hence $-4 < x \leq 0$.
Thus $0 \leq x < 4$ and $-4 < x \leq 0$, or $-4 < x < 4$.

19.4 Prove that $a^2 + b^2 > 2ab$ if a and b are real and unequal numbers.

SOLUTION

If $a^2 + b^2 > 2ab$, then $a^2 - 2ab + b^2 > 0$ or $(a - b)^2 > 0$. This last statement is true since the square of any real number different from zero is positive.

The above provides a clue as to the method of proof. Starting with $(a - b)^2 > 0$, which we know to be true if $a \neq b$, we obtain $a^2 - 2ab + b^2 > 0$ or $a^2 + b^2 > 2ab$.

Note that the proof is essentially a reversal of the steps in the first paragraph.

19.5 Prove that the sum of any positive number and its reciprocal is never less than 2.

SOLUTION

We must prove that $(a + 1/a) \geq 2$ if $a > 0$.

If $(a + 1/a) \geq 2$, then $a^2 + 1 \geq 2a$, $a^2 - 2a + 1 \geq 0$, and $(a - 1)^2 \geq 0$ which is true.

To prove the theorem we start with $(a - 1)^2 \geq 0$, which is known to be true.

Then $a^2 - 2a + 1 \geq 0$, $a^2 + 1 \geq 2a$ and $a + 1/a \geq 2$ upon division by a.

19.6 Show that $a^2 + b^2 + c^2 > ab + bc + ca$ for all real values of a,b,c unless $a = b = c$.

SOLUTION

Since $a^2 + b^2 > 2ab$, $b^2 + c^2 > 2bc$, $c^2 + a^2 > 2ca$ (see Problem 19.4), we have by addition

$$2(a^2 + b^2 + c^2) > 2(ab + bc + ca) \qquad \text{or} \qquad a^2 + b^2 + c^2 > ab + bc + ca.$$

(If $a = b = c$, then $a^2 + b^2 + c^2 = ab + bc + ca$.)

19.7 If $a^2 + b^2 = 1$ and $c^2 + d^2 = 1$, show that $ac + bd < 1$.

SOLUTION

$a^2 + c^2 > 2ac$ and $b^2 + d^2 > 2bd$; hence by addition

$$(a^2 + b^2) + (c^2 + d^2) > 2ac + 2bd \qquad \text{or} \qquad 2 > 2ac + 2bd, \text{ i.e., } 1 > ac + bd.$$

19.8 Prove that $x^3 + y^3 > x^2y + y^2x$, if x and y are real, positive and unequal numbers.

SOLUTION

If $x^3 + y^3 > x^2y + y^2x$, then $(x + y)(x^2 - xy + y^2) > xy(x + y)$. Dividing by $x + y$, which is positive.

$$x^2 - xy + y^2 > xy \qquad \text{or} \qquad x^2 - 2xy + y^2 > 0, \qquad \text{i.e., } (x - y)^2 > 0 \text{ which is true if } x \neq y.$$

The steps are reversible and supply the proof. Starting with $(x - y)^2 > 0$, $x \neq y$, obtain

$$x^2 - xy + y^2 > xy.$$

Multiplying both sides by $x + y$, we have $(x + y)(x^2 - xy + y^2) > xy(x + y)$ or $x^3 + y^3 > x^2y + y^2x$.

19.9 Prove that $a^n + b^n > a^{n-1}b + ab^{n-1}$, provided a and b are positive and unequal, and $n > 1$.

SOLUTION

If $a^n + b^n > a^{n-1}b + ab^{n-1}$, then $(a^n - a^{n-1}b) - (ab^{n-1} - b^n) > 0$ or

$$a^{n-1}(a - b) - b^{n-1}(a - b) > 0, \qquad \text{i.e., } (a^{n-1} - b^{n-1})(a - b) > 0.$$

This is true since the factors are both positive or both negative.

Reversing the steps, which are reversible, provides the proof.

19.10 Prove that

$$a^3 + \frac{1}{a^3} > a^2 + \frac{1}{a^2} \qquad \text{if} \qquad a > 0 \qquad \text{and} \qquad a \neq 1.$$

SOLUTION

Multiplying both sides of the inequality by a^3 (which is positive since $a > 0$), we have

$$a^6 + 1 > a^5 + a, \qquad a^6 - a^5 - a + 1 > 0 \qquad \text{and} \qquad (a^5 - 1)(a - 1) > 0.$$

If $a > 1$ both factors are positive, while if $0 < a < 1$ both factors are negative. In either case the product is positive. (If $a = 1$ the product is zero.)

Reversal of the steps provides the proof.

19.11 If a,b,c,d are positive numbers and

$$\frac{a}{b}>\frac{c}{d},$$

prove that

$$\frac{a+c}{b+d}>\frac{c}{d}.$$

SOLUTION

Method 1. If

$$\frac{a+c}{b+d}>\frac{c}{d},$$

then multiplying by $d(b+d)$ we obtain

$$(a+c)d>c(b+d),\ ad+cd>bc+cd,\ ad>bc$$

and, dividing by bd,

$$\frac{a}{b}>\frac{c}{d},$$

which is given as true. Reversing the steps provides the proof.

Method 2. Since

$$\frac{a}{b}>\frac{c}{d},$$

then

$$\frac{a}{b}+\frac{c}{b}>\frac{c}{d}+\frac{c}{b},\qquad \frac{a+c}{b}>\frac{c(b+d)}{bd}\qquad \text{and}\qquad \frac{a+c}{b+d}>\frac{c}{d}.$$

19.12 Prove:

(*a*) $x^2-y^2>x-y$ if $x+y>1$ and $x>y$
(*b*) $x^2-y^2<x-y$ if $x+y>1$ and $x<y$

SOLUTION

(*a*) Since $x>y$, $x-y>0$. Multiplying both sides of $x+y>1$ by the positive number $x-y$,

$$(x+y)(x-y)>(x-y)\qquad \text{or}\qquad x^2-y^2>x-y.$$

(*b*) Since $x<y$, $x-y<0$. Multiplying both sides of $x+y>1$ by the negative number $x-y$ reverses the sense of the inequality; thus

$$(x+y)(x-y)<(x-y)\qquad \text{or}\qquad x^2-y^2<x-y.$$

19.13 The arithmetic mean of two numbers a and b is $(a+b)/2$, the geometric mean is $\sqrt{ab}$, and the harmonic mean is $2ab/(a+b)$. Prove that

$$\frac{a+b}{2}>\sqrt{ab}>\frac{2ab}{a+b}$$

if a and b are positive and unequal.

SOLUTION

(*a*) If $(a+b)/2>\sqrt{ab}$, then $(a+b)^2>(2\sqrt{ab})^2$, $a^2+2ab+b^2>4ab$, $a^2-2ab+b^2>0$ and $(a-b)^2>0$ which is true if $a\neq b$. Reversing the steps, we have $(a+b)/2>\sqrt{ab}$.

(*b*) If

$$\sqrt{ab} > \frac{2ab}{a+b},$$

then

$$ab > \frac{4a^2b^2}{(a+b)^2}, \qquad (a+b)^2 > 4ab \qquad \text{and} \qquad (a-b)^2 > 0$$

which is true if $a \neq b$. Reversing the steps, we have $\sqrt{ab} > 2ab/(a+b)$.

From (*a*) and (*b*),

$$\frac{a+b}{2} > \sqrt{ab} > \frac{2ab}{a+b}.$$

19.14 Find the values of x for which (*a*) $x^2 - 7x + 12 = 0$, (*b*) $x^2 - 7x + 12 > 0$, (*c*) $x^2 - 7x + 12 < 0$.

SOLUTION

(*a*) $x^2 - 7x + 12 = (x-3)(x-4) = 0$ when $x = 3$ or 4.

(*b*) $x^2 - 7x + 12 > 0$ or $(x-3)(x-4) > 0$ when $(x-3) > 0$ and $(x-4) > 0$ simultaneously, or when $(x-3) < 0$ and $(x-4) < 0$ simultaneously.

$(x-3) > 0$ and $(x-4) > 0$ simultaneously when $x > 3$ and $x > 4$, i.e., when $x > 4$.
$(x-3) < 0$ and $(x-4) < 0$ simultaneously when $x < 3$ and $x < 4$, i.e., when $x < 3$.
Hence $x^2 - 7x + 12 > 0$ is satisfied when $x > 4$ or $x < 3$.

(*c*) $x^2 - 7x + 12 < 0$ or $(x-3)(x-4) < 0$ when $(x-3) > 0$ and $(x-4) < 0$ simultaneously, or when $(x-3) < 0$ and $(x-4) > 0$ simultaneously.

$(x-3) > 0$ and $(x-4) < 0$ simultaneously when $x > 3$ and $x < 4$, i.e., when $3 < x < 4$.
$(x-3) < 0$ and $(x-4) > 0$ simultaneously when $x < 3$ and $x > 4$, which is absurd.
Hence $x^2 - 7x + 12 < 0$ is satisfied when $3 < x < 4$.

19.15 Determine graphically the range of values of x defined by

(*a*) $x^2 + 2x - 3 = 0$

(*b*) $x^2 + 2x - 3 > 0$

(*c*) $x^2 + 2x - 3 < 0$.

SOLUTION

Figure 19-6 shows the graph of the function defined by $y = x^2 + 2x - 3$. From the graph it is clear that

(*a*) $y = 0$ when $x = 1$, $x = -3$

(*b*) $y > 0$ when $x > 1$ or $x < -3$

(*c*) $y < 0$ when $-3 < x < 1$.

19.16 Solve for x: (*a*) $|3x-6| + 2 > 9$ (*b*) $|7x-1| - 6 < 2$.

(*a*) $|3x-6| + 2 > 9$

$$|3x-6| > 7$$

$$\begin{aligned} 3x - 6 &> 7 \quad \text{or} \quad 3x - 6 < -7 \\ 3x &> 13 \quad \text{or} \quad 3x < -1 \\ x &> 13/3 \quad \text{or} \quad x < -1/3 \end{aligned}$$

The solution of $|3x-6| + 2 > 9$ is the interval $(-\infty, -1/3) \cup (13/3, \infty)$.

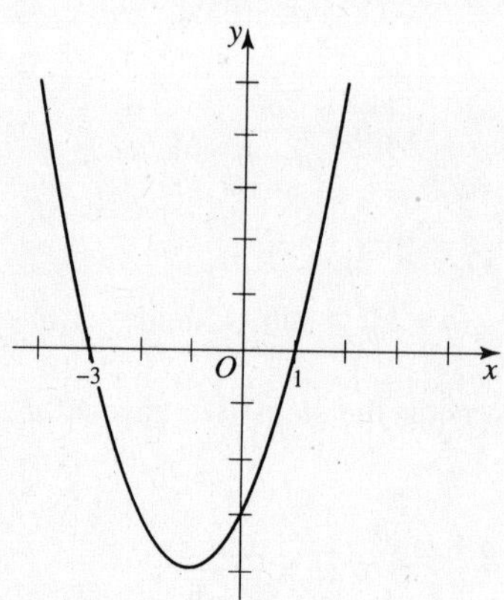

Fig. 19-6

(*b*) $|7x-1|-6<2$

$$|7x-1|<8$$
$$-8<7x-1<8$$
$$-7<7x<9$$
$$-1<x<9/7$$

The solution of $|7x-1|-6<2$ is the interval $(-1, 9/7)$.

19.17 Solve for x:

(*a*) $\dfrac{2x-1}{x+1}\le 1$ (*b*) $\dfrac{x^2-10x+21}{x^2-5x+6}\le 0.$

SOLUTION

(*a*) $\dfrac{2x-1}{x+1}\le 1$

$$\frac{2x-1}{x+1}-1\le 0$$
$$\frac{2x-1}{x+1}-\frac{x+1}{x+1}\le 0$$
$$\frac{x-2}{x+1}\le 0$$

The critical values are $x=-1$ and $x=2$. We make a sign chart (see Fig. 19-7), with a solid line through $x=2$, since it makes the fraction 0, and 0 is included in the solution, and a dashed line through $x=-1$, since it makes the fraction undefined. Next, we determine the sign of each factor in the three intervals. Finally, in intervals where an even number of factors are negative the problem is positive and in those where an odd number of factors are negative the problem is negative. The solution of

$$\frac{2x-1}{x+1}\le 1$$

is the interval $(-1, 2]$.

(*b*) $\dfrac{x^2-10x+21}{x^2-5x+6}\le 0$

$$\frac{(x-3)(x-7)}{(x-3)(x-2)}\le 0$$

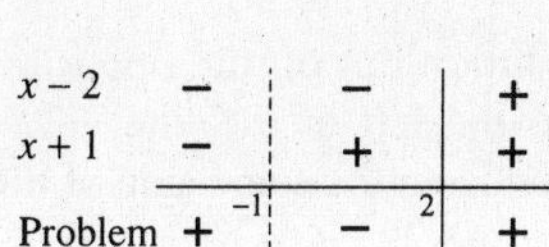

Fig. 19-7

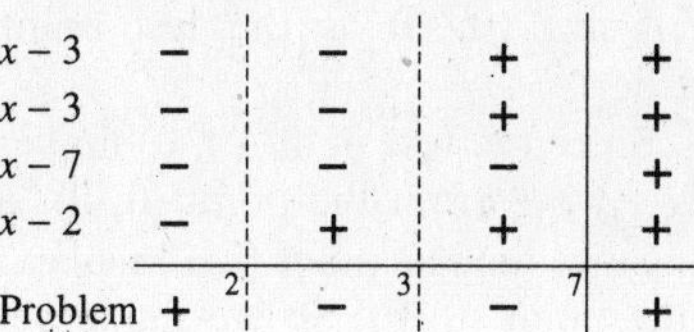

Fig. 19-8

The critical values are $x = 2$, $x = 3$, and $x = 7$. We make a sign chart (see Fig. 19-8), with dashed lines through $x = 2$ and $x = 3$ and a solid line through $x = 7$. Since $x = 3$ makes the denominator of the fraction zero it is excluded, even though it also makes the numerator 0. The signs for the factors are determined for each interval and then used to determine the sign for the problem in each interval. The factor $x - 3$ is used an even number of times in the problem and could be omitted from the sign chart, since any factor raised to an even power is always non-negative.

The solution of

$$\frac{x^2 - 10x + 21}{x^2 - 5x + 6} \leq 0$$

is the interval $(2, 3) \cup (3, 7]$.

Note 1: If we had canceled the common factor $x - 3$, we might have overlooked the fact that the problem is not defined when $x = 3$ and it cannot be in the solution set.

Note 2: When a factor appears in the problem an even number of times, it may be excluded from the sign chart and is usually omitted. When a factor appears in the problem an odd number of times it must be included in the sign chart an odd number of times, and is usually included exactly once.

19.18 Find the solution for the system of inequalities $-2x + y \geq 2$ and $2x - y \leq 6$.

SOLUTION

Graph the related equations $-2x + y = 2$ and $2x - y = 6$. Both lines are solid, since they are included in the solution.

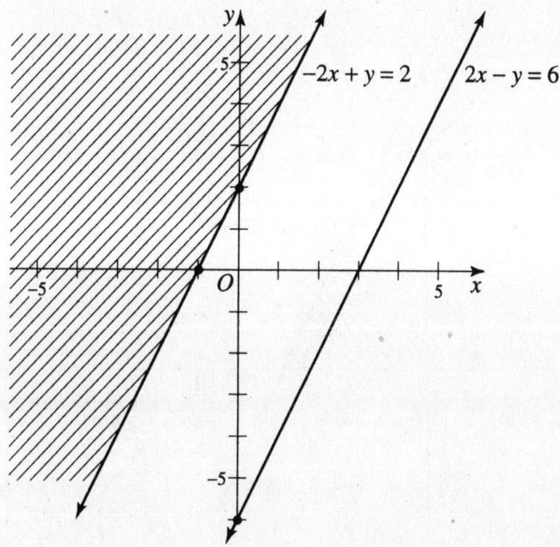

Fig. 19-9

Using (0, 0) as the test point, we get $-2(0)+0 \geq 2$, which is false, and $2(0)-0 \leq 6$, which is true.

Since the test point (0, 0) makes $-2x+y \leq 2$ false, the solution lies on the opposite side of the line $-2x+y=2$ from the point (0, 0). So we shade above and to the left of the line $-2x+y=2$.

Since the test point (0, 0) makes $2x-y \leq 6$ true, the solution is on the same side of the line $2x-y=6$ as the point (0, 0). So we shade above and to the left of the line $2x-y=6$.

The common solution is the region above and to the left of $-2x+y=2$, and is the shaded region shown in Fig. 19-9.

19.19 The Close Shave Company manufactures two types of electric shaver. One shaver is cordless, requires 4 hours to make, and sells for \$40. The other shaver is a cord type, takes 2 hours to make, and sells for \$30. The Company has only 800 work hours to use in manufacturing each day and the shipping department can pack and ship only 300 shavers per day. How many of each type of shaver should the Close Shave Company produce per day to maximize its sales revenue?

SOLUTION

Let x be the number of cordless shavers made per day and y be the number of cord-type shavers made per day.

The objective function is $R(x, y)=40x+30y$.

The stated constraints are $4x+2y \leq 800$ and $x+y \leq 300$.

The natural constraints are $x \geq 0$ and $y \geq 0$.

From Fig. 19-10, we see that the vertices of the region formed by the constraints are $A(0,0)$, $B(200,0)$, $C(100,200)$, and $D(0,300)$.

$$R(0,0)=40(0)+30(0)=0+0=0$$
$$R(200,0)=40(200)+30(0)=8000+0=8000$$
$$R(100,200)=40(100)+30(200)=4000+6000=10\,000$$
$$R(0,300)=40(0)+30(300)=0+9000=9000.$$

The Close Shave Company achieves the maximum sales revenue of \$10 000 per day by producing 100 cordless shavers and 200 cord-type shavers per day.

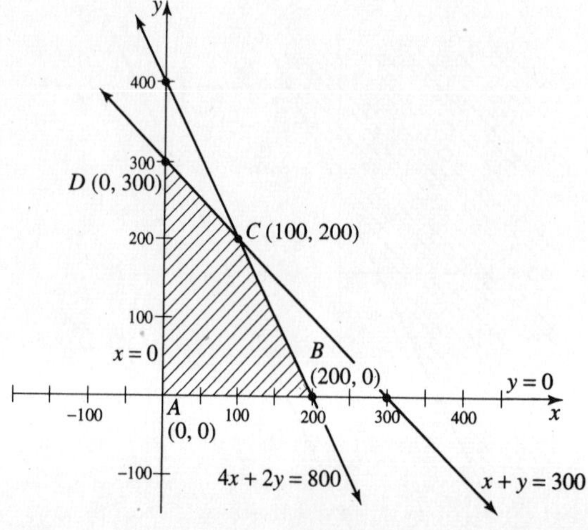

Fig. 19-10

Supplementary Problems

19.20 If $a > b$, prove that $a - c > b - c$ where c is any real number.

19.21 If $a > b$ and $k > 0$, prove that $ka > kb$.

19.22 Find the values of x for which of the following inequalities holds.

(*a*) $2(x+3) > 3(x-1) + 6$ (*b*) $\dfrac{x}{4} + \dfrac{2}{3} < \dfrac{2x}{3} - \dfrac{1}{6}$ (*c*) $\dfrac{1}{x} + \dfrac{3}{4x} > \dfrac{7}{8}$ (*d*) $x^2 > 9$

19.23 For what values of a will $(a+3) < 2(2a+1)$?

19.24 Prove that $\frac{1}{2}(a^2 + b^2) \geq b$ for all real values of a and b, the equality holding if and only if $a = b$.

19.25 Prove that

$$\frac{1}{x} + \frac{1}{y} > \frac{2}{x+y}$$

if x and y are positive and $x \neq y$.

19.26 Prove that

$$\frac{x^2 + y^2}{x+y} < x + y \quad \text{if} \quad x > 0,\ y > 0.$$

19.27 Prove that $xy + 1 \geq x + y$ if $x \geq 1$ and $y \geq 1$ or if $x \leq 1$ and $y \leq 1$.

19.28 If $a > 0$, $a \neq 1$ and n is any positive integer, prove that

$$a^{n+1} + \frac{1}{a^{n+1}} > a^n + \frac{1}{a^n}.$$

19.29 Show that $\sqrt{2} + \sqrt{6} < \sqrt{3} + \sqrt{5}$.

19.30 Determine the values of x for which each of the following inequalities holds.

(*a*) $x^2 + 2x - 24 > 0$ (*b*) $x^2 - 6 < x$ (*c*) $3x^2 - 2x < 1$ (*d*) $3x + \dfrac{1}{x} > \dfrac{7}{2}$

19.31 Determine graphically the range of values of x for which (*a*) $x^2 - 3x - 4 > 0$, (*b*) $2x^2 - 5x + 2 < 0$.

19.32 Write the solution for each inequality in interval notation.

(*a*) $|3x+3| - 15 \geq -6$ (*b*) $|2x-3| < 7$

19.33 Write the solution for each inequality in the interval notation.

(*a*) $x^2 \geq 10x - 21$ (*c*) $(x-1)(x-2)(x+3) > 0$ (*e*) $\dfrac{x-5}{x+1} \leq 3$

(*b*) $\dfrac{(x+1)(x-1)}{x} < 0$ (*d*) $\dfrac{x-1}{x+2} \leq 0$ (*f*) $\dfrac{(x-6)(x-3)}{x+2} \geq 0$

19.34 Graph each inequality and shade the solution region.

(*a*) $4x - y \leq 5$ (*b*) $y - 3x > 2$

19.35 Graph each system of inequalities and shade the solution region.

(*a*) $x + 2y \le 20$ and $3x + 10y \le 80$

(*b*) $3x + y \ge 4$, $x + y \ge 2$, $-x + y \le 4$, and $x \le 5$

19.36 Use linear programming to solve each problem.

(*a*) Ramone builds portable storage buildings. He uses 10 sheets of plywood and 15 studs in a small building and 15 sheets of plywood and 45 studs in a large building. Ramone has 60 sheets of plywood and 135 studs available for use. If Ramone makes a profit of $400 on a small building and $500 on a large building, how many of each type of building should he make to maximize his profit?

(*b*) Jean and Wesley make wind chimes and bird houses in their craft shop. Each wind chime requires 3 hours of work from Jean and 1 hour of work from Wesley. Each bird house requires 4 hours of work from Jean and 2 hours of work from Wesley. Jean cannot work more than 48 hours per week and Wesley cannot work more than 20 hours per week. If each wind chime sells for $12 and each bird house sells for $20, how many of each item should they make to maximize their revenue?

ANSWERS TO SUPPLEMENTARY PROBLEMS

19.22 (*a*) $x < 3$ (*b*) $x > 2$ (*b*) $0 < x < 2$ (*d*) $x < -3$ or $x > 3$

19.23 $a > \frac{1}{3}$

19.30 (*a*) $x > 4$ or $x < -6$ (*b*) $-2 < x < 3$ (*c*) $-\frac{1}{3} < x < 1$ (*d*) $x > \frac{2}{3}$ or $0 < x < \frac{1}{2}$

19.31 (*a*) $x > 4$ or $x < -1$ (*b*) $\frac{1}{2} < x < 2$,

19.32 (*a*) $(-\infty, -4) \cup [2, \infty)$ (*b*) $(2, 5)$

19.33 (*a*) $(-\infty, 3] \cup [7, \infty)$ (*c*) $(-3, 1) \cup (2, \infty)$ (*e*) $(-\infty, -4) \cup (-1, \infty)$
(*b*) $(-\infty, -1) \cup (0, 1)$ (*d*) $(-2, 1]$ (*f*) $(-6, 3) \cup (6, \infty)$

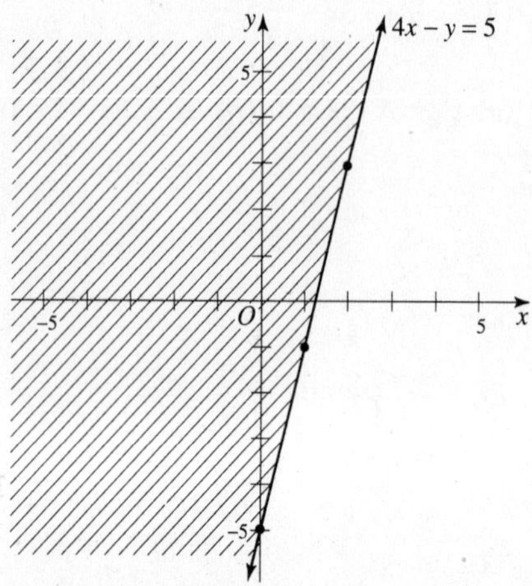

Fig. 19-11

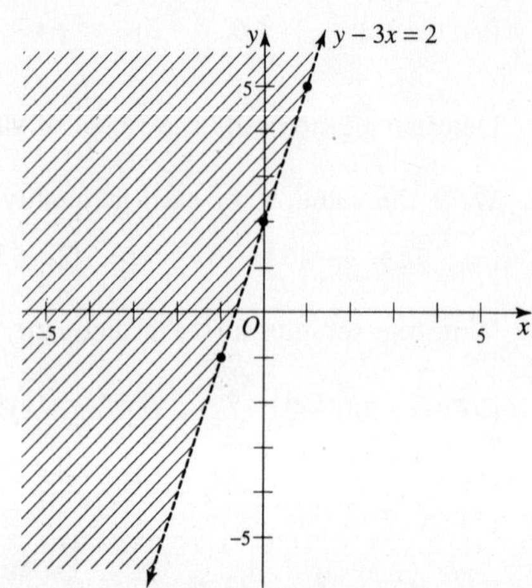

Fig. 19-12

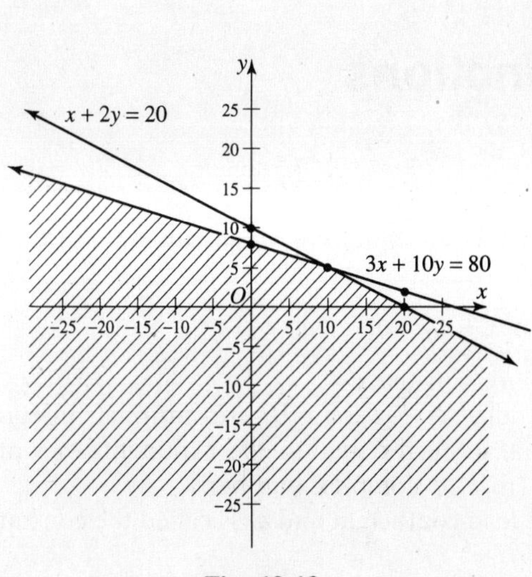

Fig. 19-13

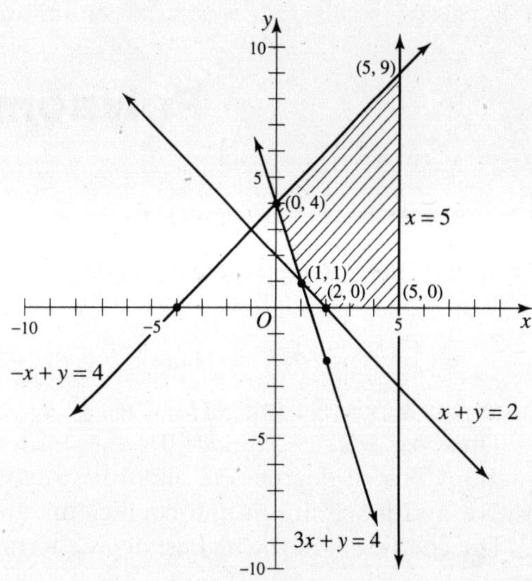

Fig. 19-14

19.34 (*a*) Figure 19-11 (*b*) Figure 19-12

19.35 (*a*) Figure 19-13 (*b*) Figure 19-14

19.36 (*a*) Ramone maximizes his profit by making 6 small buildings and 0 large buildings.
(*b*) Jean and Wesley will maximize their revenue by making 6 wind chimes and 8 bird houses.

Chapter 20

Polynomial Functions

20.1 POLYNOMIAL EQUATIONS

A rational integral equation of degree n in the variable x is an equation which can be written in the form

$$a_nx^n + a_{n-1}x^{n-1} + a_{n-2}x^{n-2} + \cdots + a_1x + a_0 = 0, \qquad a_n \neq 0$$

where n is a positive integer and $a_0, a_1, a_2, \ldots, a_{n-1}, a_n$ are constants.

Thus $4x^3 - 2x^2 + 3x - 5 = 0$, $x^2 - \sqrt{2}x + \frac{1}{4} = 0$ and $x^4 + \sqrt{-3}x - 8 = 0$ are rational integral equations in x of degree 3, 2 and 4 respectively. Note that in each equation the exponents of x are positive and integral, and the coefficients are constants (real or complex numbers).

The coefficient of the highest degree term is called the lead coefficient and a_0 is called the constant term.

In this chapter, only rational integral equations are considered.

A polynomial of degree n in the variable x is a function of x which can be written in the form

$$P(x) = a_nx^n + a_{n-1}x^{n-1} + a_{n-2}x^{n-2} + \cdots + a_1x + a_0, \qquad a_n \neq 0$$

where n is a positive integer and $a_0, a_1, a_2, \ldots, a_{n-1}, a_n$ are constants. Then $P(x) = 0$ is a rational integral equation of degree n in x.

If $P(x) = 3x^3 + x^2 + 5x - 6$, then $P(-2) = 3(-2)^3 + (-2)^2 + 5(-2) - 6 = -36$.

If $P(x) = x^2 + 2x - 8$, then $P(\sqrt{5}) = 5 + 2\sqrt{5} - 8 = 2\sqrt{5} - 3$.

Any value of x which makes $P(x)$ vanish is called a *root* of the equation $P(x) = 0$. Thus 2 is a root of the equation $P(x) = 3x^3 - 2x^2 - 5x - 6 = 0$, since $P(2) = 24 - 8 - 10 - 6 = 0$.

20.2 ZEROS OF POLYNOMIAL EQUATIONS

A. Remainder theorem If r is any constant and if a polynomial $P(x)$ is divided by $(x - r)$, the remainder is $P(r)$.

For example, if $P(x) = 2x^3 - 3x^2 - x + 8$ is divided by $x + 1$, then $r = -1$ and the remainder $= P(-1) = -2 - 3 + 1 + 8 = 4$. That is,

$$\frac{2x^3 - 3x^2 - x + 8}{x + 1} = Q(x) + \frac{4}{x + 1}, \quad \text{where } Q(x) \text{ is a polynomial in } x.$$

B. Factor theorem If r is a root of the equation $P(x) = 0$, i.e. if $P(r) = 0$, then $(x - r)$ is a factor of $P(x)$. Conversely, if $(x - r)$ is a factor of $P(x)$, then r is a root of $P(x) = 0$, or $P(r) = 0$.

Thus $1, -2, -3$ are the three roots of the equation $P(x) = x^3 + 4x^2 + x - 6 = 0$, since $P(1) = P(-2) = P(-3) = 0$. Then $(x - 1)$, $(x + 2)$ and $(x + 3)$ are factors of $x^3 + 4x^2 + x - 6$.

C. Synthetic division Synthetic division is a simplified method of dividing a polynomial $P(x)$ by $x - r$, where r is any assigned number. By this method we determine values of the coefficients of the quotient and the value of the remainder can readily be determined.

EXAMPLE 20.1. Divide $(5x + x^4 - 14x^2)$ by $(x + 4)$ using synthetic division.

Write the terms of the dividend in descending powers of the variable and fill in missing terms using zero for the coefficients; write the divisor in the form $x - a$.

$$(x^4 + 0x^3 - 14x^2 + 5x + 0) \div (x - (-4))$$

Write the constant term a from the divisor on the left in a $\underline{\quad}|$ and write the coefficients from the dividend to the right of the symbol

$$\underline{-4|}\ \ 1+0-14+5+0$$

Bring down the first term in the divisor to the third row, leaving a blank row for now.

$$\begin{array}{r|l} \underline{-4} & 1+0-14+5+0 \\ & \\ \hline & 1 \end{array}$$

Multiply the term in the quotient row (third row) by the divisor and write the product in the second row under the second term in the first row, add the numbers in the column formed, and write the sum as the second term in the quotient row.

$$\begin{array}{r|l} \underline{-4} & 1+0-14+5+0 \\ & \quad -4 \\ \hline & 1-4 \end{array}$$

Multiply the last term on the right in the quotient row by the divisor, write it under the next term in the top row, add, and write the sum in the quotient row. Continue this process until all of the terms in the top row have a number under them.

$$\begin{array}{r|l} \underline{-4} & 1+0-14+5+0 \\ & \quad -4+16-8+12 \\ \hline & 1-4+\ 2-3+12 \end{array}$$

The third row is the quotient row with the last term being the remainder. The degree of the quotient polynomial is one less than the degree of the dividend because we are dividing by a linear factor. The terms of the quotient row are the coefficients of the terms in the quotient polynomial. The degree of the quotient polynomial here is 3.

The quotient with remainder for $(5x + x^4 - 14x^2) \div (x + 4)$ is

$$1x^3 - 4x^2 + 2x - 3 + \frac{12}{x+4}$$

D. Fundamental theorem of algebra Every polynomial equation $P(x) = 0$ has at least one root, real or complex.

Thus $x^7 - 3x^5 + 2 = 0$ has at least one root.

But $f(x) = \sqrt{x} + 3 = 0$ has no root, since no number r exists such that $f(r) = 0$. Since this equation is not rational, the fundamental theorem does not apply.

E. Number of roots of an equation Every rational integral equation $P(x) = 0$ of the nth degree has exactly n roots.

Thus $2x^3 + 5x^2 - 14x - 8 = 0$ has exactly 3 roots, namely 2, $-\frac{1}{2}$, -4.

Some of the n roots may be equal. Thus the equation of the sixth degree $(x-2)^3(x-5)^2(x+4) = 0$ has 2 as a triple root, 5 as a double root, and -4 as a single root; i.e., the six roots are $2, 2, 2, 5, 5, -4$.

20.3 SOLVING POLYNOMIAL EQUATIONS

A. Complex and irrational roots

(1) If a complex number $a + bi$ is a root of the rational integral equation $P(x) = 0$ with *real coefficients*, then the conjugate complex number $a - bi$ is also a root.

It follows that every rational integral equation of odd degree with real coefficients has at least one real root.

(2) If the rational integral equation $P(x) = 0$ with *rational coefficients* has $a + \sqrt{b}$ as a root, where a and b are rational and $\sqrt{b}$ is irrational, then $a - \sqrt{b}$ is also a root.

B. Rational root theorem

If b/c, a rational fraction in lowest terms, is a root of the equation

$$a_n x^n + a_{n-1}x^{n-1} + a_{n-2}x^{n-2} + \cdots + a_1 x + a_0 = 0, \qquad a_n \neq 0$$

with integral coefficients, then b is a factor of a_0 and c is a factor of a_n.

Thus if b/c is a rational root of $6x^3 + 5x^2 - 3x - 2 = 0$, the values of b are limited to the factors of 2, which are ± 1, ± 2; and the values of c are limited to the factors of 6, which are ± 1, ± 2, ± 3, ± 6. Hence the only possible rational roots are ± 1, ± 2, $\pm 1/2$, $\pm 1/3$, $\pm 1/6$, $\pm 2/3$.

C. Integral root theorem

It follows that if an equation $P(x) = 0$ has integral coefficients and the lead coefficient is 1:

$$x^n + a_{n-1}x^{n-1} + a_{n-2}x^{n-2} + \cdots + a_1 x + a_0 = 0,$$

then any rational root of $P(x) = 0$ is an integer and a factor of a_0.

Thus the rational roots, if any, of $x^3 + 2x^2 - 11x - 12 = 0$ are limited to the integral factors of 12, which are ± 1, ± 2, ± 3, ± 4, ± 6, ± 12.

D. Intermediate value theorem

If $P(x) = 0$ is a polynomial equation with real coefficients, then approximate values of the real roots of $P(x) = 0$ may be found by obtaining the graph of $y = P(x)$ and determining the values of x at the points where the graph intersects the x-axis ($y = 0$). Fundamental in this procedure is the fact that if $P(a)$ and $P(b)$ have opposite signs then $P(x) = 0$ has at least one root between $x = a$ and $x = b$. This fact is based on the continuity of the graph of $y = P(x)$ when $P(x)$ is a polynomial with real coefficients.

EXAMPLE 20.2. For each real zero of $P(x) = 2x^3 - 5x^2 - 6x + 4$ isolate the zero between two consecutive integers.

Since $P(x) = 2x^3 - 5x^2 - 6x + 4$ has degree 3, there are at most 3 real zeros. We will look for the real zeros in the interval -5 to 5. The interval is arbitrary and may need to be expanded if the real zeros are not found here. By synthetic division, we will find the value of $P(x)$ for each integer in the interval selected. The remainders from the synthetic division are the values of $P(x)$ and are summarized in the table below.

x	-5	-4	-3	-2	-1	0	1	2	3	4	5
$P(x)$	-341	-180	-77	-20	3	4	-5	-12	-5	28	99

Note that $P(-2) = -20$ and $P(-1) = 3$ have opposite signs, so from the Intermediate Value Theorem there is a real zero between -2 and -1. Similarly, since $P(0) = 4$ and $P(1) = -5$ there is a real zero between 0 and 1, and since $P(3) = -5$ and $P(4) = 28$ there is a real zero between 3 and 4. Three real zeros have been isolated, so we have located all the real zeros of $P(x)$.

It is not always possible to locate all the real zeros this way because there could be more than one zero between two consecutive integers. When there are an even number of zeros between two consecutive integers the Intermediate Value Theorem will not reveal them when we use just integers for x. The Intermediate Value Theorem does not tell you how many real zeros are in the interval, just that there is at least one real zero in the interval.

E. Upper and lower limits for the real roots

A number a is called an *upper limit* or *upper bound* for the real roots of $P(x) = 0$ if no root is greater than a. A number b is called a *lower limit* or *lower bound* for the real roots of $P(x) = 0$

if no root is less than b. The following theorem is useful in determining upper and lower limits.

Let $P(x) = a_n x^n + a_{n-1}x^{n-1} + a_{n-2}x^{n-2} + \cdots + a_0 = 0$, where $a_0, a_1 \ldots, a_n$ are real and $a_n > 0$. Then:

(1) If upon synthetic division of $P(x)$ by $x - a$, where $a \geq 0$, all of the numbers obtained in the third row are positive or zero, then a is an upper limit for all the real roots of $P(x) = 0$.

(2) If upon synthetic division of $P(x)$ by $x - b$, where $b \leq 0$, all of the numbers obtained in the third row are alternately positive and negative (or zero), then b is a lower limit for all the real roots of $P(x) = 0$.

EXAMPLE 20.3. Find an interval that contains all the real zeros of $P(x) = 2x^3 - 5x^2 + 6$.

We will find the integer, b, that is the least upper limit of the real zeros of $P(x)$ and the integer, a, that is the great lower limit on the real zeros of $P(x)$. All real zeros will be in the interval $[a, b]$. To find a and b we use synthetic division on $P(x) = 2x^3 - 5x^2 + 6$.

$$\begin{array}{r|l} 1 & 2-5+0+6 \\ & +2-3-3 \\ \hline & 2-3-3+3 \end{array} \qquad \begin{array}{r|l} 2 & 2-5+0+6 \\ & +4-2-4 \\ \hline & 2-2-2+2 \end{array} \qquad \begin{array}{r|l} 3 & 2-5+0+\ 6 \\ & +6+3+\ 9 \\ \hline & 2+1+3+15 \end{array}$$

When we divide using 3, the quotient row is all positive, so 3 is the smallest integer that is an upper limit for the real zeros of $P(x)$. Thus $b = 3$.

$$\begin{array}{r|l} -1 & 2-5+0+6 \\ & -2+7-7 \\ \hline & 2-7+7-1 \end{array}$$

When we divide using -1, the quotient row alternates in sign, so -1 is the greatest integer that is a lower limit for the real zeros of $P(x)$. Thus, $a = -1$.

The real zeros of $P(x) = 2x^3 - 5x^2 + 6$ are in the interval $(-1, 3)$ or $-1 < x < 3$. Since $P(-1) \neq 0$ and $P(3) \neq 0$, we used interval notation that indicates that neither endpoint is a zero.

F. Descartes' Rule of Signs

If the terms of a polynomial $P(x)$ with real coefficients are arranged in order of descending powers of x, a *variation of sign* occurs when two consecutive terms differ in sign. For example, $x^3 - 2x^2 + 3x - 12$ has 3 variations of sign, and $2x^7 - 6x^5 - 4x^4 + x^2 - 2x + 4$ has 4 variations of sign.

Descartes' Rule of Signs says that the number of positive roots of $P(x) = 0$ is either equal to the number of variations of sign of $P(x)$ or is less than that number by an even integer. The number of negative roots of $P(x) = 0$ is either equal to the number of variations of sign of $P(-x)$ or is less than that number by an even integer.

Thus in $P(x) = x^9 - 2x^5 + 2x^2 - 3x + 12 = 0$ there are 4 variations of sign of $P(x)$; hence the number of positive roots of $P(x) = 0$ are 4, $(4 - 2)$ or $(4 - 4)$. Since $P(-x) = (-x)^9 - 2(-x)^5 + 2(-x)^2 - 3(-x) + 12 = -x^9 + 2x^5 + 2x^2 + 3x + 12 = 0$ has one variation of sign, then $P(x) = 0$ has exactly one negative root. Hence there are 4, 2, or 0 positive roots, 1 negative root, and at least $9 - (4 + 1) = 4$ complex roots. (There are 4, 6, or 8 complex roots. Why?)

20.4 APPROXIMATING REAL ZEROS

In solving a polynomial equation $P(x) = 0$, it is not always possible to find all the zeros by the previous methods. We have been able to determine the irrational and imaginary zeros when we were able to find quadratic factors that we could solve using the quadratic formula. If we cannot find the quadratic factors of $P(x) = 0$, we will not be able to solve for the imaginary zeros, but we can often find an approximation for some of the real zeros.

To approximate a real zero of $P(x) = 0$, we must first find an interval that contains a real zero of $P(x) = 0$. We can do this using the Intermediate Value Theorem to locate to numbers a and b such that $P(a)$ and $P(b)$ have opposite signs. We keep using the Intermediate Value Theorem until we have isolated the real zero in an interval small enough that it will be known to the desired degree of accuracy.

EXAMPLE 20.4. Find a real zero of $x^3 + 3x + 8 = 0$ correct to two decimal places.

By Descartes' Rule of Signs, $P(x) = x^3 + 3x + 8$ has no positive real zeros and 1 negative real zeros.

Using synthetic division, we find $P(-2) = -6$ and $P(-1) = 4$, so by the Intermediate Value Theorem $P(x) = x^3 + 3x + 8$ has a real zero between -2 and -1.

We now use synthetic division and the Intermediate Value Theorem to determine the tenths interval containing the zero. The results are summarized in the table below.

x	-1.0	-1.1	-1.2	-1.3	-1.4	-1.5	-1.6	-1.7	-1.8	-1.9	-2.0
$P(x)$	4	3.37	2.67	1.90	1.06	0.13	-0.80	-2.01	-3.23	-4.56	-6

We can see that $P(-1.5)$ is positive and $P(-1.6)$ is negative so the zero is between -1.6 and -1.5.

Now we check for the hundredths digit by using synthetic on the interval between -1.6 and -1.5. We do not have to find all the hundredths values, just a sign change between two consecutive values.

x	-1.50	-1.51	-1.52
$P(x)$	0.13	0.03	-0.07

We see that $P(-1.51)$ is positive and $P(-1.52)$ is negative, so by the Intermediate Value Theorem there is a real zero between -1.51 and -1.52.

Since the real zero is located between -1.51 and -1.52, we just need to determine whether it rounds off to -1.51 or -1.52. To do this we find $P(-1.515)$, which is about -0.02. This value of $P(-1.515)$ is negative and $P(-1.51)$ is positive, so we know that the zero is between -1.515 and -1.510, and all the numbers in this interval rounded to two decimals places are -1.51.

Thus, to two decimal places the only real zero of $x^3 + 3x + 8 = 0$ is -1.51.

To use a graphing calculator to approximate the real zeros of a polynomial we graph the function and use the trace and zoom features of the calculator. After we graph the function, we use the trace feature to locate an interval that contains a real zero using the Intermediate Value Theorem. We then use the zoom feature to focus in on this interval. We continue using the trace and zoom feature until we find two x values that round to the desired degree of accuracy and have function values that are opposite in sign.

Solved Problems

20.1 Prove the remainder theorem: If a polynomial $P(x)$ is divided by $(x - r)$ the remainder is $P(r)$.

SOLUTION

In the division of $P(x)$ by $(x - r)$, let $Q(x)$ be the quotient and R, a constant, the remainder. By definition $P(x) = (x - r)Q(x) + R$, an identity for all values of x. Letting $x = r$, $R(r) = R$.

20.2 Determine the remainder R after each of the following divisions.

(*a*) $(2x^3+3x^2-18x-4)\div(x-2)$. $R=P(2)=2(2^3)+3(2^2)-18(2)-4=-12$

(*b*) $(x^4-3x^3+5x+8)\div(x+1)$. $R=P(-1)=(-1)^4-3(-1)^3+5(-1)+8=1+3-5+8=7$

(*c*) $(4x^3+5x^2-1)\div\left(x+\frac{1}{2}\right)$. $R=P\left(-\frac{1}{2}\right)=4\left(-\frac{1}{2}\right)^3+5\left(-\frac{1}{2}\right)^2-1=-\frac{1}{4}$

(*d*) $(x^3-2x^2+x-4)\div x$. $R=P(0)=-4$

(*e*) $\left(\frac{8}{27}x^3-\frac{4}{9}x^2+x-\frac{3}{2}\right)\div(2x-3)$. $R=P\left(\frac{3}{2}\right)=\frac{8}{27}\left(\frac{3}{2}\right)^3-\frac{4}{9}\left(\frac{3}{2}\right)^2+\frac{3}{2}-\frac{3}{2}=0$

(*f*) $(x^8-x^5-x^3+1)\div(x+\sqrt{-1})$. $R=P(-i)=(-i)^8-(-i)^5-(-i)^3+1=i^8+i^5+i^3+1$
$=1+i-i+1=2$

20.3 Prove the factor theorem: If r is a root of the equation $P(x)=0$, then $(x-r)$ is a factor of $P(x)$; and conversely if $(x-r)$ is a factor of $P(x)$, then r is a root of $P(x)=0$.

SOLUTION

In the division of $P(x)$ by $(x-r)$, let $Q(x)$ be the quotient and R, a constant, the remainder. Then $P(x)=(x-r)Q(x)+R$ or $P(x)=(x-r)Q(x)+P(r)$ by the remainder theorem.

If r is a root of $P(x)=0$, then $P(r)=0$. Hence $P(x)=(x-r)Q(x)$, or $(x-r)$ is a factor of $P(x)$.

Conversely if $(x-r)$ is a factor of $P(x)$, then the remainder in the division of $P(x)$ by $(x-r)$ is zero. Hence $P(r)=0$, i.e., r is a root of $P(x)=0$.

20.4 Show that $(x-3)$ is a factor of the polynomial $P(x)=x^4-4x^3-7x^2+22x+24$.

SOLUTION

$P(3)=81-108-63+66+24=0$. Hence $(x-3)$ is a factor of $P(x)$, 3 is a *zero* of the polynomial $P(x)$, and 3 is a *root* of the equation $P(x)=0$.

20.5 (*a*) Is -1 a root of the equation $P(x)=x^3-7x-6=0$?

(*b*) Is 2 a root of the equation $P(y)=y^4-2y^2-y+7=0$?

(*c*) Is $2i$ a root of the equation $P(z)=2z^3+3z^2+8z+12=0$?

SOLUTION

(*a*) $P(-1)=-1+7-6=0$. Hence -1 is a root of the equation $P(x)=0$, and $[x-(-1)]=x+1$ is a factor of the polynomial $P(x)$.

(*b*) $P(2)=16-8-2+7=13$. Hence 2 is not a root of $P(y)=0$, and $(y-2)$ is not a factor of y^4-2y^2-y+7.

(*c*) $P(2i)=2(2i)^3+3(2i)^2+8(2i)+12=-16i-12+16i+12=0$. Hence $2i$ is a root of $P(z)=0$, and $(z-2i)$ is a factor of the polynomial $P(z)$.

20.6 Prove that $x-a$ is a factor of x^n-a^n, if n is any positive integer.

SOLUTION

$P(x)=x^n-a^n$; then $P(a)=a^n-a^n=0$. Since $P(a)=0$, $x-a$ is a factor of x^n-a^n.

20.7 (*a*) Show that x^5+a^5 is exactly divisible by $x+a$.

(*b*) What is the remainder when y^6+a^6 is divided by $y+a$?

SOLUTION

(*a*) $P(x) = x^5 + a^5$; then $P(-a) = (-a)^5 + a^5 = -a^5 + a^5 = 0$. Since $P(-a) = 0$, $x^5 + a^5$ is exactly divisible by $x + a$.

(*b*) $P(y) = y^6 + a^6$. Remainder $= P(-a) = (-a)^6 + a^6 = a^6 + a^6 = 2a^6$.

20.8 Show that $x + a$ is a factor of $x^n - a^n$ when n is an even positive integer, but not a factor when n is an odd positive integer. Assume $a \neq 0$.

SOLUTION

$$P(x) = x^n - a^n.$$

When n is even, $P(-a) = (-a)^n - a^n = a^n - a^n = 0$. Since $P(-a) = 0$, $x + a$ is a factor of $x^n - a^n$ when n is even.

When n is odd, $P(-a) = (-a)^n - a^n = -a^n - a^n = -2a^n$. Since $P(-a) \neq 0$, $x^n - a^n$ is not exactly divisible by $x + a$ when n is odd (the remainder being $-2a^n$).

20.9 Find the values of p for which

(*a*) $2x^3 - px^2 + 6x - 3p$ is exactly divisible by $x + 2$,

(*b*) $(x^4 - p^2x + 3 - p) \div (x - 3)$ has a remainder of 4.

SOLUTION

(*a*) The remainder is $2(-2)^3 - p(-2)^2 + 6(-2) - 3p = -16 - 4p - 12 - 3p = -28 - 7p = 0$. Then $p = -4$.

(*b*) The remainder is $3^4 - p^2(3) + 3 - p = 84 - 3p^2 - p = 4$. Then $3p^2 + p - 80 = 0$, $(p - 5)(3p + 16) = 0$ and $p = 5, -16/3$.

20.10 By synthetic division, determine the quotient and remainder in the following.

$$(3x^5 - 4x^4 - 5x^3 - 8x + 25) \div (x - 2)$$

SOLUTION

$$\begin{array}{r|l} 2 & 3 - 4 - 5 + 0 - 8 + 25 \\ & \phantom{3 - {}} 6 + 4 - 2 - 4 - 24 \\ \hline & 3 + 2 - 1 - 2 - 12 + 1 \end{array}$$

Quotient: $3x^4 + 2x^3 - x^2 - 2x - 12$

Remainder: 1

The top row of figures gives the coefficients of the dividend, with zero as the coefficient of the missing power of x ($0x^2$). The 2 at the extreme left is the second term of the divisor with the sign changed (since the coefficient of x in the divisor is 1).

The first coefficient of the top row, 3, is written first in the third row and then multiplied by the 2 of the divisor. The product 6 is placed first in the second row and added to the -4 above it to give 2, which is the next figure in the third row. This 2 is then multiplied by the 2 of the divisor. The product 4 is placed in the second row and added to the -5 above it to give the -1 in the third row; etc. The last figure in the third row is the Remainder, while all the figures to its left constitute the coefficients of the Quotient.

Since the dividend is of the 5th degree and the divisor of the 1st degree, the quotient is of the 4th degree.

The answer may be written:

$$3x^4 + 2x^3 - x^2 - 2x - 12 + \frac{1}{x - 2}.$$

20.11 $(x^4 - 2x^3 - 24x^2 + 15x + 50) \div (x + 4)$

SOLUTION

$$\begin{array}{r|l} \underline{-4} & 1 - 2 - 24 + 15 + 50 \\ & \quad - 4 + 24 - \ 0 - 60 \\ \hline & 1 - 6 + \ 0 + 15 - 10 \end{array}$$

Answer: $x^3 - 6x^2 + 15 - \dfrac{10}{x + 4}$

20.12 $(2x^4 - 17x^2 - 4) \div (x + 3)$

SOLUTION

$$\begin{array}{r|l} \underline{-3} & 2 + 0 - 17 + 0 - 4 \\ & \quad - 6 + 18 - 3 + 9 \\ \hline & 2 - 6 + \ 1 - 3 + 5 \end{array}$$

Answer: $2x^3 - 6x^2 + x - 3 + \dfrac{5}{x + 3}$

20.13 $(4x^3 - 10x^2 + x - 1) \div (x - 1/2)$

SOLUTION

$$\begin{array}{r|l} \underline{1/2} & 4 - 10 + 1 - \ 1 \\ & \quad + \ 2 - 4 - 3/2 \\ \hline & 4 - \ 8 - 3 - 5/2 \end{array}$$

Answer: $4x^2 - 8x - 3 - \dfrac{5}{2x - 1}$

20.14 Given $P(x) = x^3 - 6x^2 - 2x + 40$, compute (*a*) $P(-5)$ and (*b*) $P(4)$ using synthetic division.

SOLUTION

(*a*)
$$\begin{array}{r|l} \underline{-5} & 1 - \ 6 - \ 2 + \ 40 \\ & \quad - \ 5 + 55 - 265 \\ \hline & 1 - 11 + 53 - 225 \end{array}$$

$P(-5) = -225$

(*b*)
$$\begin{array}{r|l} \underline{4} & 1 - 6 - \ 2 + 40 \\ & \quad + 4 - \ 8 - 40 \\ \hline & 1 - 2 - 10 + \ 0 \end{array}$$

$P(4) = 0$

20.15 Given that one root of $x^3 + 2x^2 - 23x - 60 = 0$ is 5, solve the equation.

SOLUTION

$$\begin{array}{r|l} \underline{5} & 1 + 2 - 23 - 60 \\ & \quad + 5 + 35 + 60 \\ \hline & 1 + 7 + 12 + \ 0 \end{array}$$

Divide $x^3 + 2x^2 - 23x - 60$ by $x - 5$.

The depressed equation is $x^2 + 7x + 12 = 0$, whose roots are -3, -4. The three roots are $5, -3, -4$.

20.16 Two roots of $x^4 - 2x^2 - 3x - 2 = 0$ are -1 and 2. Solve the equation.

SOLUTION

$$\begin{array}{r|l} \underline{-1} & 1 + 0 - 2 - 3 - 2 \\ & \quad - 1 + 1 + 1 + 2 \\ \hline & 1 - 1 - 1 - 2 + 0 \end{array}$$

Divide $x^4 - 2x^2 - 3x - 2$ by $x + 1$.

The first depressed equation is $x^3 - x^2 - x - 2 = 0$.

$$\begin{array}{r|l} \underline{2} & 1 - 1 - 1 - 2 \\ & \quad + 2 + 2 + 2 \\ \hline & 1 + 1 + 1 + 0 \end{array}$$

Divide $x^3 - x^2 - x - 2$ by $x - 2$.

The second depressed equation is $x^2 + x + 1 = 0$, whose roots are $-\frac{1}{2} \pm \frac{1}{2}i\sqrt{3}$.

The four roots are -1, 2, $-\frac{1}{2} \pm \frac{1}{2}i\sqrt{3}$.

20.17 Determine the roots of each of the following equations.

(*a*) $(x-1)^2(x+2)(x+4) = 0$. Ans. 1 as a double root, -2, -4

(*b*) $(2x+1)(3x-2)^3(2x-5) = 0$. $-1/2$, $2/3$ as a triple root, $5/2$

(*c*) $x^3(x^2 - 2x - 15) = 0$. 0 as a triple root, 5, -3

(*d*) $(x+1+\sqrt{3})(x+1-\sqrt{3})(x-6) = 0$. $(-1-\sqrt{3})$, $(-1+\sqrt{3})$, 6

(*e*) $[(x-i)(x+i)]^3(x+1)^2 = 0$. $\pm i$ as triple roots, -1 as a double root

(*f*) $3(x+m)^4(5x-n)^2 = 0$. $-m$ as a quadruple root, $n/5$ as a double root

20.18 Write the equation having only the following roots.
(*a*) 5, 1, -3; (*b*) 2, $-1/4$, $-1/2$; (*c*) ± 2, $2 \pm \sqrt{3}$; (*d*) 0, $1 \pm 5i$.

SOLUTION

(*a*) $(x-5)(x-1)(x+3) = 0$ or $x^3 - 3x^2 - 13x + 15 = 0$.

(*b*) $(x-2)\left(x+\frac{1}{4}\right)\left(x+\frac{1}{2}\right) = 0$ or $x^3 - \frac{5x^2}{4} - \frac{11x}{8} - \frac{1}{4} = 0$ or $8x^3 - 10x^2 - 11x - 2 = 0$,

which has integral coefficients.

(*c*) $(x-2)(x+2)[x-(2-\sqrt{3})][x-(2+\sqrt{3})] = (x^2-4)[(x-2)+\sqrt{3}][(x-2)-\sqrt{3}]$
$= (x^2-4)[(x-2)^2-3] = (x^2-4)(x^2-4x+1) = 0$, or $x^4 - 4x^3 - 3x^2 + 16x - 4 = 0$.

(*d*) $x[x-(1+5i)][x-(1-5i)] = x[(x-1)-5i][(x-1)+5i] = x[(x-1)^2+25]$
$= x(x^2-2x+26) = 0$, or $x^3 - 2x^2 + 26x = 0$.

20.19 Form the equation with integral coefficients having only the following roots.

(*a*) $1, \frac{1}{2}, -\frac{1}{3}$; (*b*) $0, \frac{3}{4}, \frac{2}{3}, -1$; (*c*) $\pm 3i, \pm\frac{1}{2}\sqrt{2}$; (*d*) 2 as a triple root, -1.

SOLUTION

(*a*) $(x-1)(2x-1)(3x+1) = 0$ or $6x^3 - 7x^2 + 1 = 0$

(*b*) $x(4x-3)(3x-2)(x+1) = 0$ or $12x^4 - 5x^3 - 11x^2 + 6x = 0$

(*c*) $(x-3i)(x+3i)\left(x-\frac{1}{2}\sqrt{2}\right)\left(x+\frac{1}{2}\sqrt{2}\right) = (x^2+9)\left(x^2-\frac{1}{2}\right) = 0$, $(x^2+9)(2x^2-1) = 0$,

or $2x^4 + 17x^2 - 9 = 0$

(*d*) $(x-2)^3(x+1) = 0$ or $x^4 - 5x^3 + 6x^2 + 4x - 8 = 0$

20.20 Each given number is a root of a polynomial equation with *real coefficients*. What other number is a root? (*a*) $2i$, (*b*) $-3 \pm 2i$, (*c*) $-3 - i\sqrt{2}$.

SOLUTION

(*a*) $-2i$, (*b*) $-3-2i$, (*c*) $-3+i\sqrt{2}$

20.21 Each given number is a root of a polynomial equation with *rational coefficients*. What other number is a root? (*a*) $-\sqrt{7}$, (*b*) $-4+2\sqrt{3}$, (*c*) $5-\frac{1}{2}\sqrt{2}$.

SOLUTION

(*a*) $\sqrt{7}$, (*b*) $-4-2\sqrt{3}$, (*c*) $5+\frac{1}{2}\sqrt{2}$

20.22 Criticize the validity of each of the following conclusions.

(*a*) $x^3+7x-6i=0$ has $x=i$ as a root; hence $x=-i$ is a root.

(*b*) $x^3+(1-2\sqrt{3})x^2+(5-2\sqrt{3})x+5=0$ has $\sqrt{3}-i\sqrt{2}$ as a root; hence $\sqrt{3}+i\sqrt{2}$ is a root.

(*c*) $x^4+(1-2\sqrt{2})x^3+(4-2\sqrt{2})x^2+(3-4\sqrt{2})x+1=0$ has $x=-1+\sqrt{2}$ as a root; hence $x=-1-\sqrt{2}$ is a root.

SOLUTION

(*a*) $x=-i$ is not necessarily a root, since not all the coefficients of the given equation are *real*. By substitution it is, in fact, found that $x=-i$ is not a root.

(*b*) The conclusion is valid, since the given equation has real coefficients.

(*c*) $x=-1-2\sqrt{2}$ is not necessarily a root, since not all the coefficients of the given equation are *rational*. By substitution it is found that $x=-1-\sqrt{2}$ is not a root.

20.23 Write the polynomial equation of lowest degree with real coefficients having 2 and $1-3i$ as two of its roots.

SOLUTION

$$(x-2)[x-(1-3i)][x-(1+3i)]=(x-2)(x^2-2x+10)=0 \quad \text{or} \quad x^3-4x^2+14x-20=0$$

20.24 Form the polynomial equation of lowest degree with rational coefficients having $-1+\sqrt{5}$ and -6 as two of its roots.

SOLUTION

$$[x-(-1+\sqrt{5})][x-(-1-\sqrt{5})](x+6)=(x^2+2x-4)(x+6)=0 \quad \text{or} \quad x^3+8x^2+8x-24=0$$

20.25 Form the quartic polynomial equation with rational coefficients having as two of its roots (*a*) $-5i$ and $\sqrt{6}$, (*b*) $2+i$ and $1-\sqrt{3}$.

SOLUTION

(*a*) $(x+5i)(x-5i)(x-\sqrt{6})(x+\sqrt{6})=(x^2+25)(x^2-6)=0$ or $x^4+19x^2-150=0$

(*b*) $[x-(2+i)][x-(2-i)][x-(1-\sqrt{3})][x-(1+\sqrt{3})]=(x^2-4x+5)(x^2-2x-2)=0$ or $x^4-6x^3+11x^2-2x-10=0$

20.26 Find the four roots of $x^4+2x^2+1=0$.

SOLUTION

$x^4+2x^2+1=(x^2+1)^2=[(x+i)(x-i)]^2=0$. The roots are $i, i, -i, -i$.

20.27 Solve $x^4-3x^3+5x^2-27x-36=0$, given that one root is a pure imaginary number of the form bi where b is real.

SOLUTION

Substituting bi for x, $b^4 + 3b^3i - 5b^2 - 27bi - 36 = 0$.

Equating real and imaginary parts to zero:

$$b^4 - 5b^2 - 36 = 0,\ (b^2 - 9)(b^2 + 4) = 0 \text{ and } b = \pm 3 \text{ since } b \text{ is real;}$$

$$3b^3 - 27b = 0,\ 3b(b^2 - 9) = 0 \text{ and } b = 0,\ \pm 3.$$

The common solution is $b = \pm 3$; hence two roots are $\pm 3i$ and $(x - 3i)(x + 3i) = x^2 + 9$ is a factor of $x^4 - 3x^3 + 5x^2 - 27x - 36$. By division the other factor is $x^2 - 3x - 4 = (x - 4)(x + 1)$, and the other two roots are $4, -1$.

The four roots are $\pm 3i, 4, -1$.

20.28 Form the polynomial equation of lowest degree with *rational coefficients*, one of whose roots is (*a*) $\sqrt{3} - \sqrt{2}$, (*b*) $\sqrt{2} + \sqrt{-1}$.

SOLUTION

(*a*) Let $x = \sqrt{3} - \sqrt{2}$.

Squaring both sides, $x^2 = 3 - 2\sqrt{6} + 2 = 5 - 2\sqrt{6}$ and $x^2 - 5 = -2\sqrt{6}$.

Squaring again, $x^4 - 10x^2 + 25 = 24$ and $x^4 - 10x^2 + 1 = 0$.

(*b*) Let $x = \sqrt{2} + \sqrt{-1}$.

Squaring both sides, $x^2 = 2 + 2\sqrt{-2} - 1 = 1 + 2\sqrt{-2}$ and $x^2 - 1 = 2\sqrt{-2}$.

Squaring again, $x^4 - 2x^2 + 1 = -8$ and $x^4 - 2x^2 + 9 = 0$.

20.29 (*a*) Write the polynomial equation of lowest degree with *constant* (real or complex) coefficients having the roots 2 and $1 - 3i$. Compare with Problem 20.23 above.

(*b*) Write the polynomial equation of lowest degree with *real* coefficients having the roots -6 and $-1 + \sqrt{5}$. Compare with Problem 20.24 above.

SOLUTION

(*a*) $(x - 2)[x - (1 - 3i)] = 0$ or $x^2 - 3(1 - i)x + 2 - 6i = 0$

(*b*) $(x + 6)[x - (-1 + \sqrt{5})] = 0$ or $x^2 + (7 - \sqrt{5})x - 6(\sqrt{5} - 1) = 0$

20.30 Obtain the rational roots, if any, of each of the following polynomial equations.

(*a*) $x^4 - 2x^2 - 3x - 2 = 0$

The rational roots are limited to the integral factors of 2, which are ± 1, ± 2.

Testing these values for x in order, $+1, -1, +2, -2$, by synthetic division or by substitution, we find that the only rational roots are -1 and 2.

(*b*) $x^3 - x - 6 = 0$

The rational roots are limited to the integral factors of 6, which are ± 1, ± 2, ± 3, ± 6.

Testing these values for x in order, $+1, -1, +2, -2, +3, -3, +6, -6$, the only rational root obtained is 2.

(*c*) $2x^3 + x^2 - 7x - 6 = 0$

If b/c (in lowest terms) is a rational root, the only possible values of b are ± 1, ± 2, ± 3, ± 6; and the only possible values of c are $\pm 1, \pm 2$. Hence the possible rational roots are limited to the following numbers: ± 1, ± 2, ± 3, ± 6, $\pm 1/2$, $\pm 3/2$.

Testing these values for x, we obtain -1, 2, $-3/2$ as the rational roots.

(*d*) $2x^4 + x^2 + 2x - 4 = 0$

If b/c is a rational root, the values of b are limited to ± 1, ± 2, ± 4; and the values of c are limited to $\pm 1, \pm 2$. Hence the possible rational roots are limited to the numbers ± 1, ± 2, ± 4, $\pm 1/2$.

Testing these values for x, we find that there are no rational roots.

20.31 Solve the polynomial equation $x^3 - 2x^2 - 31x + 20 = 0$.

SOLUTION

Any rational root of this equation is an integral factor of 20. Then the possibilities for rational roots are: $\pm 1, \pm 2, \pm 4, \pm 5, \pm 10, \pm 20$.

Testing these values for x by synthetic division, we find that -5 is a root.

$$\begin{array}{r|l} -5 & 1 - 2 - 31 + 20 \\ & \quad - 5 + 35 - 20 \\ \hline & 1 - 7 + \ 4 + \ 0 \end{array}$$

The depressed equation $x^2 - 7x + 4 = 0$ has irrational roots $7/2 \pm \sqrt{33}/2$.
Hence the three roots of the given equation are -5, $7/2 \pm \sqrt{33}/2$.

20.32 Solve the polynomial equation $2x^4 - 3x^3 - 7x^2 - 8x + 6 = 0$.

SOLUTION

If b/c is a rational root, the only possible values of b are $\pm 1, \pm 2, \pm 3, \pm 6$; and the only possible values of c are $\pm 1, \pm 2$. Hence the possibilities for rational roots are $\pm 1, \pm 2, \pm 3, \pm 6, \pm 1/2, \pm 3/2$.

Testing these values of x by synthetic division, we find that 3 is a root.

$$\begin{array}{r|l} 3 & 2 - 3 - 7 - 8 + 6 \\ & \quad + 6 + 9 + 6 - 6 \\ \hline & 2 + 3 + 2 - 2 + 0 \end{array}$$

The first depressed $2x^3 + 3x^2 + 2x - 2 = 0$ is tested and 1/2 is obtained as a root.

$$\begin{array}{r|l} 1/2 & 2 + 3 + 2 - 2 \\ & \quad + 1 + 2 + 2 \\ \hline & 2 + 4 + 4 + 0 \end{array}$$

The second depressed equation $2x^2 + 4x + 4 = 0$ or $x^2 + 2x + 2 = 0$ has the complex roots $-1 \pm i$.
The four roots are 3, 1/2, $-1 \pm i$.

20.33 Prove that $\sqrt{3} + \sqrt{2}$ is an irrational number.

SOLUTION

Let $x = \sqrt{3} + \sqrt{2}$; then $x^2 = (\sqrt{3} + \sqrt{2})^2 = 3 + 2\sqrt{6} + 2 = 5 + 2\sqrt{6}$ and $x^2 - 5 = 2\sqrt{6}$.

Squaring again, $x^4 - 10x^2 + 25 = 24$ or $x^4 - 10x^2 + 1 = 0$. The only possible rational roots of this equation are ± 1. Testing these values, we find that there is no rational root. Hence $x = \sqrt{3} + \sqrt{2}$ is irrational.

20.34 Graph $P(x) = x^3 + x - 3$. From the graph determine the number of positive, negative and complex roots of $x^3 + x - 3 = 0$.

SOLUTION

x	−3	−2	−1	0	1	2	3	4
$P(x)$	−33	−13	−5	−3	−1	7	27	65

From the graph it is seen that there is one positive and no negative real root (see Fig. 20-1).
Hence there are two conjugate complex roots.

20.35 Find upper and lower limits of the real roots of (*a*) $x^3 - 3x^2 + 5x + 4 = 0$, (*b*) $x^3 + x^2 - 6 = 0$.

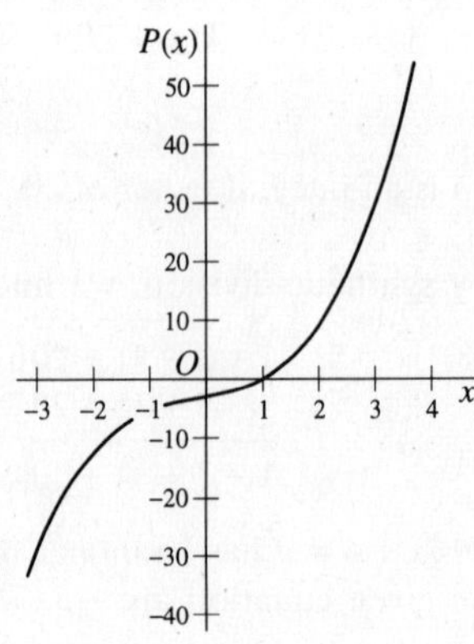

Fig. 20-1

SOLUTION

(*a*) The possible rational roots are ±1, ±2, ±4.

Testing for upper limit.

$$\begin{array}{r|l} 1 & 1-3+5+4 \\ & +1-2+3 \\ \hline & 1-2+3+7 \end{array} \qquad \begin{array}{r|l} 2 & 1-3+5+\ 4 \\ & +2-2+\ 6 \\ \hline & 1-1+3+10 \end{array} \qquad \begin{array}{r|l} 3 & 1-3+5+\ 4 \\ & +3+0+15 \\ \hline & 1+0+5+19 \end{array}$$

Since all the numbers in the third row of the synthetic division of $P(x)$ by $x-3$ are positive (or zero), an upper limit of the roots is 3, i.e. no root is greater than 3.

Testing for lower limit.

$$\begin{array}{r|l} -1 & 1-3+5+4 \\ & -1+4-9 \\ \hline & 1-4+9-5 \end{array}$$

Since the numbers in the third row are alternately positive and negative, -1 is a lower limit of the roots, i.e. no root is less than -1.

(*b*) The possible rational roots are ±1, ±2, ±3, ±6.

Testing for upper limit.

$$\begin{array}{r|l} 1 & 1+1+0-6 \\ & +1+2+2 \\ \hline & 1+2+2-4 \end{array} \qquad \begin{array}{r|l} 2 & 1+1+0-\ 6 \\ & +2+6+12 \\ \hline & 1+3+6+\ 6 \end{array}$$

Hence 2 is an upper limit of the roots.

Testing for lower limit.

$$\begin{array}{r|l} -1 & 1+1+0-6 \\ & -1-0+0 \\ \hline & 1+0+0-6 \end{array}$$

Since all the numbers of the third row are alternately positive and negative (or zero), a lower limit to the roots is -1.

20.36 Determine the rational roots of $4x^3+15x-36=0$ and thus solve the equation completely.

SOLUTION

The possible rational roots are ±1, ±2, ±3, ±4, ±6, ±9, ±12, ±18, ±36, ±1/2, ±3/2, ±9/2, ±1/4, ±3/4, ±9/4. To avoid testing all of these possibilities, find upper and lower limits of the roots.

Testing for upper limit.

$$\begin{array}{r|l} 1 & 4+0+15-36 \\ & \quad +4+\ \ 4+19 \\ \hline & 4+4+19-17 \end{array} \qquad \begin{array}{r|l} 2 & 4+0+15-36 \\ & \quad +8+16+62 \\ \hline & 4+8+31+26 \end{array}$$

Hence no (real) root is greater than or equal to 2.

Testing for lower limit.

$$\begin{array}{r|l} -1 & 4+0+15-36 \\ & \quad -4+\ \ 4-19 \\ \hline & 4-4+19-55 \end{array}$$

Hence no real root is less than or equal to -1.

The only possible rational roots greater than -1 and less than 2 are $+1$, $\pm 1/2$, $\pm 3/2$, $\pm 1/4$, $\pm 3/4$. Testing these we find that 3/2 is the only rational root.

$$\begin{array}{r|l} 3/2 & 4+0+15-36 \\ & \quad +6+\ \ 9+36 \\ \hline & 4+6+24+\ \ 0 \end{array}$$

The other roots are solutions of $4x^2 + 6x + 24 = 0$ or $2x^2 + 3x + 12 = 0$, i.e., $x = -\frac{3}{4} \pm \frac{\sqrt{87}}{4} i$.

20.37 Employing Descartes' Rule of Signs, what may be inferred as to the number of positive, negative and complex roots of the following equations?

(*a*) $2x^3 + 3x^2 - 13x + 6 = 0$
(*b*) $x^4 - 2x^2 - 3x - 2 = 0$
(*c*) $x^2 - 2x + 7 = 0$
(*d*) $2x^4 + 7x^2 + 6 = 0$
(*e*) $x^4 - 3x^2 - 4 = 0$
(*f*) $x^3 + 3x - 14 = 0$
(*g*) $x^6 + x^3 - 1 = 0$
(*h*) $x^6 - 3x^2 - 4x + 1 = 0$

SOLUTION

(*a*) There are 2 variations of sign in $P(x) = 2x^3 + 3x^2 - 13x + 6$. There is 1 variation of sign in $P(-x) = -2x^3 + 3x^2 + 13x + 6$. Hence there are at most 2 positive roots and 1 negative root.

The roots may be: (1) 2 positive, 1 negative, 0 complex; or (2) 0 positive, 1 negative, 2 complex. (Complex roots occur in conjugate pairs.)

(*b*) There is 1 variation of sign in $P(x) = x^4 - 2x^2 - 3x - 2$, and 3 variations of sign in $P(-x) = x^4 - 2x^2 + 3x - 2$. Hence there are at most 1 positive root and 3 negative roots.

The roots may be: (1) 1 positive, 3 negative, 0 complex;
or (2) 1 positive, 1 negative, 2 complex.

(*c*) There are 2 variations of sign in $P(x) = x^2 - 2x + 7$, and no variation of sign in $P(-x) = x^2 + 2x + 7$.

Hence the roots may be: (1) 2 positive, 0 negative, 0 complex;
or (2) 0 positive, 0 negative, 2 complex;

(*d*) Neither $P(x) = 2x^4 + 7x^2 + 6$ nor $P(-x) = 2x^4 + 7x^2 + 6$ has a variation of sign. Hence all 4 roots are complex, since $P(0) \neq 0$.

(*e*) There is 1 variation of sign in $P(x) = x^4 - 3x^2 - 4 = 0$, and 1 variation of sign in $P(-x) = x^4 - 3x^2 - 4$.

Hence the roots are: 1 positive, 1 negative, 2 complex.

(*f*) There is 1 variation of sign in $P(x) = x^3 + 3x - 14$, and no variation in $P(-x) = -x^3 - 3x - 14$.

Hence the roots are: 1 positive, 2 complex.

(*g*) There is 1 variation of sign in $P(x) = x^6 + x^3 - 1$, and 1 variation in $P(-x) = x^6 - x^3 - 1$.

Hence the roots are: 1 positive, 1 negative, 4 complex.

(*h*) There are 2 variations of sign in $P(x) = x^6 - 3x^2 - 4x + 1$, and 2 variations of sign in $P(-x) = x^6 - 3x^2 + 4x + 1$.

Hence the roots may be:

(1) 2 positive, 2 negative, 2 complex; (3) 0 positive, 2 negative, 4 complex;

(2) 2 positive, 0 negative, 4 complex; (4) 0 positive, 0 negative, 6 complex.

20.38 Determine the nature of the roots of $x^n - 1 = 0$ when n is a positive integer and (*a*) n is even, (*b*) n is odd.

SOLUTION

(*a*) $P(x) = x^n - 1$ has 1 variation of sign, and $P(-x) = x^n - 1$ has 1 variation of sign. Hence the roots are: 1 positive, 1 negative, $(n-2)$ complex.

(*b*) $P(x) = x^n - 1$ has 1 variation of sign, and $P(-x) = -x^n - 1$ has no variation of sign. Hence the roots are: 1 positive, 0 negative, $(n-1)$ complex.

20.39 Obtain the rational roots, if any, of each equation, making use of Descartes' Rule of Signs.

(*a*) $x^3 - x^2 + 3x - 27 = 0$, (*c*) $2x^5 + x - 66 = 0$,

(*b*) $x^3 + 2x + 12 = 0$, (*d*) $3x^4 + 7x^2 + 6 = 0$.

SOLUTION

(*a*) By Descartes' Rule of Signs, the equation has 3 or 1 positive roots and no negative root. Hence the rational roots are limited to positive integral factors of 27, which are 1,3,9,27.

Testing these values for x, the only rational root obtained is 3.

(*b*) By the rule of signs, the equation has no positive root and 1 negative root. Hence the rational roots are limited to the negative integral factors of 12, i.e., to -1, -2, -3, -4, -6, -12.

Testing these values for x, we obtain -2 as the only rational root.

(*c*) By the rule of signs, the equation has 1 positive root and no negative root. Hence the rational roots are limited to positive rational numbers of the form b/c where b is limited to integral factors of 66 and c is limited to integral factors of 2. The possible rational roots are then 1, 2, 3, 6, 11, 22, 33, 66, 1/2, 3/2, 11/2, 33/2.

Testing these values for x, we obtain 2 as the only rational root.

(*d*) The equation has no real root, since neither $P(x) = 3x^4 + 7x^2 + 6$ nor $P(-x) = 3x^4 + 7x^2 + 6$ has a variation of sign and $P(0) \neq 0$.

Hence all of its four roots are complex.

20.40 Use the Intermediate Value Theorem to isolate each of the real zeros of $P(x)$ between two consecutive integers.

(*a*) $P(x) = 3x^3 - 8x^2 - 8x + 8$ (*b*) $P(x) = 5x^3 - 4x^2 - 10x + 8$

SOLUTION

(*a*) For $P(x) = 3x^3 - 8x^2 - 8x + 8$, we find the upper and lower limits of the real zeros.

$$\begin{array}{l|l} 0 & 3-8-8+8 \\ & \quad +0+0+0 \\ \hline & 3-8-8+8 \end{array} \qquad \begin{array}{l|l} 1 & 3-8-\ 8+\ 8 \\ & \quad +3-\ 5-13 \\ \hline & 3-5-13-\ 5 \end{array} \qquad \begin{array}{l|l} 2 & 3-8-\ 8+\ 8 \\ & \quad +6-\ 4-24 \\ \hline & 3-2-12-16 \end{array}$$

$$\begin{array}{l|l} 3 & 3-8-8+\ 8 \\ & \quad +9+3-15 \\ \hline & 3+1-5-\ 7 \end{array} \qquad \begin{array}{l|l} 4 & 3-\ 8-\ 8+\ 8 \\ & \quad +12+16+32 \\ \hline & 3+\ 4+\ 8+40 \end{array}$$

Since the quotient row when synthetic division is done with 4 is all positive, the upper limit of the real zeros of $P(x)$ is 4.

$$\begin{array}{r|rrrr} -1 & 3 & -8 & -8 & +8 \\ & & -3 & +11 & -3 \\ \hline & 3 & -11 & +3 & +5 \end{array} \qquad \begin{array}{r|rrrr} -2 & 3 & -8 & -8 & +8 \\ & & -6 & +28 & -40 \\ \hline & 3 & -14 & +14 & -32 \end{array}$$

Since the quotient row when synthetic division is done with -2 alternates in sign, the lower limit of the real zeros of $P(x)$ is -2.

We now examine the interval from -2 to 4 to isolate the real zeros of $P(x)$ between consecutive integers.

Since $P(0) = 8$ and $P(1) = -5$ there is a real zero between 0 and 1. Since $P(3) = -7$ and $P(4) = 40$ there is a real zero between 3 and 4. Since $P(-1) = 5$ and $P(-2) = -32$, there is a real zero between -2 and -1.

The real zeros of $P(x) = 3x^3 - 8x^2 - 8x + 8$ are between -2 and -1, 0 and 1, and 3 and 4.

(*b*) For $P(x) = 5x^3 - 4x^2 - 10x + 8$, we find the upper and lower limits of the real zeros.

$$\begin{array}{r|rrrr} 0 & 5 & -4 & -10 & +8 \\ & & +0 & +0 & +0 \\ \hline & 5 & -4 & -10 & +8 \end{array} \qquad \begin{array}{r|rrrr} 1 & 5 & -4 & -10 & +8 \\ & & +5 & +1 & -9 \\ \hline & 5 & +1 & -9 & -1 \end{array} \qquad \begin{array}{r|rrrr} 2 & 5 & -4 & -10 & +8 \\ & & +10 & +12 & +4 \\ \hline & 5 & +6 & +2 & +12 \end{array}$$

The upper limit for the zeros of $P(x)$ is 2.

$$\begin{array}{r|rrrr} -1 & 5 & -4 & -10 & +8 \\ & & -5 & +9 & +1 \\ \hline & 5 & -9 & -1 & +9 \end{array} \qquad \begin{array}{r|rrrr} -2 & 5 & -4 & -10 & +8 \\ & & -10 & +28 & -36 \\ \hline & 5 & -14 & +18 & -28 \end{array}$$

The lower limit of the real zeros of $P(x)$ is -2.

We now examine the interval from -2 to 2 to isolate the real zeros of $P(x)$. Since $P(0) = 8$ and $P(1) = -1$, there is a real zero between 0 and 1. Since $P(1) = -1$ and $P(2) = 12$, there is a real zero between 1 and 2. Since $P(-1) = 9$ and $P(-2) = -28$, there is a real zero between -2 and -1.

Thus, the real zeros of $P(x) = 5x^3 - 4x^2 - 10x + 8$ are located between -2 and -1, 0 and 1, and 1 and 2.

20.41 Approximate a real zero of $P(x) = x^3 - x - 5$ to two decimal places.

SOLUTION

By Descartes' Rule of Signs, $P(x) = x^3 - x - 5$ has 1 positive real zero and 2 or 0 negative real zeros.

Now we determine the upper limit of real zeros of $P(x)$.

$$\begin{array}{r|rrrr} 0 & 1 & +0 & -1 & -5 \\ & & +0 & +0 & +0 \\ \hline & 1 & +0 & -1 & -5 \end{array} \qquad \begin{array}{r|rrrr} 1 & 1 & +0 & -1 & -5 \\ & & +1 & +1 & +0 \\ \hline & 1 & +1 & +0 & -5 \end{array} \qquad \begin{array}{r|rrrr} 2 & 1 & +0 & -1 & -5 \\ & & +2 & +4 & +6 \\ \hline & 1 & +2 & +3 & +1 \end{array}$$

The upper limit on the real zeros of $P(x)$ is 2.

Since $P(1) = -5$ and $P(2) = 1$, the positive real zero is between 1 and 2.

We now determine the tenths interval of the zero. Using synthetic division, we determine the tenths values until we find two values with different signs. Since $P(2) = 1$ is closer to 0 than $P(1) = -5$, we start with $x = 1.9$. Since $P(1.9) = -0.041$ and $P(2.0) = 1$, the real zero is between 1.9 and 2.0.

Since $P(1.90) = -0.41$ is closer to 0 than $P(2.0) = 1$, we look for the hundredths interval starting with $x = 1.91$. $P(1.91) = 0.579$. Since $P(1.90) = -0.041$ and $P(1.91) = 0.579$, there is a real zero between 1.90 and 1.91.

We now determine $P(1.905)$ to decide whether the zero rounds off to 1.90 or 1.91. $P(1.905) = 0.008$. Since $P(1.900)$ is negative and $P(1.905)$ is positive, the zero is between 1.900 and 1.905. When rounded to two decimal places, all numbers in this interval round to 1.90.

Thus, rounded to two decimal places, a real zero of $P(x) = x^3 - x - 5$ is 1.90.

20.42 Approximate $\sqrt[3]{3}$ to three decimal places.

SOLUTION

Let $x = \sqrt[3]{3}$, so $x^3 = 3$ and $P(x) = x^3 - 3 = 0$.

By Descartes' Rule of Signs, $P(x)$ has one positive real zero and no negative real zeros.

Since $P(1) = -2$ and $P(2) = 5$, the zero is between 1 and 2.

Since $P(1)$ is closer to 0 than $P(2)$, locate the tenths interval by evaluating $P(x)$ from $x = 1$ to $x = 2$ starting with $x = 1.1$. Once a sign change in $P(x)$ is found we stop.

x	1.0	1.1	1.2	1.3	1.4	1.5
$P(x)$	-2	-1.669	-1.272	-0.803	-0.256	0.375

Since $P(1.4)$ is negative and $P(1.5)$ is positive, the real zero is between 1.4 and 1.5.

Now determine the hundredths interval by exploring the values of $P(x)$ in the interval from $x = 1.40$ to $x = 1.50$.

x	1.40	1.41	1.42	1.43	1.44	1.45
$P(x)$	-0.256	-0.97	-0.137	-0.076	-0.014	0.49

Since $P(1.4)$ is negative and $P(1.5)$ is positive, the zero is between 1.4 and 1.5.

The next step is to determine the thousandths interval for the zero by exploring the values of $P(x)$ in the interval between $x = 1.440$ and $x = 1.450$.

x	1.440	1.441	1.442	1.443
$P(x)$	-0.014	-0.008	-0.002	0.005

Since $P(1.442)$ is negative and $P(1.443)$ is positive, the zero is between 1.442 and 1.443.

Since $P(1.4425) = 0.002$, the real zero is between 1.4420 and 1.4425. All values in the interval from 1.4420 to 1.4425 round to 1.442 to three decimal places.

Therefore, $\sqrt[3]{3} = 1.442$ to three decimal places.

Supplementary Problems

20.43 If $P(x) = 2x^3 - x^2 - x + 2$, find (*a*) $P(0)$, (*b*) $P(2)$, (*c*) $P(-1)$, (*d*) $P(\frac{1}{2})$, (*e*) $P(\sqrt{2})$.

20.44 Determine the remainder in each of the following.

(*a*) $(2x^5 - 7) \div (x + 1)$

(*b*) $(x^3 + 3x^2 - 4x + 2) \div (x - 2)$

(*c*) $(3x^3 + 4x - 4) \div (x - \frac{1}{2})$

(*d*) $(4y^3 + y + 27) \div (2y + 3)$

(*e*) $(x^{12} + x^6 + 1) \div (x - \sqrt{-1})$

(*f*) $(2x^{33} + 35) \div (x + 1)$

20.45 Prove that $x + 3$ is a factor of $x^3 + 7x^2 + 10x - 6$ and that $x = -3$ is a root of the equation $x^3 + 7x^2 + 10x - 6 = 0$.

20.46 Determine which of the following numbers are roots of the equation $y^4 + 3y^3 + 12y - 16 = 0$: (*a*) 2, (*b*) -4, (*c*) 3, (*d*) 1, (*e*) $2i$.

20.47 Find the values of k for which

(*a*) $4x^3 + 3x^2 - kx + 6k$ is exactly divisible by $x + 3$,

(*b*) $x^5 + 4kx - 4k^2 = 0$ has the root $x = 2$.

20.48 By synthetic division determine the quotient and remainder in each of the following.

(a) $(2x^3 + 3x^2 - 4x - 2) \div (x + 1)$ (c) $(y^6 - 3y^5 + 4y - 5) \div (y + 2)$

(b) $(3x^5 + x^3 - 4) \div (x - 2)$ (d) $(4x^3 + 6x^2 - 2x + 3) \div (2x + 1)$

20.49 If $P(x) = 2x^4 - 3x^3 + 4x - 4$, compute $P(2)$ and $P(-3)$ using synthetic division.

20.50 Given that one root of $x^3 - 7x - 6 = 0$ is -1, find the other two roots.

20.51 Show that $2x^4 - x^3 - 3x^2 - 31x - 15 = 0$ has roots 3, $-\frac{1}{2}$. Find the other roots.

20.52 Find the roots of each equation.

(a) $(x + 3)^2(x - 2)^3(x + 1) = 0$ (c) $(x^2 + 3x + 2)(x^2 - 4x + 5) = 0$

(b) $4x^4(x + 2)^4(x - 1) = 0$ (d) $(y^2 + 4)^2(y + 1)^2 = 0$

20.53 Form equations with integral coefficients having only the following roots.

(a) $2, -3, -\frac{1}{2}$ (b) $0, -4, 2/3, 1$ (c) $\pm 3i$, double root 2 (d) $-1 \pm 2i, 2 \pm i$

20.54 Form an equation whose only roots are $1 \pm \sqrt{2}, -1 \pm i\sqrt{3}$.

20.55 Write the equation of lowest possible degree with integral coefficients having the given roots.

(a) $1, 0, i$ (b) $2 + i$ (c) $-1 \pm \sqrt{3}, 1/3$ (d) $-2, i\sqrt{3}$ (e) $\sqrt{2}, i$ (f) $i/2, 6/5$

20.56 In the equation $x^3 + ax^2 + bx + a = 0$, a and b are real numbers. If $x = 2 + i$ is a root of the equation, find a and b.

20.57 Write an equation of lowest degree with integral coefficients having $\sqrt{2} - 1$ as a double root.

20.58 Write an equation of lowest degree with integral coefficients having $\sqrt{3} + 2i$ as a root.

20.59 Solve each equation, given the indicated root.

(a) $x^4 + x^3 - 12x^2 + 32x - 40 = 0$; $1 - i\sqrt{3}$ (c) $x^3 - 5x^2 + 6 = 0$; $3 - \sqrt{3}$

(b) $6x^4 - 11x^3 + x^2 + 33x - 45 = 0$; $1 + i\sqrt{2}$ (d) $x^4 - 4x^3 + 6x^2 - 16x + 8 = 0$; $2i$

20.60 Obtain the rational roots, if any, of each equation.

(a) $x^4 + 2x^3 - 4x^2 - 5x - 6 = 0$ (c) $2x^4 - x^3 + 2x^2 - 2x - 4 = 0$

(b) $4x^3 - 3x + 1 = 0$ (d) $3x^3 + x^2 - 12x - 4 = 0$

20.61 Solve each equation.

(a) $x^3 - x^2 - 9x + 9 = 0$ (d) $4x^4 + 8x^3 - 5x^2 - 2x + 1 = 0$

(b) $2x^3 - 3x^2 - 11x + 6 = 0$ (e) $5x^4 + 3x^3 + 8x^2 + 6x - 4 = 0$

(c) $3x^3 + 2x^2 + 2x - 1 = 0$ (f) $3x^5 + 2x^4 - 15x^3 - 10x^2 + 12x + 8 = 0$

20.62 Prove that (a) $\sqrt{5} - \sqrt{2}$ and (b) $\sqrt[5]{2}$ are irrational numbers.

20.63 If $P(x) = 2x^3 - 3x^2 + 12x - 16$, determine the number of positive, negative and complex roots.

20.64 Locate between two successive integers the real roots of $x^4 - 3x^2 - 6x - 2 = 0$. Find the least positive root of the equation accurate to two decimal places.

20.65 Find upper and lower limits for the real roots of each equation.

(*a*) $x^3 - 3x^2 + 2x - 4 = 0$ (*b*) $2x^4 + 5x^2 - 6x - 14 = 0$

20.66 Find the rational roots of $2x^3 - 5x^2 + 4x + 24 = 0$ and thus solve the equation completely.

20.67 Using Descartes' Rule of Signs, what may be inferred as to the number of positive, negative and complex roots of the following equations?

(*a*) $2x^3 + 3x^2 + 7 = 0$ (*c*) $x^5 + 4x^3 - 3x^2 - x + 12 = 0$

(*b*) $3x^3 - x^2 + 2x - 1 = 0$ (*d*) $x^5 - 3x - 2 = 0$

20.68 Given the equation $3x^4 - x^3 + x^2 - 5 = 0$, determine (*a*) the maximum number of positive roots, (*b*) the minimum number of positive roots, (*c*) the exact number of negative roots, and (*d*) the maximum number of complex roots.

20.69 Given the equation $5x^3 + 2x - 4 = 0$, how many roots are (*a*) negative, (*b*) real?

20.70 Does the equation $x^6 + 4x^4 + 3x^2 + 16 = 0$ have (*a*) 4 complex and 2 real roots, (*b*) 4 real and 2 complex roots, (*c*) 6 complex roots, or (*d*) 6 real roots?

20.71 (*a*) How many positive roots has the equation $x^6 - 7x^2 - 11 = 0$?

(*b*) How many complex roots has the equation $x^7 + x^4 - x^2 - 3 = 0$?

(*c*) Show that $x^6 + 2x^3 + 3x - 4 = 0$ has exactly 4 complex roots.

(*d*) Show that $x^4 + x^3 - x^2 - 1 = 0$ has only one negative root.

20.72 Solve completely each equation.

(*a*) $8x^3 - 20x^2 + 14x - 3 = 0$ (*c*) $4x^3 + 5x^2 + 2x - 6 = 0$

(*b*) $8x^4 - 14x^3 - 9x^2 + 11x - 2 = 0$ (*d*) $2x^4 - x^3 - 23x^2 + 18x + 18 = 0$

20.73 Approximate the indicated root of each equation to the specified accuracy.

(*a*) $2x^3 + 3x^2 - 9x - 7 = 0$; positive root, to the nearest tenth

(*b*) $x^3 + 9x^2 + 27x - 50 = 0$; positive root, to the nearest hundredth

(*c*) $x^3 - 3x^2 - 3x + 18 = 0$; negative root, to the nearest tenth

(*d*) $x^3 + 6x^2 + 9x + 17 = 0$; negative root, to the nearest tenth

(*e*) $x^5 + x^4 - 27x^3 - 83x^2 + 50x + 162 = 0$; root between 5 and 6, to the nearest hundredth

(*f*) $x^4 - 3x^3 + x^2 - 7x + 12 = 0$; root between 1 and 2, to the nearest hundredth

20.74 In finding the maximum deflection of a beam of given length loaded in a certain way, it is necessary to solve the equation $4x^3 - 150x^2 + 1500x - 2871 = 0$. Find correct to the nearest tenth the root of the equation lying between 2 and 3.

20.75 The length of a rectangular box is twice its width, and its depth is one foot greater than its width. If its volume is 64 cubic feet, find its width to the nearest tenth of a foot.

20.76 Find $\sqrt[3]{20}$ correct to the nearest hundredth.

ANSWERS TO SUPPLEMENTARY PROBLEMS

20.43 (*a*) 2 (*b*) 12 (*c*) 0 (*d*) 3/2 (*e*) $3\sqrt{2}$

20.44 (*a*) −9 (*b*) 14 (*c*) −13/8 (*d*) 12 (*e*) 1 (*f*) 33

20.46 -4, 1 and $2i$ are roots

20.47 (*a*) $k = 9$ (*b*) $k = 4, -2$

20.48 (*a*) $2x^2 + x - 5 + \dfrac{3}{x+1}$ (*c*) $y^5 - 5y^4 + 10y^3 - 20y^2 + 40y - 76 + \dfrac{147}{y+2}$

(*b*) $3x^4 + 6x^3 + 13x^2 + 26x + 52 + \dfrac{100}{x-2}$ (*d*) $2x^2 + 2x - 2 + \dfrac{5}{2x+1}$

20.49 12, 227

20.50 3, -2

20.51 $-1 \pm 2i$

20.52 (*a*) double root -3, triple root 2, -1 (*c*) $-1, -2, 2 \pm i$
(*b*) quadruple root 0, quadruple root -2, 1 (*d*) double roots $\pm 2i$, double root -1

20.53 (*a*) $2x^3 + 3x^2 - 11x - 6 = 0$ (*c*) $x^4 - 4x^3 + 13x^2 - 36x + 36 = 0$
(*b*) $3x^4 + 7x^3 - 18x^2 + 8x = 0$ (*d*) $x^4 - 2x^3 + 2x^2 - 10x + 25 = 0$

20.54 $x^4 - x^2 - 10x - 4 = 0$

20.55 (*a*) $x^4 - x^3 + x^2 - x = 0$ (*d*) $x^3 + 2x^2 + 3x + 6 = 0$
(*b*) $x^2 - 4x + 5 = 0$ (*e*) $x^4 - x^2 - 2 = 0$
(*c*) $3x^3 + 5x^2 - 8x + 2 = 0$ (*f*) $20x^3 - 24x^2 + 5x - 6 = 0$

20.56 $a = -5$, $b = 9$

20.57 $x^4 + 4x^3 + 2x^2 - 4x + 1 = 0$

20.58 $x^4 + 2x^2 + 49 = 0$

20.59 (*a*) $1 \pm i\sqrt{3}, -5, 2$ (*b*) $1 \pm i\sqrt{2}, -5/3, 3/2$ (*c*) $3 \pm \sqrt{3}, -1$ (*d*) $\pm 2i, 2 \pm \sqrt{2}$

20.60 (*a*) $-3, 2$ (*b*) $1/2, 1/2, -1$ (*c*) no rational root (*d*) $-1/3, \pm 2$

20.61 (*a*) $1, \pm 3$ (*d*) $\pm\frac{1}{2}, -1 \pm \sqrt{2}$
(*b*) $3, -2, 1/2$ (*e*) $-1, 2/5, \pm\sqrt{2}i$
(*c*) $1/3, -\frac{1}{2} + \frac{1}{2}\sqrt{3}i$ (*f*) $\pm 1, \pm 2, -2/3$

20.63 1 positive root, 0 negative roots, 2 complex roots

20.64 Positive root between 2 and 3; negative root between -1 and 0; pos. root = 2.41 approx.

20.65 (*a*) Upper limit 3, lower limit -1 (*b*) Upper limit 2, lower limit -2

20.66 $-3/2, 2 \pm 2i$

20.67 (*a*) 1 negative, 2 complex
(*b*) 3 positive or 1 positive, 2 complex
(*c*) 1 negative, 2 positive, 2 complex or 1 negative, 4 complex
(*d*) 1 positive, 2 negative, 2 complex or 1 positive, 4 complex

20.68 (*a*) 3 (*b*) 1 (*c*) 1 (*d*) 2

20.69 (*a*) none (*b*) one

20.70 (*c*)

20.71 (*a*) one (*b*) four or six

20.72 (*a*) 1/2, 1/2, 3/2 (*b*) 2, −1, 1/4, 1/2 (*c*) 3/4, $-1 \pm i$ (*d*) 3, 3/2, $-2 \pm \sqrt{2}$

20.73 (*a*) 1.9 (*b*) 1.25 (*c*) −2.2 (*d*) −4.9 (*e*) 5.77 (*f*) 1.38

20.74 2.5

20.75 2.9 ft

20.76 2.71

Chapter 21

Rational Functions

21.1 RATIONAL FUNCTIONS

A rational function is the ratio of two polynomial functions. If $P(x)$ and $Q(x)$ are polynomials, then a function of the form $R(x) = P(x)/Q(x)$ is a rational function where $Q(x) \neq 0$.

The domain of $R(x)$ is the intersection of the domains of $P(x)$ and $Q(x)$.

21.2 VERTICAL ASYMPTOTES

If $R(x) = P(x)/Q(x)$, then values of x that make $Q(x) = 0$ result in vertical asymptotes if $P(x) \neq 0$. However, if for some value $x = a$, $P(a) = 0$ and $Q(a) = 0$, then $P(x)$ and $Q(x)$ have a common factor of $x - a$. If $R(x)$ is then reduced to lowest terms, the graph of $R(x)$ has a hole in it where $x = a$.

A vertical asymptote for $R(x)$ is a vertical line $x = k$, k is a constant, that the graph of $R(x)$ approaches but does not touch. $R(k)$ is not defined because $Q(k) = 0$ and $P(k) \neq 0$. The domain of $R(x)$ is separated in distinct intervals by the vertical asymptotes of $R(x)$.

EXAMPLE 21.1. What are the vertical asymptotes of

$$R(x) = \frac{2x-3}{x^2-4}?$$

Since

$$R(x) = \frac{2x-3}{x^2-4}$$

is undefined when $x^2 - 4 = 0$, $x = 2$ and $x = -2$ could result in vertical asymptotes. When $x = 2$, $2x - 3 \neq 0$ and when $x = -2$, $2x - 3 \neq 0$. Thus, the graph of $R(x)$ has vertical asymptotes of $x = 2$ and $x = -2$.

21.3 HORIZONTAL ASYMPTOTES

A rational function $R(x) = P(x)/Q(x)$ has a horizontal asymptote $y = a$ if, as $|x|$ increases without limit, $R(x)$ approaches a. $R(x)$ has at most one horizontal asymptote. The horizontal asymptote of $R(x)$ may be found from a comparison of the degree of $P(x)$ and the degree of $Q(x)$.

(1) If the degree of $P(x)$ is less than the degree of $Q(x)$, then $R(x)$ has a horizontal asymptote of $y = 0$.

(2) If the degree of $P(x)$ is equal to the degree of $Q(x)$, then $R(x)$ has a horizontal asymptote of $y = a_n/b_n$, where a_n is the lead coefficient (coefficient of the highest degree term) of $P(x)$ and b_n is the lead coefficient of $Q(x)$.

(3) If the degree of $P(x)$ is greater than the degree of $Q(x)$, then $R(x)$ does not have a horizontal asymptote.

The graph of $R(x)$ may cross a horizontal asymptote in the interior of its domain. This is possible since we are only concerned with how $R(x)$ behaves as $|x|$ increases without limit in determining the horizontal asymptote.

EXAMPLE 21.2. What are the horizontal asymptotes of each rational function $R(x)$?

(*a*) $R(x) = \dfrac{3x^3}{x^2 - 1}$ (*b*) $R(x) = \dfrac{x}{x^2 - 4}$ (*c*) $R(x) = \dfrac{2x + 1}{3 + 5x}$

(*a*) For

$$R(x) = \frac{3x^3}{x^2 - 1},$$

the degree of the numerator $3x^3$ is 3 and the degree of the denominator is 2. Since the numerator exceeds the degree of the denominator, $R(x)$ does not have a horizontal asymptote.

(*b*) The degree of the numerator of

$$R(x) = \frac{x}{x^2 - 1}$$

is 1 and the degree of the denominator is 2, so $R(x)$ has a horizontal asymptote of $y = 0$.

(*c*) The numerator and denominator of

$$R(x) = \frac{2x + 1}{3 + 5x}$$

each have degree 1. Since the lead coefficient of the numerator is 2 and the lead coefficient of the denominator is 5, $R(x)$ has a horizontal asymptote of $y = \frac{2}{5}$.

21.4 GRAPHING RATIONAL FUNCTIONS

To graph a rational function $R(x) = P(x)/Q(x)$, we first determine the holes: values of x for which both $P(x)$ and $Q(x)$ are zero. After any holes are located, we reduce $R(x)$ to lowest terms. The value of the reduced form of $R(x)$ for an x that yields a hole is the y coordinate of the point that corresponds to the hole.

Once $R(x)$ is in lowest terms, we determine the asymptotes, symmetry, zeros, and y intercept if they exist. We graph the asymptotes as dashed lines, plot the zeros and y intercept, and plot several other points to determine how the graph approaches the asymptotes. Finally, we sketch the graph through the plotted points and approaching the asymptotes.

EXAMPLE 21.3. Sketch a graph of each rational function $R(x)$.

(*a*) $R(x) = \dfrac{3}{x^2 - 1}$ (*b*) $R(x) = \dfrac{x^2}{4 - x^2}$

(*a*)

$$R(x) = \frac{3}{x^2 - 1}$$

has vertical asymptotes at $x = 1$, and at $x = -1$, a horizontal asymptote of $y = 0$, and no holes.

Since the numerator of $R(x)$ is a constant, it does not have any zeros. Since $R(0) = -3$, $R(x)$ has a y intercept of $(0, -3)$.

Plot the y intercept and graph the asymptotes as dashed lines. We determine some values of $R(x)$ in each interval of the domain $(-\infty, -1)$, and $(-1, 1)$, and $(1, \infty)$. $R(-x) = R(x)$, so $R(x)$ is symmetric with respect to the y axis.

$$R(2) = R(-2) = \frac{3}{2^2 - 1} = 1, \ R(0.5) = R(-0.5) = \frac{3}{(0.5)^2 - 1} = -4$$

Plot $(2, 1)$, $(-2, 1)$, $(0.5, -4)$, and $(-0.5, -4)$. Using the asymptotes as a boundary, we sketch the graph. The graph of

$$R(x) = \frac{3}{x^2 - 1}$$

is shown in Fig. 21-1.

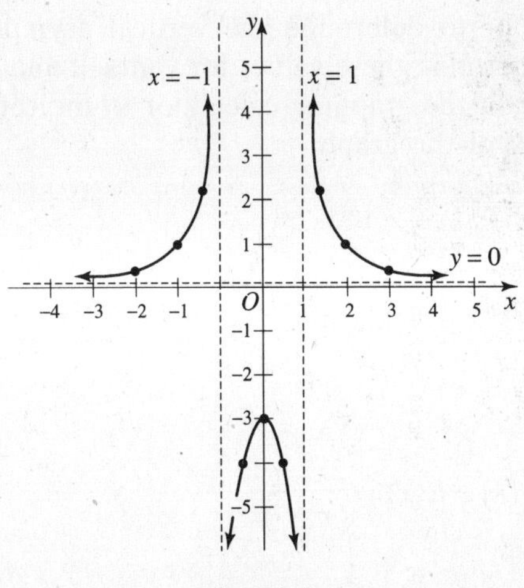

Fig. 21-1

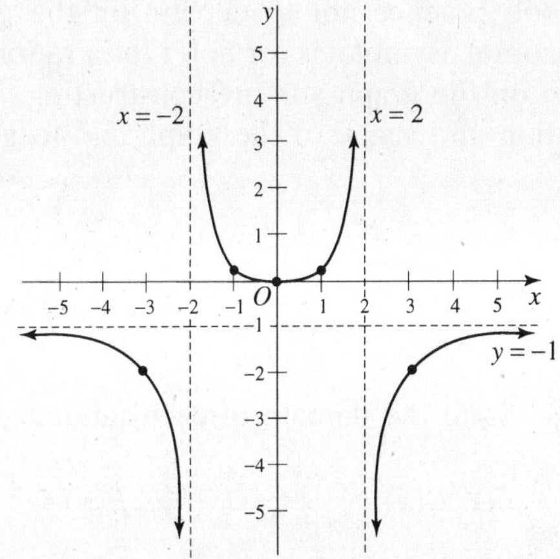

Fig. 21-2

(*b*)
$$R(x) = \frac{x^2}{4 - x^2}$$

has vertical asymptotes at $x = 2$ and $x = -2$, a horizontal asymptote of $y = -1$, and no holes.

The zeros of $R(x)$ are for $x = 0$. Since when $x = 0$, $R(0) = 0$ and $(0, 0)$ is the zero and the y intercept.

Plot the point (0, 0) and the vertical and horizontal asymptotes.

We determine some values of $R(x)$ in each interval of its domain $(-\infty, -2)$, $(-2, 2)$, and $(2, \infty)$. Since $R(-x) = R(x)$, the graph is symmetric with respect to the y axis.

$$R(3) = R(-3) = \frac{3^2}{4 - 3^2} = \frac{9}{-5} = \frac{-9}{5}; \qquad R(1) = R(-1) = \frac{1^2}{4 - 1^2} = \frac{1}{3}$$

Plot $(3, -9/5)$, $(-3, -9/5)$, $(1, 1/3)$, and $(1, -1/3)$.

Using the asymptotes as boundary lines, sketch the graph of $R(x)$.

The graph of

$$R(x) = \frac{x^2}{4 - x^2}$$

is shown in Fig. 21-2.

21.5 GRAPHING RATIONAL FUNCTIONS USING A GRAPHING CALCULATOR

The graphing features of a graphing calculator allow for easy graphing of rational functions, However, unless the graphing calculator specifically plots the x values for the asymptotes, it connects the distinct branches of the rational function. You need to determine the vertical asymptotes, then set the scale on the x axis of the graphing window so that the values of the vertical asymptotes are used.

Horizontal asymptotes must be read from the graph itself, since they are not drawn or labeled and only appear as a characteristic of the graph in the calculator display window.

Holes are based on factors that can be canceled out from the rational function. These are difficult to locate from the display on a graphing calculator.

In general, when using a graphing calculator to aid in the production of a graph on paper, it is good practice not to depend on the graphing calculator to determine the vertical asymptotes, horizontal asymptotes, or holes for a rational function. Determine these values for yourself and place them on the graph you are constructing. Use the display on the graphing calculator to indicate the location and shape of the graph and to guide your sketch of the graph.

Solved Problems

21.1 State the domain of each rational function $R(x)$.

(a) $R(x) = \dfrac{3x}{x+2}$ (b) $R(x) = \dfrac{x^3 - 2x^2 - 3x}{x}$ (c) $R(x) = \dfrac{3x^2 - 1}{x^3 - x}$

SOLUTION

(a) For $R(x) = 3x/(x+2)$, set $x + 2 = 0$ and see that $R(x)$ is not defined for $x = -2$. The domain of $R(x)$ is {all real numbers except -2} or domain $= (-\infty, -2) \cup (-2, \infty)$.

(b) For $R(x) = (x^3 - 2x^2 - 3x)/x$, we see that $R(x)$ is not defined for $x = 0$. Thus, the domain of $R(x)$ is {all real numbers except 0} or domain $= (-\infty, 0) \cup (0, \infty)$.

(c) For $R(x) = (3x^2 - 1)/(x^3 - x)$, we set $x^3 - x = 0$ and determine that for $x = 0$, $x = 1$, and $x = -1$ $R(x)$ is undefined. The domain of $R(x)$ is {all real numbers except -1, 0, 1} or domain $= (-\infty, -1) \cup (-1, 0) \cup (0, 1) \cup (1, \infty)$.

21.2 Determine the vertical asymptotes, horizontal asymptotes, and holes for each rational function $R(x)$.

(a) $R(x) = \dfrac{3x}{x+2}$ (b) $R(x) = \dfrac{x^3 - 2x^2 - 3x}{x}$ (c) $R(x) = \dfrac{3x^2 - 1}{x^3 - x}$

SOLUTION

(a) Values that make the denominator zero but do not make the numerator zero yield asymptotes. In $R(x) = 3x/(x+2)$, $x = -2$ makes the denominator $x + 2 = 0$ but does not make the numerator $3x = 0$. Thus, $x = -2$ is a vertical asymptote.

Since the degree of the numerator $3x$ is 1 and the degree of the denominator $x + 2$ is 1, $R(x)$ has a horizontal asymptote of $y = 3/1 = 3$, where 3 is the lead coefficient of the numerator and 1 is the lead coefficient of the denominator.

Since $x = 0$ is the only value that makes the numerator 0 and $x = 0$ does not make the denominator 0, $R(x)$ has no holes in its graph.

(b) Only $x = 0$ makes the denominator of $R(x) = (x^3 - 2x^2 - 3x)/x$ zero. Since $x = 0$ also makes the numerator zero, $R(x)$ has no vertical asymptotes.

Since the degree of the numerator of $R(x)$ is 3 and the degree of the denominator is 1, the degree of the numerator exceeds the degree of the denominator and there is no horizontal asymptote.

Since $x = 0$ makes both the numerator and denominator of $R(x)$ zero, there is a hole in the graph of $R(x)$ when $x = 0$. We reduce $R(x)$ to lowest terms and get $R(x) = x^2 - 2x - 3$ when $x \neq 0$. The graph of this reduced form would have the value of -3 if x were 0, so the graph of $R(x)$ has a hole at $(0, -3)$.

(c) Since $x^3 - x = 0$ has solutions of $x = 0$, $x = 1$, and $x = -1$, the vertical asymptotes of $R(x) = (3x^2 - 1)/(x^3 - x)$ are $x = 0$, $x = 1$, and $x = -1$.

The degree of the numerator of $R(x)$ is less than the degree of the denominator, so $y = 0$ is the horizontal asymptote of $R(x)$.

The numerator is not zero for any values, $x = -1$, $x = 0$, and $x = 1$, that make the denominator zero, so the graph of $R(x)$ does not have any holes in it.

21.3 What are the zeros and y intercept of each rational function $R(x)$?

(a) $R(x) = \dfrac{3x}{x+2}$ (b) $R(x) = \dfrac{x^3 - 2x^2 - 3x}{x}$ (c) $R(x) = \dfrac{3x^2 - 1}{x^3 - x}$

SOLUTION

(a) For $R(x) = 3x/(x+2)$, the numerator $3x$ is zero if $x = 0$. Since $x = 0$ does not make the denominator zero, there is a zero when $x = 0$. Thus, (0, 0) is the zero of $R(x)$. The y intercept is the value of y when $x = 0$. Thus, (0, 0) is the point for the y intercept.

(b) For $R(x) = (x^3 - 2x^2 - 3x)/x$, the numerator $x^3 - 2x^2 - 3x$ is zero when $x = 0$, $x = -1$, and $x = 3$. However, $x = 0$ makes the denominator zero, so it will not yield a zero of $R(x)$. The zeros of $R(x)$ are $(3, 0)$ and $(-1, 0)$.

From Problem 21.1(b), we know that $x = 0$ is not in the domain of $R(x)$. Thus, $R(x)$ does not have a y intercept.

(c) For $R(x) = (3x^2 - 1)/(x^3 - x)$, the numerator $3x^2 - 1 = 0$ has solutions $x = \sqrt{3}/3$ and $x = -\sqrt{3}/3$. Thus the zeros of $R(x)$ are $(\sqrt{3}/3, 0)$ and $(-\sqrt{3}/3, 0)$.

From Problem 21.1(c), we know that the domain of $R(x)$ does not contain $x = 0$, so $R(x)$ does not have a y intercept.

21.4 Sketch a graph of each rational function $R(x)$.

(a) $R(x) = \dfrac{3x}{x+2}$ (b) $R(x) = \dfrac{x^3 - 2x^2 - 3x}{x}$ (c) $R(x) = \dfrac{3x^2 - 1}{x^3 - x}$

SOLUTION

From Problems 21.1, 21.2, and 21.3, we know the domains, vertical asymptotes, horizontal asymptotes, holes, zeros, and y intercepts for these three rational functions. We will use this information in graphing each of these rational functions.

(a) $R(x) = 3x/(x+2)$ has a vertical asymptote of $x = -2$, a horizontal asymptote of $y = 3$, and (0, 0) is the point for the zero and y intercept. We draw dashed lines for the asymptotes and plot (0, 0). We select some x values and determine the points, then plot them. For $x = -4$, -3, -1, and 2, we get the points $(-4, 6)$, $(-3, 9)$, $(-1, -3)$, and $(2, 1.5)$. Now we sketch the graph of $R(x)$ going through the plotted points and approaching the asymptotes. Since the domain of $R(x)$ is separated into two parts, we will have two parts to the graph of $R(x)$. See Fig. 21-3.

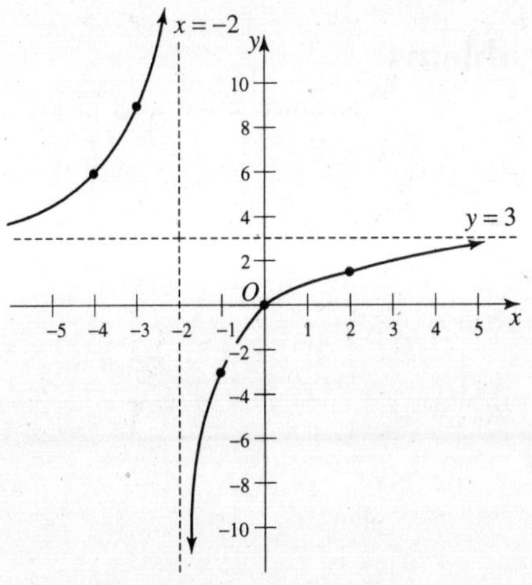

Fig. 21-3

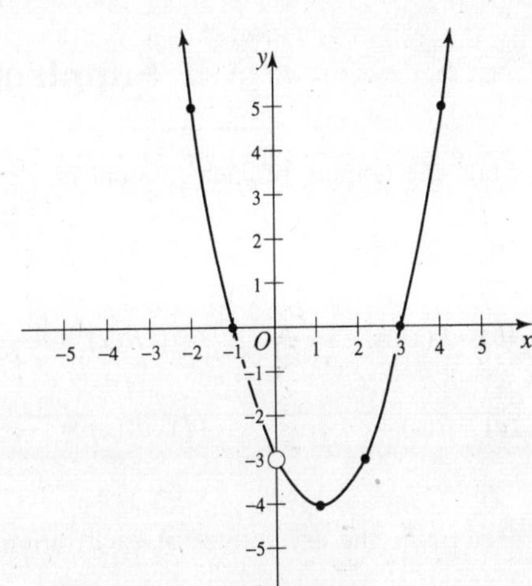

Fig. 21-4

(b) $R(x) = (x^3 - 2x^2 - 3x)/x = x^2 - 2x - 3$ when $x \neq 0$, and there is a hole at $(0, -3)$. There are no asymptotes for the graph of $R(x)$, there is no y intercept, but there are zeros at $(3, 0)$ and $(-1, 0)$. We plot the zeros and place an open circle O around the point $(0, -3)$ to indicate the hole in the graph. Now we select values of x, determine the corresponding points, and plot them. For $x = -2$, 1, 2, and 4, we get the points $(-2, 5)$, $(1, -4)$, $(2, 3)$, and $(4, 5)$. Since the domain of $R(x)$ is separated into two parts by the x value for the hole of $R(x)$, the graph of $R(x)$ is separated into two parts by the hole at $(0, -3)$. See Fig. 21-4.

(c) $R(x) = (3x^2 - 1)/(x^3 - x)$ has vertical asymptotes of $x = -1$, $x = 0$, and $x = 1$, a horizontal asymptote of $y = 0$, and zeros of $(\sqrt{3}/3, 0)$ and $(-\sqrt{3}/3, 0)$. We approximate the zeros to be $(0.6, 0)$ and $(-0.6, 0)$ and we plot these points and graph the asymptotes. To be sure that we graph all the parts of $R(x)$, we select x values from each interval in the domain. For $x = -2, -1.5, -0.75, -0.25, 0.25, 0.75, 1.5$, and 2, we get the points $(-2, -1.8)$, $(-1.5, -3.1)$, $(-0.75, 2.1)$, $(-0.25, -3.5)$, $(0.25, 3.5)$, $(0.75, -2.1)$, $(1.5, 3.1)$, and $(2, 1.8)$. Since the domain of $R(x)$ is separated into four parts the graph of $R(x)$ is in four separate parts. See Fig. 21-5.

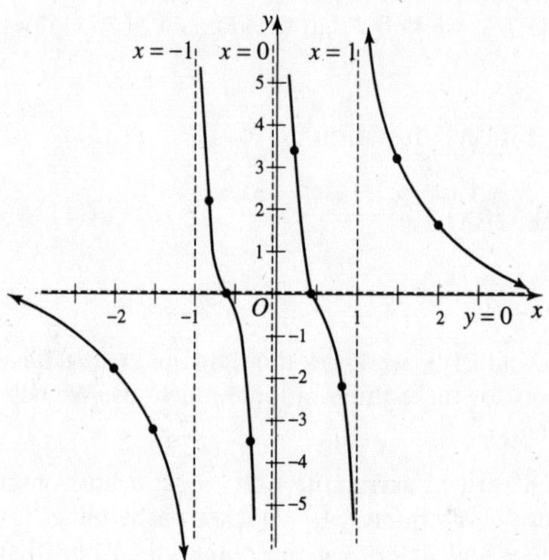

Fig. 21-5

Supplementary Problems

21.5 State the domain of each rational function.

(a) $R(x) = \dfrac{4}{x+2}$ (d) $R(x) = \dfrac{4}{x^2 + x - 2}$ (g) $R(x) = \dfrac{x^3 + 2}{x^2}$

(b) $R(x) = -\dfrac{1}{x-2}$ (e) $R(x) = \dfrac{6-x}{x+3}$ (h) $R(x) = \dfrac{x^2 + 4}{27x^3 - 3x}$

(c) $R(x) = -\dfrac{x}{x^2 - 4}$ (f) $R(x) = \dfrac{2x-5}{x+4}$ (i) $R(x) = \dfrac{-x^2 + x}{x^2 - 5x + 6}$

21.6 Determine the asymptotes of each rational function.

(a) $R(x) = \dfrac{4}{x-3}$ (b) $R(x) = \dfrac{x}{x^2 - 16}$ (c) $R(x) = \dfrac{3x+6}{x-1}$

(*d*) $R(x) = \frac{x^2 - 6x + 9}{x}$ (*f*) $R(x) = \frac{-x^2 + 2x}{x + 3}$ (*h*) $R(x) = \frac{2}{x^2 - 7x + 10}$

(*e*) $R(x) = \frac{2x^2 - 5}{x + 2}$ (*g*) $R(x) = -\frac{3}{x + 4}$ (*i*) $R(x) = \frac{x + 5}{5 - x}$

21.7 Determine the zeros and y intercept for each rational function.

(*a*) $R(x) = \frac{3}{x + 2}$ (*d*) $R(x) = \frac{x^3 - 27}{x^2}$ (*g*) $R(x) = \frac{x^2 - 5x + 6}{x^2 + 6x + 9}$

(*b*) $R(x) = -\frac{x}{x^2 - 4}$ (*e*) $R(x) = \frac{x^2}{x - 3}$ (*h*) $R(x) = \frac{x^3 - 1}{x}$

(*c*) $R(x) = \frac{2x + 8}{x + 3}$ (*f*) $R(x) = \frac{x^2 - 4}{x^2 - 1}$ (*i*) $R(x) = \frac{x + 3}{x^2 + 2x + 1}$

21.8 Graph each rational function.

(*a*) $R(x) = \frac{2}{x}$ (*d*) $R(x) = \frac{2x - 6}{x + 1}$ (*g*) $R(x) = \frac{4 - x^2}{x^2 - 9}$

(*b*) $R(x) = \frac{3}{x - 2}$ (*e*) $R(x) = \frac{x + 2}{x - 3}$ (*h*) $R(x) = -\frac{x}{x^2 - 4}$

(*c*) $R(x) = -\frac{1}{x^2 - 1}$ (*f*) $R(x) = \frac{x}{x^2 - 9}$ (*i*) $R(x) = \frac{x^2 - 5x + 4}{x^2 + 7x + 6}$

21.9 Graph each rational function.

(*a*) $R(x) = \frac{x + 2}{x^2 - 4}$ (*b*) $R(x) = \frac{3x}{x^2 - 2x}$

21.10 Graph each rational function.

(*a*) $R(x) = \frac{2}{x^2 + 4}$ (*b*) $R(x) = \frac{x^2 + 2}{x^3 - x}$

ANSWERS TO SUPPLEMENTARY PROBLEMS

21.5 (*a*) $(-\infty, -2) \cup (-2, \infty)$
(*b*) $(-\infty, 2) \cup (2, \infty)$
(*c*) $(-\infty, -2) \cup (-2, 2) \cup (2, \infty)$
(*d*) $(-\infty, -2) \cup (-2, 1) \cup (1, \infty)$
(*e*) $(-\infty, -3) \cup (-3, \infty)$
(*f*) $(-\infty, -4) \cup (-4, \infty)$
(*g*) $(-\infty, 0) \cup (0, \infty)$
(*h*) $(-\infty, -1/3) \cup (-1/3, 0) \cup (0, 1/3) \cup (1/3, \infty)$
(*i*) $(-\infty, 2) \cup (2, 3) \cup (3, \infty)$

21.6

	Vertical asymptotes	Horizontal asymptotes
(*a*)	$x = 3$	$y = 0$
(*b*)	$x = 4,\ x = -4$	$y = 0$
(*c*)	$x = 1$	$y = 3$
(*d*)	$x = 0$	none
(*e*)	$x = -2$	none
(*f*)	$x = -3$	none
(*g*)	$x = -4$	$y = 0$
(*h*)	$x = 5,\ x = 2$	$y = 0$
(*i*)	$x = 5$	$y = -1$

21.7

Zeros	y intercept
(*a*) none	$(0, 3/2)$
(*b*) $(0, 0)$	$(0, 0)$
(*c*) $(-4, 0)$	$(0, 8/3)$
(*d*) $(3, 0)$	none
(*e*) $(0, 0)$	$(0, 0)$
(*f*) $(2, 0)$, $(-2, 0)$	$(0, 4)$
(*g*) $(3, 0)$, $(2, 0)$	$(0, 2/3)$
(*h*) $(1, 0)$	none
(*i*) $(-3, 0)$	$(0, 3)$

21.8 (*a*) Fig. 21-6 (*d*) Fig. 21-9 (*g*) Fig. 21-12
(*b*) Fig. 21-7 (*e*) Fig. 21-10 (*h*) Fig. 21-13
(*c*) Fig. 21-8 (*f*) Fig. 21-11 (*i*) Fig. 21-14

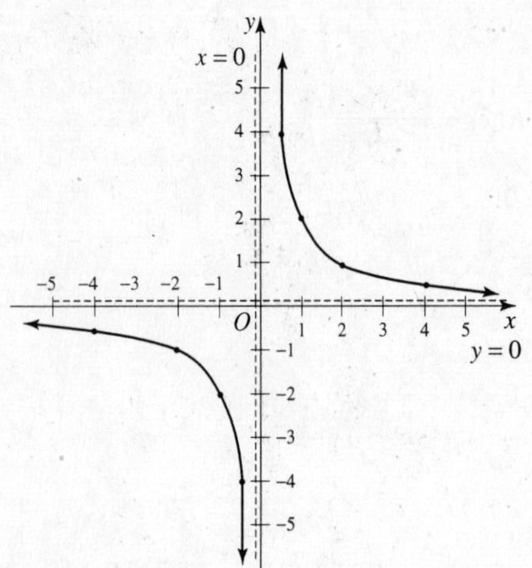

Fig. 21-6

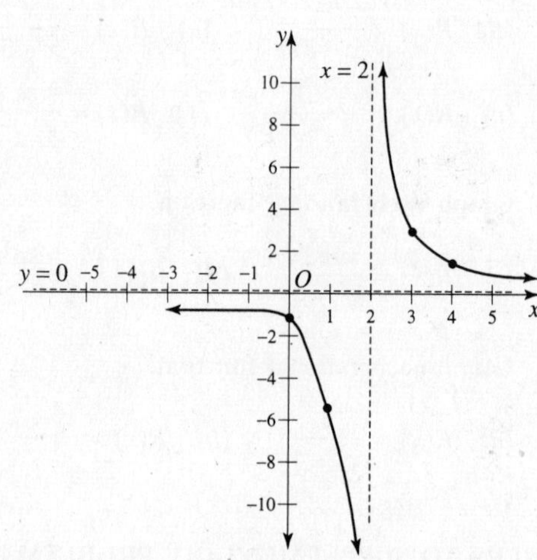

Fig. 21-7

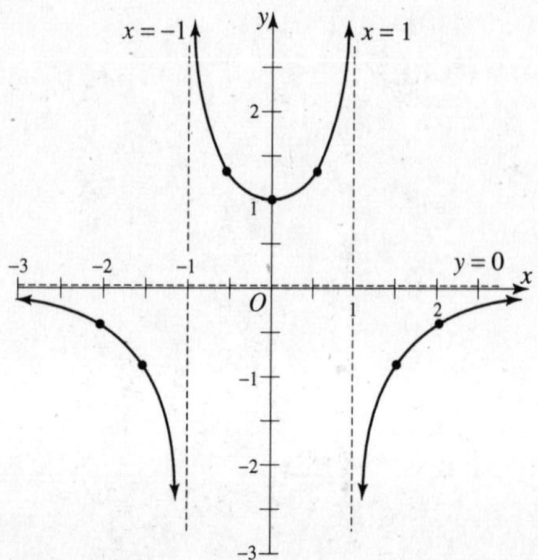

Fig. 21-8

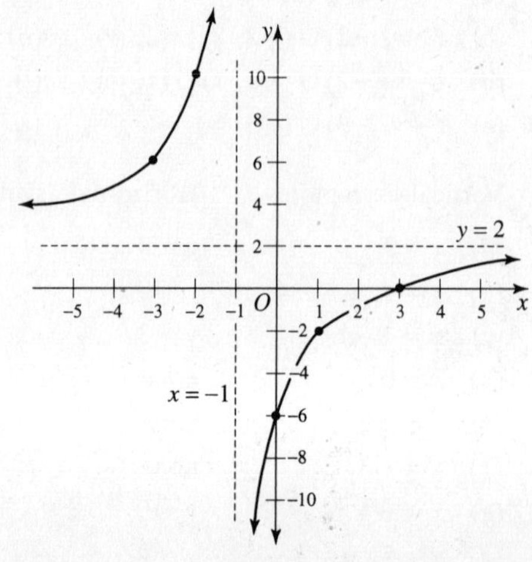

Fig. 21-9

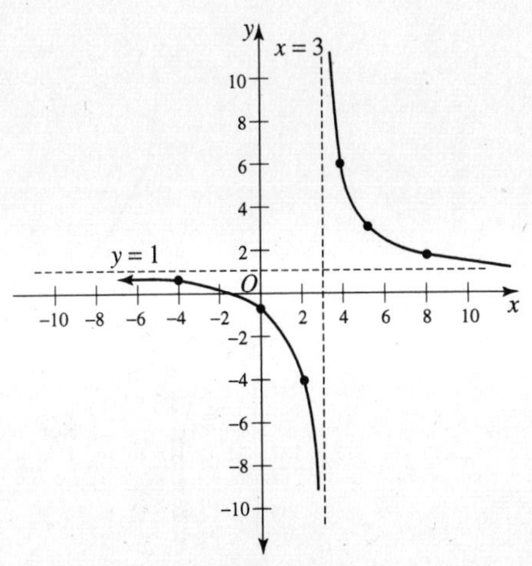

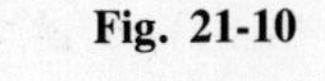

Fig. 21-10

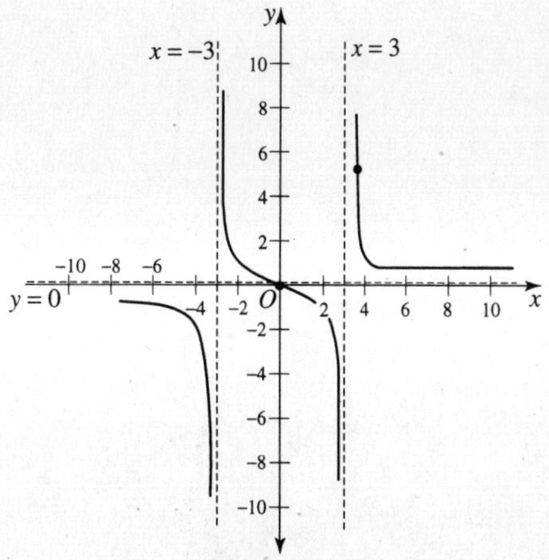

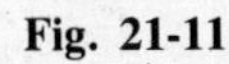

Fig. 21-11

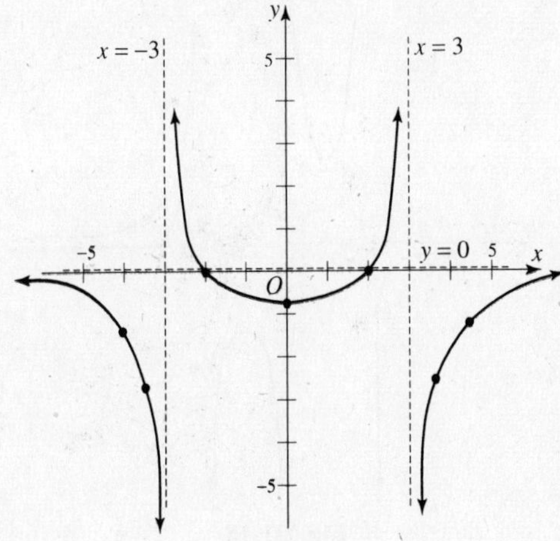

Fig. 21-12

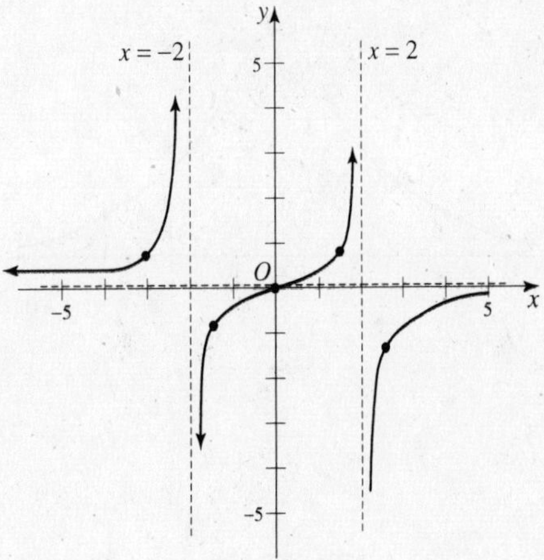

Fig. 21-13

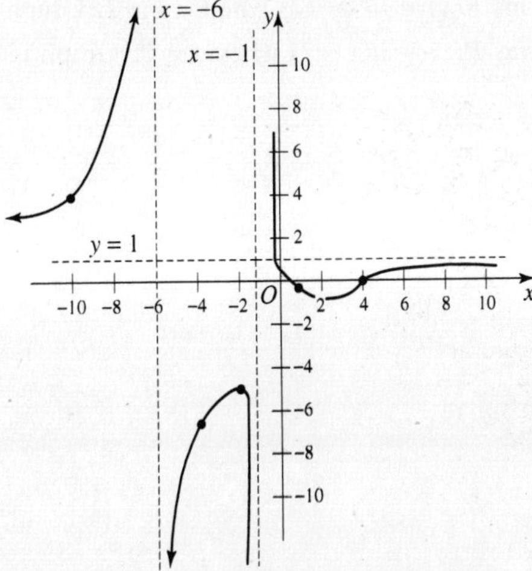

Fig. 21-14

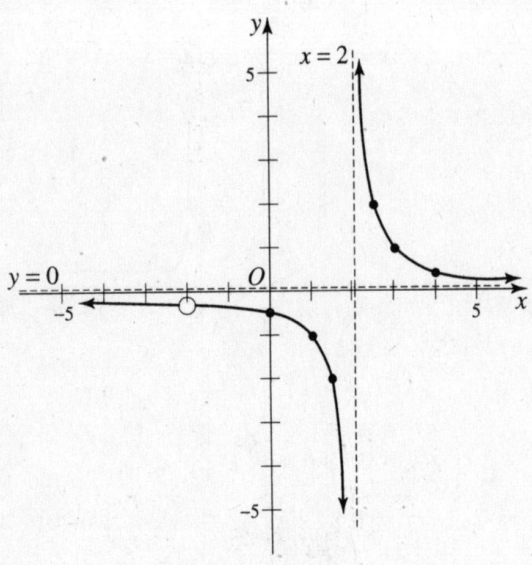

Fig. 21-15

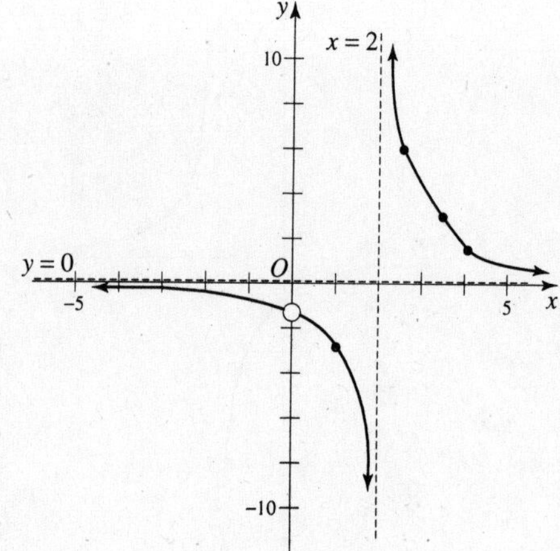

Fig. 21-16

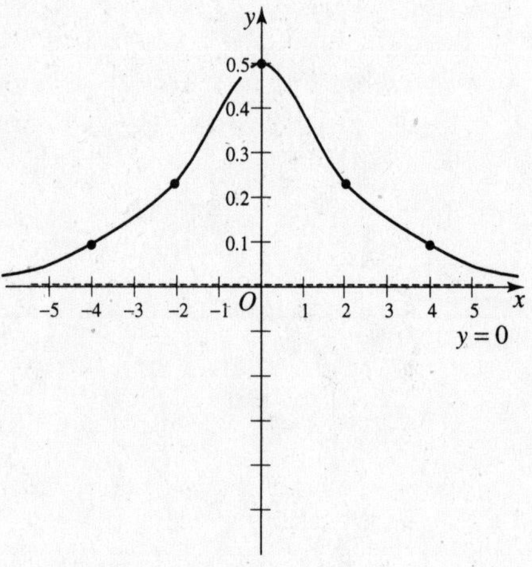

Fig. 21-17

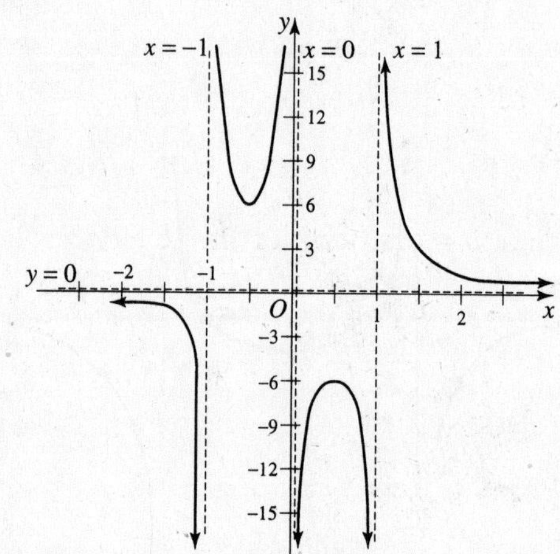

Fig. 21-18

21.9 (*a*) Reduced to lowest terms $R(x) = 1/(x - 2)$ when $x \neq -2$. Graph is Fig. 21-15.
(*b*) Reduced to lowest terms $R(x) = 3/(x - 2)$ when $x \neq 0$. Graph is Fig. 21-16.

21.10 (*a*) Fig. 21-17 (*b*) Fig. 21-18

Chapter 22

Sequences and Series

22.1 SEQUENCES

A sequence of numbers is a function defined on the set of positive integers. The numbers in the sequence are called *terms*. A series is the sum of the terms of a sequence.

22.2 ARITHMETIC SEQUENCES

A. *An arithmetic sequence* is a sequence of numbers each of which, after the first, is obtained by adding to the preceding number a constant number called the *common difference*.

Thus $3, 7, 11, 15, 19, \ldots$ is an arithmetic sequence because each term is obtained by adding 4 to the preceding number. In the arithmetic sequence $50, 45, 40, \ldots$ the common difference is $45 - 50 = 40 - 45 = -5$.

B. *Formulas for arithmetic sequences*

(1) The nth term, or last term: $l = a + (n-1)d$

(2) The sum of the first n terms: $S = \dfrac{n}{2}(a + l) = \dfrac{n}{2}\,[2a + (n-1)d]$

where a = first term of the sequence; d = common difference;
n = number of terms; l = nth term, or last term;
S = sum of first n terms.

EXAMPLE 22.1. Consider the arithmetic sequence $3, 7, 11, \ldots$ where $a = 3$ and $d = 7 - 3 = 11 - 7 = 4$. The sixth term is $l = a + (n-1)d = 3 + (6-1)4 = 23$.

The sum of the first six terms is

$$S = \frac{n}{2}(a + l) = \frac{6}{2}(3 + 23) = 78 \quad \text{or} \quad S = \frac{n}{2}[2a + (n-1)d] = \frac{6}{2}[2(3) + (6-1)4] = 78.$$

22.3 GEOMETRIC SEQUENCES

A. *A geometric sequence* is a sequence of numbers each of which, after the first, is obtained by multiplying the preceding number by a constant number called the *common ratio*.

Thus $5, 10, 20, 40, 80, \ldots$ is a geometric sequence because each number is obtained by multiplying the preceding number by 2. In the geometric sequence $9, -3, 1 -\frac{1}{3}, \frac{1}{9}, \ldots$ the common ratio is

$$\frac{-3}{9} = \frac{1}{-3} = \frac{-1/3}{1} = \frac{1/9}{-1/3} = -\frac{1}{3}.$$

B. *Formulas for geometric sequences.*

(1) The nth term, or last term: $l = ar^{n-1}$

(2) The sum of the first n terms: $S = \dfrac{a(r^n - 1)}{r - 1} = \dfrac{rl - a}{r - 1}, \; r \neq 1$

where a = first term; r = common ratio; n = number of terms;
l = nth term, or last term; S = sum of first n terms.

EXAMPLE 22.2. Consider the geometric sequence $5, 10, 20, \ldots$ where $a = 5$ and

$$r = \frac{10}{5} = \frac{20}{10} = 2$$

The seventh term is $l = ar^{n-1} = 5(2^{7-1}) = 5(2^6) = 320$.

The sum of the first seven terms is

$$S = \frac{a(r^n - 1)}{r - 1} = \frac{5(2^7 - 1)}{2 - 1} = 635.$$

22.4 INFINITE GEOMETRIC SERIES

The sum to infinity (S_∞) of any geometric sequence in which the common ratio r is numerically less than 1 is given by

$$S_\infty = \frac{a}{1 - r}, \qquad \text{where } |r| < 1.$$

EXAMPLE 22.3. Consider the infinite geometric series

$$1 - \frac{1}{2} + \frac{1}{4} - \frac{1}{8} + \cdots$$

where $a = 1$ and $r = -\frac{1}{2}$. Its sum to infinity is

$$S_\infty = \frac{a}{1 - r} = \frac{1}{1 - (-1/2)} = \frac{1}{3/2} = \frac{2}{3}.$$

22.5 HARMONIC SEQUENCES

A *harmonic sequence* is a sequence of numbers whose reciprocals form an arithmetic sequence.

Thus

$$\frac{1}{2}, \frac{1}{4}, \frac{1}{6}, \frac{1}{8}, \frac{1}{10}, \ldots$$

is a harmonic sequence because 2,4,6,8,10,... is an arithmetic sequence.

22.6 MEANS

The terms between any two given terms of sequence are called the *means* between these *two* terms.

Thus in the arithmetic sequence $3, 5, 7, 9, 11, \ldots$ the arithmetic mean between 3 and 7 is 5, and *four* arithmetic means between 3 and 13 are $5, 7, 9, 11$.

In the geometric sequence $2, -4, 8, -16, \ldots$ *two* geometric means between 2 and -16 are $-4, 8$.

In the harmonic sequence

$$\frac{1}{2}, \frac{1}{3}, \frac{1}{4}, \frac{1}{5}, \frac{1}{6}, \ldots$$

the harmonic mean between $\frac{1}{2}$ and $\frac{1}{4}$ is $\frac{1}{3}$, and *three* harmonic means between $\frac{1}{2}$ and $\frac{1}{6}$ are $\frac{1}{3}, \frac{1}{4}, \frac{1}{5}$.

Solved Problems

22.1 Which of the following sequences are arithmetic sequences?

(*a*) $1, 6, 11, 16, \ldots$ Yes, since $6 - 1 = 11 - 6 = 16 - 11 = 5$. $(d = 5)$

(*b*) $\frac{1}{3}, 1, \frac{5}{3}, \frac{7}{3}, \ldots$ Yes, since $1 - \frac{1}{3} = \frac{5}{3} - 1 = \frac{7}{3} - \frac{5}{3} = \frac{2}{3}$. $(d = \frac{2}{3})$

(*c*) $4, -1, -6, -11, \ldots$ Yes, since $-1 - 4 = -6 - (-1) = -11 - (-6) = -5$. $(d = -5)$

(*d*) $9, 12, 16, \ldots$ No, since $12 - 9 \neq 16 - 12$.

(*e*) $\frac{1}{2}, \frac{1}{3}, \frac{1}{4}, \ldots$ No, since $\frac{1}{3} - \frac{1}{2} \neq \frac{1}{4} - \frac{1}{3}$.

(*f*) $7, 9 + 3p, 11 + 6p, \ldots$ Yes, with $d = 2 + 3p$.

22.2 Prove the formula $S = (n/2)(a + l)$ for the sum of the first n terms of an arithmetic sequence.

SOLUTION

The sum of the first n terms of an arithmetic sequence may be written

$$S = a + (a + d) + (a + 2d) + \cdots + l \qquad (n \text{ terms})$$

or

$$S = l + (l - d) + (l - 2d) + \cdots + a \qquad (n \text{ terms})$$

where the sum is written in reversed order.

Adding, $\quad 2S = (a + l) + (a + l) + (a + l) + \cdots + (a + l) \quad$ to n terms.

Hence $\quad 2S = n(a + l)$ and $S = \dfrac{n}{2}(a + l)$.

22.3 Find the 16th term of the arithmetic sequence: $4, 7, 10, \ldots$.

SOLUTION

Here $a = 4$, $n = 16$, $d = 7 - 4 = 10 - 7 = 3$, and $l = a + (n - 1)d = 4 + (16 - 1)3 = 49$.

22.4 Determine the sum of the first 12 terms of the arithmetic sequence: $3, 8, 13, \ldots$

SOLUTION

Here $a = 3$, $d = 8 - 3 = 13 - 8 = 5$, $n = 12$, and

$$S = \frac{n}{2}[2a + (n - 1)d] = \frac{12}{2}[2(3) + (12 - 1)5] = 366.$$

Otherwise: $\quad l = a + (n - 1)d = 3 + (12 - 1)5 = 58 \quad$ and

$$S = \frac{n}{2}(a + l) = \frac{12}{2}(3 + 58) = 366.$$

22.5 Find the 40th term and the sum of the first 40 terms of the arithmetic sequence: $10, 8, 6, \ldots$

SOLUTION

Here $d = 8 - 10 = 6 - 8 = -2$, $a = 10$, $n = 40$.

Then $\quad l = a + (n - 1)d = 10 + (40 - 1)(-2) = -68 \quad$ and

$$S = \frac{n}{2}(a + l) = \frac{40}{2}(10 - 68) = -1160.$$

22.6 Which term of the sequence 5, 14, 23, ... is 239?

SOLUTION

$l = a + (n-1)d, \quad 239 = 5 + (n-1)9, \quad 9n = 243$ and the required term is $n = 27$.

22.7 Compute the sum of the first 100 positive integers exactly divisible by 7.

SOLUTION

The sequence is 7, 14, 21, ... an arithmetic sequence in which $a = 7$, $d = 7$, $n = 100$.

Hence $$S = \frac{n}{2}[2a + (n-1)d] = \frac{100}{2}[2(7) + (100-1)7] = 35\,350.$$

22.8 How many consecutive integers, beginning with 10, must be taken for their sum to equal 2035?

SOLUTION

The sequence is 10, 11, 12, ... an arithmetic sequence in which $a = 10$, $d = 1$, $S = 2035$.

Using $$S = \frac{n}{2}[2a + (n-1)d],$$

we obtain $$2035 = \frac{n}{2}[20 + (n-1)1], \quad 2035 = \frac{n}{2}(n+19), \quad n^2 + 19n - 4070 = 0,$$

$$(n-55)(n+74) = 0, \quad n = 55, -74.$$

Hence 55 integers must be taken.

22.9 How long will it take to pay off a debt of \$880 if \$25 is paid the first month, \$27 the second month, \$29 the third month, etc.?

SOLUTION

From $$S = \frac{n}{2}[2a + (n-1)d],$$

we obtain $$880 = \frac{n}{2}[2(25) + (n-1)2], \quad 880 = 24n + n^2, \quad n^2 + 24n - 880 = 0,$$

$$(n-20)(n+44) = 0, \; n = 20, -44.$$

The debt will be paid off in 20 months.

22.10 How many terms of the arithmetic sequence 24, 22, 20, ... are needed to give the sum of 150? Write the terms.

SOLUTION

$$150 = \frac{n}{2}[48 + (n-1)(-2)], \quad n^2 - 25n + 150 = 0, \quad (n-10)(n-15) = 0, \quad n = 10, 15.$$

For $n = 10$: 24, 22, 20, 18, 16, 14, 12, 10, 8, 6.
For $n = 15$: 24, 22, 20, 18, 16, 14, 12, 10, 8, 6, 4, 2, 0, −2, −4.

22.11 Determine the arithmetic sequence whose sum to n terms is $n^2 + 2n$.

SOLUTION

The nth term = sum to n terms − sum to $n - 1$ terms

$$= n^2 + 2n - [(n-1)^2 + 2(n-1)] = 2n + 1.$$

Thus the arithmetic sequence is 3,5,7,9, . . .

22.12 Show that the sum of n consecutive odd integers beginning with 1 equals n^2.

SOLUTION

We are to find the sum of the arithmetic sequence 1, 3, 5, . . . to n terms.

Then $a = 1,\ d = 2,\ n = n$ and $S = \frac{n}{2}[2a + (n-1)d] = \frac{n}{2}[2(1) + (n-1)2] = n^2.$

22.13 Find three numbers in an arithmetic sequence such that the sum of the first and third is 12 and the product of the first and second is 24.

SOLUTION

Let the numbers in the arithmetic sequence be $(a - d),\ a,\ (a + d)$. Then $(a - d) + (a + d) = 12$ or $a = 6$.

Since $(a - d)a = 24$, $(6 - d)6 = 24$ or $d = 2$. Hence the numbers are 4, 6, 8.

22.14 Find three numbers in an arithmetic sequence whose sum is 21 and whose product is 280.

SOLUTION

Let the numbers be $(a - d),\ a,\ (a + d)$. Then $(a - d) + a + (a + d) = 21$ or $a = 7$.

Since $(a - d)(a)(a + d) = 280$, $a(a^2 - d^2) = 7(49 - d^2) = 280$ and $d = \pm 3$.

The required numbers are 4,7,10 or 10,7,4.

22.15 Three numbers are in the ratio of 2:5:7. If 7 is subtracted from the second, the resulting numbers form an arithmetic sequence. Determine the original numbers.

SOLUTION

Let the original numbers be $2x, 5x, 7x$. The resulting numbers in the arithmetic sequence are $2x$, $(5x - 7), 7x$.

Then $(5x - 7) - 2x = 7x - (5x - 7)$ or $x = 14$. Hence the original numbers are 28, 70, 98.

22.16 Compute the sum of all integers between 100 and 800 that are divisible by 3.

SOLUTION

The arithmetic sequence is 102, 105, 108, . . ., 798. Then $l = a + (n-1)d$, $798 = 102 + (n-1)3$, $n = 233$, and

$$S = \frac{n}{2}(a + l) = \frac{233}{2}(102 + 798) = 104\,850.$$

22.17 A slide of uniform grade is to be built on a level surface and is to have 10 supports equidistant from each other. The heights of the longest and shortest supports will be $42\frac{1}{2}$ feet and 2 feet respectively. Determine the required height of each support.

SOLUTION

From $l = a + (n-1)d$ we have $42\frac{1}{2} = 2 + (10-1)d$ and $d = 4\frac{1}{2}$ ft.

Thus the heights are 2, $6\frac{1}{2}$, 11, $15\frac{1}{2}$, 20, $24\frac{1}{2}$, 29, $33\frac{1}{2}$, 38, $42\frac{1}{2}$ feet respectively.

22.18 A freely falling body, starting from rest, falls 16 ft during the first second, 48 ft during the second second, 80 ft during the third second, etc. Calculate the distance it falls during the fifteenth second and the total distance it falls in 15 seconds from rest.

SOLUTION

Here $d = 48 - 16 = 80 - 48 = 32$.

During the 15th second it falls a distance $l = a + (n-1)d = 16 + (15-1)32 = 464$ ft.

Total distance covered during 15 sec is $S = \frac{n}{2}(a + l) = \frac{15}{2}(16 + 464) = 3600$ ft.

22.19 In a potato race, 8 potatoes are placed 6 ft apart on a straight line, the first being 6 ft from the basket. A contestant starts from the basket and puts one potato at a time into the basket. Find the total distance she must run in order to finish the race.

SOLUTION

Here $a = 2\cdot 6 = 12$ ft and $l = 2(6\cdot 8) = 96$ ft. Then $S = \frac{n}{2}(a + l) = \frac{8}{2}(12 + 96) = 432$ ft.

22.20 Show that if the sides of a right triangle are in an arithmetic sequence, their ratio is 3:4:5.

SOLUTION

Let the sides be $(a-d)$, a, $(a+d)$, where the hypotenuse is $(a+d)$.

Then $(a+d)^2 = a^2 + (a-d)^2$ or $a = 4d$. Hence $(a-d):a:(a+d) = 3d:4d:5d = 3:4:5$.

22.21 Derive the formula for the arithmetic mean (x) between two numbers p and q.

SOLUTION

Since p, x, q are in an arithmetic sequence, we have $x - p = q - x$ or $x = \frac{1}{2}(p + q)$.

22.22 Find the arithmetic mean between each of the following pairs of numbers.

(*a*) 4 and 56. Arithmetic mean $= \dfrac{4 + 56}{2} = 30$.

(*b*) $3\sqrt{2}$ and $-6\sqrt{2}$. Arithmetic mean $= \dfrac{3\sqrt{2} + (-6\sqrt{2})}{2} = -\dfrac{3\sqrt{2}}{2}$.

(*c*) $a + 5d$ and $a - 3d$. Arithmetic mean $= \dfrac{(a + 5d) + (a - 3d)}{2} = a + d$.

22.23 Insert 5 arithmetic means between 8 and 26.

SOLUTION

We require an arithmetic sequence of the form 8,—,—,—,—,—,26; thus $a = 8$, $l = 26$ and $n = 7$.

Then $l = a + (n-1)d, 26 = 8 + (7-1)d, d = 3$.

The five arithmetic means are 11, 14, 17, 20, 23.

22.24 Insert between 1 and 36 a number of arithmetic means so that the sum of the resulting arithmetic sequence will be 148.

SOLUTION

$$S = \tfrac{1}{2}n(a+l), \quad 148 = \tfrac{1}{2}n(1+36), \quad 37n = 296 \quad \text{and} \quad n = 8.$$
$$l = a + (n-1)d, \quad 36 = 1 + (8-1)d, \quad 7d = 35 \quad \text{and} \quad d = 5.$$

The complete arithmetic sequence is $1, 6, 11, 16, 21, 26, 31, 36$.

22.25 Which of the following sequences are geometric sequences?

(*a*) $3, 6, 12, \ldots$ Yes, since $\dfrac{6}{3} = \dfrac{12}{6} = 2.$ $(r = 2)$

(*b*) $16, 12, 9, \ldots$ Yes, since $\dfrac{12}{16} = \dfrac{9}{12} = \dfrac{3}{4}.$ $\left(r = \dfrac{3}{4}\right)$

(*c*) $-1, 3, -9, \ldots$ Yes, since $\dfrac{3}{-1} = \dfrac{-9}{3} = -3.$ $(r = -3)$

(*d*) $1, 4, 9, \ldots$ No, since $\dfrac{4}{1} \neq \dfrac{9}{4}.$

(*e*) $\dfrac{1}{2}, \dfrac{1}{3}, \dfrac{2}{9}, \ldots$ Yes, since $\dfrac{1/3}{1/2} = \dfrac{2/9}{1/3} = \dfrac{2}{3}.$ $\left(r = \dfrac{2}{3}\right)$

(*f*) $2h, \dfrac{1}{h}, \dfrac{1}{2h^3}, \ldots$ Yes, since $\dfrac{1/h}{2h} = \dfrac{1/2h^3}{1/h} = \dfrac{1}{2h^2}.$ $\left(r = \dfrac{1}{2h^2}\right)$

22.26 Prove the formula

$$S = \frac{a(r^n - 1)}{r - 1}$$

for the sum of the first n terms of a geometric sequence.

SOLUTION

The sum of the first n terms of a geometric sequence may be written

$$(1) \quad S = a + ar + ar^2 + ar^3 + \cdots + ar^{n-1} \quad (n \text{ terms}).$$

Multiplying (1) by r, we obtain

$$(2) \quad rS = ar + ar^2 + ar^3 + \cdots + ar^{n-1} + ar^n \quad (n \text{ terms}).$$

Subtracting (1) from (2),

$$rS - S = ar^n - a, \quad (r-1)S = a(r^n - 1) \quad \text{and} \quad S = \frac{a(r^n - 1)}{r - 1}.$$

22.27 Find the 8th term and the sum of the first eight terms of the sequence $4, 8, 16, \ldots$

SOLUTION

Here $a = 4$, $r = 8/4 = 16/8 = 2$, $n = 8$.

The 8th term is $l = ar^{n-1} = 4(2)^{8-1} = 4(2^7) = 4(128) = 512$.

The sum of the first eight terms is

$$S = \frac{a(r^n - 1)}{r - 1} = \frac{4(2^8 - 1)}{2 - 1} = \frac{4(256 - 1)}{1} = 1020.$$

22.28 Find the 7th term and the sum of the first seven terms of the sequence $9, -6, 4, \ldots$.

SOLUTION

Here $$a = 9, \qquad r = \frac{-6}{9} = \frac{4}{-6} = -\frac{2}{3}.$$

Then the 7th term is $$l = ar^{n-1} = 9\left(-\frac{2}{3}\right)^{7-1} = \frac{64}{81}.$$

$$S = \frac{a(r^n - 1)}{r - 1} = \frac{a(1 - r^n)}{1 - r} = \frac{9[1 - (-2/3)^7]}{1 - (-2/3)} = \frac{9[1 - (-128/2187)]}{5/3} = \frac{463}{81}$$

22.29 The second term of a geometric sequence is 3 and the fifth term is 81/8. Find the eighth term.

SOLUTION

$$5\text{th term} = ar^4 = \frac{81}{8}, \quad 2\text{nd term} = ar = 3. \quad \text{Then } \frac{ar^4}{ar} = \frac{81/8}{3}, \quad r^3 = \frac{27}{8} \quad \text{and} \quad r = \frac{3}{2}.$$

$$\text{Hence the 8th term} = ar^7 = (ar^4)r^3 = \frac{81}{8}\left(\frac{27}{8}\right) = \frac{2187}{64}.$$

22.30 Find three numbers in a geometric sequence whose sum is 26 and whose product is 216.

SOLUTION

Let the numbers in geometric sequence be a/r, a, ar. Then $(a/r)(a)(ar) = 216$, $a^3 = 216$ and $a = 6$.
Also $a/r + a + ar = 26$, $6/r + 6 + 6r = 26$, $6r^2 - 20r + 6 = 0$ and $r = 1/3, 3$.
For $r = 1/3$, the numbers are $18, 6, 2$; for $r = 3$, the numbers are $2, 6, 18$.

22.31 The first term of a geometric sequence is 375 and the fourth term is 192. Find the common ratio and the sum of the first four terms.

SOLUTION

1st term $= a = 375$, 4th term $= ar^3 = 192$. Then $375r^3 = 192$, $r^3 = 64/125$ and $r = 4/5$.
The sum of the first four terms is

$$S = \frac{a(1 - r^n)}{1 - r} = \frac{375[1 - (4/5)^4]}{1 - 4/5} = 1107.$$

22.32 The first term of a geometric sequence is 160 and the common ratio is 3/2. How many consecutive terms must be taken to give a sum of 2110?

SOLUTION

$$S = \frac{a(r^n - 1)}{r - 1}, \quad 2110 = \frac{160[(3/2)^n - 1]}{3/2 - 1}, \quad \left(\frac{3}{2}\right)^n - 1 = \frac{211}{32}, \quad \left(\frac{3}{2}\right)^n = \frac{243}{32} = \left(\frac{3}{2}\right)^5, \qquad n = 5.$$

The five consecutive terms are $160, 240, 360, 540, 810$.

22.33 In a geometric sequence consisting of four terms in which the ratio is positive, the sum of the first two terms is 8 and the sum of the last two terms is 72. Find the sequence.

SOLUTION

The four terms are a, ar, ar^2, ar^3. Then $a + ar = 8$ and $ar^2 + ar^3 = 72$.

$$\text{Hence } \frac{ar^2 + ar^3}{a + ar} = \frac{ar^2(1+r)}{a(1+r)} = r^2 = \frac{72}{8} = 9, \qquad \text{so that } r = 3.$$

Since $a + ar = 8$, $a = 2$ and the sequence is $2, 6, 18, 54$.

22.34 Prove that x, $x+3$, $x+6$ cannot be a geometric sequence.

SOLUTION

If x, $x+3$, $x+6$ is a geometric sequence then

$$r = \frac{x+3}{x} = \frac{x+6}{x+3}, \qquad x^2 + 6x + 9 = x^2 + 6x \quad \text{or} \quad 9 = 0.$$

Since this equality can never be true, x, $x+3$, $x+6$ cannot be a geometric sequence.

22.35 A boy agrees to work at the rate of one cent the first day, two cents the second day, four cents the third day, eight cents the fourth day, etc. How much would he receive at the end of 12 days?

SOLUTION

Here $a = 1$, $r = 2$, $n = 12$.

$$S = \frac{a(r^n - 1)}{r - 1} = 2^{12} - 1 = 4096 - 1 = 4095¢ = \$40.95.$$

22.36 It is estimated that the population of a certain town will increase 10% each year for four years. What is the percentage increase in population after four years?

SOLUTION

Let p denote the initial population. After one year the population is $1.10p$, after two years $(1.10)^2p$, after three years $(1.10)^3p$, after four years $(1.10)^4p = 1.46p$. Thus the population increases 46%.

22.37 From a tank filled with 240 gallons of alcohol, 60 gallons are drawn off and the tank is filled up with water. Then 60 gallons of the mixture are removed and replaced with water, etc. How many gallons of alcohol remain in the tank after 5 drawings of 60 gallons each are made?

SOLUTION

After the first drawing, $240 - 60 = 180$ gal of alcohol remain in the tank.

After the second drawing,

$$180\left(\frac{240 - 60}{240}\right) = 180\left(\frac{3}{4}\right) \text{ gal}$$

of alcohol remain; etc.

The geometric sequence for the number of gallons of alcohol remaining in the tank after successive drawings is

$$180, 180\left(\frac{3}{4}\right), 180\left(\frac{3}{4}\right)^2, \ldots \quad \text{where } a = 180,\ r = \frac{3}{4}.$$

After the fifth drawing ($n = 5$):

$$l = ar^{n-1} = 180\left(\frac{3}{4}\right)^4 = 57 \text{ gal}$$

of alcohol remain.

22.38 A sum of \$400 is invested today at 6% per year. To what amount will it accumulate in five years if interest is compounded (*a*) annually, (*b*) semiannually, (*c*) quarterly?

SOLUTION

Let P = initial principal, i = interest rate per period. S = compound amount after n periods.
At end of 1st period: interest = Pi, new amount = $P + Pi = P(1+i)$.
At end of 2nd period: interest = $P(1+i)i$, new amount = $P(1+i) + P(1+i)i = P(1+i)^2$.
Compound amount at end of n periods is $S = P(1+i)^n$.

(*a*) Since there is 1 interest period per year, $n = 5$ and $i = 0.06$.

$$S = P(1+i)^n = 400(1+0.06)^5 = 400(1.3382) = \$535.28$$

(*b*) Since there are 2 interest periods per year, $n = 2(5) = 10$ and $i = \frac{1}{2}(0.06) = 0.03$.

$$S = P(1+i)^n = 400(1+0.03)^{10} = 400(1.3439) = \$537.56.$$

(*c*) Since there are 4 interest periods per year, $n = 4(5) = 20$ and $i = \frac{1}{4}(0.06) = 0.015$.

$$S = P(1+i)^n = 400(1+0.015)^{20} = 400(1.3469) = \$538.76.$$

22.39 What sum (P) should be invested in a loan association at 4% per annum compounded semiannually, so that the compound amount (S) will be \$500 at the end of $3\frac{1}{2}$ years?

SOLUTION

Since there are 2 interest periods per year, $n = 2(3\frac{1}{2}) = 7$ (periods) and the interest rate per period is $i = \frac{1}{2}(0.04) = 0.02$.
Then $S = P(1+i)^n$ or $P = S(1+i)^{-n} = 500(1+0.02)^{-7} = 500(0.870\,56) = \$435.28$.

22.40 Derive the formula for the geometric mean, G, between two numbers p and q.

SOLUTION

Since p, G, q are in geometric sequence, we have $G/p = q/G$, $G^2 = pq$ and $G = \pm\sqrt{pq}$.

It is customary to take $G = \sqrt{pq}$ if p and q are positive.
and $G = -\sqrt{pq}$ if p and q are negative.

22.41 Find the geometric mean between each of the following pairs of numbers.

(*a*) 4 and 9. $G = \sqrt{4(9)} = 6$

(*b*) -2 and -8. $G = -\sqrt{(-2)(-8)} = -4$

(*c*) $\sqrt{7}+\sqrt{3}$ and $\sqrt{7}-\sqrt{3}$. $G = \sqrt{(\sqrt{7}+\sqrt{3})(\sqrt{7}-\sqrt{3})} = \sqrt{7-3} = 2$

22.42 Show that the arithmetic mean A of two positive numbers p and q is greater than or equal to their geometric mean G.

SOLUTION

Arithmetic mean of p and q is $A = \frac{1}{2}(p+q)$. Geometric mean of p and q is $G = \sqrt{pq}$.
Then $A - G = \frac{1}{2}(p+q) - \sqrt{pq} = \frac{1}{2}(p - 2\sqrt{pq} + q) = \frac{1}{2}(\sqrt{p} - \sqrt{q})^2$.
Now $\frac{1}{2}(\sqrt{p} - \sqrt{q})^2$ is always positive or zero; hence $A \geq G$. ($A = G$ if and only if $p = q$.)

22.43 Insert two geometric means between 686 and 2.

SOLUTION

We require a geometric sequence of the form 686, —, —, 2 where $a = 686$, $l = 2$, $n = 4$.

Then $l = ar^{n-1}, 2 = 686r^3$, $r^3 = 1/343$ and $r = 1/7$.

Thus the geometric sequence is 686, 98, 14, 2 and the means are 98, 14.

Note. Actually, $r^3 = 1/343$ is satisfied by three different values of r, one of the roots being real and two imaginary. It is customary to exclude geometric sequences with imaginary numbers.

22.44 Insert five geometric means between 9 and 576.

SOLUTION

We require a geometric sequence of the form 9, —, —, —, —, —, 576 where $a = 9$, $l = 576$, $n = 7$.

Then $l = ar^{n-1}, 576 = 9r^6$, $r^6 = 64$, $r^3 = \pm 8$ and $r = \pm 2$.

Thus the sequences are 9, 18, 36, 72, 144, 288, 576 and 9, −18, 36, −72, 144, −288, 576; and the corresponding means are 18, 36, 72, 144, 288 and −18, 36, −72, 144, −288.

22.45 Find the sum of the infinite geometric series.

(*a*) $2 + 1 + \frac{1}{2} + \frac{1}{4} + \cdots$ $\qquad S_\infty = \frac{a}{1-r} = \frac{2}{1-1/2} = 4$

(*b*) $\frac{1}{3} - \frac{2}{9} + \frac{4}{27} - \frac{8}{81} + \cdots$ $\qquad S_\infty = \frac{a}{1-r} = \frac{1/3}{1-(-2/3)} = \frac{1}{5}$

(*c*) $1 + \frac{1}{1.04} + \frac{1}{(1.04)^2} + \cdots$ $\qquad S_\infty = \frac{a}{1-r} = \frac{1}{1-1/1.04} = \frac{1.04}{1.04-1} = \frac{104}{4} = 26$

22.46 Express each of the following repeating decimals as a rational fraction.

(*a*) 0.444... (*b*) 0.4272727... (*c*) 6.305305... (*d*) 0.78367836...

SOLUTION

(*a*) $0.444\ldots = 0.4 + 0.04 + 0.004 + \ldots$, where $a = 0.4$, $r = 0.1$.

$$S_\infty = \frac{a}{1-r} = \frac{0.4}{1-0.1} = \frac{0.4}{0.9} = \frac{4}{9}$$

(*b*) $0.4272727\ldots = 0.4 + 0.0272727\ldots$

$0.0272727\ldots = 0.027 + 0.00027 + 0.0000027 + \ldots$, where $a = 0.027$, $r = 0.01$.

$$S_\infty = 0.4 + \frac{a}{1-r} = 0.4 + \frac{0.027}{1-0.01} = 0.4 + \frac{27}{990} = \frac{4}{10} + \frac{3}{110} = \frac{47}{110}$$

(*c*) $6.305305\ldots = 6 + 0.305305\ldots$

$0.305305\ldots = 0.305 + 0.000305 + \ldots$, where $a = 0.305$, $r = 0.001$.

$$S_\infty = 6 + \frac{a}{1-r} = 6 + \frac{0.305}{1-0.001} = 6 + \frac{305}{999} = 6\frac{305}{999}$$

(*d*) $0.78367836\ldots = 0.7836 + 0.00007836 + \ldots$, where $a = 0.7836$, $r = 0.0001$.

$$S_\infty = \frac{a}{1-r} = \frac{0.7836}{1-0.0001} = \frac{7836}{9999} = \frac{2612}{3333}$$

22.47 The distances passed over by a certain pendulum bob in succeeding swings form the geometric sequence 16, 12, 9, ... inches respectively. Calculate the total distance traversed by the bob before coming to rest.

SOLUTION

$$S_\infty = \frac{a}{1-r} = \frac{16}{1-3/4} = \frac{16}{1/4} = 64 \text{ inches}$$

22.48 Find the least number of terms of the series $\frac{1}{3}+\frac{1}{6}+\frac{1}{12}+\cdots$ that should be taken so that their sum will differ from their sum to infinity by less than 1/1000.

SOLUTION

Let S_∞ = sum to infinity, S_n = sum to n terms. Then

$$S_\infty - S_n = \frac{a}{1-r} - \frac{a(1-r^n)}{1-r} = \frac{ar^n}{1-r}.$$

It is required that

$$\frac{ar^n}{1-r} < \frac{1}{1000}, \quad \text{where } a = 1/3,\ r = 1/2.$$

Then

$$\frac{(1/3)(1/2)^n}{1-1/2} < \frac{1}{1000}, \quad \frac{1}{3(2^n)} < \frac{1}{2000}, \quad 3(2^n) > 2000, \quad 2^n > 666\frac{2}{3}.$$

When $n = 9, 2^n < 666\frac{2}{3}$; when $n = 10, 2^n > 666\frac{2}{3}$. Thus at least 10 terms should be taken.

22.49 Which of the following sequences are harmonic sequences?

(*a*) $\frac{1}{3}, \frac{1}{5}, \frac{1}{7}, \ldots$ is a harmonic sequence since $3, 5, 7, \ldots$ is an arithmetic sequence.

(*b*) $2, 4, 6, \ldots$ is not a harmonic sequence since $\frac{1}{2}, \frac{1}{4}, \frac{1}{6}, \ldots$ is not an arithmetic sequence.

(*c*) $\frac{1}{12}, \frac{2}{15}, \frac{1}{3}, \ldots$ is a harmonic sequence since $12, \frac{15}{2}, 3, \ldots$ is an arithmetic sequence.

22.50 Compute the 15th term of the harmonic sequence $\frac{1}{4}, \frac{1}{7}, \frac{1}{10}, \ldots$.

SOLUTION

The corresponding arithmetic sequence is $4, 7, 10, \ldots$; its 15th term is $l = a + (n-1)d = 4 + (15-1)3 = 46$.

Hence the 15th term of the harmonic progression is $\frac{1}{46}$.

22.51 Derive the formula for the harmonic mean, H, between two numbers p and q.

SOLUTION

Since p, H, q is a harmonic sequence, $\frac{1}{p}, \frac{1}{H}, \frac{1}{q}$ is an arithmetic sequence.

Then $\quad \frac{1}{H} - \frac{1}{p} = \frac{1}{q} - \frac{1}{H}, \quad \frac{2}{H} = \frac{1}{p} + \frac{1}{q} = \frac{p+q}{pq} \quad$ and $\quad H = \frac{2pq}{p+q}.$

Another method.

Harmonic mean between p and q = reciprocal of the arithmetic mean between $\frac{1}{p}$ and $\frac{1}{q}$.

Arithmetic mean between $\frac{1}{p}$ and $\frac{1}{q} = \frac{1}{2}\left(\frac{1}{p} + \frac{1}{q}\right) = \frac{p+q}{2pq}$.

Hence the harmonic mean between p and $q = \frac{2pq}{p+q}$.

22.52 What is the harmonic mean between 3/8 and 4?

SOLUTION

Arithmetic mean between $\frac{8}{3}$ and $\frac{1}{4} = \frac{1}{2}\left(\frac{8}{3}+\frac{1}{4}\right) = \frac{35}{24}$.

Hence the harmonic mean between $\frac{3}{8}$ and $4 = 24/35$.

Or, by formula, harmonic mean $= \dfrac{2pq}{p+q} = \dfrac{2(3/8)(4)}{3/8+4} = \dfrac{24}{35}$.

22.53 Insert four harmonic means between 1/4 and 1/64.

SOLUTION

To insert four arithmetic means between 4 and 64: $l = a + (n-1)d$, $64 = 4 + (6-1)d$, $d = 12$. Thus the four arithmetic means between 4 and 64 are $16, 28, 40, 52$.

Hence the four harmonic means between $\frac{1}{4}$ and $\frac{1}{64}$ are $\frac{1}{16}, \frac{1}{28}, \frac{1}{40}, \frac{1}{52}$.

22.54 Insert three harmonic means between 10 and 20.

SOLUTION

To insert three arithmetic means between $\frac{1}{10}$ and $\frac{1}{20}$:

$$l = a + (n-1)d, \quad \frac{1}{20} = \frac{1}{10} + (5-1)d, \quad d = -\frac{1}{80}.$$

Thus the three arithmetic means between $\frac{1}{10}$ and $\frac{1}{20}$ are $\frac{7}{80}, \frac{6}{80}, \frac{5}{80}$.

Hence the three harmonic means between 10 and 20 are $\frac{80}{7}, \frac{40}{3}, 16$.

22.55 Determine whether the sequence $-1, -4, 2$ is in arithmetic, geometric or harmonic sequence.

SOLUTION

Since $-4-(-1) \neq 2-(-4)$, it is not in arithmetic sequence.

Since $\frac{-4}{-1} \neq \frac{2}{-4}$, it is not in geometric sequence.

Since $\frac{1}{-1}, \frac{1}{-4}, \frac{1}{2}$ are in arithmetic sequence, i.e., $\frac{1}{-4} - (-1) = \frac{1}{2} - \left(\frac{1}{-4}\right)$, the given sequence is in harmonic sequence.

Supplementary Problems

22.56 Find the nth term and the sum of the first n terms of each arithmetic sequence for the indicated value of n.

(*a*) $1, 7, 13, \ldots$ $n = 100$

(*b*) $2, 5\frac{1}{2}, 9, \ldots$ $n = 23$

(*c*) $-26, -24, -22, \ldots$ $n = 40$

(*d*) $2, 6, 10, \ldots$ $n = 16$

(*e*) $3, 4\frac{1}{2}, 6, \ldots$ $n = 37$

(*f*) $x-y, x, x+y, \ldots$ $n = 30$

22.57 Find the sum of the first n terms of each arithmetic sequence.

(*a*) $1, 2, 3, \ldots$ (*b*) $2, 8, 14, \ldots$ (*c*) $1\frac{1}{2}, 5, 8\frac{1}{2}, \ldots$

22.58 An arithmetic sequence has first term 4 and last term 34. If the sum of its terms is 247, find the number of terms and their common difference.

22.59 An arithmetic sequence consisting of 49 terms has last term 28. If the common difference of its terms is 1/2, find the first term and the sum of the terms.

22.60 Find the sum of all even integers between 17 and 99.

22.61 Find the sum of all integers between 84 and 719 which are exactly divisible by 5.

22.62 How many terms of the arithmetic sequence $3, 7, 11, \ldots$ are needed to yield the sum 1275?

22.63 Find three numbers in an arithmetic sequence whose sum is 48 and such that the sum of their squares is 800.

22.64 A ball starting from rest rolls down an inclined plane and passes over 3 in. during the 1st second, 5 in. during the 2nd second, 7 in. during the 3rd second, etc. In what time from rest will it cover 120 inches?

22.65 If 1¢ is saved the 1st day, 2¢ the 2nd day, 3¢ the third day, etc., find the sum that will accumulate at the end of 365 days.

22.66 The sum of 40 terms of a certain arithmetic sequence is 430, while the sum of 60 terms is 945. Determine the nth term of the arithmetic sequence.

22.67 Find an arithmetic sequence which has the sum of its first n terms equal to $2n^2 + 3n$.

22.68 Determine the arithmetic mean between (*a*) 15 and 41, (*b*) -16 and 23, (*c*) $2 - \sqrt{3}$ and $4 + 3\sqrt{3}$, (*d*) $x - 3y$ and $5x + 2y$.

22.69 (*a*) Insert 4 arithmetic means between 9 and 24.
(*b*) Insert 2 arithmetic means between -1 and 11.
(*c*) Insert 3 arithmetic means between $x + 2y$ and $x + 10y$.
(*d*) Insert between 5 and 26 a number of arithmetic means such that the sum of the resulting arithmetic sequence will be 124.

22.70 Find the nth term and the sum of the first n terms of each geometric sequence for the indicated value of n.

(*a*) $2, 3, 9/2, \ldots$ $n = 5$ (*d*) $1, 3, 9, \ldots$ $n = 8$
(*b*) $6, -12, 24, \ldots$ $n = 9$ (*e*) $8, 4, 2, \ldots$ $n = 12$
(*c*) $1, 1/2, 1/4, \ldots$ $n = 10$ (*f*) $\sqrt{3}, 3, 3\sqrt{3}, \ldots$ $n = 8$

22.71 Find the sum of the first n terms of each geometric sequence.

(*a*) $1, 1/3, 1/9, \ldots$ (*b*) $4/3, 2, 3, \ldots$ (*c*) $1, -2, 4, \ldots$

22.72 A geometric sequence has first term 3 and last term 48. If each term is twice the previous term, find the number of terms and the sum of the geometric sequence.

22.73 Prove that the sum S of the terms of a geometric sequence in which the first term is a, the last term is l and the common ratio is r is given by

$$S = \frac{rl - a}{r - 1}.$$

22.74 In a geometric sequence the second term exceeds the first term by 4, and the sum of the second and third terms is 24. Show that there are two possible geometric sequences satisfying these conditions and find the sum of the first 5 terms of each geometric sequence.

22.75 In a geometric sequence consisting of four terms, in which the ratio is positive, the sum of the first two terms is 10 and the sum of the last two terms is $22\frac{1}{2}$. Find the sequence.

22.76 The first two terms of a geometric sequence are $b/(1 + c)$ and $b/(1 + c)^2$. Show that the sum of n terms of this sequence is given by the formula

$$S = b\left(\frac{1 - (1 + c)^{-n}}{c}\right)$$

22.77 Find the sum of the first n terms of the geometric sequence: $a - 2b$, $ab^2 - 2b^3$, $ab^4 - 2b^5, \ldots$

22.78 The third term of a geometric sequence is 6 and the fifth term is 81 times the first term. Write the first five terms of the progression, assuming the terms are positive.

22.79 Find three numbers in a geometric sequence whose sum is 42 and whose product is 512.

22.80 The third term of a geometric sequence is 144 and the sixth term is 486. Find the sum of the first five terms of the geometric sequence.

22.81 A tank contains a salt water solution in which is dissolved 972 lb of salt. One third of the solution is drawn off and the tank is filled with pure water. After stirring so that the solution is uniform, one third of the mixture is again drawn off and the tank is again filled with water. If this process is performed four times, what weight of salt remains in the tank?

22.82 The sum of the first three terms of a geometric sequence is 26 and the sum of the first six terms is 728. What is the nth term of the geometric sequence?

22.83 The sum of three numbers in geometric sequence is 14. If the first two terms are each increased by 1 and the third term decreased by 1, the resulting numbers are in an arithmetic sequence. Find the geometric sequence.

22.84 Determine the geometric mean between:

(*a*) 2 and 18, (*b*) 4 and 6, (*c*) -4 and -16, (*d*) $a + b$ and $4a + 4b$.

22.85 (*a*) Insert two geometric means between 3 and 192.

(*b*) Insert four geometric means between $\sqrt{2}$ and 8.

(*c*) The geometric mean of two numbers is 8. If one of the numbers is 6, find the other.

22.86 The first term of an arithmetic sequence is 2; and the first, third and eleventh terms are also the first three terms of a geometric sequence. Find the sum of the first eleven terms of the arithmetic sequence.

22.87 How many terms of the arithmetic sequence $9, 11, 13, \ldots$ must be added in order that the sum should equal the sum of nine terms of the geometric sequence $3, -6, 12, -24, \ldots$?

22.88 In a set of four numbers, the first three are in a geometric sequence and the last three are in an arithmetic sequence with a common difference of 6. If the first number is the same as the fourth, find the four numbers.

22.89 Find two numbers whose difference is 32 and whose arithmetic mean exceeds the geometric mean by 4.

22.90 Find the sum of the infinite geometric series.

(a) $3 + 1 + 1/3 + \cdots$ (c) $1 + 1/2^2 + 1/2^4 + \cdots$ (e) $4 - 8/3 + 16/9 - \cdots$
(b) $4 + 2 + 1 + \cdots$ (d) $6 - 2 + 2/3 - \cdots$ (f) $1 + 0.1 + 0.01 + \cdots$

22.91 The sum of the first two terms of a decreasing geometric sequence is 5/4 and the sum to infinity is 9/4. Write the first three terms of the geometric series.

22.92 The sum of an infinite number of terms of a decreasing geometric sequence is 3, and the sum of their squares is also 3. Write the first three terms of the series.

22.93 The successive distances traversed by a swinging pendulum bob are respectively 36,24,16, ... inches. Find the distance which the bob will travel before coming to rest.

22.94 Express each repeating decimal as a rational fraction.

(a) $0.121212\cdots$ (c) $0.270270\cdots$ (e) $0.1363636\cdots$
(b) $0.090909\cdots$ (d) $1.424242\cdots$ (f) $0.428571428571428\cdots$

22.95 (a) Find the 8th term of the harmonic sequence $2/3, 1/2, 2/5, \ldots$
(b) Find the 10th term of the harmonic sequence $5, 30/7, 15/4, \ldots$
(c) What is the nth term of the harmonic sequence $10/3, 2, 10/7, \ldots$?

22.96 Find the harmonic mean between each pair of numbers.

(a) 3 and 6 (b) 1/2 and 1/3 (c) $\sqrt{3}$ and $\sqrt{2}$ (d) $a + b$ and $a - b$

22.97 (a) Insert two harmonic means between 5 and 10.
(b) Insert four harmonic means between 3/2 and 3/7.

22.98 An object moves at uniform speed a from A to B and then travels at uniform speed b from B to A. Show that the average speed in making the round trip is $2ab/(a + b)$, the harmonic mean between a and b. Calculate the average speed if $a = 30$ ft/sec and $b = 60$ ft/sec.

ANSWERS TO SUPPLEMENTARY PROBLEMS

22.56 (a) $l = 595,\ S = 29\,800$ (c) $l = 52,\ S = 520$ (e) $l = 57,\ S = 1110$
(b) $l = 79,\ S = 931\frac{1}{2}$ (d) $l = 62,\ S = 512$ (f) $l = x + 28y,\ S = 30x + 405y$

22.57 (a) $\dfrac{n(n+1)}{2}$ (b) $n(3n - 1)$ (c) $\dfrac{n(7n-1)}{4}$

22.58 $n = 13,\ d = 5/2$

22.59 $a = 4,\ S = 784$

22.60 2378

22.61 50 800

22.62 25

22.63 12, 16, 20

22.64 10 sec

22.65 \$667.95

22.66 $\dfrac{n+1}{2}$

22.67 5, 9, 13, 17, . . . nth term $= 4n+1$

22.68 (*a*) 28, (*b*) 7/2, (*c*) $3+\sqrt{3}$, (*d*) $3x - y/2$

22.69 (*a*) 12, 15, 18, 21 (*b*) 3, 7 (*c*) $x+4y,\ x+6y,\ x+8y$ (*d*) The arithmetic sequence is 5, 8, 11, 14, 17, 20, 23, 26

22.70 (*a*) $l = 81/8,\ S = 211/8$ (*b*) $l = 1536,\ S = 1026$ (*c*) $l = 1/512,\ S = 1023/512$ (*d*) $l = 2187,\ S = 3280$ (*e*) $l = 1/256,\ S = 4095/256$ (*f*) $l = 81,\ S = 120 + 40\sqrt{3}$

22.71 (*a*) $\dfrac{3}{2}\left[1-\left(\dfrac{1}{3}\right)^n\right]$ (*b*) $\dfrac{8}{3}\left[\left(\dfrac{3}{2}\right)^n - 1\right]$ (*c*) $\dfrac{1-(-2)^n}{3}$

22.72 $n = 5,\ S = 93$

22.74 2, 6, 18, . . . and $S = 242$; 4, 8, 16, . . . and $S = 124$

22.75 4, 6, 9, 27/2

22.77 $\dfrac{(a-2b)(b^{2n}-1)}{b^2-1}$

22.78 2/3, 2, 6, 18, 54

22.79 2, 8, 32

22.80 844

22.81 192 lb

22.82 $2 \cdot 3^{n-1}$

22.83 2, 4, 8

22.84 (*a*) 6 (*b*) $2\sqrt{6}$ (*c*) -8 (*d*) $2a+2b$

22.85 (*a*) 12, 48 (*b*) $2, 2\sqrt{2}, 4, 4\sqrt{2}$ (*c*) 32/3

22.86 187 or 22

22.87 19

22.88 $8, -4, 2, 8$

22.89 $18, 50$

22.90 (*a*) 9/2 (*b*) 8 (*c*) 4/3 (*d*) 9/2 (*e*) 12/5 (*f*) 10/9

22.91 3/4, 1/2, 1/3

22.92 3/2, 3/4, 3/8

22.93 108 in.

22.94 (*a*) 4/33 (*b*) 1/11 (*c*) 10/37 (*d*) 47/33 (*e*) 3/22 (*f*) 3/7

22.95 (*a*) 1/5 (*b*) 2 (*c*) $\dfrac{10}{2n+1}$

22.96 (*a*) 4 (*b*) 2/5 (*c*) $6\sqrt{2} - 4\sqrt{3}$ (*d*) $\dfrac{a^2 - b^2}{a}$

22.97 (*a*) 6, 15/2 (*b*) 1, 3/4, 3/5, 1/2

22.98 40 ft/sec

Chapter 23

Logarithms

23.1 DEFINITION OF A LOGARITHM

If $b^x = N$, where N is a positive number and b is a positive number different from 1, then the exponent x is the logarithm of N to the base b and is written $x = \log_b N$.

EXAMPLE 23.1. Write $3^2 = 9$ using logarithmic notation.
Since $3^2 = 9$, then 2 is the logarithm of 9 to the base 3, i.e., $2 = \log_3 9$.

EXAMPLE 23.2. Evaluate $\log_2 8$.
$\log_2 8$ is that number x to which the base 2 must be raised in order to yield 8, i.e., $2^x = 8$, $x = 3$. Hence $\log_2 8 = 3$.

Both $b^x = N$ and $x = \log_b N$ are equivalent relationships; $b^x = N$ is called the *exponential form* and $x = \log_b N$ the *logarithmic form* of the relationship. As a consequence, corresponding to *laws of exponents* there are *laws of logarithms*.

23.2 LAWS OF LOGARITHMS

I. The logarithm of the product of two positive numbers M and N is equal to the sum of the logarithms of the numbers, i.e.,

$$\log_b MN = \log_b M + \log_b N.$$

II. The logarithm of the quotient of two positive numbers M and N is equal to the difference of the logarithms of the numbers, i.e.,

$$\log_b \frac{M}{N} = \log_b M - \log_b N.$$

III. The logarithm of the pth power of a positive number M is equal to p mutiplied by the logarithm of the number, i.e.,

$$\log_b M^p = p \log_b M.$$

EXAMPLES 23.3. Apply the laws of logarithms to each expression.

(*a*) $\log_2 3(5)$ (*b*) $\log_{10} \frac{17}{24}$ (*c*) $\log_7 5^3$ (*d*) $\log_{10} \sqrt[3]{2}$

(*a*) $\log_2 3(5) = \log_2 3 + \log_2 5$

(*b*) $\log_{10} \frac{17}{24} = \log_{10} 17 - \log_{10} 24$

(*c*) $\log_7 5^3 = 3 \log_7 5$

(*d*) $\log_{10} \sqrt[3]{2} = \log_{10} 2^{1/3} = \frac{1}{3} \log_{10} 2$

23.3 COMMON LOGARITHMS

The system of logarithms whose base is 10 is called the common logarithm system. When the base is omitted, it is understood that base 10 is to be used. Thus $\log 25 = \log_{10} 25$.

Consider the following table.

Number N	0.0001	0.001	0.01	0.1	1	10	100	1000	10 000
Exponential form of N	10^{-4}	10^{-3}	10^{-2}	10^{-1}	10^{0}	10^{1}	10^{2}	10^{3}	10^{4}
$\log N$	−4	−3	−2	−1	0	1	2	3	4

It is obvious that $10^{1.5377}$ will give some number greater than 10 (which is 10^1) but smaller than 100 (which is 10^2). Actually, $10^{1.5377} = 34.49$; hence $\log 34.49 = 1.5377$.

The digit before the decimal point is the *characteristic* of the log, and the decimal fraction part is the *mantissa* of the log. In the above example, the characteristic is 1 and the mantissa is .5377.

The mantissa of the log of a number is found in tables, ignoring the decimal point of the number. Each mantissa in the tables is understood to have a decimal point preceding it, and the mantissa is always considered positive.

The characteristic is determined by inspection from the number itself according to the following rules.

(1) For a number greater than 1, the characteristic is positive and is one *less* than the number of digits before the decimal point. For example:

Number	5297	348	900	34.8	60	5.764	3
Characteristic	3	2	2	1	1	0	0

(2) For a number less than 1, the characteristic is negative and is one *more* than the number of zeros immediately following the decimal point. The negative sign of the characteristic is written in either of these two ways: (*a*) above the characteristic as $\bar{1}$, $\bar{2}$, etc.; (*b*) as $9-10$, $8-10$, etc. Thus the characteristic of 0.3485 is $\bar{1}$ or $9-10$, of 0.0513 is $\bar{2}$ or $8-10$, and of 0.0024 is $\bar{3}$ or $7-10$.

23.4 USING A COMMON LOGARITHM TABLE

To find the common logarithm of a positive number use the table of common logarithms in Appendix A.

Suppose it is required to find the log of the number 728. In the table of common logarithms glance down the N column to 72, then horizontally to the right to column 8 and note the figure 8621, which is the required mantissa. Since the characteristic is 2, $\log 728 = 2.8621$. (This means that $728 = 10^{2.8621}$.)

The mantissa for $\log 72.8$, for $\log 7.28$, for $\log 0.728$, for $\log 0.0728$, etc., is .8621, but the characteristics differ. Thus:

$$\begin{aligned} \log 728 &= 2.8621 & \log 0.728 &= \bar{1}.8621 \text{ or } 9.8621-10 \\ \log 72.8 &= 1.8621 & \log 0.0728 &= \bar{2}.8621 \text{ or } 8.8621-10 \\ \log 7.28 &= 0.8621 & \log 0.00728 &= \bar{3}.8621 \text{ or } 7.8621-10 \end{aligned}$$

When the number contains four digits, interpolate using the method of proportional parts.

EXAMPLE 23.4. Find $\log 4.638$.

The characteristic is 0. The mantissa is found as follows.

$$\begin{aligned} \text{Mantissa of } \log 4640 &= .6665 \\ \text{Mantissa of } \log 4630 &= \underline{.6656} \\ \text{Tabular difference} &= .0009 \end{aligned}$$

$.8 \times$ tabular difference $= .000\,72$ or $.0007$ to four decimal places.
Mantissa of $\log 4638 = .6656 + .0007 = .6663$ to four digits.
Hence $\log 4.638 = 0.6663$.

The mantissa for $\log 4638$, for $\log 463.8$, for $\log 46.38$, etc., is .6663, but the characteristics differ. Thus:

$$\begin{aligned} \log 4638 &= 3.6663 & \log 0.4638 &= \bar{1}.6663 \text{ or } 9.6663 - 10 \\ \log 463.8 &= 2.6663 & \log 0.046\,38 &= \bar{2}.6663 \text{ or } 8.6663 - 10 \\ \log 46.38 &= 1.6663 & \log 0.004\,638 &= \bar{3}.6663 \text{ or } 7.6663 - 10 \\ \log 4.638 &= 0.6663 & \log 0.000\,463\,8 &= \bar{4}.6663 \text{ or } 6.6663 - 10 \end{aligned}$$

The antilogarithm is the number corresponding to a given logarithm. "The antilog of 3" means "the number whose log is 3"; that number is obviously 1000.

EXAMPLES 23.5. Find the value of N.

(*a*) $\log N = 1.9058$ (*b*) $\log N = 7.8657 - 10$ (*c*) $\log N = 9.3842 - 10$.

(*a*) In the table the mantissa .9058 corresponds to the number 805. Since the characteristic of $\log N$ is 1, the number must have two digits before the decimal point; thus $N = 80.5$ (or antilog $1.9058 = 80.5$).

(*b*) In the table the mantissa .8657 corresponds to the number 734. Since the characteristic is $7 - 10$, the number must have two zeros immediately following the decimal point; thus $N = 0.00734$ (or antilog $7.8657 - 10 = 0.00734$).

(*c*) Since the mantissa .3842 is not found in the tables, interpolation must be used.

$$\begin{aligned} \text{Mantissa of } \log 2430 &= .3856 & \text{Given mantissa} &= .3842 \\ \text{Mantissa of } \log 2420 &= \underline{.3838} & \text{Next smaller mantissa} &= \underline{.3838} \\ \text{Tabular difference} &= .0018 & \text{Difference} &= .0004 \end{aligned}$$

Then $2420 + \frac{4}{18}(2430 - 2420) = 2422$ to four digits, and $N = 0.2422$.

23.5 NATURAL LOGARITHMS

The system of logarithms whose base is the constant e is called the natural logarithm system. When we want to indicate the base of a logarithm is e we write ln. Thus, $\ln 25 = \log_e 25$.

The exponential form of $\ln a = b$ is $e^b = a$. The number e is an irrational number that has a decimal expansion $e = 2.718\,281\,828\,450\,45\ldots$.

23.6 USING A NATURAL LOGARITHM TABLE

To find the natural logarithm of a positive number use the table of natural logarithms in Appendix B.

To find the natural logarithm of a number from 1 to 10, such as 5.26, we go down the N column to 5.2, then across to the right to the column headed by .06 to get the value 1.6601. Thus, $\ln 5.26 = 1.6601$. This means that $5.26 = e^{1.6601}$.

If we want to find the natural logarithm of a number greater than 10 or less than one, we write the number in scientific notation, apply the laws of logarithms, and use the natural logarithm table and the fact that $\ln 10 = 2.3026$.

EXAMPLES 23.6. Find the natural logarithm of each number.

(*a*) 346 (*b*) 0.0217

(*a*)
$$\begin{aligned} \ln 346 &= \ln(3.46 \times 10^2) \\ &= \ln 3.46 + \ln 10^2 \\ &= \ln 3.46 + 2 \ln 10 \\ &= 1.2413 + 2(2.3026) \\ &= 1.2413 + 4.6052 \\ \ln 346 &= 5.8465 \end{aligned}$$

(*b*)
$$\begin{aligned} \ln 0.0217 &= \ln(2.17 \times 10^{-2}) \\ &= \ln 2.17 + \ln 10^{-2} \\ &= \ln 2.17 - 2 \ln 10 \\ &= 0.7747 - 2(2.3026) \\ &= 0.7747 - 4.6052 \\ \ln 0.0217 &= -3.8305 \end{aligned}$$

The value of $\ln 4.638$ cannot be found directly from the natural logarithm table since it has four significant digits, but we can interpolate to find it.

$$\begin{aligned} \ln 4.640 &= 1.5347 \\ \ln 4.630 &= \underline{1.5326} \\ \text{tabular difference} &= 0.0021 \end{aligned}$$

$0.8 \times$ tabular difference $= 0.8 \times 0.0021 = 0.001\,68$ or 0.0017 to four decimal places.

Thus, $\ln 4.638 = \ln 4.630 + 0.0017 = 1.5324 + 0.0017 = 1.5343$.

The antilogarithm of a natural logarithm is the number that has the given logarithm. The procedure for finding the antilogarithm of a natural logarithm less than 0 or greater than 2.3026, requires us to add or subtract multiples of $\ln 10 = 2.3026$ to bring the natural logarithm into the range 0 to 2.3026 that can be found from the table in Appendix B.

EXAMPLES 23.7 Find the value of N.

(*a*) $\ln N = 2.1564$ (*b*) $\ln N = -4.9705$ (*c*) $\ln N = 1.8869$

(*a*) $\ln N = 2.1564$ is between 0 and 2.3026, so we look in the natural logarithm table for 2.1564. It is in the table, so we get N from the sum of the numbers that head the row and column for 1.1564. Thus, $N =$ antilogarithm $2.1564 = 8.64$.

(*b*) Since $\ln N = -4.9705$ is less than 0, we must rewrite it as a number between 0 and 2.3026 minus a multiple of $2.3026 = \ln 10$. Since if we add 3 times 2.3026 to -4.9705 we get a positive number between 0 and 2.3026, we will rewrite -4.9705 as $1.9373 - 3(2.3026)$.

$$\begin{aligned} \ln N &= -4.9705 \\ &= 1.9373 - 3(2.3026) \\ &= \ln 6.94 - 3 \ln 10 \qquad \text{Note: } \ln 6.94 = 1.9373 \text{ and } \ln 10 = 2.3026 \\ &= \ln 6.94 + \ln 10^{-3} \\ &= \ln(6.94 \times 10^{-3}) \\ \ln N &= \ln 0.006\,94 \\ N &= 0.006\,94 \end{aligned}$$

(*c*) Since $\ln N = 1.8869$ is between 0 and 2.3026, we look for 1.8869 in the natural logarithm table, but it is not there. We will have to use interpolation to find N.

$$\begin{aligned} \ln 6.600 &= 1.8871 & \ln N &= 1.8869 \\ \ln 6.590 &= \underline{1.8856} & \ln 6.590 &= \underline{1.8856} \\ \text{tabular difference} &= 0.0015 & \text{difference} &= 0.0013 \end{aligned}$$

$$N = 6.590 + \frac{13}{15}(6.600 - 6.590) = 6.590 + 0.0009 = 6.599$$

23.7 FINDING LOGARITHMS USING A CALCULATOR

If the number we want to find the logarithm of a number with four or more significant digits, we can round the number to four significant digits and use the logarithm tables and interpolation or we can use a scientific or graphing calculator to find the logarithm for the given number. The use of the calculator will yield a more accurate result.

A scientific calculator can be used to find logarithms and antilogarithms to base 10 or base e. Scientific calculators have keys for the log and ln functions, and the inverse of these yields the antilogarithms.

Much of the computation once done using logarithms can be done directly on a scientific calculator. The advantages of doing a problem on the calculator are that the numbers rarely have to be rounded and that the problem can be worked quickly and accurately.

Solved Problems

23.1 Express each of the following exponential forms in logarithmic form:

(*a*) $p^q = r$, (*b*) $2^3 = 8$, (*c*) $4^2 = 16$, (*d*) $3^{-2} = \frac{1}{9}$, (*e*) $8^{-2/3} = \frac{1}{4}$.

SOLUTION

(*a*) $q = \log_p r$, (*b*) $3 = \log_2 8$, (*c*) $2 = \log_4 16$, (*d*) $-2 = \log_3 \frac{1}{9}$, (*e*) $-\frac{2}{3} = \log_8 \frac{1}{4}$

23.2 Express each of the following logarithmic forms in exponential form:

(*a*) $\log_5 25 = 2$, (*b*) $\log_2 64 = 6$, (*c*) $\log_{1/4} \frac{1}{16} = 2$, (*d*) $\log_a a^3 = 3$, (*e*) $\log_r 1 = 0$.

SOLUTION

(*a*) $5^2 = 25$, (*b*) $2^6 = 64$, (*c*) $\left(\frac{1}{4}\right)^2 = \frac{1}{16}$, (*d*) $a^3 = a^3$, (*e*) $r^0 = 1$

23.3 Determine the value of each of the following.

(*a*) $\log_4 64$. Let $\log_4 64 = x$; then $4^x = 64 = 4^3$ and $x = 3$.

(*b*) $\log_3 81$. Let $\log_3 81 = x$; then $3^x = 81 = 3^4$ and $x = 4$.

(*c*) $\log_{1/2} 8$. Let $\log_{1/2} 8 = x$; then $(\frac{1}{2})^x = 8$, $(2^{-1})^x = 2^3$, $2^{-x} = 2^3$ and $x = -3$.

(*d*) $\log \sqrt[3]{10} = x$, $10^x = \sqrt[3]{10} = 10^{1/3}$, $x = 1/3$

(*e*) $\log_5 125\sqrt{5} = x$, $5^x = 125\sqrt{5} = 5^3 \cdot 5^{1/2} = 5^{7/2}$, $x = 7/2$

23.4 Solve each of the following equations.

(*a*) $\log_3 x = 2$, $3^2 = x$, $x = 9$

(*b*) $\log_4 y = -\frac{3}{2}$, $4^{-3/2} = y$, $y = \frac{1}{8}$

(*c*) $\log_x 25 = 2$, $x^2 = 25$, $x = \pm 5$. Since bases are positive, the solution is $x = 5$.

(*d*) $\log_y \frac{9}{4} = -\frac{2}{3}$, $y^{-2/3} = \frac{9}{4}$, $y^{2/3} = \frac{4}{9}$, $y = \left(\frac{4}{9}\right)^{3/2} = \frac{8}{27}$ is the required solution.

(*e*) $\log(3x^2 + 2x - 4) = 0$, $10^0 = 3x^2 + 2x - 4$, $3x^2 + 2x - 5 = 0$, $x = 1, -5/3$

23.5 Prove the laws of logarithms.

SOLUTION

Let $M = b^x$ and $N = b^y$; then $x = \log_b M$ and $y = \log_b N$.

I. Since $MN = b^x \cdot b^y = b^{x+y}$, then $\log_b MN = x + y = \log_b M + \log_b N$.

II. Since $\frac{M}{N} = \frac{b^x}{b^y} = b^{x-y}$, then $\log_b \frac{M}{N} = x - y = \log_b M - \log_b N$.

III. Since $M^p = (b^x)^p = b^{px}$, then $\log_b M^p = px = p \log_b M$.

23.6 Express each of the following as an algebraic sum of logarithms, using the laws I, II, III.

(*a*) $\log_b UVW = \log_b (UV)W = \log_b UV + \log_b W = \log_b U + \log_b V + \log_b W$

(*b*) $\log_b \frac{UV}{W} = \log_b UV - \log_b W = \log_b U + \log_b V - \log_b W$

(*c*) $\log \frac{XYZ}{PQ} = \log XYZ - \log PQ = \log X + \log Y + \log Z - (\log P + \log Q)$
$= \log X + \log Y + \log Z - \log P - \log Q$

(*d*) $\log \frac{U^2}{V^3} = \log U^2 - \log V^3 = 2 \log U - 3 \log V$

(*e*) $\log \frac{U^2 V^3}{W^4} = \log U^2 V^3 - \log W^4 = \log U^2 + \log V^3 - \log W^4$
$= 2 \log U + 3 \log V - 4 \log W$

(*f*) $\log \frac{U^{1/2}}{V^{2/3}} = \log U^{1/2} - \log V^{2/3} = \frac{1}{2} \log U - \frac{2}{3} \log V$

(*g*) $\log_e \frac{\sqrt{x^3}}{\sqrt[4]{y^3}} = \log_e \frac{x^{3/2}}{y^{3/4}} = \log_e x^{3/2} - \log_e y^{3/4} = \frac{3}{2} \log_e x - \frac{3}{4} \log_e y$

(*h*) $\log \sqrt[4]{a^2 b^{-3/4} c^{1/3}} = \frac{1}{4}\left\{2 \log a - \frac{3}{4} \log b + \frac{1}{3} \log c\right\}$
$= \frac{1}{2} \log a - \frac{3}{16} \log b + \frac{1}{12} \log c$

23.7 Given that $\log 2 = 0.3010$, $\log 3 = 0.4771$, $\log 5 = 0.6990$, $\log 7 = 0.8451$ (all base 10) accurate to four decimal places, evaluate the following.

(*a*) $\log 105 = \log (3 \cdot 5 \cdot 7) = \log 3 + \log 5 + \log 7 = 0.4771 + 0.6990 + 0.8451 = 2.0212$

(*b*) $\log 108 = \log (2^2 \cdot 3^3) = 2 \log 2 + 3 \log 3 = 2(0.3010) + 3(0.4771) = 2.0333$

(*c*) $\log \sqrt[3]{72} = \log \sqrt[3]{3^2 \cdot 2^3} = \log (3^{2/3} \cdot 2) = \frac{2}{3} \log 3 + \log 2 = 0.6191$

(*d*) $\log 2.4 = \log \frac{24}{10} = \log \frac{3 \cdot 2^3}{10} = \log 3 + 3 \log 2 - \log 10$
$= 0.4771 + 3(0.3010) - 1 = 0.3801$

(*e*) $\log 0.0081 = \log \frac{81}{10^4} = \log 81 - \log 10^4 = \log 3^4 - \log 10^4$
$= 4 \log 3 - 4 \log 10 = 4(0.4771) - 4 = -2.0916$ or $7.9084 - 10$

Note. In exponential form this means $10^{-2.0916} = 0.0081$.

23.8 Express each of the following as a single logarithm (base is 10 unless otherwise indicated).

(a) $\log 2 - \log 3 + \log 5 = \log \frac{2}{3} + \log 5 = \log \frac{2}{3}(5) = \log \frac{10}{3}$

(b) $3 \log 2 - 4 \log 3 = \log 2^3 - \log 3^4 = \log \frac{2^3}{3^4} = \log \frac{8}{81}$

(c) $\frac{1}{2}\log 25 - \frac{1}{3}\log 64 + \frac{2}{3}\log 27 = \log 25^{1/2} - \log 64^{1/3} + \log 27^{2/3}$

$= \log 5 - \log 4 + \log 9 = \log \frac{5}{4} + \log 9 = \log \frac{5}{4}(9) = \log \frac{45}{4}$

(d) $\log 5 - 1 = \log 5 - \log 10 = \log \frac{5}{10} = \log \frac{1}{2}$

(e) $2 \log 3 + 4 \log 2 - 3 = \log 3^2 + \log 2^4 - 3 \log 10 = \log 9 + \log 16 - \log 10^3$

$= \log (9 \cdot 16) - \log 10^3 = \log \frac{9 \cdot 16}{10^3} = \log 0.144$

(f) $3 \log_a b - \frac{1}{2}\log_a c = \log_a b^3 + \log_a c^{-1/2} = \log_a (b^3 c^{-1/2})$

23.9 In each of the following equations, solve for the indicated letter in terms of the other quantities.

(a) $\log_2 x = y + c : x$. $\quad x = 2^{y+c}$

(b) $\log a = 2 \log b : a$. $\quad \log a = \log b^2,\ a = b^2$

(c) $\log_e I = \log_e I_0 - t : I$. $\quad \log_e I = \log_e I_0 - t \log_e e = \log_e I_0 + \log_e e^{-t}$

$= \log_e I_0 e^{-t},\ I = I_0 e^{-t}$

(d) $2 \log x + 3 \log y = 4 \log z - 2 : y$.

Solving for $\log y$, $3 \log y = 4 \log z - 2 - 2 \log x$ and

$$\log y = \frac{4}{3}\log z - \frac{2}{3} - \frac{2}{3}\log x = \log z^{4/3} + \log 10^{-2/3} + \log x^{-2/3} = \log z^{4/3} 10^{-2/3} x^{-2/3}.$$

Hence $y = 10^{-2/3} x^{-2/3} z^{4/3}$.

(e) $\log (x + 3) = \log x + \log 3 : x$. $\quad \log (x + 3) = \log 3x,\quad x + 3 = 3x,\quad x = 3/2$

23.10 Determine the characteristic of the common logarithm of each of the following numbers.

(a) 57 (c) 5.63 (e) 982.5 (g) 186 000 (i) 0.7314 (k) 0.0071
(b) 57.4 (d) 35.63 (f) 7824 (h) 0.71 (j) 0.0325 (l) 0.0003

SOLUTION

(a) 1 (c) 0 (e) 2 (g) 5 (i) 9 −10 (k) 7 −10
(b) 1 (d) 1 (f) 3 (h) 9 −10 (j) 8 −10 (l) 6 −10

23.11 Verify each of the following common logarithms.

(a) $\log 87.2 = 1.9405$ (e) $\log 0.382 = 9.5821 - 10$
(b) $\log 37\,300 = 4.5717$ (f) $\log 0.001\,59 = 7.2014 - 10$
(c) $\log 753 = 2.8768$ (g) $\log 0.0256 = 8.4082 - 10$
(d) $\log 9.21 = 0.9643$ (h) $\log 6.753 = 0.8295\ (8293 + 2)$

(*i*) $\log 183.2 = 2.2630$ (2625 + 5)
(*j*) $\log 43.15 = 1.6350$ (6345 + 5)
(*k*) $\log 876\,400 = 5.9427$ (9425 + 2)
(*l*) $\log 0.2548 = 9.4062 - 10$ (4048 + 14)
(*m*) $\log 0.043\,72 = 8.6407 - 10$ (6405 + 2)
(*n*) $\log 0.009\,848 = 7.9933 - 10$ (9930 + 3)

23.12 Verify each of the following.

(*a*) Antilog 3.8531 = 7130
(*b*) Antilog 1.4997 = 31.6
(*c*) Antilog 9.8267 − 10 = 0.671
(*d*) Antilog 7.7443 − 10 = 0.005 55
(*e*) Antilog 0.1875 = 1.54
(*f*) Antilog $\bar{2}.3927$ = 0.0247
(*g*) Antilog 4.9360 = 86 300
(*h*) Antilog 2.6715 = 469.3 ($3/9 \times 10 = 3$ approx.)
(*i*) Antilog 4.1853 = 15 320 ($6/28 \times 10 = 2$ approx.)
(*j*) Antilog 0.9245 = 8.404 ($2/5 \times 10 = 4$)
(*k*) Antilog $\bar{1}.6089$ = 0.4064 ($4/11 \times 10 = 4$ approx.)
(*l*) Antilog 8.8907 − 10 = 0.077 75 ($3/6 \times 10 = 5$)
(*m*) Antilog 1.2000 = 15.85 ($13/27 \times 10 = 5$ approx.)
(*n*) Antilog 7.2409 − 10 = 0.001 742 ($4/25 \times 10 = 2$ approx.)

23.13 Write each of the following numbers as a power of 10: (*a*) 893, (*b*) 0.358.

SOLUTION

(*a*) We require x such that $10^x = 893$. Then $x = \log 893 = 2.9509$ and $893 = 10^{2.9509}$.

(*b*) We require x such that $10^x = 0.358$.

Then $x = \log 0.358 = 9.5539 - 10 = -0.4461$ and $0.358 = 10^{-0.4461}$.

Calculate each of the following using logarithms.

23.14 $P = 3.81 \times 43.4$

SOLUTION

$$\log P = \log 3.81 + \log 43.4$$

$$\begin{aligned} \log 3.81 &= 0.5809 \\ (+)\ \log 43.4 &= \underline{1.6375} \\ \log P &= 2.2184 \end{aligned}$$

Hence $P = \text{antilog}\, 2.2184 = 165.3$.

Note the exponential significance of the computation. Thus

$$\begin{aligned} 3.81 \times 43.4 &= 10^{0.5809} \times 10^{1.6375} \\ &= 10^{0.5809+1.6375} = 10^{2.2184} = 165.3. \end{aligned}$$

23.15 $P = 73.42 \times 0.004\,62 \times 0.5143$

SOLUTION

$$\log P = \log 73.42 + \log 0.004\,62 + \log 0.5143$$

$$\begin{aligned} \log 73.42 &= 1.8658 \\ (+)\ \log 0.004\,62 &= 7.6646 - 10 \\ (+)\ \log 0.5143 &= \underline{9.7112 - 10} \\ \log P &= 19.2416 - 20 = 9.2416 - 10. \end{aligned}$$

Hence $P = 0.1744$.

23.16 $P = \dfrac{784.6 \times 0.0431}{28.23}$

SOLUTION

$$\log P = \log 784.6 + \log 0.0431 - \log 28.23$$

$$\begin{aligned} \log 784.6 &= 2.8947 \\ (+)\ \log 0.0431 &= \underline{8.6345 - 10} \\ & 11.5292 - 10 \\ (-)\ \log 28.23 &= \underline{1.4507} \\ \log P &= 10.0785 - 10 = 0.0785 \\ P &= 1.198 \end{aligned}$$

23.17 $P = \dfrac{0.4932 \times 653.7}{0.072\,13 \times 8456}$

SOLUTION

Numerator N

$$\begin{aligned} \log 0.4932 &= 9.6930 - 10 \\ (+)\ \log 653.7 &= \underline{2.8154} \\ \log N &= 12.5084 - 10 \\ (-)\ \log D &= \underline{2.7853} \\ \log P &= 9.7231 - 10 \\ P &= 0.5286 \end{aligned}$$

Denominator D

$$\begin{aligned} \log 0.072\,13 &= 8.8581 - 10 \\ (+)\ \log 8456 &= \underline{3.9272} \\ \log D &= 12.7853 - 10 \\ &= 2.7853 \end{aligned}$$

23.18 $P = (7.284)^5$

SOLUTION

$\log P = 5 \log 7.284 = 5(0.8623) = 4.3115$ and $P = 20\,490$.

23.19 $P = \dfrac{(63.28)^3(0.008\,43)^2(0.4623)}{(412.3)(2.184)^5}$

SOLUTION

$$\log P = 3\log 63.28 + 2\log 0.008\,43 + \log 0.4623 - (\log 412.3 + 5\log 2.184)$$

Numerator N

$$\begin{aligned} 3\log 63.28 &= 3(1.8013) = 5.4039 \\ (+)\ 2\log 0.008\,43 &= 2(7.9258 - 10) = 15.8516 - 20 \\ (+)\ \log 0.4623 &= \underline{9.6649 - 10} \\ \log N &= 30.9204 - 30 \\ (-)\ \log D &= \underline{4.3117} \\ \log P &= 26.6087 - 30 = 6.6087 - 10 \\ P &= 0.000\,406\,2 \ (\text{or } 4.062 \times 10^{-4}) \end{aligned}$$

Denominator D

$$\begin{aligned} \log 412.3 &= 2.6152 \\ (+)\ 5\log 2.184 &= \underline{1.6965} \\ \log D &= 4.3117 \end{aligned}$$

23.20 $P = \sqrt[5]{0.8532}$

SOLUTION

$\log P = \frac{1}{5}\log 0.8532 = \frac{1}{5}(9.9310 - 10) = \frac{1}{5}(49.9310 - 50) = 9.9862 - 10$ and $P = 0.9688$.

23.21 $P = \dfrac{(78.41)^3 \sqrt{142.3}}{\sqrt[4]{0.1562}}$

SOLUTION

$$\log P = 3 \log 78.41 + \frac{1}{2} \log 142.3 - \frac{1}{4} \log 0.1562.$$

Numerator N

$$\begin{aligned} 3\log 78.41 &= 3(1.8944) = 5.6832 \\ (+)\ \tfrac{1}{2}\log 142.3 &= \tfrac{1}{2}(2.1532) = 1.0766 \\ \log N &= 6.7598 = 16.7598 - 10 \\ (-)\ \log D &= 9.7984 - 10 \\ \log P &= 6.9614 \\ P &= 9\,150\,000 \quad \text{or} \quad 9.15 \times 10^6 \end{aligned}$$

Denominator D

$$\begin{aligned} \tfrac{1}{4}\log 0.1562 &= \tfrac{1}{4}(9.1937 - 10) \\ &= \tfrac{1}{4}(39.1937 - 40) \\ \log D &= 9.7984 - 10 \end{aligned}$$

23.22 The period T of a simple pendulum of length l is given by the formula $T = 2\pi\sqrt{l/g}$, where g is the acceleration due to gravity. Find T (in seconds) if $l = 281.3$ cm and $g = 981.0$ cm/sec^2. Take $2\pi = 6.283$.

SOLUTION

$$T = 2\pi\sqrt{\frac{l}{g}} = 6.283\sqrt{\frac{281.3}{981.0}}$$

$$\log T = \log 6.283 + \tfrac{1}{2}(\log 281.3 - \log 981.0)$$

$$\begin{aligned} \log 6.283 &= \phantom{\tfrac{1}{2}(2.4492)} = 0.7982 \\ (+)\ \tfrac{1}{2}\log 281.3 &= \tfrac{1}{2}(2.4492) = 1.2246 \\ & \phantom{= \tfrac{1}{2}(2.4492) = 1}2.0228 \\ (-)\ \tfrac{1}{2}\log 981.0 &= \tfrac{1}{2}(2.9917) = 1.4959 \\ \log T &= 0.5269 \\ T &= 3.365 \text{ seconds} \end{aligned}$$

23.23 Solve for x: $5^{2x+2} = 3^{5x-1}$.

SOLUTION

Taking logarithms, $(2x + 2)\log 5 = (5x - 1)\log 3$.

Then $2x\log 5 - 5x\log 3 = -\log 3 - 2\log 5$,

$x(2\log 5 - 5\log 3) = -\log 3 - 2\log 5$,

and $x = \dfrac{\log 3 + 2\log 5}{5\log 3 - 2\log 5} = \dfrac{0.4771 + 2(0.6990)}{5(0.4771) - 2(0.6990)} = \dfrac{1.8751}{0.9875}.$

$$\begin{aligned} \log 1.875 &= 10.2730 - 10 \\ (-)\ \log 0.9875 &= 9.9946 - 10 \\ \log x &= 0.2784 \\ x &= 1.898 \end{aligned}$$

23.24 Find the value of each of these natural logarithms.

(*a*) ln 5.78 (*c*) ln 3.456 (*e*) ln 190 (*g*) ln 2839

(*b*) ln 8.62 (*d*) ln 4.643 (*f*) ln 0.0084 (*h*) ln 0.014 85

SOLUTION

(a) $\ln 5.78 = 1.7544$ from natural logarithm table

(b) $\ln 8.62 = 2.1541$ from natural logarithm table

(c)
$$\begin{aligned}\ln 3.456 &= \ln 3.45 + 0.6(\ln 3.46 - \ln 3.45)\\ &= 1.2384 + 0.6(1.2413 - 1.2384)\\ &= 1.2384 + 0.6(0.0029)\\ &= 1.2384 + 0.0017\\ \ln 3.456 &= 1.2401\end{aligned}$$

(d)
$$\begin{aligned}\ln 4.643 &= \ln 4.64 + 0.3(\ln 4.65 - \ln 4.64)\\ &= 1.5347 + 0.3(1.5369 - 1.5347)\\ &= 1.5347 + 0.3(0.0022)\\ &= 1.5347 + 0.0007\\ \ln 4.643 &= 1.5354\end{aligned}$$

(e)
$$\begin{aligned}\ln 190 &= \ln(1.90 \times 10^2)\\ &= \ln 1.90 + \ln 10^2\\ &= \ln 1.90 + 2 \ln 10\\ &= 0.6419 + 2(2.3026)\\ &= 0.6419 + 4.6052\\ \ln 190 &= 5.2471\end{aligned}$$

(f)
$$\begin{aligned}\ln 0.0084 &= \ln(8.40 \times 10^{-3})\\ &= \ln 8.40 + \ln 10^{-3}\\ &= \ln 8.40 - 3 \ln 10\\ &= 2.1282 - 3(2.3026)\\ &= 2.1282 - 6.9078\\ \ln 0.0084 &= -4.7796\end{aligned}$$

(g)
$$\begin{aligned}\ln 2839 &= \ln(2.839 \times 10^3)\\ &= \ln 2.839 + \ln 10^3\\ &= [\ln 2.83 + 0.9(\ln 2.84 - \ln 2.83)] + 3 \ln 10\\ &= [1.0403 + 0.9(1.0438 - 1.0403)] + 3(2.3026)\\ &= [1.0403 + 0.9(0.0035)] + 6.9078\\ &= [1.0403 + 0.0032] + 6.9078\\ &= 1.0435 + 6.9078\\ \ln 2839 &= 7.9513\end{aligned}$$

(h)
$$\begin{aligned}\ln 0.014\,85 &= \ln(1.485 \times 10^{-2})\\ &= \ln 1.485 + \ln 10^{-2}\\ &= [\ln 1.48 + 0.5(\ln 1.49 - \ln 1.48)] - 2 \ln 10\\ &= [0.3920 + 0.5(0.3988 - 0.3920)] - 2\,(2.3026)\\ &= [0.3920 + 0.5(0.0068)] - 4.6052\\ &= [0.3920 + 0.0034] - 4.6052\\ &= 0.3954 - 4.6052\\ \ln 0.014\,85 &= -4.2098\end{aligned}$$

23.25 Find the value of N.

(a) $\ln N = 2.4146$ (b) $\ln N = 0.9847$ (c) $\ln N = 4.1482$ (d) $\ln N = -1.7654$

SOLUTION

(a)
$$\begin{aligned}\ln N &= 2.4146\\ &= 0.1120 + 2.3026\\ &= \ln\left(1.11 + \frac{0.1120 - 0.1044}{0.1133 - 0.1044}(1.12 - 1.11)\right) + \ln 10\\ &= \ln\left(1.11 + \frac{0.0076}{0.0089}(0.01)\right) + \ln 10\end{aligned}$$

$$= \ln(1.11 + 0.009) + \ln 10$$
$$= \ln 1.119 + \ln 10$$
$$= \ln(1.119 \times 10)$$
$$\ln N = \ln 11.19$$
$$N = 11.19$$

(*b*) $\ln N = 0.9847$

$$= \ln\left(2.67 + \frac{0.9847 - 0.9821}{0.9858 - 0.9821}(2.68 - 2.67)\right)$$
$$= \ln\left((2.67 + \frac{0.0026}{0.0037}(0.01)\right)$$
$$= \ln(2.67 + 0.007)$$
$$\ln N = \ln 2.677$$
$$N = 2.677$$

(*c*) $\ln N = 4.1482$

$$= 1.8456 + 2.3026$$
$$= \ln\left(6.33 + \frac{1.8456 - 1.8453}{1.8469 - 1.8453}(6.34 - 6.33)\right) + \ln 10$$
$$= \ln\left(6.33 + \frac{0.0003}{0.0016}(0.01)\right) + \ln 10$$
$$= \ln(6.33 + 0.002) + \ln 10$$
$$= \ln 6.332 + \ln 10$$
$$= \ln(6.332 \times 10)$$
$$\ln N = \ln 63.32$$
$$N = 63.32$$

(*d*) $\ln N = -1.7654$

$$= 0.5372 - 2.3026$$
$$= \ln\left(1.71 + \frac{0.5372 - 0.5365}{0.5423 - 0.5365}(1.72 - 1.71)\right) - \ln 10$$
$$= \ln\left(1.71 + \frac{0.0007}{0.0058}(0.01)\right) + \ln 10^{-1}$$
$$= \ln(1.71 + 0.001) + \ln 10^{-1}$$
$$= \ln 1.711 + \ln 10^{-1}$$
$$= \ln(1.711 \times 10^{-1})$$
$$\ln N = \ln 0.1711$$
$$N = 0.1711$$

Supplementary Problems

23.26 Evaluate: (*a*) $\log_2 32$, (*b*) $\log \sqrt[4]{10}$, (*c*) $\log_3 1/9$, (*d*) $\log_{1/4} 16$, (*e*) $\log_e e^x$, (*f*) $\log_8 4$.

23.27 Solve each equation for the unknown.

(*a*) $\log_2 x = 3$ (*c*) $\log_x 8 = -3$ (*e*) $\log_4 x^3 = 3/2$

(*b*) $\log y = -2$ (*d*) $\log_3 (2x + 1) = 1$ (*f*) $\log_{(x-1)}(4x - 4) = 2$

23.28 Express as an algebraic sum of logarithms.

(*a*) $\log \dfrac{U^3 V^2}{W^5}$ (*b*) $\log \sqrt{\dfrac{2x^3 y}{z^7}}$ (*c*) $\ln \sqrt[3]{x^{1/2} y^{-1/2}}$ (*d*) $\log \dfrac{xy^{-3/2} z^3}{a^2 b^{-4}}$

23.29 Solve each equation for the indicated letter in terms of the other quantities.

(*a*) $2\log x = \log 16$; x
(*b*) $3\log y + 2\log 2 = \log 32$; y
(*c*) $\log_3 F = \log_3 4 - 2\log_3 x$; F
(*d*) $\ln(30 - U) = \ln 30 - 2t$; U

23.30 Prove that if a and b are positive and $\neq 1$, $(\log_a b)(\log_b a) = 1$.

23.31 Prove that $10^{\log N} = N$ where $N > 0$.

23.32 Determine the characteristic of the common logarithm of each number.

(*a*) 248 (*b*) 2.48 (*c*) 0.024 (*d*) 0.162 (*e*) 0.0006 (*f*) 18.36 (*g*) 1.06 (*h*) 6000 (*i*) 4 (*j*) 40.60 (*k*) 237.63 (*l*) 146.203 (*m*) 7 000 000 (*n*) 0.000 007

23.33 Find the common logarithm of each number.

(*a*) 237 (*b*) 28.7 (*c*) 1.26 (*d*) 0.263 (*e*) 0.086 (*f*) 0.007 (*g*) 10 400 (*h*) 0.00 607 (*i*) 0.000 000 728 (*j*) 6 000 000 (*k*) 23.70 (*l*) 6.03 (*m*) 1 (*n*) 1000

23.34 Find the antilogarithm of each of the following.

(*a*) 2.8802 (*b*) 1.6590 (*c*) 0.6946 (*d*) $\bar{2}.9042$ (*e*) 8.3160 − 10 (*f*) 7.8549 − 10 (*g*) 4.6618 (*h*) 0.4216 (*i*) $\bar{1}.9484$ (*j*) 9.8344 − 10

23.35 Find the common logarithm of each number by interpolation.

(*a*) 1463 (*b*) 810.6 (*c*) 86.27 (*d*) 8.106 (*e*) 0.6041 (*f*) 0.046 22 (*g*) 1.006 (*h*) 300.6 (*i*) 460.3 (*j*) 0.003 001

23.36 Find the antilogarithm of each of the following by interpolation.

(*a*) 2.9060 (*b*) $\bar{1}.4860$ (*c*) 1.6600 (*d*) $\bar{1}.9840$ (*e*) 3.7045 (*f*) 8.9266 − 10 (*g*) 2.2500 (*h*) 0.8003 (*i*) $\bar{1}.4700$ (*j*) 1.2925

23.37 Write each number as a power of 10: (*a*) 45.4, (*b*) 0.005 278.

23.38 Evaluate.

(*a*) $(42.8)(3.26)(8.10)$

(*b*) $\dfrac{(0.148)(47.6)}{284}$

(*c*) $\dfrac{(1.86)(86.7)}{(2.87)(1.88)}$

(*d*) $\dfrac{2453}{(67.2)(8.55)}$

(*e*) $\dfrac{5608}{(0.4536)(11\,000)}$

(*f*) $\dfrac{(3.92)^3(72.16)}{\sqrt[4]{654}}$

(*g*) $3.14\sqrt{11.65/32}$

(*h*) $\sqrt{\dfrac{906}{(3.142)(14.6)}}$

(*i*) $\sqrt{\dfrac{(1600)(310.6)^2}{7290}}$

(*j*) $\sqrt[3]{\dfrac{(5.52)(2610)}{(7.36)(3.142)}}$

23.39 Solve the following hydraulics equation:

$$\frac{20.0}{14.7} = \left(\frac{0.0613}{x}\right)^{1.32}.$$

23.40 The formula

$$D = \sqrt[3]{\frac{W}{0.5236(A - G)}}$$

gives the diameter of a spherical balloon required to lift a weight W. Find D if $A = 0.0807$, $G = 0.0056$ and $W = 1250$.

23.41 Given the formula $T = 2\pi\sqrt{l/g}$, find l if $T = 2.75$, $\pi = 3.142$ and $g = 32.16$.

23.42 Solve for x.

(a) $3^x = 243$ (c) $2^{x+2} = 64$ (e) $x^{-3/4} = 8$ (g) $7x^{-1/2} = 4$ (i) $5^{x-2} = 1$
(b) $5^x = 1/125$ (d) $x^{-2} = 16$ (f) $x^{-2/3} = 1/9$ (h) $3^x = 1$ (j) $2^{2x+3} = 1$

23.43 Solve each exponential equation: (a) $4^{2x-1} = 5^{x+2}$, (b) $3^{x-1} = 4 \cdot 5^{1-3x}$.

23.44 Find the natural logarithms.

(a) ln 2.367 (b) ln 8.532 (c) ln 4875 (d) ln 0.000 189 4

23.45 Find N, the antilogarithm of the given number.

(a) $\ln N = 0.7642$ (b) $\ln N = 1.8540$ (c) $\ln N = 8.4731$ (d) $\ln N = -6.2691$

ANSWERS TO SUPPLEMENTARY PROBLEMS

23.26 (a) 5 (b) 1/4 (c) −2 (d) −2 (e) x (f) 2/3

23.27 (a) 8 (b) 0.01 (c) 1/2 (d) 1 (e) 2 (f) 5

23.28 (a) $3\log U + 2\log V - 5\log W$ (c) $\frac{1}{6}\ln x - \frac{1}{6}\ln y$
(b) $\frac{1}{2}\log 2 + \frac{3}{2}\log x + \frac{1}{2}\log y - \frac{7}{2}\log z$ (d) $\log x - \frac{3}{2}\log y + 3\log z - 2\log a + 4\log b$

23.29 (a) 4 (b) 2 (c) $F = 4/x^2$ (d) $U = 30(1 - e^{-2t})$

23.32 (a) 2 (c) $\bar{2}$ (e) $\bar{4}$ (g) 0 (i) 0 (k) 2 (m) 6
(b) 0 (d) $\bar{1}$ (f) 1 (h) 3 (j) 1 (l) 2 (n) $\bar{6}$

23.33 (a) 2.3747 (d) $\bar{1}.4200$ (g) 4.0170 (j) 6.7782 (m) 0.0000
(b) 1.4579 (e) $\bar{2}.9345$ (h) $\bar{3}.7832$ (k) 1.3747 (n) 3.0000
(c) 0.1004 (f) 7.8451 − 10 (i) $\bar{7}.8621$ (l) 0.7803

23.34 (a) 759 (c) 4.95 (e) 0.0207 (g) 45 900 (i) 0.888
(b) 45.6 (d) 0.0802 (f) 0.007 16 (h) 2.64 (j) 0.683

23.35 (a) 3.1653 (c) 1.9359 (e) $\bar{1}.7811$ (g) 0.0026 (i) 2.6631
(b) 2.9088 (d) 0.9088 (f) 8.6648 − 10 (h) 2.4780 (j) 7.4773 − 10

23.36 (a) 805.4 (c) 45.71 (e) 5064 (g) 177.8 (i) 0.2951
(b) 0.3062 (d) 0.9638 (f) 0.084 45 (h) 6.314 (j) 19.61

23.37 (*a*) $10^{1.6571}$ (*b*) $10^{-2.2776}$

23.38 (*a*) 1130 (*c*) 29.9 (*e*) 1.124 (*g*) 1.90 (*i*) 145.5
(*b*) 0.0248 (*d*) 4.27 (*f*) 860 (*h*) 4.44 (*j*) 8.54

23.39 0.0486

23.40 31.7

23.41 6.16

23.42 (*a*) 5 (*c*) 4 (*e*) 1/16 (*g*) 49/16 (*i*) 2
(*b*) −3 (*d*) ±1/4 (*f*) ±27 (*h*) 0 (*j*) −3/2

23.43 (*a*) 3.958 (*b*) 0.6907

23.44 (*a*) 0.8616 (*b*) 2.1438 (*c*) 8.4919 (*d*) −8.5717

23.45 (*a*) 2.147 (*b*) 6.385 (*c*) 4784 (*d*) 0.001 894

Chapter 24

Applications of Logarithms and Exponents

24.1 INTRODUCTION

Logarithms have their major use in solving exponential equations and solving equations in which the variables are related logarithmically. To solve equations in which the variable is in the exponent, we generally start by changing the expression from exponential form to logarithmic form.

24.2 SIMPLE INTEREST

Interest is money paid for the use of a sum of money called the principal. The interest is usually paid at the ends of specified equal time intervals, such as monthly, quarterly, semiannually, or annually. The sum of the principal and the interest is called the amount.

The simple interest, I, on the principal, P, for a time in years, t, at an interest rate per year, r, is given by the formula $I = Prt$, and the amount, A, is found by $A = P + Prt$ or $A = P(1 + rt)$.

EXAMPLE 24.1. If an individual borrows \$800 at 8% per year for two and one-half years, how much interest must be paid on the loan?

$$I = Prt$$
$$I = \$800(0.08)(2.5)$$
$$I = \$160$$

EXAMPLE 24.2. If a person invests \$3000 at 6% per year for five years, how much will the investment be worth at the end of the five years?

$$A = P + Prt$$
$$A = \$3000 + \$3000(0.06)(5)$$
$$A = \$3000 + \$900$$
$$A = \$3900$$

24.3 COMPOUND INTEREST

Compound interest means that the interest is paid periodically over the term of the loan which results in a new principal at the end of each interval of time.

If a principal, P, is invested for t years at an annual interest rate, r, compounded n times per year, then the amount, A, or ending balance is given by:

$$A = P\left(1 + \frac{r}{n}\right)^{nt}$$

EXAMPLE 24.3. Find the amount of an investment if \$20 000 is invested at 6% compounded monthly for three years.

$$A = P\left(1 + \frac{r}{n}\right)^{nt}$$

$$A = 20\,000\left(1 + \frac{0.06}{12}\right)^{12(3)}$$

$$\begin{aligned}
A &= 20\,000(1 + 0.005)^{36}\\
A &= 20\,000(1.005)^{36}\\
\log A &= \log 20\,000(1.005)^{36}\\
\log A &= \log 20\,000 + 36\ \log 1.005\\
\log A &= 4.3010 + 36(0.002\,15)\\
\log A &= 4.3010 + 0.0774\\
\log A &= 4.3784\\
A &= \text{antilog}\,4.3784\\
A &= 2.39 \times 10^4 \qquad \log 2.39 = 0.3784 \text{ and } \log 10^4 = 4\\
A &= \$23\,900
\end{aligned}$$

When the interest is compounded more and more frequently, we get to a situation of continuously compounded interest. If a principal, P, is invested for t years at an annual interest rate, r, compounded continuously, then the amount, A, or ending balance, is given by:

$$A = Pe^{rt}$$

EXAMPLE 24.4. Find the amount of an investment if \$20 000 is invested at 6% compounded continuously for three years.

$$\begin{aligned}
A &= Pe^{rt}\\
A &= 20\,000\mathrm{e}^{0.06(3)}\\
A &= 20\,000\mathrm{e}^{0.18}\\
\ln A &= \ln 20\,000\mathrm{e}^{0.18}\\
\ln A &= \ln 20\,000 + \ln \mathrm{e}^{0.18}\\
\ln A &= \ln(2.00 \times 10^4) + 0.18\ \ln \mathrm{e}\\
\ln A &= \ln 2.00 + 4\ \ln 10 + 0.18(1) \qquad \ln \mathrm{e} = 1\\
\ln A &= 0.6931 + 4(2.3026) + 0.18 \qquad \ln 2.00 = 0.6931 \text{ and } \ln 10 = 2.3026\\
\ln A &= 10.0835\\
\ln A &= 0.8731 + 4(2.3026)\\
\ln A &= \ln\left(2.39 + \frac{0.8731 - 0.8713}{0.8755 - 0.8713}(2.40 - 2.39)\right) + 4\ln 10\\
\ln A &= \ln\left(2.39 + \frac{0.0018}{0.0042}(0.01)\right) + \ln 10^4\\
\ln A &= \ln(2.39 + 0.004) + \ln 10^4\\
\ln A &= \ln 2.394 + \ln 10^4\\
\ln A &= \ln(2.394 \times 10^4)\\
\ln A &= \ln 23\,940\\
A &= \$23\,940
\end{aligned}$$

In doing Examples 24.3 and 24.4 we found the answers to four significant digits. However, using the logarithm tables and doing interpolation results in some error. Also, we may have a problem if the interest is compounded daily, because when we divide r by n the result could be zero when rounded to thousandths. To deal with this problem and to get greater accuracy, we can use five-place logarithm tables, calculators, or computers. Generally, banks and other businesses use computers or calculators to get the accuracy they need.

EXAMPLE 24.5. Use a scientific or graphing calculator to find the amount of an investment if \$20 000 is invested at 6% compounded monthly for three years.

$$A = P\left(1 + \frac{r}{n}\right)^{nt}$$

$$A = \$20\,000\left(1 + \frac{0.06}{12}\right)^{12(3)}$$

$$A = \$20\,000(1.005)^{36} \qquad \text{use the power key to compute } (1.005)^{36}$$

$$A = \$23\,933.61$$

To the nearest cent, the amount has been increased by \$33.61 from the amount found in Example 24.3. It is possible to compute the answer to the nearest cent here, while we were able to compute the result to the nearest ten dollars in Example 24.3.

EXAMPLE 24.6. Use a scientific or graphing calculator to find the amount of an investment if \$20 000 is invested at 6% compounded continuously for three years.

$$A = Pe^{rt}$$

$$A = \$20\,000e^{0.06(3)}$$

$$A = \$20\,000e^{0.18} \qquad \text{use the inverse of } \ln x \text{ to compute } e^{0.18}$$

$$A = \$23\,944.35$$

To the nearest cent, the amount has increased by \$4.35 from the amount found in Example 24.4. The greater accuracy was possible because the calculator computes with more decimal places in each operation and then the answer is rounded. In our examples, we rounded to hundredths because cents are the smallest units of money that have a general usefulness. Most calculators compute with 8, 10, or 12 significant digits in doing the operations.

24.4 APPLICATIONS OF LOGARITHMS

The loudness, L, of a sound (in decibels) perceived by the human ear depends on the ratio of the intensity, I, of the sound to the threshold, I_0, of hearing for the average human ear.

$$L = 10 \log\left(\frac{I}{I_0}\right)$$

EXAMPLE 24.7. Find the loudness of a sound that has an intensity 10 000 times the threshold of hearing for the average human ear.

$$L = 10 \log\left(\frac{I}{I_0}\right)$$

$$L = 10 \log\left(\frac{10\,000 I_0}{I_0}\right)$$

$$L = 10 \log 10\,000$$

$$L = 10\ (4)$$

$$L = 40 \text{ decibels}$$

Chemists use the hydrogen potential, pH, of a solution to measure its acidity or basicity. The pH of distilled water is about 7. If the pH of a solution exceeds 7, it is called an acid, but if its pH is less than 7 it is called a base. If $[H^+]$ is the concentration of hydrogen ions in moles per liter, the pH is given by the formula:

$$\text{pH} = -\log[\text{H}^+]$$

EXAMPLE 24.8. Find the pH of the solution whose concentration of hydrogen ions is 5.32×10^{-5} moles per liter.

$$\begin{aligned}
\text{pH} &= -\log[\text{H}^+] \\
\text{pH} &= -\log(5.32 \times 10^{-5}) \\
\text{pH} &= -[\log 5.32 + \log 10^{-5}] \\
\text{pH} &= -\log 5.32 - (-5)\log 10 \qquad \log 10 = 1 \\
\text{pH} &= -\log 5.32 + 5(1) \\
\text{pH} &= -0.7259 + 5 \\
\text{pH} &= 4.2741 \\
\text{pH} &= 4.3
\end{aligned}$$

Seismologists use the Richter scale to measure and report the magnitude of earthquakes. The magnitude or Richter number of an earthquake depends on the ratio of the intensity, I, of an earthquake to the reference intensity, I_0, which is the smallest earth movement that can be recorded on a seismograph. Richter numbers are usually rounded to the nearest tenth or hundredth. The Richter number is given by the formula:

$$R = \log\left(\frac{I}{I_0}\right)$$

EXAMPLE 24.9. If the intensity of an earthquake is determined to be 50 000 times the reference intensity, what is its reading on the Richter scale?

$$\begin{aligned}
R &= \log\left(\frac{I}{I_0}\right) \\
R &= \log\left(\frac{50\,000 I_0}{I_0}\right) \\
R &= \log 50\,000 \\
R &= 4.6990 \\
R &= 4.70
\end{aligned}$$

24.5 APPLICATIONS OF EXPONENTS

The number e is involved in many functions occurring in nature. The growth curve of many materials can be described by the exponential growth equation:

$$A = A_0 e^{rt}$$

where A_0 is the initial amount of the material, r is the annual rate of growth, t is the time in years, and A is the amount of the material at the ending time.

EXAMPLE 24.10. The population of a country was 2 400 000 in 1990 and it has an annual growth rate of 3%. If the growth is exponential, what will its population be in 2000?

$$A = A_0 e^{rt}$$

$$A = 2\,400\,000\mathrm{e}^{(0.03)(10)}$$
$$A = 2\,400\,000\mathrm{e}^{0.3} \qquad N = \mathrm{e}^{0.3}$$
$$A = 2\,400\,000(1.350) \qquad \ln N = 0.3 \ln \mathrm{e}$$
$$A = 3\,240\,000 \qquad \ln N = 0.3$$
$$N = 1.350$$

The decay or decline equation is similar to the growth except the exponent is negative.

$$A = A_0\mathrm{e}^{-rt}$$

where A_0 is the initial amount, r is the annual rate of decay or decline, t is the time in years, and A is the ending amount.

EXAMPLE 24.11. A piece of wood is found to contain 100 grams of carbon-14 when it was removed from a tree. If the rate of decay of carbon-14 is 0.0124% per year, how much carbon-14 will be left in the wood after 200 years?

$$A = A_0\mathrm{e}^{-rt}$$
$$A = 100\mathrm{e}^{-0.000\,124\,(200)} \qquad N = \mathrm{e}^{-0.0248}$$
$$A = 100\mathrm{e}^{-0.0248} \qquad \ln N = -0.0248$$
$$A = 100(0.9755) \qquad \ln N = \ln 2.2778 - 2.3026$$
$$A = 97.55 \text{ grams} \qquad \ln N = \ln 9.755 - \ln 10$$
$$\ln N = \ln(9.755 \times 10^{-1})$$
$$\ln N = \ln 0.9755$$
$$N = 0.9755$$

Solved Problems

24.1 A woman borrows \$400 for 2 years at a simple interest rate of 3%. Find the amount required to repay the loan at the end of 2 years.

SOLUTION

Interest I $= Prt = 400(0.03)(2) = \$24$. Amount A = principal P + interest $I = \$424$.

24.2 Find the interest I and amount A for

(*a*) \$600 for 8 months (2/3 yr) at 4%,

(*b*) \$1562.60 for 3 years, 4 months (10/3 yr) at $3\frac{1}{2}$%.

SOLUTION

(*a*) $I = Prt = 600(0.04)(2/3) = \$16.$ $\qquad A = P + I = \$616.$

(*b*) $I = Prt = 1562.60(0.035)(10/3) = \$182.30.$ $\qquad A = P + I = \$1744.90.$

24.3 What principal invested at 4% for 5 years will amount to \$1200?

SOLUTION

$$A = P(1 + rt) \qquad \text{or} \qquad P = \frac{A}{1 + rt} = \frac{1200}{1 + (0.04)(5)} = \frac{1200}{1.2} = \$1000.$$

The principal of \$1000 is called the present value of \$1200. This means that \$1200 to be paid *5 years from now* is worth \$1000 *now* (the interest rate being 4%).

24.4 What rate of interest will yield \$1000 on a principal of \$800 in 5 years?

SOLUTION

$$A = P(1 + rt) \qquad \text{or} \qquad r = \frac{A - P}{Pt} = \frac{1000 - 800}{800(5)} = 0.05 \text{ or } 5\%.$$

24.5 A man wishes to borrow \$200. He goes to the bank where he is told that the interest rate is 5%, interest payable in advance, and that the \$200 is to be paid back at the end of one year. What interest rate is he actually paying?

SOLUTION

The simple interest on \$200 for 1 year at 5% is $I = 200(0.05)(1) = \$10$. Thus he receives $\$200 - \$10 = \$190$. Since he must pay back \$200 after a year, $P = \$190$, $A = \$200$, $t = 1$ year. Thus

$$r = \frac{A - P}{Pt} = \frac{200 - 190}{190(1)} = 0.0526,$$

i.e., the effective interest rate is 5.26%.

24.6 A merchant borrows \$4000 under the condition that she pay at the end of every 3 months \$200 on the principal plus the simple interest of 6% on the principal outstanding at the time. Find the total amount she must pay.

SOLUTION

Since \$4000 is to be paid (excluding interest) at the rate of \$200 every 3 months, it will take $4000/200(4) = 5$ years, i.e., 20 payments.

$$\begin{aligned}
&\text{Interest paid at 1st payment (for first 3 months)} = 4000(0.06)(\tfrac{1}{4}) = \$60.00.\\
&\text{Interest paid at 2nd payment} \qquad\qquad\qquad = 3800(0.06)(\tfrac{1}{4}) = \$57.00\\
&\text{Interest paid at 3rd payment} \qquad\qquad\qquad = 3600(0.06)(\tfrac{1}{4}) = \$54.00.\\
&\quad\vdots\\
&\text{Interest paid at 20th payment} \qquad\qquad\quad\;\; = 200(0.06)(\tfrac{1}{4}) = \$\ 3.00.
\end{aligned}$$

The total interest is $60 + 57 + 54 + \cdots + 9 + 6 + 3$: an arithmetic sequence having sum given by $S = (n/2)(a + l)$, where a = 1st term, l = last term, n = number of terms.

Then $S = (20/2)(60 + 3) = \$630$, and the total amount she must pay is \$4630.

24.7 What will \$500 deposited in a bank amount to in 2 years if interest is compounded semiannually at 2%?

SOLUTION

Method 1. Without formula.

$$\begin{aligned}
\text{At end of 1st half year, interest} &= 500(0.02)(\tfrac{1}{2}) = \$5.00\\
\text{At end of 2nd half year, interest} &= 505(0.02)(\tfrac{1}{2}) = \$5.05.\\
\text{At end of 3rd half year, interest} &= 510.05(0.02)(\tfrac{1}{2}) = \$5.10.\\
\text{At end of 4th half year, interest} &= 515.15(0.02)(\tfrac{1}{2}) = \$5.15.\\
\text{Total interest} &= \$20.30. \qquad \text{Total amount} = \$520.30.
\end{aligned}$$

Method 2. Using formula.
$P = \$500$, i = rate per period $= 0.02/2 = 0.01$. n = number of periods $= 4$.

$$A = P(1+i)^n = 500(1.01)^4 = 500(1.0406) = \$520.30.$$

Note. $(1.01)^4$ may be evaluated by the binomial formula, logarithms or tables.

24.8 Find the compound interest and amount of \$2800 in 8 years at 5% compounded quarterly.

SOLUTION

$$A = P(1+i)^n = 2800(1+0.05/4)^{32} = 2800(1.0125)^{32} = 2800(1.4881) = \$4166.68.$$
$$\text{Interest} = A - P = \$4166.68 - \$2800 = \$1366.68.$$

24.9 A man expects to receive \$2000 in 10 years. How much is that money worth now considering interest at 6% compounded quarterly? What is the discount?

SOLUTION

We are asked for the present value P which will amount to $A = \$2000$ in 10 years.

$$A = P(1+i)^n \quad \text{or} \quad P = \frac{A}{(1+i)^n} = \frac{2000}{(1.015)^{40}} = \$1102.52 \text{ using tables.}$$

The discount is $\$2000 - \$1102.52 = \$897.48$.

24.10 What rate of interest compounded annually is the same as the rate of interest of 6% compounded semiannually?

SOLUTION

Amount from principal P in 1 year at rate $r = P(1+r)$.

Amount from principal P in 1 year at rate 6% compounded semiannually $= P(1+0.03)^2$.

The amounts are equal if $P(1+r) = P(1.03)^2$, $1+r = (1.03)^2$, $r = 0.0609$ or 6.09%.

The rate of interest i per year compounded a given number of times per year is called the *nominal rate*. The rate of interest r which, if compounded annually, would result in the same amount of interest is called the *effective rate*. In this example, 6% compounded semiannually is the nominal rate and 6.09% is the effective rate.

24.11 Derive a formula for the effective rate in terms of the nominal rate.

SOLUTION

Let r = effective interest rate,
i = interest rate per annum compounded k times per year, i.e. nominal rate.

Amount from principal P in 1 year at rate $r = P(1+r)$.
Amount from principal P in 1 year at rate i compounded k times per year $= P(1+i/k)^k$.
The amounts are equal if $P(1+r) = P(1+i/k)^k$.
Hence $r = (1+i/k)^k - 1$.

24.12 When US Saving Bonds were introduced, a bond that cost \$18.75 could be cashed in 10 years later for \$25. If the interest was compounded annually, what was the interest rate?

SOLUTION

$$A = P\left(1 + \frac{r}{n}\right)^{nt}$$
$$25.00 = 18.75(1 + r)^{10} \qquad n = 1$$
$$4/3 = (1 + r)^{10}$$
$$\log 4 - \log 3 = 10 \log(1 + r)$$
$$0.6021 - 0.3771 = 10 \log(1 + r)$$
$$0.125 = 10 \log(1 + r)$$
$$0.0125 = \log(1 + r)$$
$$1 + r = \text{antilog}\, 0.0125$$
$$1 + r = 1.02 + \frac{39}{42}(0.01)$$
$$1 + r = 1.02 + 0.009$$
$$1 + r = 1.029$$
$$r = 0.029$$
$$r = 2.9\%$$

24.13 The population in a country grows at a rate of 4% compounded annually. At this rate, how long will it take the population to double?

SOLUTION

$$A = P\left(1 + \frac{r}{n}\right)^{nt}$$
$$2P = P(1 + 0.04)^t \qquad n = 1$$
$$2 = (1.04)^t$$
$$\log 2 = t \log(1.04)$$
$$t = \frac{\log 2}{\log 1.04}$$
$$t = \frac{0.3010}{0.0170}$$
$$t = 17.7 \text{ years}$$

24.14 If \$1000 is invested at 10% compounded continuously, how long will it take the investment to triple?

SOLUTION

$$A = Pe^{rt}$$
$$3000 = 1000e^{0.10t}$$
$$3 = e^{0.10t}$$
$$\ln 3 = 0.10 \qquad \ln e = 1$$
$$t = \frac{\ln 3}{0.10}$$
$$t = \frac{1.0986}{0.10}$$
$$t = 10.986$$
$$t = 11.0$$

24.15 If \$5000 is invested at 9% compounded continuously, what will be the value of the investment in 5 years?

SOLUTION

$$A = Pe^{rt}$$
$$A = 5000e^{(0.09)(5)} \qquad N = e^{0.45}$$
$$A = 5000e^{0.45} \qquad \ln N = 0.45$$
$$A = 5000(1.568) \qquad \ln N = \ln\left(1.56 + \frac{0.4500 - 0.4447}{0.4500 - 0.4447}(0.01)\right)$$
$$A = \$7840$$
$$\ln N = \ln(1.56 + 0.008)$$
$$\ln N = \ln 1.568$$
$$N = 1.568$$

24.16 Find the pH of blood if the concentration of hydrogen ions is 3.98×10^{-8}.

SOLUTION

$$\text{pH} = -\log[\text{H}^+]$$
$$\text{pH} = -\log(3.98 \times 10^{-8})$$
$$\text{pH} = -\log 3.98 - (-8)\log 10$$
$$\text{pH} = -0.5999 + 8$$
$$\text{pH} = 7.4001$$
$$\text{pH} = 7.40$$

24.17 An earthquake in San Francisco in 1989 was reported to have a Richter number of 6.90. How does the intensity of the earthquake compare with the reference intensity?

SOLUTION

$$R = \log\frac{I}{I_0}$$
$$6.90 = \log\frac{I}{I_0}$$
$$\log\frac{I}{I_0} = 6.90$$
$$\frac{I}{I_0} = \text{antilog}\,6.90 \qquad \text{antilog}\,0.9000 = 7.94 + \frac{0.9000 - 0.8998}{0.9004 - 0.8998}(0.01)$$
$$I = (\text{antilog}\,6.90)\,I_0 \qquad = 7.94 + \frac{0.0002}{0.0006}(0.01)$$
$$I = (7.943 \times 10^6)\,I_0 \qquad = 7.94 + 0.003$$
$$I = 7\,943\,000 I_0 \qquad \text{antilog}\,0.9000 = 7.943$$
$$\text{antilog}\,6.9000 = 7.943 \times 10^6$$

24.18 A sound that causes pain has an intensity 10^{14} times the threshold intensity. What is the decibel level of this sound?

SOLUTION

$$L = 10\log\left(\frac{I}{I_0}\right)$$
$$L = 10\log\left(\frac{10^{14}I_0}{I_0}\right)$$
$$L = 10\log 10^{14}$$
$$L = 10(14\log 10)$$
$$L = 10(14)(1)$$
$$L = 140 \text{ decibels}$$

24.19 If the rate of decay of carbon-14 is 0.0124% per year, how long, rounded to 3 significant digits, will it take for the carbon-14 to diminish to 1% of the original amount after the death of the plant or animal?

SOLUTION

$$A = A_0 e^{-rt}$$
$$0.01A_0 = A_0 e^{-0.000124t}$$
$$0.01 = e^{-0.000124t}$$
$$\ln 0.01 = -0.000\,124t \ln e$$
$$\ln(1\times 10^{-2}) = -0.000\,124t(1)$$
$$0 - 2(2.3026) = -0.000\,124t$$
$$-4.6052 = -0.000\,124t$$
$$37\,139 = t$$
$$t = 37\,100 \text{ years}$$

24.20 The population of the world increased from 2.5 billion in 1950 to 5.0 billion in 1987. If the growth was exponential, what was the annual growth rate?

SOLUTION

$$A = A_0 e^{rt}$$
$$5.0 = 2.5e^{r(37)}$$
$$2 = e^{37r}$$
$$\ln 2 = 37r \ln e$$
$$0.6931 = 37r(1)$$
$$0.6931 = 37r$$
$$0.018\,73 = r$$
$$r = 0.0187$$
$$r = 1.87\%$$

24.21 In Nigeria the rate of deforestation is 5.25% per year. If the decrease in forests in Nigeria is exponential, how long will it take until only 25% of the current forests are left?

SOLUTION

$$A = A_0 e^{-rt}$$
$$0.25A_0 = A_0 e^{-0.0525t}$$
$$0.25 = e^{-0.0525t}$$

$$\ln 0.25 = -0.0525t \ln \mathrm{e}$$
$$\ln(2.5 \times 10^{-1}) = -0.0525t(1)$$
$$\ln 2.5 - \ln 10 = -0.0525t$$
$$0.9163 - 2.3026 = -0.0525t$$
$$-1.3863 = -0.0525t$$
$$26.405 = t$$
$$t = 26.41 \text{ years}$$

Supplementary Problems

24.22 If \$51.30 interest is earned in two years on a deposit of \$95, then what is the simple annual interest rate?

24.23 If \$500 is borrowed for one month and \$525 must be paid back at the end of the month, what is the simple annual interest?

24.24 If \$4000 is invested at a bank that pays 8% interest compounded quarterly, how much will the investment be worth in 6 years?

24.25 If \$8000 is invested in an account that pays 12% interest compounded monthly, how much will the investment be worth after 10 years?

24.26 A bank tried to attract new, large, long-term investments by paying 9.75% interest compounded continuously if at least \$30 000 was invested for at least 5 years. If \$30 000 is invested for 5 years at this bank, how much would the investment be worth at the end of the 5 years?

24.27 What interest will be earned if \$8000 is invested for 4 years at 10% compounded semiannually?

24.28 What interest will be earned if \$3500 is invested for 5 years at 8% compounded quarterly?

24.29 What interest will be earned if \$4000 is invested for 6 years at 8% compounded continuously?

24.30 Find the amount that will result if \$9000 is invested for 2 years at 12% compounded monthly.

24.31 Find the amount that will result if \$9000 is invested for 2 years at 12% compounded continuously.

24.32 In 1990 an earthquake in Iran was said to have about 6 times the intensity of the 1989 San Francisco earthquake, which had a Richter number of 6.90. What is the Richter number of the Iranian earthquake?

24.33 Find the Richter number of an earthquake if its intensity is 3 160 000 times as great as the reference intensity.

24.34 An earthquake in Alaska in 1964 measured 8.50 on the Richter scale. What is the intensity of this earthquake compared with the reference intensity?

24.35 Find the intensity as compared with the reference intensity of the 1906 San Francisco earthquake if it has a Richter number of 8.25.

24.36 Find the Richter number of an earthquake with an intensity 20 000 times greater than the reference intensity.

24.37 Find the pH of each substance with the given concentration of hydrogen ions.

(*a*) beer: $[H^+] = 6.31 \times 10^{-5}$
(*b*) orange juice: $[H^+] = 1.99 \times 10^{-4}$
(*c*) vinegar: $[H^+] = 6.3 \times 10^{-3}$
(*d*) tomato juice: $[H^+] = 7.94 \times 10^{-5}$

24.38 Find the approximate hydrogen ions concentration, $[H^+]$, for the substances with the given pH.

(*a*) apples: pH = 3.0 (*b*) eggs: pH = 7.8

24.39 If gastric juices in your stomach have a hydrogen ion concentration of 1.01×10^{-1} moles per liter, what is the pH of the gastric juices?

24.40 A relatively quiet room has a background noise level of 32 decibels. How many times the hearing threshold intensity is the intensity of a relatively quiet room?

24.41 If the intensity of an argument is about 3 980 000 times the hearing threshold intensity, what is the decibel level of the argument?

24.42 The population of the world compounds continuously. If in 1987 the growth rate was 1.63% annually and an initial population of 5 million people, what will the world population be in the year 2000?

24.43 During the Black Plague the world population declined by about 1 million from 4.7 million to about 3.7 million during the 50-year period from 1350 to 1400. If world population decline was exponential, what was the annual rate of decline?

24.44 If the world population grew exponentially from 1.6 billion in 1900 to 5.0 billion in 1987, what was the annual rate of population growth?

24.45 If the deforestation of El Salvador continues at the current rate for 20 more years only 53% of the present forests will be left. If the decline of the forests is exponential, what is the annual rate of deforestation for El Salvador?

24.46 A bone is found to contain 40% of the carbon-14 that it contained when it was part of a living animal. If the decay of carbon-14 is exponential with an annual rate of decay of 0.0124%, how long ago did the animal die?

24.47 Radioactive strontium-90 is used in nuclear reactors and decays exponentially with an annual rate of decay of 2.48%. How much of 50 grams of strontium-90 will be left after 100 years?

24.48 How long does it take 12 grams of carbon-14 to decay to 10 grams when the decay is exponential with an annual rate of decay of 0.0124%?

24.49 How long does it take for 10 grams of strontium-90 to decay to 8 grams if the decay is exponential and the annual rate of decay is 2.48%?

ANSWERS TO SUPPLEMENTARY PROBLEMS

Note: The tables in Appendices A and B were used in computing these answers. If a calculator is used your answers may vary.

24.22 2.7%

24.23 60%

24.24 \$6436

24.25 \$26 248

24.26 \$48 840

24.27 \$3824

24.28 \$1701

24.29 \$2464

24.30 \$11 412

24.31 \$11 439

24.32 7.68

24.33 6.50

24.34 316 200 000 I_0

24.35 177 800 000 I_0

24.36 4.30

24.37 (*a*) pH = 4.2 (*b*) pH = 2.2 (*c*) pH = 3.7 (*d*) pH = 4.1

24.38 (*a*) $[H^+] = 0.001$ or 1.00×10^{-3} (*b*) $[H^+] = 1.585 \times 10^8$

24.39 1.0

24.40 1.585 I_0

24.41 66 decibels

24.42 6.18 billion

24.43 0.48% per year

24.44 1.31% per year

24.45 3.17% per year

24.46 7390 years

24.47 4.2 grams

24.48 1471 years

24.49 8.998 years

Chapter 25

Permutations and Combinations

25.1 FUNDAMENTAL COUNTING PRINCIPLE

If one thing can be done in m different ways and, when it is done in any one of these ways, a second thing can be done in n different ways, then the two things in succession can be done in mn different ways.

For example, if there are 3 candidates for governor and 5 for mayor, then the two offices may be filled in $3 \cdot 5 = 15$ ways.

In general, if a_1 can be done in x_1 ways, a_2 can be done in x_2 ways, a_3 can be done in x_3 ways, ..., and a_n can be done in x_n ways, then the event $a_1 a_2 a_3 \cdots a_n$ can be done in $x_1 \cdot x_2 \cdot x_3 \cdot \cdots \cdot x_n$ ways.

EXAMPLE 25.1 A man has 3 jackets, 10 shirts, and 5 pairs of slacks. If an outfit consists of a jacket, a shirt, and a pair of slacks, how many different outfits can the man make?

$$x_1 \cdot x_2 \cdot x_3 = 3 \cdot 10 \cdot 5 = 150 \text{ outfits}$$

25.2 PERMUTATIONS

A permutation is an arrangement of all or part of a number of things in a definite order.

For example, the permutations of the three letters a, b, c taken all at a time are *abc*, *acb*, *bca*, *bac*, *cba*, *cab*. The permutations of the three letters a, b, c taken two at a time are *ab*, *ac*, *ba*, *bc*, *ca*, *cb*.

For a natural number n, n factorial, denoted by $n!$, is the product of the first n natural numbers. That is, $n! = n \cdot (n-1) \cdot (n-2) \cdot \cdots \cdot 2 \cdot 1$. Also, $n! = n \cdot (n-1)!$

Zero factorial is defined to be $1 : 0! = 1$.

EXAMPLES 25.2. Evaluate each factorial.

(*a*) 7! (*b*) 5! (*c*) 1! (*d*) 2! (*e*) 4!

(*a*) $7! = 7 \cdot 6 \cdot 5 \cdot 4 \cdot 3 \cdot 2 \cdot 1 = 5040$

(*b*) $5! = 5 \cdot 4 \cdot 3 \cdot 2 \cdot 1 = 120$

(*c*) $1! = 1$

(*d*) $2! = 2 \cdot 1 = 2$

(*e*) $4! = 4 \cdot 3 \cdot 2 \cdot 1 = 24$

The symbol ${}_nP_r$ represents the number of permutations (arrangements, orders) of n things taken r at a time.

Thus ${}_8P_3$ denotes the number of permutations of 8 things taken 3 at a time, and ${}_5P_5$ denotes the number of permutations of 5 things taken 5 at a time.

Note. The symbol $P(n, r)$ having the same meaning as ${}_nP_r$ is sometimes used.

A. Permutations of n different things taken r at a time

$${}_nP_r = n(n-1)(n-2) \cdots (n-r+1) = \frac{n!}{(n-r)!}$$

When $r = n$, ${}_nP_r = {}_nP_n = n(n-1)(n-2) \cdots 1 = n!$.

EXAMPLES 25.3.

$$_5P_1 = 5,\ _5P_2 = 5\cdot 4 = 20,\ _5P_3 = 5\cdot 4\cdot 3 = 60,\ _5P_4 = 5\cdot 4\cdot 3\cdot 2 = 120,\ _5P_5 = 5! = 5\cdot 4\cdot 3\cdot 2\cdot 1 = 120,$$
$$_{10}P_7 = 10\cdot 9\cdot 8\cdot 7\cdot 6\cdot 5\cdot 4 = 604\,800.$$

The number of ways in which 4 persons can take their places in a cab having 6 seats is $_6P_4 = 6\cdot 5\cdot 4\cdot 3 = 360$.

B. Permutations with some things alike, taken all at a time

The number of permutations P of n things taken all at a time, of which n_1 are alike, n_2 others are alike, n_3 others are alike, etc., is

$$P = \frac{n!}{n_1!\,n_2!\,n_3!\cdots} \qquad \text{where } n_1 + n_2 + n_3 + \cdots = n.$$

For example, the number of ways 3 dimes and 7 quarters can be distributed among 10 boys, each to receive one coin, is

$$\frac{10!}{3!7!} = \frac{10\cdot 9\cdot 8}{1\cdot 2\cdot 3} = 120.$$

C. Circular permutations

The number of ways of arranging n different objects around a circle is $(n-1)!$ ways.

Thus 10 persons may be seated at a round table in $(10-1)! = 9!$ ways.

25.3 COMBINATIONS

A combination is a grouping or selection of all or part of a number of things without reference to the arrangement of the things selected.

Thus the combinations of the three letters a, b, c taken 2 at a time are ab, ac, bc. Note that ab and ba are 1 combination but 2 permutations of the letters a, b.

The symbol $_nC_r$ represents the number of combinations (selections, groups) of n things taken r at a time.

Thus $_9C_4$ denotes the number of combinations of 9 things taken 4 at a time.

Note. The symbol $C(n, r)$ having the same meaning as $_nC_r$ is sometimes used.

A. Combinations of n different things taken r at a time

$$_nC_r = \frac{_nP_r}{r!} = \frac{n!}{r!(n-r)!} = \frac{n(n-1)(n-2)\cdots(n-r+1)}{r!}$$

For example, the number of handshakes that may be exchanged among a party of 12 students if each student shakes hands once with each other student is

$$_{12}C_2 = \frac{12!}{2!(12-2)!} = \frac{12!}{2!\,10!} = \frac{12\cdot 11}{1\cdot 2} = 66.$$

The following formula is very useful in simplifying calculations:

$$_nC_r = {}_nC_{n-r}.$$

This formula indicates that the number of selections of r out of n things is the same as the number of selections of $n-r$ out of n things.

EXAMPLES 25.4.

$$_5C_1 = \frac{5}{1} = 5, \qquad _5C_2 = \frac{5\cdot 4}{1\cdot 2} = 10, \qquad _5C_5 = \frac{5!}{5!} = 1$$

$$_9C_7 = {_9C_{9-7}} = {_9C_2} = \frac{9\cdot 8}{1\cdot 2} = 36, \qquad _{25}C_{22} = {_{25}C_3} = \frac{25\cdot 24\cdot 23}{1\cdot 2\cdot 3} = 2300$$

Note that in each case the numerator and denominator have the same number of factors.

B. Combinations of different things taken any number at a time

The total number of combinations C of n different things taken $1, 2, 3 \ldots, n$ at a time is

$$C = 2^n - 1.$$

For example, a woman has in her pocket a quarter, a dime, a nickel, and a penny. The total number of ways she can draw a sum of money from her pocket is $2^4 - 1 = 15$.

25.4 USING A CALCULATOR

Scientific and graphing calculators have keys for factorials, $n!$; permutations, $_nP_r$, and combinations, $_nC_r$. As factorials get larger, the results are displayed in scientific notation. Many calculators have only two digits available for the exponent, which limits the size of the factorial that can be displayed. Thus, 69! can be displayed and 70! can not, because 70! needs more than two digits for the exponent in scientific notation. When the calculator can perform an operation, but it can not display the result an error message is displayed instead of an answer.

The values of $_nP_r$ and $_nC_r$ can often be computed on the calculator when $n!$ cannot be displayed. This can be done because the internal procedure does not require the result to be displayed, just used.

Solved Problems

25.1 Evaluate $_{20}P_2$, $_8P_5$, $_7P_5$, $_7P_7$.

SOLUTION

$_{20}P_2 = 20\cdot 19 = 380$ $\qquad$ $_7P_5 = 7\cdot 6\cdot 5\cdot 4\cdot 3 = 2520$

$_8P_5 = 8\cdot 7\cdot 6\cdot 5\cdot 4 = 6720$ $\qquad$ $_7P_7 = 7! = 7\cdot 6\cdot 5\cdot 4\cdot 3\cdot 2\cdot 1 = 5040$

25.2 Find n if (*a*) $7\cdot {_nP_3} = 6\cdot {_{n+1}P_3}$, (*b*) $3\cdot {_nP_4} = {_{n-1}P_5}$.

SOLUTION

(*a*) $7n(n-1)(n-2) = 6(n+1)(n)(n-1)$.
Since $n \neq 0, 1$ we may divide by $n(n-1)$ to obtain $7(n-2) = 6(n+1)$, $n = 20$.

(*b*) $3n(n-1)(n-2)(n-3) = (n-1)(n-2)\ (n-3)(n-4)(n-5)$.
Since $n \neq 1, 2, 3$ we may divide by $(n-1)(n-2)(n-3)$ to obtain

$$3n = (n-4)(n-5), \qquad n^2 - 12n + 20 = 0, \qquad (n-10)(n-2) = 0.$$

Thus $n = 10$.

25.3 A student has a choice of 5 foreign languages and 4 sciences. In how many ways can he choose 1 language and 1 science?

SOLUTION

He can choose a language in 5 ways, and with each of these choices there are 4 ways of choosing a science.

Hence the required number of ways $= 5 \cdot 4 = 20$ ways.

25.4 In how many ways can 2 different prizes be awarded among 10 contestants if both prizes (*a*) may not be given to the same person, (*b*) may be given to the same person?

SOLUTION

(*a*) The first prize can be awarded in 10 different ways and, when it is awarded, the second prize can be given in 9 ways, since both prizes may not be given to the same contestant.

Hence the required number of ways $= 10 \cdot 9 = 90$ ways.

(*b*) The first prize can be awarded in 10 ways, and the second prize also in 10 ways, since both prizes may be given to the same contestant.

Hence the required number of ways $= 10 \cdot 10 = 100$ ways.

25.5 In how many ways can 5 letters be mailed if there are 3 mailboxes available?

SOLUTION

Each of the 5 letters may be mailed in any of the 3 mailboxes.

Hence the required number of ways $= 3 \cdot 3 \cdot 3 \cdot 3 \cdot 3 = 3^5 = 243$ ways.

25.6 There are 4 candidates for president of a club, 6 for vice-president and 2 for secretary. In how many ways can these three positions be filled?

SOLUTION

A president may be selected in 4 ways, a vice-president in 6 ways, and a secretary in 2 ways. Hence the required number of ways $= 4 \cdot 6 \cdot 2 = 48$ ways.

25.7 In how many different orders may 5 persons be seated in a row?

SOLUTION

The first person may take any one of 5 seats, and after the first person is seated, the second person may take any one of the remaining 4 seats, etc. Hence the required number of orders $= 5 \cdot 4 \cdot 3 \cdot 2 \cdot 1 = 120$ orders.

Otherwise. Number of orders = number of arrangements of 5 persons taken all at a time

$$= {}_5P_5 = 5! = 5 \cdot 4 \cdot 3 \cdot 2 \cdot 1 = 120 \text{ orders.}$$

25.8 In how many ways can 7 books be arranged on a shelf?

SOLUTION

Number of ways = number of arrangements of 7 books taken all at a time

$$= {}_7P_7 = 7! = 7 \cdot 6 \cdot 5 \cdot 4 \cdot 3 \cdot 2 \cdot 1 = 5040 \text{ ways.}$$

25.9 Twelve different pictures are available, of which 4 are to be hung in a row. In how many ways can this be done?

SOLUTION

The first place may be occupied by any one of 12 pictures, the second place by any one of 11, the third place by any one of 10, and the fourth place by any one of 9.

Hence the required number of ways $= 12 \cdot 11 \cdot 10 \cdot 9 = 11\,880$ ways.

Otherwise. Number of ways = number of arrangements of 12 pictures taken 4 at a time

$$= {}_{12}P_4 = 12 \cdot 11 \cdot 10 \cdot 9 = 11\,880 \text{ ways.}$$

25.10 It is required to seat 5 men and 4 women in a row so that the women occupy the even places. How many such arrangements are possible?

SOLUTION

The men may be seated in ${}_5P_5$ ways, and the women in ${}_4P_4$ ways. Each arrangement of the men may be associated with each arrangement of the women.

Hence the required number of arrangements $= {}_5P_5 \cdot {}_4P_4 = 5!4! = 120 \cdot 24 = 2880$.

25.11 In how many orders can 7 different pictures be hung in a row so that 1 specified picture is (*a*) at the center, (*b*) at either end?

SOLUTION

(*a*) Since 1 given picture is to be at the center, 6 pictures remain to be arranged in a row. Hence the number of orders $= {}_6P_6 = 6! = 720$ orders.

(*b*) After the specified picture is hung in any one of 2 ways, the remaining 6 can be arranged in ${}_6P_6$ ways.

Hence the number of orders $= 2 \cdot {}_6P_6 = 1440$ orders.

25.12 In how many ways can 9 different books be arranged on a shelf so that (*a*) 3 of the books are always together, (*b*) 3 of the books are never all 3 together?

SOLUTION

(*a*) The specified 3 books can be arranged among themselves in ${}_3P_3$ ways. Since the specified 3 books are always together, they may be considered as 1 thing. Then together with the other 6 books (things) we have a total of 7 things which can be arranged in ${}_7P_7$ ways.

Total number of ways $= {}_3P_3 \cdot {}_7P_7 = 3!7! = 6 \cdot 5040 = 30\,240$ ways.

(*b*) Number of ways in which 9 books can be arranged on a shelf if there are no restrictions $= 9! = 362\,880$ ways.

Number of ways in which 9 books can be arranged on a shelf when 3 specified books are always together (from (*a*) above) $= 3!7! = 30\,240$ ways.

Hence the number of ways in which 9 books can be arranged on a shelf so that 3 specified books are never all 3 together $= 362\,880 - 30\,240 = 332\,640$ ways.

25.13 In how many ways can n women be seated in a row so that 2 particular women will not be next to each other?

SOLUTION

With no restrictions, n women may be seated in a row in ${}_nP_n$ ways. If 2 of the n women must always sit next to each other, the number of arrangements $= 2!({}_{n-1}P_{n-1})$.

Hence the number of ways n women can be seated in a row if 2 particular women may never sit together $= {}_nP_n - 2({}_{n-1}P_{n-1}) = n! - 2(n-1)! = n(n-1)! - 2(n-1)! = (n-2) \cdot (n-1)!$

25.14 Six different biology books, 5 different chemistry books and 2 different physics books are to be arranged on a shelf so that the biology books stand together, the chemistry books stand together, and the physics books stand together. How many such arrangements are possible?

SOLUTION

The biology books can be arranged among themselves in 6! ways, the chemistry books in 5! ways, the physics books in 2! ways, and the three groups in 3! ways.

Required number of arrangements = 6!5!2!3! = 1 036 800.

25.15 Determine the number of different words of 5 letters each that can be formed with the letters of the word *chromate* (*a*) if each letter is used not more than once, (*b*) if each letter may be repeated in any arrangement. (These words need not have meaning.)

SOLUTION

(*a*) Number of words = arrangements of 8 different letters taken 5 at a time
$= {}_8P_5 = 8 \cdot 7 \cdot 6 \cdot 5 \cdot 4 = 6720$ words.

(*b*) Number of words $= 8 \cdot 8 \cdot 8 \cdot 8 \cdot 8 = 8^5 = 32\,768$ words.

25.16 How many numbers may be formed by using 4 out of the 5 digits 1,2,3,4,5 (*a*) if the digits must not be repeated in any number, (*b*) if they may be repeated? If the digits must not be repeated, how many of the 4-digit numbers (*c*) begin with 2, (*d*) end with 25?

SOLUTION

(*a*) Numbers formed $= {}_5P_4 = 5 \cdot 4 \cdot 3 \cdot 2 = 120$ numbers.

(*b*) Numbers formed $= 5 \cdot 5 \cdot 5 \cdot 5 = 5^4 = 625$ numbers.

(*c*) Since the first digit of each number is specified, there remain 4 digits to be arranged in 3 places.
Numbers formed $= {}_4P_3 = 4 \cdot 3 \cdot 2 = 24$ numbers.

(*d*) Since the last two digits of every number are specified, there remain 3 digits to be arranged in 2 places.
Numbers formed $= {}_3P_2 = 3 \cdot 2 = 6$ numbers.

25.17 How many 4-digit numbers may be formed with the 10 digits 0,1,2,3,...,9 (*a*) if each digit is used only once in each number? (*b*) How many of these numbers are odd?

SOLUTION

(*a*) The first place may be filled by any one of the 10 digits except 0, i.e., by any one of 9 digits. The 9 digits remaining may be arranged in the 3 other places in ${}_9P_3$ ways.
Numbers formed $= 9 \cdot {}_9P_3 = 9(9 \cdot 8 \cdot 7) = 4536$ numbers.

(*b*) The last place may be filled by any one of the 5 odd digits, 1,3,5,7.9. The first place may be filled by any one of the 8 digits, i.e., by the remaining 4 odd digits and the even digits, 2,4,6,8. The 8 remaining digits may be arranged in the 2 middle positions in ${}_8P_2$ ways.
Numbers formed $= 5 \cdot 8 \cdot {}_8P_2 = 5 \cdot 8 \cdot 8 \cdot 7 = 2240$ odd numbers.

25.18 (*a*) How many 5-digit numbers can be formed from the 10 digits 0,1,2,3,...,9, repetitions allowed? How many of these numbers (*b*) begin with 40, (*c*) are even, (*d*) are divisible by 5?

SOLUTION

(*a*) The first place may be filled by any one of 9 digits (any of the 10 except 0). Each of the other 4 places may be filled by any one of the 10 digits whatever.

Numbers formed $= 9 \cdot 10 \cdot 10 \cdot 10 \cdot 10 = 9 \cdot 10^4 = 90\,000$ numbers.

(*b*) The first 2 places may be filled in 1 way, by 40. The other 3 places may be filled by any one of the 10 digits whatever.

Numbers formed $= 1 \cdot 10 \cdot 10 \cdot 10 = 10^3 = 1000$ numbers.

(*c*) The first place may be filled in 9 ways, and the last place in 5 ways $(0, 2, 4, 6, 8)$. Each of the other 3 places may be filled by any one of the 10 digits whatever.

Even numbers $= 9 \cdot 10 \cdot 10 \cdot 10 \cdot 5 = 45\,000$ numbers.

(*d*) The first place may be filled in 9 ways, the last place in 2 ways $(0, 5)$, and the other 3 places in 10 ways each.

Numbers divisible by $5 = 9 \cdot 10 \cdot 10 \cdot 10 \cdot 2 = 18\,000$ numbers.

25.19 How many numbers between 3000 and 5000 can be formed by using the 7 digits 0,1,2,3,4,5,6 if each digit must not be repeated in any number?

SOLUTION

Since the numbers are between 3000 and 5000, they consist of 4 digits. The first place may be filled in 2 ways, i.e., by digits 3, 4. Then the remaining 6 digits may be arranged in the 3 other places in ${}_6P_3$ ways.

Numbers formed $= 2 \cdot {}_6P_3 = 2(6 \cdot 5 \cdot 4) = 240$ numbers.

25.20 From 11 novels and 3 dictionaries, 4 novels and 1 dictionary are to be selected and arranged on a shelf so that the dictionary is always in the middle. How many such arrangements are possible?

SOLUTION

The dictionary may be chosen in 3 ways. The number of arrangements of 11 novels taken 4 at a time is ${}_{11}P_4$.

Required number of arrangements $= 3 \cdot {}_{11}P_4 = 3(11 \cdot 10 \cdot 9 \cdot 8) = 23\,760$.

25.21 How many signals can be made with 5 different flags by raising them any number at a time?

SOLUTION

Signals may be made by raising the flags 1, 2, 3, 4, and 5 at a time. Hence the total number of signals is

$${}_5P_1 + {}_5P_2 + {}_5P_3 + {}_5P_4 + {}_5P_5 = 5 + 20 + 60 + 120 + 120 = 325 \text{ signals.}$$

25.22 Compute the sum of the 4-digit numbers which can be formed with the four digits 2,5,3,8 if each digit is used only once in each arrangement.

SOLUTION

The number of arrangements, or numbers, is ${}_4P_4 = 4! = 4 \cdot 3 \cdot 2 \cdot 1 = 24$.

The sum of the digits $= 2 + 5 + 3 + 8 = 18$, and each digit will occur $24/4 = 6$ times each in the units, tens, hundreds, and thousands positions. Hence the sum of all the numbers formed is

$$1(6 \cdot 18) + 10(6 \cdot 18) + 100(6 \cdot 18) + 1000(6 \cdot 18) = 119\,988.$$

25.23 (*a*) How many arrangements can be made from the letters of the word *cooperator* when all are taken at a time? How many of such arrangements (*b*) have the three *o*s together, (*c*) begin with the two *r*s?

SOLUTION

(*a*) The word *cooperator* consists of 10 letters: 3 *o*'s, 2 *r*'s, and 5 different letters.

$$\text{Number of arrangements} = \frac{10!}{3!2!} = \frac{10\cdot9\cdot8\cdot7\cdot6\cdot5\cdot4\cdot3\cdot2\cdot1}{(1\cdot2\cdot3)(1\cdot2)} = 302\,400.$$

(*b*) Consider the 3 *o*'s as 1 letter. Then we have 8 letters of which 2 *r*'s are alike.

$$\text{Number of arrangements} = \frac{8!}{2!} = 20\,160.$$

(*c*) The number of arrangements of the remaining 8 letters, of which 3 *o*'s are alike, $= 8!/3! = 6720$.

25.24 There are 3 copies each of 4 different books. In how many different ways can they be arranged on a shelf?

SOLUTION

There are $3\cdot4 = 12$ books of which 3 are alike, 3 others alike, etc.

$$\text{Number of arrangements} = \frac{(3\cdot4)!}{3!3!3!3!} = \frac{12!}{(3!)^4} = 369\,600.$$

25.25 (*a*) In how many ways can 5 persons be seated at a round table?

(*b*) In how many ways can 8 persons be seated at a round table if 2 particular persons must always sit together?

SOLUTION

(*a*) Let 1 of them be seated anywhere. Then the 4 persons remaining can be seated in 4! ways. Hence there are $4! = 24$ ways of arranging 5 persons in a circle.

(*b*) Consider the two particular persons as one person. Since there are 2! ways of arranging 2 persons among themselves and 6! ways of arranging 7 persons in a circle, the required number of ways $= 2!6! = 2\cdot720 = 1440$ ways.

25.26 In how many ways can 4 men and 4 women be seated at a round table if each woman is to be between two men?

SOLUTION

Consider that the men are seated first. Then the men can be arranged in 3! ways, and the women in 4! ways.

Required number of circular arrangements $= 3!4! = 144$.

25.27 By stringing together 9 differently colored beads, how many different bracelets can be made?

SOLUTION

There are 8! arrangements of the beads on the bracelet, but half of these can be obtained from the other half simply by turning the bracelet over.

Hence there are $\frac{1}{2}(8!) = 20\,160$ different bracelets.

25.28 In each case, find n: (*a*) ${}_nC_{n-2} = 10$, (*b*) ${}_nC_{15} = {}_nC_{11}$, (*c*) ${}_nP_4 = 30 \cdot {}_nC_5$.

SOLUTION

(*a*) $${}_nC_{n-2} = {}_nC_2 = \frac{n(n-1)}{2!} = \frac{n^2-n}{2} = 10, \qquad n^2 - n - 20 = 0, \qquad n = 5$$

(*b*) $${}_nC_r = {}_nC_{n-r}, \qquad {}_nC_{15} = {}_nC_{n-11}, \qquad 15 = n - 11, \qquad n = 26$$

(*c*) $$30 \cdot {}_nC_5 = 30\left(\frac{{}_nP_5}{5!}\right) = \frac{30 \cdot {}_nP_4 \cdot (n-4)}{5!}$$

Then $${}_nP_4 = \frac{30 \cdot {}_nP_4 \cdot (n-4)}{5!}, \qquad 1 = \frac{30(n-4)}{120}, \qquad n = 8.$$

25.29 Given ${}_nP_r = 3024$ and ${}_nC_r = 126$, find r.

SOLUTION

$${}_nP_r = r!({}_nC_r), \qquad r! = \frac{{}_nP_r}{{}_nC_r} = \frac{3024}{126} = 24, \qquad r = 4$$

25.30 How many different sets of 4 students can be chosen out of 17 qualified students to represent a school in a mathematics contest?

SOLUTION

Number of sets = number of combinations of 4 out of 17 students

$$= {}_{17}C_4 = \frac{17 \cdot 16 \cdot 15 \cdot 14}{1 \cdot 2 \cdot 3 \cdot 4} = 2380 \text{ sets of 4 students.}$$

25.31 In how many ways can 5 styles be selected out of 8 styles?

SOLUTION

Number of ways = number of combinations of 5 out of 8 styles

$$= {}_8C_5 = {}_8C_3 = \frac{8 \cdot 7 \cdot 6}{1 \cdot 2 \cdot 3} = 56 \text{ ways.}$$

25.32 In how many ways can 12 books be divided between A and B so that one may get 9 and the other 3 books?

SOLUTION

In each separation of 12 books into 9 and 3, A may get the 9 and B the 3, or A may get the 3 and B the 9.

$$\text{Hence the number of ways} = 2 \cdot {}_{12}C_9 = 2 \cdot {}_{12}C_3 = 2\left(\frac{12 \cdot 11 \cdot 10}{1 \cdot 2 \cdot 3}\right) = 440 \text{ ways.}$$

25.33 Determine the number of different triangles which can be formed by joining the six vertices of a hexagon, the vertices of each triangle being on the hexagon.

SOLUTION

Number of triangles = number of combinations of 3 out of 6 points

$$= {}_6C_3 = \frac{6 \cdot 5 \cdot 4}{1 \cdot 2 \cdot 3} = 20 \text{ triangles.}$$

25.34 How many angles less than 180° are formed by 12 straight lines which terminate in a point, if no two of them are in the same straight line?

SOLUTION

$$\text{Number of angles} = \text{number of combinations of 2 out of 12 lines}$$
$$= {}_{12}C_2 = \frac{12\cdot 11}{1\cdot 2} = 66 \text{ angles.}$$

25.35 How many diagonals has an octagon?

SOLUTION

$$\text{Lines formed} = \text{number of combinations of 2 out of 8 corners (points)} = {}_8C_2 = \frac{8\cdot 7}{2} = 28.$$

Since 8 of these 28 lines are the sides of the octagon, the number of diagonals = 20.

25.36 How many parallelograms are formed by a set of 4 parallel lines intersecting another set of 7 parallel lines?

SOLUTION

Each combination of 2 lines out of 4 can intersect each combination of 2 lines out of 7 to form a parallelogram.

Number of parallelograms = ${}_4C_2\cdot{}_7C_2 = 6\cdot 21 = 126$ parallelograms.

25.37 There are 10 points in a plane. No three of these points are in a straight line, except 4 points which are all in the same straight line. How many straight lines can be formed by joining the 10 points?

SOLUTION

$$\text{Number of lines formed if no 3 of the 10 points were in a straight line} = {}_{10}C_2 = \frac{10\cdot 9}{2} = 45.$$

$$\text{Number of lines formed by 4 points, no 3 of which are collinear} = {}_4C_2 = \frac{4\cdot 3}{2} = 6.$$

Since the 4 points are collinear, they form 1 line instead of 6 lines.
Required number of lines = $45 - 6 + 1 = 40$ lines.

25.38 In how many ways can 3 women be selected out of 15 women

(*a*) if 1 of the women is to be included in every selection,

(*b*) if 2 of the women are to be excluded from every selection,

(*c*) if 1 is always included and 2 are always excluded?

SOLUTION

(*a*) Since 1 is always included, we must select 2 out of 14 women.

$$\text{Hence the number of ways} = {}_{14}C_2 = \frac{14\cdot 13}{2} = 91 \text{ ways.}$$

(*b*) Since 2 are always excluded, we must select 3 out of 13 women.

$$\text{Hence the number of ways} = {}_{13}C_3 = \frac{13\cdot 12\cdot 11}{3!} = 286 \text{ ways.}$$

(c) Number of ways $= {}_{15-1-2}C_{3-1} = {}_{12}C_2 = \dfrac{12 \cdot 11}{2} = 66$ ways.

25.39 An organization has 25 members, 4 of whom are doctors. In how many ways can a committee of 3 members be selected so as to include at least 1 doctor?

SOLUTION

Total number of ways in which 3 can be selected out of $25 = {}_{25}C_3$.

Number of ways in which 3 can be selected so that no doctor is included $= {}_{25-4}C_3 = {}_{21}C_3$.

Then the number of ways in which 3 members can be selected so that at least 1 doctor is included is

$$ {}_{25}C_3 - {}_{21}C_3 = \frac{25 \cdot 24 \cdot 23}{3!} - \frac{21 \cdot 20 \cdot 19}{3!} = 970 \text{ ways.} $$

25.40 From 6 chemists and 5 biologists, a committee of 7 is to be chosen so as to include 4 chemists. In how many ways can this be done?

SOLUTION

Each selection of 4 out of 6 chemists can be associated with each selection of 3 out of 5 biologists.

Hence the number of ways $= {}_6C_4 \cdot {}_5C_3 = {}_6C_2 \cdot {}_5C_2 = 15 \cdot 10 = 150$ ways.

25.41 Given 8 consonants and 4 vowels, how many 5-letter words can be formed, each word consisting of 3 different consonants and 2 different vowels?

SOLUTION

The 3 different consonants can be selected in ${}_8C_3$ ways, the 2 different vowels in ${}_4C_2$ ways, and the 5 different letters (3 consonants, 2 vowels) can be arranged among themselves in ${}_5P_5 = 5!$ ways.

Hence the number of words $= {}_8C_3 \cdot {}_4C_2 \cdot 5! = 56 \cdot 6 \cdot 120 = 40\,320$.

25.42 From 7 capitals, 3 vowels and 5 consonants, how many words of 4 letters each can be formed if each word begins with a capital and contains at least 1 vowel, all the letters of each word being different?

SOLUTION

The first letter, or capital, may be selected in 7 ways.

The remaining 3 letters may be

(a) 1 vowel and 2 consonants, which may be selected in ${}_3C_1 \cdot {}_5C_2$ ways,

(b) 2 vowels and 1 consonant, which may be selected in ${}_3C_2 \cdot {}_5C_1$ ways, and

(c) 3 vowels, which may be selected in ${}_3C_3 = 1$ way.

Each of these selections of 3 letters may be arranged among themselves in ${}_3P_3 = 3!$ ways.

$$ \begin{aligned} \text{Hence the number of words} &= 7 \cdot 3!({}_3C_1 \cdot {}_5C_2 + {}_3C_2 \cdot {}_5C_1 + 1) \\ &= 7 \cdot 6(3 \cdot 10 + 3 \cdot 5 + 1) = 1932 \text{ words.} \end{aligned} $$

25.43 A has 3 maps and B has 9 maps. Determine the number of ways in which they can exchange maps if each keeps his initial number of maps.

SOLUTION

A can exchange 1 map with B in ${}_3C_1 \cdot {}_9C_1 = 3 \cdot 9 = 27$ ways.
A can exchange 2 maps with B in ${}_3C_2 \cdot {}_9C_2 = 3 \cdot 36 = 108$ ways.
A can exchange 3 maps with B in ${}_3C_3 \cdot {}_9C_3 = 1 \cdot 84 = 84$ ways.
Total number of ways $= 27 + 108 + 84 = 219$ ways.

Another method. Consider that A and B put their maps together. Then the problem is to find the number of ways A can select 3 maps out of 12, not including the selection by A of his original three maps.

$$\text{Hence, } {}_{12}C_3 - 1 = \frac{12 \cdot 11 \cdot 10}{1 \cdot 2 \cdot 3} - 1 = 219 \text{ ways.}$$

25.44 (*a*) In how many ways can 12 books be divided among 3 students so that each receives 4 books?

(*b*) In how many ways can 12 books be divided into 3 groups of 4 each?

SOLUTION

(*a*) The first student can select 4 out of 12 books in ${}_{12}C_4$ ways.
The second student can select 4 of the remaining 8 books in ${}_8C_4$ ways.
The third student can select 4 of the remaining 4 books in 1 way.

$$\text{Number of ways} = {}_{12}C_4 \cdot {}_8C_4 \cdot 1 = 495 \cdot 70 \cdot 1 = 34\,650 \text{ ways.}$$

(*b*) The 3 groups could be distributed among the students in $3! = 6$ ways.

$$\text{Hence the number of groups} = 34\,650/3! = 5775 \text{ groups.}$$

25.45 In how many ways can a person choose 1 or more of 4 electrical appliances?

SOLUTION

Each appliance may be dealt with in 2 ways, as it can be chosen or not chosen. Since each of the 2 ways of dealing with an appliance is associated with 2 ways of dealing with each of the other appliances, the number of ways of dealing with the 4 appliances $= 2 \cdot 2 \cdot 2 \cdot 2 = 2^4$ ways. But 2^4 ways includes the case in which no appliance is chosen.

Hence the required number of ways $= 2^4 - 1 = 16 - 1 = 15$ ways.

Another method. The appliances may be chosen singly, in twos, etc. Hence the required number of ways $= {}_4C_1 + {}_4C_2 + {}_4C_3 + {}_4C_4 = 4 + 6 + 4 + 1 = 15$ ways.

25.46 How many different sums of money can be drawn from a wallet containing one bill each of 1, 2, 5, 10, 20 and 50 dollars?

SOLUTION

$$\text{Number of sums} = 2^6 - 1 = 63 \text{ sums.}$$

25.47 In how many ways can 2 or more ties be selected out of 8 ties?

SOLUTION

One or more ties may be selected in $(2^8 - 1)$ ways. But since 2 or more must be chosen, the required number of ways $= 2^8 - 1 - 8 = 247$ ways.

Another method. 2, 3, 4, 5, 6, 7, or 8 ties may be selected in

$$\begin{aligned} {}_8C_2 + {}_8C_3 + {}_8C_4 + {}_8C_5 + {}_8C_6 + {}_8C_7 + {}_8C_8 &= {}_8C_2 + {}_8C_3 + {}_8C_4 + {}_8C_3 + {}_8C_2 + {}_8C_1 + 1 \\ &= 28 + 56 + 70 + 56 + 28 + 8 + 1 = 247 \text{ ways.} \end{aligned}$$

25.48 There are available 5 different green dyes, 4 different blue dyes, and 3 different red dyes. How many selections of dyes can be made, taking at least 1 green and 1 blue dye?

SOLUTION

The green dyes can be chosen in $(2^5 - 1)$ ways, the blue dyes in $(2^4 - 1)$ ways, and the red dyes in 2^3 ways.

$$\text{Number of selections} = (2^5 - 1)(2^4 - 1)(2^3) = 31 \cdot 15 \cdot 8 = 3720 \text{ selections.}$$

Supplementary Problems

25.49 Evaluate: ${}_{16}P_3$, ${}_7P_4$, ${}_5P_5$, ${}_{12}P_1$.

25.50 Find n if (*a*) $10 \cdot {}_nP_2 = {}_{n+1}P_4$, (*b*) $3 \cdot {}_{2n+4}P_3 = 2 \cdot {}_{n+4}P_4$.

25.51 In how many ways can six people be seated on a bench?

25.52 With four signal flags of different colors, how many different signals can be made by displaying two flags one above the other?

25.53 With six signal flags of different colors, how many different signals can be made by displaying three flags one above the other?

25.54 In how many ways can a club consisting of 12 members choose a president, a secretary, and a treasurer?

25.55 If no two books are alike, in how many ways can 2 red, 3 green, and 4 blue books be arranged on a shelf so that all the books of the same color are together?

25.56 There are 4 hooks on a wall. In how many ways can 3 coats be hung on them, one coat on a hook?

25.57 How many two-digit numbers can be formed with the digits 0, 3, 5, 7 if no repetition in any of the numbers is allowed?

25.58 How many even numbers of two different digits can be formed from the digits 3, 4, 5, 6, 8?

25.59 How many three-digit numbers can be formed from the digits 1, 2, 3, 4, 5 if no digit is repeated in any number?

25.60 How many numbers of three digits each can be written with the digits 1, 2, . . . , 9 if no digit is repeated in any number?

25.61 How many three-digit numbers can be formed from the digits 3, 4, 5, 6, 7 if digits are allowed to be repeated?

25.62 How many odd numbers of three digits each can be formed, without the repetition of any digit in a number, from the digits (*a*) 1, 2, 3, 4, (*b*) 1, 2, 4, 6, 8?

25.63 How many even numbers of four different digits each can be formed from the digits 3, 5, 6, 7, 9?

25.64 How many different numbers of 5 digits each can be formed from the digits 2, 3, 5, 7, 9 if no digit is repeated?

25.65 How many integers are there between 100 and 1000 in which no digit is repeated?

25.66 How many integers greater than 300 and less than 1000 can be made with the digits 1, 2, 3, 4, 5 if no digit is repeated in any number?

25.67 How many numbers between 100 and 1000 can be written with the digits 0, 1, 2, 3, 4 if no digit is repeated in any number?

25.68 How many four-digit numbers greater than 2000 can be formed with the digits 1, 2, 3, 4 if repetitions (*a*) are not allowed, (*b*) are allowed?

25.69 How many of the arrangements of the letters of the word *logarithm* begin with a vowel and end with a consonant?

25.70 In a telephone system four different letters P, R, S, T and the four digits 3, 5, 7, 8 are used. Find the maximum number of "telephone numbers" the system can have if each consists of a letter followed by a four-digit number in which the digits may be repeated.

25.71 In how many ways can 3 girls and 3 boys be seated in a row, if no two girls and no two boys are to occupy adjacent seats?

25.72 How many Morse code characters could be made by using three dots and two dashes in each character?

25.73 In how many ways can three dice fall?

25.74 How many fraternities can be named with the 24 letters of the Greek alphabet if each has three letters and none is repeated in any name?

25.75 How many signals can be shown with 8 flags of which 2 are red, 3 white and 3 blue, if they are all strung up on a vertical pole at once?

25.76 In how many ways can 4 men and 4 women sit at a round table so that no two men are adjacent?

25.77 How many different arrangements are possible with the factors of the term $a^2b^4c^5$ written at full length?

25.78 In how many ways can 9 different prizes be awarded to two students so that one receives 3 and the other 6?

25.79 How many different radio stations can be named with 3 different letters of the alphabet? How many with 4 different letters in which W must come first?

25.80 In each case find n: (*a*) $4 \cdot {}_nC_2 = {}_{n+2}C_3$, (*b*) ${}_{n+2}C_n = 45$, (*c*) ${}_nC_{12} = {}_nC_8$.

25.81 If $5 \cdot {}_nP_3 = 24 \cdot {}_nC_4$, find n.

25.82 Evaluate (*a*) ${}_7C_7$, (*b*) ${}_5C_3$, (*c*) ${}_7C_2$, (*d*) ${}_7C_5$, (*e*) ${}_7C_6$, (*f*) ${}_8C_7$, (*g*) ${}_8C_5$, (*h*) ${}_{100}C_{98}$.

25.83 How many straight lines are determined by (*a*) 6, (*b*) n points, no three of which lie in the same straight line?

25.84 How many chords are determined by seven points on a circle?

25.85 A student is allowed to choose 5 questions out of 9. In how many ways can she choose them?

25.86 How many different sums of money can be formed by taking two of the following: a cent, a nickel, a dime, a quarter, a half-dollar?

25.87 How many different sums of money can be formed from the coins of Problem 25.86?

25.88 A baseball league is made up of 6 teams. If each team is to play each of the other teams (*a*) twice, (*b*) three times, how many games will be played?

25.89 How many different committees of two men and one woman can be formed from (*a*) 7 men and 4 women, (*b*) 5 men and 3 women?

25.90 In how many ways can 5 colors be selected out of 8 different colors including red, blue, and green

(*a*) if blue and green are always to be included,

(*b*) if red is always excluded,

(*c*) if red and blue are always included but green excluded?

25.91 From 5 physicists, 4 chemists, and 3 mathematicians a committee of 6 is to be chosen so as to include 2 physicists, 2 chemists, and 1 mathematician. In how many ways can this be done?

25.92 In Problem 25.91, in how many ways can the committee of 6 be chosen so that

(*a*) 2 members of the committee are mathematicians.

(*b*) at least 3 members of the committee are physicists?

25.93 How many words of 2 vowels and 3 consonants may be formed (considering any set a word) from the letters of the word (*a*) *stenographic*, (*b*) *facetious*?

25.94 In how many ways can a picture be colored if 7 different colors are available for use?

25.95 In how many ways can 8 women form a committee if at least 3 women are to be on the committee?

25.96 A box contains 7 red cards, 6 white cards and 4 blue cards. How many selections of three cards can be made so that (*a*) all three are red, (*b*) none is red?

25.97 How many baseball nines can be chosen from 13 candidates if A, B, C, D are the only candidates for two positions and can play no other position?

25.98 How many different committees including 3 Democrats and 2 Republicans can be chosen from 8 Republicans and 10 Democrats?

25.99 At a meeting, after everyone had shaken hands once with everyone else, it was found that 45 handshakes were exchanged. How many were at the meeting?

25.100 Find the number of (*a*) combinations and (*b*) permutations of four letters each that can be made from the letters of the word TENNESSEE.

ANSWERS TO SUPPLEMENTARY PROBLEMS

25.49 3360, 840, 120, 12

25.50 (*a*) 4, (*b*) 6

25.51 720

25.52 12

25.53 120

25.54 1320

25.55 1728
25.56 24
25.57 9
25.58 12
25.59 60
25.60 504
25.61 125
25.62 (*a*) 12, (*b*) 12
25.63 24
25.64 120
25.65 648
25.66 36
25.67 48
25.68 (*a*) 18, (*b*) 192
25.69 90 720
25.70 1024
25.71 72
25.72 10
25.73 216
25.74 12, 144
25.75 560
25.76 144
25.77 1260
25.78 168
25.79 15 600; 13 800
25.80 (*a*) 2, 7, (*b*) 8, (*c*) 20
25.81 8
25.82 (*a*) 1, (*b*) 10, (*c*) 21, (*d*) 21, (*e*) 7, (*f*) 8, (*g*) 56, (*h*) 4950
25.83 (*a*) 15, (*b*) $\dfrac{n(n-1)}{2}$
25.84 21
25.85 126
25.86 10
25.87 31
25.88 (*a*) 30, (*b*) 45
25.89 (*a*) 84, (*b*) 30
25.90 (*a*) 20, (*b*) 21, (*c*) 10
25.91 180
25.92 (*a*) 378, (*b*) 462
25.93 (*a*) 40 320, (*b*) 4800
25.94 127
25.95 219
25.96 (*a*) 35, (*b*) 120
25.97 216
25.98 3360
25.99 10
25.100 (*a*) 17, (*b*) 163

Chapter 26

The Binomial Theorem

26.1 COMBINATORIAL NOTATION

The number of combinations of n objects selected r at a time, ${}_nC_r$, can be written in the form

$$\binom{n}{r}$$

which is called combinatorial notation.

$${}_nC_r = \frac{n!}{(n-r)!r!} = \binom{n}{r},$$

where n and r are integers and $r \leq n$.

EXAMPLES 26.1. Evaluate each expression.

(a) $\binom{7}{3}$ (b) $\binom{8}{7}$ (c) $\binom{9}{9}$ (d) $\binom{5}{0}$

(a) $\binom{7}{3} = \dfrac{7!}{(7-3)!3!} = \dfrac{7!}{4!3!} = \dfrac{7 \cdot 6 \cdot 5 \cdot 4!}{4!3 \cdot 2 \cdot 1} = 7 \cdot 5 = 35$

(b) $\binom{8}{7} = \dfrac{8!}{(8-7)!7!} = \dfrac{8!}{1!7!} = \dfrac{8 \cdot 7!}{17!} = 8$

(c) $\binom{9}{9} = \dfrac{9!}{(9-9)!9!} = \dfrac{9!}{0!9!} = \dfrac{1}{0!} = \dfrac{1}{1} = 1$

(d) $\binom{5}{0} = \dfrac{5!}{(5-0)!0!} = \dfrac{5!}{5!0!} = \dfrac{1}{0!} = \dfrac{1}{1} = 1$

26.2 EXPANSION OF $(a + x)^n$

If n is a positive integer we expand $(a + x)^n$ as shown below:

$$(a+x)^n = a^n + na^{n-1}x + \frac{n(n-1)}{2!}a^{n-2}x^2 + \frac{n(n-1)(n-2)}{3!}a^{n-3}x^3$$
$$+ \cdots + \frac{n(n-1)(n-2)\cdots(n-r+2)}{(r-1)!}a^{n-r+1}x^{r-1} + \cdots + x^n$$

This equation is called the binomial theorem, or binomial formula.

Other forms of the binomial theorem exist and some use combinations to express the coefficients. The relationship between the coefficients and combinations are shown below.

$$\frac{5 \cdot 4}{2!} = \frac{5 \cdot 4 \cdot 3 \cdot 2 \cdot 1}{3 \cdot 2 \cdot 1 \cdot 2!} = \frac{5!}{3!2!} = \frac{5!}{(5-2)!2!} = \binom{5}{2}$$

$$\frac{n(n-1)(n-2)}{3!} = \frac{n(n-1)(n-2)\cdots 2 \cdot 1}{(n-3)!3!} = \frac{n!}{(n-3)!3!} = \binom{n}{3}$$

So

$$(a+x)^n = a^n + \frac{n!}{(n-1)!1!}a^{n-1}x + \frac{n!}{(n-2)!2!}a^{n-2}x^2 + \cdots$$
$$+ \frac{n!}{(n-[r-1])!(r-1)!}a^{n-r+1}x^{r-1} + \cdots + x^n$$

and $$(a+x)^n = a^n + \binom{n}{1}a^{n-1}x + \binom{n}{2}a^{n-2}x^2 + \cdots + \binom{n}{r-1}a^{n-r+1}x^{r-1} + \cdots + x^n$$

The rth term of the expansion of $(a+x)^n$ is

$$r\text{th term} = \frac{n(n-1)(n-2)\cdots(n-r+2)}{(r-1)!}a^{n-r+1}x^{r-1}.$$

The rth term formula for the expansion of $(a+x)^n$ can be expressed in terms of combinations.

$$r\text{th term} = \frac{n(n-1)(n-2)\cdots(n-r+2)}{(r-1)!}a^{n-r+1}x^{r-1}$$
$$= \frac{n(n-1)(n-2)\cdots(n-r+2)(n-r+1)\cdots 2\cdot 1}{(n-r+1)(n-r)\cdots 2\cdot 1(r-1)!}a^{n-r+1}x^{r-1}$$
$$r\text{th term} = \frac{n!}{(n-[r-1])!(r-1)!}\, a^{n-r+1}x^{r-1}$$
$$r\text{th term} = \binom{n}{r-1}a^{n-r+1}x^{r-1}$$

Solved Problems

26.1 Evaluate each expression.

(*a*) $\binom{10}{2}$ (*b*) $\binom{10}{8}$ (*c*) $\binom{12}{10}$ (*d*) $\binom{170}{170}$

SOLUTION

(*a*) $\binom{10}{2} = \frac{10!}{(10-2)!2!} = \frac{10!}{8!2!} = \frac{10\cdot 9\cdot 8!}{8!\cdot 2\cdot 1} = 45$

(*b*) $\binom{10}{8} = \frac{10!}{(10-8)!8!} = \frac{10!}{2!8!} = \frac{10\cdot 9\cdot 8!}{2\cdot 1\cdot 8!} = 45$

(*c*) $\binom{12}{10} = \frac{12!}{(12-10)!10!} = \frac{12!}{2!\cdot 10!} = \frac{12\cdot 11\cdot 10!}{2\cdot 1\cdot 10!} = 66$

(*d*) $\binom{170}{170} = \frac{170}{(170-170)!170!} = \frac{170!}{0!\cdot 170!} = \frac{1}{0!} = \frac{1}{1} = 1$

Expand by the binomial formula.

26.2 $(a+x)^3 = a^3 + 3a^2x + \frac{3\cdot 2}{1\cdot 2}ax^2 + \frac{3\cdot 2\cdot 1}{1\cdot 2\cdot 3}x^3 = a^3 + 3a^2x + 3ax^2 + x^3$

26.3 $(a+x)^4 = a^4 + 4a^3x + \frac{4\cdot 3}{1\cdot 2}a^2x^2 + \frac{4\cdot 3\cdot 2}{1\cdot 2\cdot 3}ax^3 + \frac{4\cdot 3\cdot 2\cdot 1}{1\cdot 2\cdot 3\cdot 4}x^4 = a^4 + 4a^3x + 6a^2x^2 + 4ax^3 + x^4$

26.4 $(a+x)^5 = a^5 + 5a^4x + \frac{5\cdot 4}{1\cdot 2}a^3x^2 + \frac{5\cdot 4\cdot 3}{1\cdot 2\cdot 3}a^2x^3 + \frac{5\cdot 4\cdot 3\cdot 2}{1\cdot 2\cdot 3\cdot 4}ax^4 + x^5 = a^5 + 5a^4x + 10a^3x^2 + 10a^2x^3 + 5ax^4 + x^5$

Note that in the expansion of $(a+x)^n$:

(1) The exponent of a + the exponent of $x = n$ (i.e., the degree of each term is n).

(2) The number of terms is $n+1$, when n is a positive integer.

(3) There are *two* middle terms when n is an odd positive integer.

(4) There is only *one* middle term when n is an even positive integer.

(5) The coefficients of the terms which are equidistant from the ends are the same. It is interesting to note that these coefficients may be arranged as follows.

$$\begin{array}{ll}
(a+x)^0 & 1 \\
(a+x)^1 & 1\quad 1 \\
(a+x)^2 & 1\quad 2\quad 1 \\
(a+x)^3 & 1\quad 3\quad 3\quad 1 \\
(a+x)^4 & 1\quad 4\quad 6\quad 4\quad 1 \\
(a+x)^5 & 1\quad 5\quad 10\quad 10\quad 5\quad 1 \\
(a+x)^6 & 1\quad 6\quad 15\quad 20\quad 15\quad 6\quad 1 \\
\text{etc.} &
\end{array}$$

This array of numbers is known as *Pascal's Triangle*. The first and last number n in each row are 1, while any other number in the array can be obtained by adding the two numbers to the right and left of it in the preceding row.

26.5
$$(x-y^2)^6 = x^6 + 6x^5(-y^2) + \frac{6\cdot 5}{1\cdot 2}x^4(-y^2)^2 + \frac{6\cdot 5\cdot 4}{1\cdot 2\cdot 3}x^3(-y^2)^3 + \frac{6\cdot 5\cdot 4\cdot 3}{1\cdot 2\cdot 3\cdot 4}x^2(-y^2)^4$$
$$+ \frac{6\cdot 5\cdot 4\cdot 3\cdot 2}{1\cdot 2\cdot 3\cdot 4\cdot 5}x(-y^2)^5 + (-y^2)^6$$
$$= x^6 - 6x^5y^2 + 15x^4y^4 - 20x^3y^6 + 15x^2y^8 - 6xy^{10} + y^{12}$$

In the expansion of a binomial of the form $(a-b)^n$, where n is a positive integer, the terms are alternately + and −.

26.6
$$(3a^3-2b)^4 = (3a^3)^4 + 4(3a^3)^3(-2b) + \frac{4\cdot 3}{1\cdot 2}(3a^3)^2(-2b)^2 + \frac{4\cdot 3\cdot 2}{1\cdot 2\cdot 3}(3a^3)(-2b)^3 + (-2b)^4$$
$$= 81a^{12} - 216a^9b + 216a^6b^2 - 96a^3b^3 + 16b^4$$

26.7
$$(x-1)^7 = x^7 + 7x^6(-1) + \frac{7\cdot 6}{1\cdot 2}x^5(-1)^2 + \frac{7\cdot 6\cdot 5}{1\cdot 2\cdot 3}x^4(-1)^3 + \frac{7\cdot 6\cdot 5\cdot 4}{1\cdot 2\cdot 3\cdot 4}x^3(-1)^4 + \frac{7\cdot 6\cdot 5\cdot 4\cdot 3}{1\cdot 2\cdot 3\cdot 4\cdot 5}x^2(-1)^5$$
$$+ \frac{7\cdot 6\cdot 5\cdot 4\cdot 3\cdot 2}{1\cdot 2\cdot 3\cdot 4\cdot 5\cdot 6}x(-1)^6 + (-1)^7$$
$$= x^7 - 7x^6 + 21x^5 - 35x^4 + 35x^3 - 21x^2 + 7x - 1$$

26.8 $(1-2x)^5 = 1 + 5(-2x) + \frac{5\cdot 4}{1\cdot 2}(-2x)^2 + \frac{5\cdot 4\cdot 3}{1\cdot 2\cdot 3}(-2x)^3 + \frac{5\cdot 4\cdot 3\cdot 2}{1\cdot 2\cdot 3\cdot 4}(-2x)^4 + (-2x)^5$

$= 1 - 10x + 40x^2 - 80x^3 + 80x^4 - 32x^5$

26.9 $\left(\frac{x}{3}+\frac{2}{y}\right)^4 = \left(\frac{x}{3}\right)^4 + 4\left(\frac{x}{3}\right)^3\left(\frac{2}{y}\right) + \frac{4\cdot 3}{1\cdot 2}\left(\frac{x}{3}\right)^2\left(\frac{2}{y}\right)^2 + \frac{4\cdot 3\cdot 2}{1\cdot 2\cdot 3}\left(\frac{x}{3}\right)\left(\frac{2}{y}\right)^3 + \left(\frac{2}{y}\right)^4$

$= \frac{x^4}{81} + \frac{8x^3}{27y} + \frac{8x^2}{3y^2} + \frac{32x}{3y^3} + \frac{16}{y^4}$

26.10 $(\sqrt{x}+\sqrt{y})^6 = (x^{1/2})^6 + 6(x^{1/2})^5(y^{1/2}) + \frac{6\cdot 5}{1\cdot 2}(x^{1/2})^4(y^{1/2})^2 + \frac{6\cdot 5\cdot 4}{1\cdot 2\cdot 3}(x^{1/2})^3(y^{1/2})^3$

$+ \frac{6\cdot 5\cdot 4\cdot 3}{1\cdot 2\cdot 3\cdot 4}(x^{1/2})^2(y^{1/2})^4 + \frac{6\cdot 5\cdot 4\cdot 3\cdot 2}{1\cdot 2\cdot 3\cdot 4\cdot 5}(x^{1/2})(y^{1/2})^5 + (y^{1/2})^6$

$= x^3 + 6x^{5/2}y^{1/2} + 15x^2y + 20x^{3/2}y^{3/2} + 15xy^2 + 6x^{1/2}y^{5/2} + y^3$

26.11 $\left(\sqrt{x} - \frac{1}{\sqrt{x}}\right)^4 = (x^{1/2} - x^{-1/2})^4$

$= (x^{1/2})^4 + 4(x^{1/2})^3(-x^{-1/2}) + \frac{4\cdot 3}{1\cdot 2}(x^{1/2})^2(-x^{-1/2})^2$

$+ \frac{4\cdot 3\cdot 2}{1\cdot 2\cdot 3}(x^{1/2})(-x^{-1/2})^3 + (-x^{-1/2})^4 = x^2 - 4x + 6 - 4x^{-1} + x^{-2}$

26.12 $(a^{-2} + b^{3/2})^4 = (a^{-2})^4 + 4(a^{-2})^3(b^{3/2}) + \frac{4\cdot 3}{1\cdot 2}(a^{-2})^2(b^{3/2})^2 + \frac{4\cdot 3\cdot 2}{1\cdot 2\cdot 3}(a^{-2})(b^{3/2})^3 + (b^{3/2})^4$

$= a^{-8} + 4a^{-6}b^{3/2} + 6a^{-4}b^3 + 4a^{-2}b^{9/2} + b^6$

26.13 $(e^x - e^{-x})^7 = (e^x)^7 + 7(e^x)^6(-e^{-x}) + \frac{7\cdot 6}{1\cdot 2}(e^x)^5(-e^{-x})^2 + \frac{7\cdot 6\cdot 5}{1\cdot 2\cdot 3}(e^x)^4(-e^{-x})^3$

$+ \frac{7\cdot 6\cdot 5\cdot 4}{1\cdot 2\cdot 3\cdot 4}(e^x)^3(-e^{-x})^4 + \frac{7\cdot 6\cdot 5\cdot 4\cdot 3}{1\cdot 2\cdot 3\cdot 4\cdot 5}(e^x)^2(-e^{-x})^5$

$+ \frac{7\cdot 6\cdot 5\cdot 4\cdot 3\cdot 2}{1\cdot 2\cdot 3\cdot 4\cdot 5\cdot 6}(e^x)(-e^{-x})^6 + (-e^{-x})^7$

$= e^{7x} - 7e^{5x} + 21e^{3x} - 35e^x + 35e^{-x} - 21e^{-3x} + 7e^{-5x} - e^{-7x}$

26.14 $(a+b-c)^3 = [(a+b)-c]^3 = (a+b)^3 + 3(a+b)^2(-c) + \frac{3\cdot 2}{1\cdot 2}(a+b)(-c)^2 + (-c)^3$

$= a^3 + 3a^2b + 3ab^2 + b^3 - 3a^2c - 6abc - 3b^2c + 3ac^2 + 3bc^2 - c^3$

26.15 $(x^2+x-3)^3 = [x^2 + (x-3)]^3 = (x^2)^3 + 3(x^2)^2(x-3) + \frac{3\cdot 2}{1\cdot 2}(x^2)(x-3)^2 + (x-3)^3$

$= x^6 + (3x^5 - 9x^4) + (3x^4 - 18x^3 + 27x^2) + (x^3 - 9x^2 + 27x - 27)$

$= x^6 + 3x^5 - 6x^4 - 17x^3 + 18x^2 + 27x - 27$

In Problems 26.16–26.21, write the indicated term of each expansion, using the formula

$$r\text{th term of } (a+x)^n = \frac{n(n-1)(n-2)\cdots(n-r+2)}{(r-1)!}a^{n-r+1}x^{r-1}.$$

26.16 Sixth term of $(x+y)^{15}$.

SOLUTION

$$n=15,\ r=6,\ n-r+2=11,\ r-1=5,\ n-r+1=10$$

$$6\text{th term} = \frac{15\cdot 14\cdot 13\cdot 12\cdot 11}{1\cdot 2\cdot 3\cdot 4\cdot 5}x^{10}y^5 = 3003x^{10}y^5$$

26.17 Fifth term of $(a-\sqrt{b})^9$.

SOLUTION

$$n=9,\ r=5,\ n-r+2=6,\ r-1=4,\ n-r+1=5$$

$$5\text{th term} = \frac{9\cdot 8\cdot 7\cdot 6}{1\cdot 2\cdot 3\cdot 4}a^5(-\sqrt{b})^4 = 126a^5b^2$$

26.18 Fourth term of $(x^2-y^2)^{11}$.

SOLUTION

$$n=11,\ r=4,\ n-r+2=9,\ r-1=3,\ n-r+1=8$$

$$4\text{th term} = \frac{11\cdot 10\cdot 9}{1\cdot 2\cdot 3}(x^2)^8(-y^2)^3 = -165x^{16}y^6$$

26.19 Ninth term of $\left(\dfrac{x}{2}+\dfrac{1}{x}\right)^{12}$.

SOLUTION

$$n=12,\ r=9,\ n-r+2=5,\ r-1=8,\ n-r+1=4$$

$$9\text{th term} = \frac{12\cdot 11\cdot 10\cdot 9\cdot 8\cdot 7\cdot 6\cdot 5}{1\cdot 2\cdot 3\cdot 4\cdot 5\cdot 6\cdot 7\cdot 8}\left(\frac{x}{2}\right)^4\left(\frac{1}{x}\right)^8 = \frac{495}{16x^4}$$

26.20 Eighteenth term of $\left(1-\dfrac{1}{x}\right)^{20}$.

SOLUTION

$$n=20,\ r=18,\ n-r+2=4,\ r-1=17,\ n-r+1=3$$

$$18\text{th term} = \frac{20\cdot 19\cdot 18\cdot 17\cdots 4}{1\cdot 2\cdot 3\cdot 4\cdots 17}\left(-\frac{1}{x}\right)^{17} = -\frac{20\cdot 19\cdot 18}{1\cdot 2\cdot 3x^{17}} = -\frac{1140}{x^{17}}$$

26.21 Middle (4th) term of $(x^{1/3}-\frac{1}{2}x^{-2})^6$.

SOLUTION

$$n=6,\ r=4,\ n-r+2=4,\ r-1=3,\ n-r+1=3$$

$$4\text{th term} = \frac{6\cdot 5\cdot 4}{1\cdot 2\cdot 3}(x^{1/3})^3(-\tfrac{1}{2}x^{-2})^3 = 20(x)(-\tfrac{1}{8}x^{-6}) = -\frac{5}{2x^5}$$

26.22 Find the term involving x^2 in the expansion of

$$\left(x^3 + \frac{a}{x}\right)^{10}.$$

SOLUTION

From $(x^3)^{10-r+1}(x^{-1})^{r-1} = x^2$ we obtain $3(10 - r + 1) - 1(r - 1) = 2$ or $r = 8$.
For the 8th term: $n = 10$, $r = 8$, $n - r + 2 = 4$, $r - 1 = 7$, $n - r + 1 = 3$.

$$8\text{th term} = \frac{10\cdot 9\cdot 8\cdot 7\cdot 6\cdot 5\cdot 4}{1\cdot 2\cdot 3\cdot 4\cdot 5\cdot 6\cdot 7}(x^3)^3\left(\frac{a}{x}\right)^7 = 120a^7x^2$$

26.23 Find the term independent of x in the expansion of

$$\left(x^2 - \frac{1}{x}\right)^9.$$

SOLUTION

From $(x^2)^{9-r+1}(x^{-1})^{r-1} = x^0$ we obtain $2(9 - r + 1) - 1(r - 1) = 0$ or $r = 7$.
For the 7th term: $n = 9$, $r = 7$, $n - r + 2 = 4$, $r - 1 = 6$, $n - r + 1 = 3$.

$$7\text{th term} = \frac{9\cdot 8\cdot 7\cdot 6\cdot 5\cdot 4}{1\cdot 2\cdot 3\cdot 4\cdot 5\cdot 6}(x^2)^3(-x^{-1})^6 = 84$$

26.24 Evaluate $(1.03)^{10}$ to five significant figures.

SOLUTION

$$\begin{aligned}(1.03)^{10} = (1 + 0.03)^{10} &= 1 + 10(0.03) + \frac{10\cdot 9}{1\cdot 2}(0.03)^2 + \frac{10\cdot 9\cdot 8}{1\cdot 2\cdot 3}(0.03)^3 + \frac{10\cdot 9\cdot 8\cdot 7}{1\cdot 2\cdot 3\cdot 4}(0.03)^4 + \cdots \\ &= 1 + 0.3 + 0.0405 + 0.003\,24 + 0.000\,17 + \cdots = 1.3439\end{aligned}$$

Note that all 11 terms of the expansion of $(0.03 + 1)^{10}$ would be required in order to evaluate $(1.03)^{10}$.

26.25 Evaluate $(0.99)^{15}$ to four decimal places.

SOLUTION

$$\begin{aligned}(0.99)^{15} = (1 - 0.01)^{15} &= 1 + 15(-0.01) + \frac{15\cdot 14}{1\cdot 2}(-0.01)^2 + \frac{15\cdot 14\cdot 13}{1\cdot 2\cdot 3}(-0.01)^3 \\ &\quad + \frac{15\cdot 14\cdot 13\cdot 12}{1\cdot 2\cdot 3\cdot 4}(-0.01)^4 + \cdots \\ &= 1 - 0.15 + 0.0105 - 0.000\,455 + 0.000\,014 - \cdots = 0.8601\end{aligned}$$

26.26 Find the sum of the coefficients in the expansion of (*a*) $(1+x)^{10}$, (*b*) $(1-x)^{10}$.

SOLUTION

(*a*) If, $1, c_1, c_2, \ldots, c_{10}$ are the coefficients, we have the identity

$$(1+x)^{10} = 1 + c_1 x + c_2 x^2 + \cdots + c_{10} x^{10}. \qquad \text{Let } x = 1.$$

Then $\quad (1+1)^{10} = 1 + c_1 + c_2 + \cdots + c_{10} = \text{sum of coefficients} = 2^{10} = 1024.$

(*b*) Let $x = 1$. Then $(1-x)^{10} = (1-1)^{10} = 0 = $ sum of coefficients.

Supplementary Problems

26.27 Expand by the binomial formula.

(*a*) $(x+\frac{1}{2})^6$ (*b*) $(x-2)^5$ (*c*) $(y+3)^4$ (*d*) $\left(x+\dfrac{1}{x}\right)^5$ (*e*) $(x^2-y^3)^4$ (*f*) $(a-2b)^6$ (*g*) $\left(\dfrac{x}{2}+\dfrac{3}{y}\right)^4$ (*h*) $(y^{1/2}+y^{-1/2})^6$

26.28 Write the indicated term in the expansion of each of the following.

(*a*) Fifth term of $(a-b)^7$

(*b*) Seventh term of $\left(x^2-\dfrac{1}{x}\right)^9$

(*c*) Middle term of $\left(y-\dfrac{1}{y}\right)^8$

(*d*) Eighth term of $\left(\dfrac{x^2}{2}-2y\right)^{16}$

(*e*) Seventh term of $\left(a-\dfrac{1}{\sqrt{a}}\right)^{10}$

(*f*) Sixteenth term of $(2-1/x)^{18}$

(*g*) Sixth term of $(x^2-2y)^{11}$

(*h*) Eleventh term of $\left(x+\dfrac{1}{\sqrt{x}}\right)^{14}$

26.29 Find the term independent of x in the expansion of

$$\left(\sqrt{x}+\frac{1}{3x^2}\right)^{10}.$$

26.30 Find the term involving x^3 in the expansion of

$$\left(x^2+\frac{1}{x}\right)^{12}.$$

26.31 Evaluate $(0.98)^6$ correct to five decimal places.

26.32 Evaluate $(1.1)^{10}$ correct to the nearest hundredth.

ANSWERS TO SUPPLEMENTARY PROBLEMS

26.27 (*a*) $x^6 + 3x^5 + \dfrac{15}{4}x^4 + \dfrac{5}{2}x^3 + \dfrac{15}{16}x^2 + \dfrac{3}{16}x + \dfrac{1}{64}$

(*b*) $x^5 - 10x^4 + 40x^3 - 80x^2 + 80x - 32$

(*c*) $y^4 + 12y^3 + 54y^2 + 108y + 81$

(*d*) $x^5 + 5x^3 + 10x + \dfrac{10}{x} + \dfrac{5}{x^3} + \dfrac{1}{x^5}$

(*e*) $x^8 - 4x^6y^3 + 6x^4y^6 - 4x^2y^9 + y^{12}$

(*f*) $a^6 - 12a^5b + 60a^4b^2 - 160a^3b^3 + 240a^2b^4 - 192ab^5 + 64b^6$

(*g*) $\dfrac{x^4}{16} + \dfrac{3x^3}{2y} + \dfrac{27x^2}{2y^2} + \dfrac{54x}{y^3} + \dfrac{81}{y^4}$

(*h*) $y^3 + 6y^2 + 15y + 20 + 15y^{-1} + 6y^{-2} + y^{-3}$

26.28 (*a*) $35a^3b^4$ (*b*) 84 (*c*) 70 (*d*) $-2860x^{18}y^7$ (*e*) $210a$ (*f*) $-\dfrac{6528}{x^{15}}$ (*g*) $-14\,784x^{12}y^5$ (*h*) $\dfrac{1001}{x}$

26.29 5 **26.30** $792x^3$ **26.31** 0.885 84 **26.32** 2.59

Chapter 27

Probability

27.1 SIMPLE PROBABILITY

Suppose that an event can happen in h ways and fail to happen in f ways, all these $h+f$ ways supposed equally likely. Then the probability of the occurrence of the event (called its success) is

$$p = \frac{h}{h+f} = \frac{h}{n},$$

and the probability of the non-occurrence of the event (called its failure) is

$$q = \frac{f}{h+f} = \frac{f}{n},$$

where $n = h + f$.

It follows that $p + q = 1$, $p = 1 - q$, and $q = 1 - p$.

The odds in favor of the occurrence of the event are $h:f$ or h/f; the odds against its happening are $f:h$ or f/h.

If p is the probability that an event will occur, the odds in favor of its happening are $p:q = p:(1-p)$ or $p/(1-p)$; the odds against its happening are $q:p = (1-p):p$ or $(1-p)/p$.

27.2 COMPOUND PROBABILITY

Two or more events are said to be independent if the occurrence or non-occurrence of any one of them does not affect the probabilities of occurrence of any of the others.

Thus if a coin is tossed four times and it turns up a head each time, the fifth toss may be head or tail and is not influenced by the previous tosses.

The probability that two or more independent events will happen is equal to the product of their separate probabilities.

Thus the probability of getting a head on both the fifth and sixth tosses is $\frac{1}{2}(\frac{1}{2}) = \frac{1}{4}$.

Two or more events are said to be dependent if the occurrence or non-occurrence of one of the events affects the probabilities of occurrence of any of the others.

Consider that two or more events are dependent. If p_1 is the probability of a first event, p_2 the probability that after the first has happened the second will occur, p_3 the probability that after the first and second have happened the third will occur, etc., then the probability that all events will happen in the given order is the product $p_1 \cdot p_2 \cdot p_3 \cdots$.

For example, a box contains 3 white balls and 2 black balls. If a ball is drawn at random, the probability that it is black is $\dfrac{2}{3+2} = \dfrac{2}{5}$. If this ball is not replaced and a second ball is drawn, the probability that it also is black is $\dfrac{1}{3+1} = \dfrac{1}{4}$. Thus the probability that both will be black is $\dfrac{2}{5}\left(\dfrac{1}{4}\right) = \dfrac{1}{10}$.

Two or more events are said to be mutually exclusive if the occurrence of any one of them excludes the occurrence of the others.

The probability of occurrence of some one of two or more mutually exclusive events is the *sum* of the probabilities of the individual events.

EXAMPLE 27.1. If a die is thrown, what is the probability of getting a 5 or a 6? Getting a 5 and getting a 6 are mutually exclusive so

$$P(5 \text{ or } 6) = P(5) + P(6) = \frac{1}{6} + \frac{1}{6} = \frac{2}{6} = \frac{1}{3}$$

Two events are said to be overlapping if the events have at least one outcome in common, hence they can happen at the same time.

The probability of occurrence of some one of two overlapping events is the sum of the probabilities of the two individual events minus the probability of their common outcomes.

EXAMPLE 27.2. If a die is thrown, what is the probability of getting a number less than 4 or an even number?

The numbers less than 4 on a die are 1, 2, and 3. The even numbers on a die are 2, 4, and 6. Since these two events have a common outcome, 2, they are overlapping events.

$$\begin{aligned} P(\text{less than 4 or even}) &= P(\text{less than 4}) + P(\text{even}) - P(\text{less than 4 and even}) \\ &= \frac{3}{6} + \frac{3}{6} - \frac{1}{6} \\ &= \frac{5}{6} \end{aligned}$$

27.3 MATHEMATICAL EXPECTATION

If p is the probability that a person will receive a sum of money m, the value of his expectation is $p \cdot m$.

Thus if the probability of your winning a \$10 prize is 1/5, your expectation is $\frac{1}{5}(\$10) = \$2$.

27.4 BINOMIAL PROBABILITY

If p is the probability that an event will happen in any single trial and $q = 1 - p$ is the probability that it will fail to happen in any single trial, then the probability of its happening exactly r times in n trials is ${}_nC_r p^r q^{n-r}$. (See Problems 27.22 and 27.23.)

The probability that an event will happen at least r times in n trials is

$$p^n + {}_nC_1 p^{n-1} q + {}_nC_2 p^{n-2} q^2 + \cdots + {}_nC_r p^r q^{n-r}.$$

This expression is the sum of the first $n - r + 1$ terms of the binomial expansion of $(p + q)^n$. (See Problems 27.24–27.26.)

27.5 CONDITIONAL PROBABILITY

The probability that a second event will occur given that the first event has occurred is called conditional probability. To find the probability that the second event will occur given that the first event occurred, divide the probability that both events occurred by the probability of the first event. The probability of event B given that event A has occurred is denoted by $P(B|A)$.

EXAMPLE 27.3. A box contains black chips and red chips. A person draws two chips without replacement. If the probability of selecting a black chip and a red chip is 15/56 and the probability of drawing a black chip on the first draw is 3/4, what is the probability of drawing a red chip on the second draw, if you know the first chip drawn was black?

If B is the event drawing a black chip and R is the event drawing a red chip, then $P(R|B)$ is the probability of drawing a red chip on the second draw given that a black chip was drawn on the first draw.

$$\begin{aligned} P(R|B) &= \frac{P(R \text{ and } B)}{P(B)} \\ &= \frac{15/56}{3/4} \\ &= \frac{15}{56} \cdot \frac{4}{3} \\ &= \frac{5}{14} \end{aligned}$$

Thus, the probability of drawing a red chip on the second draw given that a black chip was drawn on the first draw is 5/14.

Solved Problems

27.1 One ball is drawn at random from a box containing 3 red balls, 2 white balls, and 4 blue balls. Determine the probability p that it is (a) red, (b) not red, (c) white, (d) red or blue.

SOLUTION

(a) $p = \dfrac{\text{ways of drawing 1 out of 3 red balls}}{\text{ways of drawing 1 out of } (3+2+4) \text{ balls}} = \dfrac{3}{3+2+4} = \dfrac{3}{9} = \dfrac{1}{3}$

(b) $p = 1 - \dfrac{1}{3} = \dfrac{2}{3}$ (c) $p = \dfrac{2}{9}$ (d) $p = \dfrac{3+4}{9} = \dfrac{7}{9}$

27.2 One bag contains 4 white balls and 2 black balls; another bag contains 3 white balls and 5 black balls. If one ball is drawn from each bag, determine the probability p that (a) both are white, (b) both are black, (c) 1 is white and 1 is black.

SOLUTION

(a) $p = \left(\dfrac{4}{4+2}\right)\left(\dfrac{3}{3+5}\right) = \dfrac{1}{4}$ (b) $p = \left(\dfrac{2}{4+2}\right)\left(\dfrac{5}{3+5}\right) = \dfrac{5}{24}$

(c) Probability that first ball is white and second black $= \dfrac{4}{6}\left(\dfrac{5}{8}\right) = \dfrac{5}{12}$.

Probability that first ball is black and second white $= \dfrac{2}{6}\left(\dfrac{3}{8}\right) = \dfrac{1}{8}$.

These are mutually exclusive; hence the required probability $p = \dfrac{5}{12} + \dfrac{1}{8} = \dfrac{13}{24}$.

Another method. $p = 1 - \dfrac{1}{4} - \dfrac{5}{24} = \dfrac{13}{24}$.

27.3 Determine the probability of throwing a total of 8 in a single throw with two dice, each of whose faces is numbered from 1 to 6.

SOLUTION

Each of the faces of one die can be associated with any of the 6 faces of the other die; thus the total number of possible cases = $6 \cdot 6 = 36$ cases.

There are 5 ways of throwing an 8: 2, 6; 3, 5; 4, 4; 5, 3; 6, 2.

$$\text{Required probability} = \frac{\text{number of favorable cases}}{\text{possible number of cases}} = \frac{5}{36}.$$

27.4 What is the probability of getting at least 1 one in 2 throws of a die?

SOLUTION

The probability of not getting a one in any single throw $= 1 - 1/6 = 5/6$.

The probability of not getting a one in 2 throws $= (5/6)(5/6) = 25/36$.

Hence the probability of getting at least 1 one in 2 throws $= 1 - 25/36 = 11/36$.

27.5 The probability of A's winning a game of chess against B is 1/3. What is the probability that A will win at least 1 of a total of 3 games?

SOLUTION

The probability of A's losing any single game $= 1 - 1/3 = 2/3$, and the probability of A losing all 3 games $= (2/3)^3 = 8/27$.

Hence the probability of A winning at least 1 game $= 1 - 8/27 = 19/27$.

27.6 Three cards are drawn from a pack of 52, each card being replaced before the next one is drawn. Compute the probability p that all are (*a*) spades, (*b*) aces, (*c*) red cards.

SOLUTION

A pack of 52 cards includes 13 spades, 4 aces, and 26 red cards.

$$(a)\quad p = \left(\frac{13}{52}\right)^3 = \frac{1}{64} \qquad (b)\quad p = \left(\frac{4}{52}\right)^3 = \frac{1}{2197} \qquad (c)\quad p = \left(\frac{26}{52}\right)^3 = \frac{1}{8}$$

27.7 The odds are 23 to 2 against a person winning a \$500 prize. What is her mathematical expectation?

SOLUTION

$$\text{Expectation} = \text{probability of winning} \times \text{sum of money} = \left(\frac{2}{23+2}\right)(\$500) = \$40.$$

27.8 Nine tickets, numbered from 1 to 9, are in a box. If 2 tickets are drawn at random, determine the probability p that (*a*) both are odd, (*b*) both are even, (*c*) one is odd and one is even, (*d*) they are numbered 2, 5.

SOLUTION

There are 5 odd and 4 even numbered tickets.

(a) $p = \dfrac{\text{number of selections of 2 out of 5 odd tickets}}{\text{number of selections of 2 out of 9 tickets}} = \dfrac{{}_5C_2}{{}_9C_2} = \dfrac{5}{18}$

(b) $p = \dfrac{{}_4C_2}{{}_9C_2} = \dfrac{1}{6}$ (c) $p = \dfrac{{}_5C_1 \cdot {}_4C_1}{{}_9C_2} = \dfrac{5 \cdot 4}{36} = \dfrac{5}{9}$ (d) $p = \dfrac{{}_2C_2}{{}_9C_2} = \dfrac{1}{36}$

27.9 A bag contains 6 red, 4 white, and 8 blue balls. If 3 balls are drawn at random, determine the probability p that (*a*) all 3 are red, (*b*) all 3 are blue, (*c*) 2 are white and 1 is red, (*d*) at least 1 is red, (*e*) 1 of each color is drawn, (*f*) the balls are drawn in the order red, white, blue.

SOLUTION

(a) $p = \dfrac{\text{number of selections of 3 out of 6 red balls}}{\text{number of selections of 3 out of 18 balls}} = \dfrac{{}_6C_3}{{}_{18}C_3} = \dfrac{5}{204}$

(b) $p = \dfrac{{}_8C_3}{{}_{18}C_3} = \dfrac{7}{102}$

(c) $p = \dfrac{{}_4C_2 \cdot {}_6C_1}{{}_{18}C_3} = \dfrac{3}{68}$

(d) Probability that none is red $= \dfrac{{}_{(4+8)}C_3}{{}_{18}C_3} = \dfrac{{}_{12}C_3}{{}_{18}C_3} = \dfrac{55}{204}$.

Hence the probability that at least 1 is red $= 1 - \dfrac{55}{204} = \dfrac{149}{204}$.

(e) $p = \dfrac{6 \cdot 4 \cdot 8}{{}_{18}C_3} = \dfrac{6 \cdot 4 \cdot 8}{18 \cdot 17 \cdot 16/6} = \dfrac{4}{17}$

(f) $p = \dfrac{4}{17} \cdot \dfrac{1}{3!} = \dfrac{4}{17} \cdot \dfrac{1}{6} = \dfrac{2}{51}$ or $p = \dfrac{6 \cdot 4 \cdot 8}{{}_{18}P_3} = \dfrac{6 \cdot 4 \cdot 8}{18 \cdot 17 \cdot 16} = \dfrac{2}{51}$

27.10 Three cards are drawn from a pack of 52 cards. Determine the probability p that (*a*) all are aces, (*b*) all are aces and drawn in the order spade, club, diamond, (*c*) all are spades, (*d*) all are of the same suit, (*e*) no two are of the same suit.

SOLUTION

(a) There are ${}_{52}C_3$ selections of 3 out of 52 cards, and ${}_4C_3$ selections of 3 out of 4 aces.

Hence $p = \dfrac{{}_4C_3}{{}_{52}C_3} = \dfrac{{}_4C_1}{{}_{52}C_3} = \dfrac{1}{5525}$.

(b) There are ${}_{52}P_3$ orders of drawing 3 out of 52 cards, one of which is the given order.

Hence $p = \dfrac{1}{{}_{52}P_3} = \dfrac{1}{52 \cdot 51 \cdot 50} = \dfrac{1}{132\,600}$.

(c) There are ${}_{13}C_3$ selections of 3 out of 13 spades.

Hence $p = \dfrac{{}_{13}C_3}{{}_{52}C_3} = \dfrac{11}{850}$.

(d) There are 4 suits, each consisting of 13 cards. Hence there are 4 ways of selecting a suit, and ${}_{13}C_3$ ways of selecting 3 cards from a given suit.

Hence $$p = \frac{4 \cdot {}_{13}C_3}{{}_{52}C_3} = \frac{22}{425}.$$

(*e*) There are ${}_4C_3 = {}_4C_1 = 4$ ways of selecting 3 out of 4 suits, and 13·13·13 ways of selecting 1 card from each of 3 given suits.

Hence $$p = \frac{4 \cdot 13 \cdot 13 \cdot 13}{{}_{52}C_3} = \frac{169}{425}.$$

27.11 What is the probability that any two different cards of a well-shuffled deck of 52 cards will be together in the deck, if their suit is not considered?

SOLUTION

Consider the probability that, for example, an ace and a king are together. There are 4 aces and 4 kings in a deck. Hence an ace can be chosen in 4 ways, and when that is done a king can be chosen in 4 ways. Thus an ace and then a king can be selected in $4 \cdot 4 = 16$ ways. Similarly, a king and then an ace can be selected in 16 ways. Then an ace and a king can be together in $2 \cdot 16 = 32$ ways.

For every one way the combination (ace, king) occurs, the remaining 50 cards and the (ace, king) combination can be permuted in 51! ways. The number of favorable arrangements is thus 32(51!). Since the total number of arrangements of all the cards in the deck is 52!, the required probability is

$$\frac{32(51!)}{52!} = \frac{32}{52} = \frac{8}{13}.$$

27.12 A man holds 2 of a total of 20 tickets in a lottery. If there are 2 winning tickets, determine the probability that he has (*a*) both, (*b*) neither, (*c*) exactly one.

SOLUTION

(*a*) There are ${}_{20}C_2$ ways of selecting 2 out of 20 tickets.

Hence the probability of his winning both prizes $= \dfrac{1}{{}_{20}C_2} = \dfrac{1}{190}.$

Another method. The probability of winning the first prize = 2/20 = 1/10. After winning the first prize (he has 1 ticket left and there remain 19 tickets from which to choose the second prize) the probability of winning the second prize is 1/19.

Hence the probability of winning both prizes $= \dfrac{1}{10}\left(\dfrac{1}{19}\right) = \dfrac{1}{190}.$

(*b*) There are 20 tickets, 18 of which are losers.

Hence the probability of winning neither prize $= \dfrac{{}_{18}C_2}{{}_{20}C_2} = \dfrac{153}{190}.$

Another method. The probability of not winning the first prize = 1 − 2/20 = 9/10. If he does not win the first prize (he still has 2 tickets), the probability of not winning the second prize = 1 − 2/19 = 17/19.

Hence the probability of winning neither prize $= \dfrac{9}{10}\left(\dfrac{17}{19}\right) = \dfrac{153}{190}.$

(*c*) Probability of winning exactly one prize

$$= 1 - \text{probability of winning neither} - \text{probability of winning both}$$

$$= 1 - \frac{153}{190} - \frac{1}{190} = \frac{36}{190} = \frac{18}{95}.$$

Another method.

$$\text{Probability of winning first but not second prize} = \frac{2}{20}\left(\frac{18}{19}\right) = \frac{9}{95}.$$

$$\text{Probability of not winning first but winning second} = \frac{18}{20}\left(\frac{2}{19}\right) = \frac{9}{95}.$$

$$\text{Hence the probability of winning exactly 1 prize} = \frac{9}{95} + \frac{9}{95} = \frac{18}{95}.$$

27.13 A box contains 7 tickets, numbered from 1 to 7 inclusive. If 3 tickets are drawn from the box, one at a time, determine the probability that they are alternately either odd, even, odd or even, odd, even.

SOLUTION

The probability that the first drawn is odd (4/7), then the second even (3/6) and then the third odd (3/5) is $\frac{4}{7}\left(\frac{3}{6}\right)\left(\frac{3}{5}\right) = \frac{6}{35}$.

The probability that the first drawn is even (3/7), then the second odd (4/6), and then the third even (2/5) is $\frac{3}{7}\left(\frac{4}{6}\right)\left(\frac{2}{5}\right) = \frac{4}{35}$.

Hence the required probability $= \frac{6}{35} + \frac{4}{35} = \frac{2}{7}$.

Another method. Possible orders of 7 numbers taken 3 at a time $= {}_7P_3 = 7 \cdot 6 \cdot 5 = 210$.
Orders where numbers are alternately odd, even, odd $= 4 \cdot 3 \cdot 3 = 36$.
Orders where numbers are alternately even, odd, even $= 3 \cdot 4 \cdot 2 = 24$.

Hence the required probability $= \dfrac{36 + 24}{210} = \dfrac{60}{210} = \dfrac{2}{7}$.

27.14 The probability that A can solve a given problem is 4/5, that B can solve it is 2/3, and that C can solve it is 3/7. If all three try, compute the probability that the problem will be solved.

SOLUTION

The probability that A will fail to solve it $= 1 - 4/5 = 1/5$, that B will fail $= 1 - 2/3 = 1/3$, and that C will fail $= 1 - 3/7 = 4/7$.

The probability that all three fail $= \frac{1}{5}\left(\frac{1}{3}\right)\left(\frac{4}{7}\right)$.

Hence the probability that all three will not fail, i.e., that at least one will solve it, is

$$1 - \frac{1}{5}\left(\frac{1}{3}\right)\left(\frac{4}{7}\right) = 1 - \frac{4}{105} = \frac{101}{105}.$$

27.15 The probability that a certain man will be alive 25 years hence is 3/7, and the probability that his wife will be alive 25 years hence is 4/5. Determine the probability that, 25 years hence, (*a*) both will be alive, (*b*) at least one of them will be alive, (*c*) only the man will be alive.

SOLUTION

(*a*) The probability that both will be alive $= \dfrac{3}{7}\left(\dfrac{4}{5}\right) = \dfrac{12}{35}$.

(*b*) The probability that both will die within 25 years $= \left(1 - \dfrac{3}{7}\right)\left(1 - \dfrac{4}{5}\right) = \dfrac{4}{7}\left(\dfrac{1}{5}\right) = \dfrac{4}{35}$.

Hence the probability that at least one will be alive $= 1 - \dfrac{4}{35} = \dfrac{31}{35}$.

(*c*) The probability that the man will be alive = 3/7, and the probability that his wife will not be alive = 1 − 4/5 = 1/5.

Hence the probability that only the man will be alive $= \dfrac{3}{7}\left(\dfrac{1}{5}\right) = \dfrac{3}{35}$.

27.16 There are three candidates, *A*, *B*, and *C*, for an office. The odds that *A* will win are 7 to 5, and the odds that *B* will win are 1 to 3. (*a*) What is the probability that either *A* or *B* will win? (*b*) What are the odds in favor of *C*?

SOLUTION

(*a*) Probability that *A* will win: $\dfrac{7}{7+5} = \dfrac{7}{12}$,

that *B* will win: $\dfrac{1}{1+3} = \dfrac{1}{4}$.

Hence the probability that either *A* or *B* will win $= \dfrac{7}{12} + \dfrac{1}{4} = \dfrac{5}{6}$.

(*b*) Probability that *C* will win: $1 - \dfrac{5}{6} = \dfrac{1}{6}$.

Hence the odds in favor of *C* are 1 to 5.

27.17 One purse contains 5 dimes and 2 quarters, and a second purse contains 1 dime and 3 quarters. If a coin is taken from one of the two purses at random, what is the probability that it is a quarter?

SOLUTION

The probability of selecting the first purse (1/2) and of then drawing a quarter from it (2/7) is (1/2)(2/7) = 1/7. The probability of selecting the second purse (1/2) and of then drawing a quarter from it (3/4) is (1/2)(3/4) = 3/8.

Hence the required probability $= \dfrac{1}{7} + \dfrac{3}{8} = \dfrac{29}{56}$.

27.18 A bag contains 2 white balls and 3 black balls. Four persons, *A*,*B*,*C*,*D*, in the order named, each draw one ball and do not replace it. The first to draw a white ball receives $10. Determine their expectations.

SOLUTION

A's probability of winning $= \dfrac{2}{5}$, and the expectation $= \dfrac{2}{5}(\$10) = \$4$.

To find *B*'s expectation: The probability that *A* fails = 1 − 2/5 = 3/5. If *A* fails, the bag contains 2 white and 2 black balls. Thus the probability that if A fails B will win $= 2/4 = 1/2$. Hence B's probability of winning $= (3/5)(1/2) = 3/10$, and the expectation is \$3.

To find C's expectation: The probability that A fails $= 3/5$, and the probability that B fails $= 1 - 1/2 = 1/2$. If A and B both fail, the bag contains 2 white balls and 1 black ball. Thus the probability that, if A and B both fail, C will win $= 2/3$. Hence C's probability of winning $= \frac{3}{5}\left(\frac{1}{2}\right)\left(\frac{2}{3}\right) = \frac{1}{5}$, and the expectation $= \frac{1}{5}(\$10) = \$2$.

If A, B, and C fail, only white balls remain and D must win. Hence D's probability of winning $= \frac{3}{5}\left(\frac{1}{2}\right)\left(\frac{1}{3}\right)\left(\frac{1}{1}\right) = \frac{1}{10}$, and the expectation $= \frac{1}{10}(\$10) = \$1$.

Check. $\$4 + \$3 + \$2 + \$1 = \$10$, and $\frac{2}{5} + \frac{3}{10} + \frac{1}{5} + \frac{1}{10} = 1$.

27.19 Eleven books, consisting of 5 engineering books, 4 mathematics books, and 2 chemistry books, are placed on a shelf at random. What is the probability p that the books of each kind are all together?

SOLUTION

When the books of each kind are all together, the engineering books could be arranged in 5! ways, the mathematics books in 4! ways, the chemistry books in 2! ways, and the 3 groups in 3! ways.

$$p = \frac{\text{ways in which books of each kind are together}}{\text{total number of ways of arranging 11 books}} = \frac{5!4!2!3!}{11!} = \frac{1}{1155}.$$

27.20 Five red blocks and 4 white blocks are placed at random in a row. What is the probability p that the extreme blocks are both red?

SOLUTION

Total possible arrangements of 5 red and 4 white blocks $= \frac{(5+4)!}{5!4!} = \frac{9!}{5!4!} = 126$.

Arrangements where extreme blocks are both red $= \frac{(9-2)!}{(5-2)!4!} = \frac{7!}{3!4!} = 35$.

Hence the required probability $p = \frac{35}{126} = \frac{5}{18}$.

27.21 One purse contains 6 copper coins and 1 silver coin; a second purse contains 4 copper coins. Five coins are drawn from the first purse and put into the second, and then 2 coins are drawn from the second and put into the first. Determine the probability that the silver coin is in (*a*) the second purse, (*b*) the first purse.

SOLUTION

Initially, the first purse contains 7 coins. When 5 coins are drawn from the first purse and put into the second, the probability that the silver coin is put into the second purse is 5/7, and the probability that it remains in the first purse is 2/7.

The second purse now contains $5 + 4 = 9$ coins. Finally, after 2 of these 9 coins are put into the first purse, the probability that the silver coin is in the second purse $= \frac{5}{7}\left(\frac{7}{9}\right) = \frac{5}{9}$, and the probability that it is in the first purse $= \frac{2}{7} + \frac{5}{7}\left(\frac{2}{9}\right) = \frac{4}{9} \left(\text{or } 1 - \frac{5}{9} = \frac{4}{9}\right)$.

27.22 Compute the probability that a single throw with 9 dice will result in exactly 2 ones.

SOLUTION

The probability that a certain pair of the 9 dice thrown will yield ones $= \frac{1}{6}\left(\frac{1}{6}\right) = \left(\frac{1}{6}\right)^2$. The probability that the other 7 dice will not yield ones $= \left(1 - \frac{1}{6}\right)^7 = \left(\frac{5}{6}\right)^7$. Since ${}_9C_2$ different pairs may be selected from the 9 dice, the probability that exactly 1 pair will be aces is ${}_9C_2\left(\frac{1}{6}\right)^2\left(\frac{5}{6}\right)^7 = \frac{78\,125}{279\,936}$.

Or, by formula: Probability $= {}_nC_r\, p^r q^{n-r} = {}_9C_2\left(\frac{1}{6}\right)^2\left(\frac{5}{6}\right)^7 = \frac{78\,125}{279\,936}$.

27.23 What is the probability of getting a 9 exactly once in 3 throws with a pair of dice?

SOLUTION

A 9 can occur in 4 ways: 3, 6; 4, 5; 5, 4; 6, 3.

In any throw with a pair of dice, the probability of getting a $9 = 4/(6\cdot 6) = \frac{1}{9}$, and the probability of not getting a $9 = 1 - \frac{1}{9} = \frac{8}{9}$. The probability that any given throw with a pair of dice is a 9 and that the other two throws are not $= \left(\frac{1}{9}\right)\left(\frac{8}{9}\right)^2$. Since there are ${}_3C_1 = 3$ different ways in which one throw is a 9 and the other two throws are not, the probability of throwing a 9 exactly once in 3 throws $=$ ${}_3C_1\left(\frac{1}{9}\right)\left(\frac{8}{9}\right)^2 = \frac{64}{243}$.

Or, by formula: Probability $= {}_nC_r\, p^r q^{n-r} = {}_3C_1\left(\frac{1}{9}\right)\left(\frac{8}{9}\right)^2 = \frac{64}{243}$.

27.24 If the probability that the average freshman will not complete four years of college is 1/3, what is the probability p that of 4 freshmen at least 3 will complete four years of college?

SOLUTION

Probability that 3 will complete and 1 will not $= {}_4C_3\left(\frac{2}{3}\right)^3\left(\frac{1}{3}\right) = {}_4C_1\left(\frac{2}{3}\right)^3\left(\frac{1}{3}\right)$.

Probability that 4 will complete $= \left(\frac{2}{3}\right)^4$. Hence $p = \left(\frac{2}{3}\right)^4 + {}_4C_1\left(\frac{2}{3}\right)^3\left(\frac{1}{3}\right) = \frac{16}{27}$.

Or, by formula: $p =$ first 2 $(n - r + 1 = 4 - 3 + 1)$ terms of the expansion of $\left(\frac{2}{3} + \frac{1}{3}\right)^4$

$$= \left(\frac{2}{3}\right)^4 + {}_4C_1\left(\frac{2}{3}\right)^3\left(\frac{1}{3}\right) = \frac{16}{81} + \frac{32}{81} = \frac{16}{27}.$$

27.25 A coin is tossed 6 times. What is the probability p of getting at least 3 heads? What are the odds in favor of getting at least 3 heads?

SOLUTION

On each toss, probability of a head = probability of a tail = 1/2.

The probability that certain 3 of the 6 tosses will give heads $= (1/2)^3$. The probability that none of the other 3 tosses will be heads $= (1/2)^3$. Since ${}_6C_3$ different selections of 3 can be made from the 6 tosses, the probability that exactly 3 will be heads is

$$_6C_3\left(\frac{1}{2}\right)^3\left(\frac{1}{2}\right)^3 = {_6C_3}\left(\frac{1}{2}\right)^6.$$

Similarly, the probability of exactly 4 heads $= {_6C_4}(1/2)^6 = {_6C_2}(1/2)^6$,
the probability of exactly 5 heads $= {_6C_5}(1/2)^6 = {_6C_1}(1/2)^6$,
the probability of exactly 6 heads $= (1/2)^6$.

$$\text{Hence } p = \left(\frac{1}{2}\right)^6 + {_6C_1}\left(\frac{1}{2}\right)^6 + {_6C_2}\left(\frac{1}{2}\right)^6 + {_6C_3}\left(\frac{1}{2}\right)^6$$

$$= \left(\frac{1}{2}\right)^6(1 + {_6C_1} + {_6C_2} + {_6C_3}) = \left(\frac{1}{2}\right)^6(1 + 6 + 15 + 20) = \frac{21}{32}.$$

The odds in favor of getting at least 3 heads is 21:11 or 21/11.

Or, by formula: $p =$ first 4 $(n - r + 1 = 6 - 3 + 1)$ terms of the expansion of $\left(\frac{1}{2} + \frac{1}{2}\right)^6$

$$= \left(\frac{1}{2}\right)^6 + {_6C_1}\left(\frac{1}{2}\right)^6 + {_6C_2}\left(\frac{1}{2}\right)^6 + {_6C_3}\left(\frac{1}{2}\right)^6 = \frac{21}{32}.$$

27.26 Determine the probability p that in a family of 5 children there will be at least 2 boys and 1 girl. Assume that the probability of a male birth is 1/2.

SOLUTION

The three favorable cases are: 2 boys, 3 girls; 3 boys, 2 girls; 4 boys, 1 girl.

$$p = \left(\frac{1}{2}\right)^5({_5C_2} + {_5C_3} + {_5C_4}) = \frac{1}{32}(10 + 10 + 5) = \frac{25}{32}.$$

27.27 The probability that a student takes chemistry and is on the honor rolls is 0.042. The probability that a student is on the honor roll is 0.21. What is the probability that the student is taking chemistry, given that the student is on the honor roll?

SOLUTION

$$P(\text{taking chemistry}|\text{on honor roll}) = \frac{P(\text{taking chemistry and on honor roll})}{P(\text{on honor roll})}$$

$$= \frac{0.042}{0.21}$$

$$= 0.2$$

27.28 At the Pine Valley Country Club, 32% of the members play golf and are female. Also, 80% of the members play golf. If a member of the club is selected at random, find the probability that the member is female given that the member plays golf.

SOLUTION

$$P(\text{female}\,|\,\text{golfer}) = \frac{P(\text{female and golfer})}{P(\text{golfer})}$$

$$= \frac{0.32}{0.8}$$

$$= 0.4$$

Supplementary Problems

27.29 Determine the probability that a digit chosen at random from the digits 1, 2, 3, . . ., 9 will be (*a*) odd, (*b*) even, (*c*) a multiple of 3.

27.30 A coin is tossed three times. If H = head and T = tail, what is the probability of the tosses coming up in the order (*a*) HTH, (*b*) THH, (*c*) HHH?

27.31 If three coins are tossed, what is the probability of obtaining (*a*) three heads, (*b*) two heads and a tail?

27.32 Find the probability of throwing a total of 7 in a single throw with two dice.

27.33 What is the probability of throwing a total of 8 or 11 in a single throw with two dice?

27.34 A die is thrown twice. What is the probability of getting a 4 or 5 on the first throw and a 2 or 3 on the second throw? What is the probability of not getting a one on either throw?

27.35 What is the probability that a coin will turn up heads at least once in six tosses of a coin?

27.36 Five discs in a bag are numbered 1, 2, 3, 4, 5. What is the probability that the sum of the numbers on three discs chosen at random is greater than 10?

27.37 Three balls are drawn at random from a box containing 5 red, 8 black, and 4 white balls. Determine the probability that (*a*) all three are white, (*b*) two are black and one red, (*c*) one of each color is selected.

27.38 Four cards are drawn from a pack of 52 cards. Find the probability that (*a*) all are kings, (*b*) two are kings and two are aces, (*c*) all are of the same suit, (*d*) all are clubs.

27.39 A woman will win $3.20 if in 5 tosses of a coin she gets either of the sequences HTHTH or THTHT where H = head and T = tail. Determine her expectation.

27.40 In a plane crash it was reported that three persons out of the total of twenty passengers were injured. Three newspapermen were in this plane. What is the probability that the three reported injured were the newspapermen?

27.41 A committee of three is to be chosen from a group consisting of 5 men and 4 women. If the selection is made at random, find the probability that (*a*) all three are women, (*b*) two are men.

27.42 Six persons seat themselves at a round table. What is the probability that two given persons are adjacent?

27.43 *A* and *B* alternately toss a coin. The first one to turn up a head wins. If no more than five tosses each are allowed for a single game, find the probability that the person who tosses first will win the game. What are the odds against *A*'s losing if she goes first?

27.44 Six red blocks and 4 white blocks are placed at random in a row. Find the probability that the two blocks in the middle are of the same color.

27.45 In 8 tosses of a coin determine the probability of (*a*) exactly 4 heads, (*b*) at least 2 tails, (*c*) at most 5 heads, (*d*) exactly 3 tails.

27.46 In 2 throws with a pair of dice determine the probability of getting (*a*) an 11 exactly once, (*b*) a 10 twice.

27.47 What is the probability of getting at least one 11 in 3 throws with a pair of dice?

27.48 In ten tosses of a coin, what is the probability of getting not less than 3 heads and not more than 6 heads?

27.49 The probability that an automobile will be stolen and found within one week is 0.0006. The probability that an automobile will be stolen is 0.0015. What is the probability that a stolen automobile will be found in one week?

27.50 In the Pizza Palace, 95% of the customers order pizza. If 65% of the customers who order pizza also order breadsticks, find the probability that a customer who orders a pizza will also order breadsticks.

27.51 In a large shopping mall, a marketing agency conducted a survey of 100 people about a ban on smoking in the mall. Of the 60 non-smokers surveyed, 48 preferred a smoking ban. Of the 40 smokers surveyed, 32 preferred a smoking ban. What is the probability that a person selected at random from the group surveyed prefers a smoking ban given that the person is a non-smoker?

27.52 In a new subdivision, 35% of the houses have a family room and a fireplace, while 70% have family rooms. What is the probability that a house selected at random in this subdivision has a fireplace given that it has a family room?

ANSWERS TO SUPPLEMENTARY PROBLEMS

27.29 (*a*) 5/9 (*b*) 4/9 (*c*) 1/3

27.30 (*a*) 1/8 (*b*) 1/8 (*c*) 1/8

27.31 (*a*) 1/8 (*b*) 3/8

27.32 1/6

27.33 7/36

27.34 1/9, 25/36

27.35 63/64

27.36 1/5

27.37 (*a*) $\frac{1}{170}$ (*b*) $\frac{7}{34}$ (*c*) $\frac{4}{17}$

27.38 (*a*) $\frac{1}{270\,725}$ (*b*) $\frac{36}{270\,725}$ (*c*) $\frac{44}{4165}$ (*d*) $\frac{11}{4165}$

27.39 20 cents

27.40 1/1140

27.41 (*a*) 1/21 (*b*) 10/21

27.42 2/5

27.43 $\frac{21}{32}$, 21:11

27.44 $\frac{7}{15}$

27.45 (*a*) $\frac{35}{128}$ (*b*) $\frac{247}{256}$ (*c*) $\frac{219}{256}$ (*d*) $\frac{7}{32}$

27.46 (*a*) $\frac{17}{162}$ (*b*) $\frac{1}{144}$

27.47 $\frac{919}{5832}$

27.48 $\frac{99}{128}$

27.49 0.4

27.50 $\frac{13}{19}$ (about 68%)

27.51 0.8

27.52 0.5

Chapter 28

Determinants and Systems of Linear Equations

28.1 DETERMINANTS OF SECOND ORDER

The symbol

$$\begin{vmatrix} a_1 & b_1 \\ a_2 & b_2 \end{vmatrix}$$

consisting of the four numbers a_1, b_1, a_2, b_2 arranged in two rows and two columns, is called a *determinant of second order* or *determinant of order two*. The four numbers are called *elements* of the determinant. By definition,

$$\begin{vmatrix} a_1 & b_1 \\ a_2 & b_2 \end{vmatrix} = a_1 b_2 - b_1 a_2.$$

Thus $$\begin{vmatrix} 2 & 3 \\ -1 & -2 \end{vmatrix} = (2)(-2) - (3)(-1) = -4 + 3 = -1.$$

Here the elements 2 and 3 are in the first row, the elements -1 and -2 are in the second row. Elements 2 and -1 are in the first column, and elements 3 and -2 are in the second column. A determinant is a number.

A determinant of order one is the number itself.

28.2 CRAMER'S RULE

Systems of two linear equations in two unknowns may be solved by use of second-order determinants. Given the system of equations

$$\begin{cases} a_1 x + b_1 y = c_1 \\ a_2 x + b_2 y = c_2 \end{cases} \tag{1}$$

it is possible by any of the methods of Chapter 15 to obtain

$$x = \frac{c_1 b_2 - b_1 c_2}{a_1 b_2 - b_1 a_2}, \qquad y = \frac{a_1 c_2 - c_1 a_2}{a_1 b_2 - b_1 a_2} \qquad (a_1 b_2 - b_1 a_2 \neq 0).$$

These values for x and y may be written in terms of second-order determinants as follows:

$$x = \frac{\begin{vmatrix} c_1 & b_1 \\ c_2 & b_2 \end{vmatrix}}{\begin{vmatrix} a_1 & b_1 \\ a_2 & b_2 \end{vmatrix}}, \qquad y = \frac{\begin{vmatrix} a_1 & c_1 \\ a_2 & c_2 \end{vmatrix}}{\begin{vmatrix} a_1 & b_1 \\ a_2 & b_2 \end{vmatrix}} \tag{2}$$

The form involving determinants is easy to remember if one keeps in mind the following:

(*a*) The denominators in (2) are given by the determinant

$$\begin{vmatrix} a_1 & b_1 \\ a_2 & b_2 \end{vmatrix}$$

in which the elements are the coefficients of x and y arranged as in the given equations (1).

This determinant, usually denoted by D, is called the *determinant of the coefficients*.

(*b*) The numerator in the solution for either unknown is the same as the determinant of the coefficients D with the exception that the column of coefficients of the unknown to be determined is replaced by the column of constants on the right side of equations (1). When the column of coefficients for the variable x in determinant D is replaced with the column of constants, we call the new determinant D_x. When the column of y coefficients in determinant D is replaced with the column of constants, we call the new determinant D_y.

EXAMPLE 28.1 Solve the system $\begin{cases} 2x + 3y = 8 \\ x - 2y = -3. \end{cases}$

The denominator for both x and y is $D = \begin{vmatrix} 2 & 3 \\ 1 & -2 \end{vmatrix} = 2(-2) - 3(1) = -7.$

$$D_x = \begin{vmatrix} 8 & 3 \\ -3 & -2 \end{vmatrix} = 8(-2) - 3(-3) = -7, \qquad D_y = \begin{vmatrix} 2 & 8 \\ 1 & -3 \end{vmatrix} = 2(-3) - 8(1) = -14$$

$$x = \frac{D_x}{D} = \frac{-7}{-7} = 1, \qquad y = \frac{D_y}{D} = \frac{-14}{-7} = 2$$

Thus, the solution of the system is $(1, 2)$.

The method of solution of linear equations by determinants is called *Cramer's Rule*. If the determinant $D = 0$, then Cramer's Rule can not be used to solve the system.

28.3 DETERMINANTS OF THIRD ORDER

The symbol

$$\begin{vmatrix} a_1 & b_1 & c_1 \\ a_2 & b_2 & c_2 \\ a_3 & b_3 & c_3 \end{vmatrix}$$

consisting of nine numbers arranged in three rows and three columns is called a *determinant of third order*. By definition, the value of this determinant is given by

$$a_1 b_2 c_3 + b_1 c_2 a_3 + c_1 a_2 b_3 - c_1 b_2 a_3 - a_1 c_2 b_3 - b_1 a_2 c_3$$

and is called the expansion of the determinant.

In order to remember this definition, the following scheme is given. Rewrite the first two columns on the right of the determinant as follows:

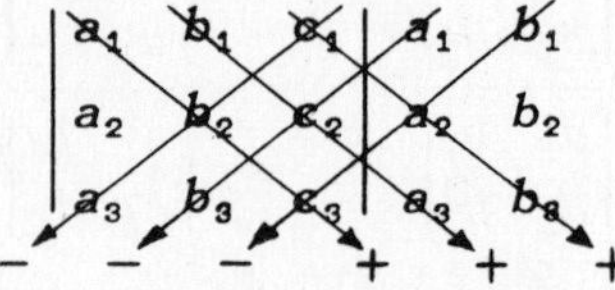

(*a*) Form the products of the elements in each of the 3 diagonals shown which run down from left to right, and precede each of these 3 terms by a positive sign.

(*b*) Form the products of the elements in each of the 3 diagonals shown which run down from right to left, and precede each of these 3 terms by a negative sign.

(*c*) The algebraic sum of the six products of (1) and (2) is the required expansion of the determinant.

EXAMPLE 28.2

Evaluate
$$\begin{vmatrix} 3 & -2 & 2 \\ 6 & 1 & -1 \\ -2 & -3 & 2 \end{vmatrix}$$

Rewriting,

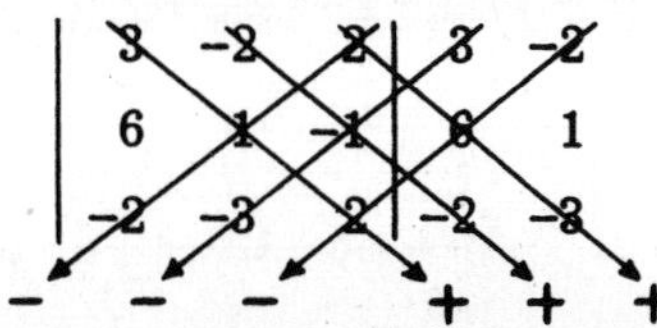

The value of the determinant is

$$(3)(1)(2) + (-2)(-1)(-2) + (2)(6)(-3) - (2)(1)(-2) - (3)(-1)(-3) - (-2)(6)(2) = -15.$$

Cramer's Rule for linear equations in 3 unknowns is a method of solving the following equations for x, y, z

$$\begin{cases} a_1x + b_1y + c_1z = d_1 \\ a_2x + b_2y + c_2z = d_2 \\ a_3x + b_3y + c_3z = d_3 \end{cases} \tag{3}$$

by determinants. It is an extension of Cramer's Rule for linear equations in two unknowns. If we solve equations (3) by the methods of Chapter 12, we obtain

$$x = \frac{d_1b_2c_3 + c_1d_2b_3 + b_1c_2d_3 - c_1b_2d_3 - b_1d_2c_3 - d_1c_2b_3}{a_1b_2c_3 + b_1c_2a_3 + c_1a_2b_3 - c_1b_2a_3 - b_1a_2c_3 - a_1c_2b_3}$$

$$y = \frac{a_1d_2c_3 + c_1a_2d_3 + d_1c_2a_3 - c_1d_2a_3 - d_1a_2c_3 - a_1c_2d_3}{a_1b_2c_3 + b_1c_2a_3 + c_1a_2b_3 - c_1b_2a_3 - b_1a_2c_3 - a_1c_2b_3}$$

$$z = \frac{a_1b_2d_3 + d_1a_2b_3 + b_1d_2a_3 - d_1b_2a_3 - b_1a_2d_3 - a_1d_2b_3}{a_1b_2c_3 + b_1c_2a_3 + c_1a_2b_3 - c_1b_2a_3 - b_1a_2c_3 - a_1c_2b_3}$$

These may be written in terms of determinants as follows

$$D = \begin{vmatrix} a_1 & b_1 & c_1 \\ a_2 & b_2 & c_2 \\ a_3 & b_3 & c_3 \end{vmatrix} \qquad D_x = \begin{vmatrix} d_1 & b_1 & c_1 \\ d_2 & b_2 & c_2 \\ d_3 & b_3 & c_3 \end{vmatrix} \qquad D_y = \begin{vmatrix} a_1 & d_1 & c_1 \\ a_2 & d_2 & c_2 \\ a_3 & d_3 & c_3 \end{vmatrix} \qquad D_z = \begin{vmatrix} a_1 & b_1 & d_1 \\ a_2 & b_2 & d_2 \\ a_3 & b_3 & d_3 \end{vmatrix}$$

$$x = \frac{D_x}{D}, \qquad y = \frac{D_y}{D}, \qquad z = \frac{D_z}{D} \tag{4}$$

D is the determinant of the coefficients of x, y, z in equations (3) and is assumed not equal to zero. If D is zero, Cramer's Rule cannot be used to solve the system of equations.

The form involving determinants is easy to remember if one keeps in mind the following:

(*a*) The denominators in (4) are given by the determinant D in which the elements are the coefficients of x, y, and z arranged as in the given equations (3).

(*b*) The numerator in the solution for any unknown is the same as the determinant of the coefficients D with the exception that the column of coefficients of the unknown to be determined is replaced by the column of constants on the right side of equations (3).

(*c*) The solution of the system is (x, y, z) where $x = \dfrac{D_x}{D}$, $y = \dfrac{D_y}{D}$, and $z = \dfrac{D_z}{D}$.

EXAMPLE 28.3 Solve the system

$$\begin{cases} x + 2y - z = -3 \\ 3x + y + z = 4 \\ x - y + 2z = 6. \end{cases}$$

$$D = \begin{vmatrix} 1 & 2 & -1 \\ 3 & 1 & 1 \\ 1 & -1 & 2 \end{vmatrix} = 2 + 2 + 3 + 1 + 1 - 12 = -3$$

$$D_x = \begin{vmatrix} -3 & 2 & -1 \\ 4 & 1 & 1 \\ 6 & -1 & 2 \end{vmatrix} = -6 + 12 + 4 + 6 - 3 - 16 = -3$$

$$D_y = \begin{vmatrix} 1 & -3 & -1 \\ 3 & 4 & 1 \\ 1 & 6 & 2 \end{vmatrix} = 8 - 3 - 18 + 4 - 6 + 18 = 3$$

$$D_z = \begin{vmatrix} 1 & 2 & -3 \\ 3 & 1 & 4 \\ 1 & -1 & 6 \end{vmatrix} = 6 + 8 + 9 + 3 + 4 - 36 = -6$$

$$x = \frac{D_x}{D} = \frac{-3}{-3} = 1, \qquad y = \frac{D_y}{D} = \frac{3}{-3} = -1, \qquad z = \frac{D_z}{D} = \frac{-6}{-3} = 2$$

The solution of the system is $(1, -1, 3)$.

Solved Problems

28.1 Evaluate the following determinants.

(*a*) $\begin{vmatrix} 3 & 2 \\ 1 & 4 \end{vmatrix} = (3)(4) - (2)(1) = 12 - 2 = 10$

(*b*) $\begin{vmatrix} 3 & -1 \\ 6 & -2 \end{vmatrix} = (3)(-2) - (-1)(6) = -6 + 6 = 0$

(*c*) $\begin{vmatrix} 0 & 3 \\ 2 & -5 \end{vmatrix} = (0)(-5) - (3)(2) = 0 - 6 = -6$

(*d*) $\begin{vmatrix} x & x^2 \\ y & y^2 \end{vmatrix} = xy^2 - x^2y$

(*e*) $\begin{vmatrix} x+2 & 2x+5 \\ 3x-1 & x-3 \end{vmatrix} = (x+2)(x-3) - (2x+5)(3x-1) = -5x^2 - 14x - 1$

28.2 (*a*) Show that if the rows and columns of a determinant of order two are interchanged the value of the determinant is the same.

(*b*) Show that if the elements of one row (or column) are proportional respectively to the elements of the other row (or column), the determinant is equal to zero.

SOLUTION

(*a*) Let the determinant be $\begin{vmatrix} a_1 & b_1 \\ a_2 & b_2 \end{vmatrix} = a_1b_2 - a_2b_1$.

The determinant with rows and columns interchanged so that 1st row becomes 1st column and 2nd row becomes 2nd column is $\begin{vmatrix} a_1 & a_2 \\ b_1 & b_2 \end{vmatrix} = a_1b_2 - a_2b_1$.

(*b*) The determinant with proportional rows is $\begin{vmatrix} a_1 & b_1 \\ ka_1 & kb_1 \end{vmatrix} = a_1kb_1 - b_1ka_1 = 0$.

28.3 Find the values of x for which $\begin{vmatrix} 2x-1 & 2x+1 \\ x+1 & 4x+2 \end{vmatrix} = 0$.

SOLUTION

$$\begin{vmatrix} 2x-1 & 2x+1 \\ x+1 & 4x+2 \end{vmatrix} = (2x-1)(4x+2) - (2x+1)(x+1) = 6x^2 - 3x - 3 = 0.$$

Then $2x^2 - x - 1 = (x-1)(2x+1) = 0$ so that $x = 1, -1/2$.

28.4 Solve for the unknowns in each of the following systems.

(*a*) $\begin{cases} 4x + 2y = 5 \\ 3x - 4y = 1 \end{cases}$

SOLUTION

$$D = \begin{vmatrix} 4 & 2 \\ 3 & -4 \end{vmatrix} = -22, \quad D_x = \begin{vmatrix} 5 & 2 \\ 1 & -4 \end{vmatrix} = -22, \quad D_y = \begin{vmatrix} 4 & 5 \\ 3 & 1 \end{vmatrix} = -11$$

$$x = \frac{D_x}{D} = \frac{-22}{-22} = 1, \quad y = \frac{D_y}{D} = \frac{-11}{-22} = \frac{1}{2}$$

The solution of the system is $(1, 1/2)$.

(*b*) $\begin{cases} 3u + 2v = 18 \\ -5u - v = 12 \end{cases}$

SOLUTION

$$D = \begin{vmatrix} 3 & 2 \\ -5 & -1 \end{vmatrix} = -7, \quad D_u = \begin{vmatrix} 18 & 2 \\ 12 & -1 \end{vmatrix} = -42, \quad D_v = \begin{vmatrix} 3 & 18 \\ -5 & 12 \end{vmatrix} = 126$$

$$u = \frac{D_u}{D} = \frac{-42}{7} = -6, \quad v = \frac{D_v}{D} = \frac{126}{7} = 18$$

The solution of the system is $(-6, 18)$.

(*c*) $\begin{cases} 5x - 2y - 14 = 0 \\ 2x + 3y + 3 = 0 \end{cases}$

SOLUTION

Rewrite as $\begin{cases} 5x - 2y = 14 \\ 2x + 3y = -3. \end{cases}$

$$D = \begin{vmatrix} 5 & -2 \\ 2 & 3 \end{vmatrix} = 19, \qquad D_x = \begin{vmatrix} 14 & -2 \\ -3 & 3 \end{vmatrix} = 36, \qquad D_y = \begin{vmatrix} 5 & 14 \\ 2 & -3 \end{vmatrix} = -43$$

$$x = \frac{D_x}{D} = \frac{36}{19}, \qquad y = \frac{D_y}{D} = \frac{-43}{19}$$

The solution of the system is (36/19, −43/19).

28.5 Solve for x and y.

(*a*) $\begin{cases} \dfrac{3x-2}{5} + \dfrac{7y+1}{10} = 10 \quad (1) \\ \dfrac{x+3}{2} - \dfrac{2y-5}{3} = 3 \quad (2) \end{cases}$ Multiply (1) by 10: $6x + 7y = 103$. Multiply (2) by 6: $3x - 4y = -1$.

$$D = \begin{vmatrix} 6 & 7 \\ 3 & -4 \end{vmatrix} = -45, \qquad D_x = \begin{vmatrix} 103 & 7 \\ -1 & -4 \end{vmatrix} = -405, \qquad D_y = \begin{vmatrix} 6 & 103 \\ 3 & -1 \end{vmatrix} = -315$$

$$x = \frac{D_x}{D} = \frac{-405}{-45} = 9, \qquad y = \frac{D_y}{D} = \frac{-315}{-45} = 7$$

The solution of the system is (9, 7).

(*b*) $\begin{cases} \dfrac{2}{y+1} - \dfrac{3}{x+1} = 0 \quad (1) \\ \dfrac{2}{x-7} + \dfrac{3}{2y-3} = 0 \quad (2) \end{cases}$ Multiply (1) by $(x+1)(y+1)$: $2x - 3y = 1$. Multiply (2) by $(x-7)(2y-3)$: $3x + 4y = 27$.

$$D = \begin{vmatrix} 2 & -3 \\ 3 & 4 \end{vmatrix} = 17, \qquad D_x = \begin{vmatrix} 1 & -3 \\ 27 & 4 \end{vmatrix} = 85, \qquad D_y = \begin{vmatrix} 2 & 1 \\ 3 & 27 \end{vmatrix} = 51$$

$$x = \frac{D_x}{D} = \frac{85}{17} = 5, \qquad y = \frac{D_y}{D} = \frac{51}{17} = 3$$

The solution of the system is (5, 3).

28.6 Solve the following systems of equations.

(*a*) $\begin{cases} \dfrac{3}{x} - \dfrac{6}{y} = \dfrac{1}{6} \\ \dfrac{2}{x} + \dfrac{3}{y} = \dfrac{1}{2} \end{cases}$ These are linear equations in $\dfrac{1}{x}$ and $\dfrac{1}{y}$.

$$D = \begin{vmatrix} 3 & -6 \\ 2 & 3 \end{vmatrix} = 21, \qquad D_{1/x} = \begin{vmatrix} 1/6 & -6 \\ 1/2 & 3 \end{vmatrix} = \frac{7}{2}, \qquad D_{1/y} = \begin{vmatrix} 3 & 1/6 \\ 2 & 1/2 \end{vmatrix} = \frac{7}{6}$$

$$\frac{1}{x} = \frac{D_{1/x}}{D} = \frac{7/2}{21} = \frac{1}{6}, \qquad \frac{1}{y} = \frac{D_{1/y}}{D} = \frac{7/6}{21} = \frac{1}{18}$$

$$x = \frac{1}{1/x} = \frac{1}{1/6} = 6, \qquad y = \frac{1}{1/y} = \frac{1}{1/18} = 18$$

The solution of the system is (6, 18).

(*b*) $\begin{cases} \dfrac{3}{2x} + \dfrac{8}{5y} = 3 \\ \dfrac{4}{3y} - \dfrac{1}{x} = 1 \end{cases}$ can be written $\begin{cases} \dfrac{3}{2}\left(\dfrac{1}{x}\right) + \dfrac{8}{5}\left(\dfrac{1}{y}\right) = 3 \\ -\left(\dfrac{1}{x}\right) + \dfrac{4}{3}\left(\dfrac{1}{y}\right) = 1. \end{cases}$

$$D = \begin{vmatrix} 3/2 & 8/5 \\ -1 & 4/3 \end{vmatrix} = \frac{18}{5}, \qquad D_{1/x} = \begin{vmatrix} 3 & 8/5 \\ 1 & 4/3 \end{vmatrix} = \frac{12}{5}, \qquad D_{1/y} = \begin{vmatrix} 3/2 & 3 \\ -1 & 1 \end{vmatrix} = \frac{9}{2}$$

$$\frac{1}{x} = \frac{D_{1/x}}{D} = \frac{12/5}{18/5} = \frac{2}{3}, \qquad \frac{1}{y} = \frac{D_{1/y}}{D} = \frac{9/2}{18/5} = \frac{5}{4}$$

$$x = \frac{1}{1/x} = \frac{1}{2/3} = \frac{3}{2}, \qquad y = \frac{1}{1/y} = \frac{1}{5/4} = \frac{4}{5}$$

The solution of the system is (3/2, 4/5).

28.7 Evaluate each of the following determinants.

(*a*) $\begin{vmatrix} 3 & -2 & 2 \\ 1 & 4 & 5 \\ 6 & -1 & 2 \end{vmatrix}$

Repeat the first two columns:

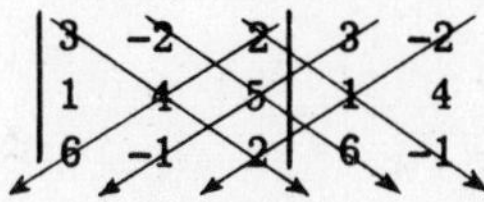

$$(3)(4)(2) + (-2)(5)(6) + (2)(1)(-1) - (2)(4)(6) - (3)(5)(-1) - (-2)(1)(2) = -67$$

(*b*) $\begin{vmatrix} -1 & 2 & -3 \\ 5 & -3 & 2 \\ 1 & -1 & -3 \end{vmatrix} = 29$

(*c*) $\begin{vmatrix} 2 & 3 & 2 \\ 0 & -2 & 1 \\ -1 & 4 & 0 \end{vmatrix} = -15$

(*d*) $\begin{vmatrix} a & b & c \\ c & a & b \\ b & c & a \end{vmatrix} = a^3 + b^3 + c^3 - 3abc$

(*e*) $\begin{vmatrix} (x-2) & (y+3) & (z-2) \\ -2 & 3 & 4 \\ 1 & -2 & 1 \end{vmatrix} = 11x + 6y + z - 6$

28.8 (*a*) Show that if two rows (or two columns) of a third-order determinant have their corresponding elements proportional, the value of the determinant is zero.

(*b*) Show that if the elements of any row (or column) are multiplied by any given constant and added to the corresponding elements of any other row (or column), the value of the determinant is unchanged.

SOLUTION

(*a*) We must show that

$$\begin{vmatrix} a_1 & b_1 & c_1 \\ ka_1 & kb_1 & kc_1 \\ a_3 & b_3 & c_3 \end{vmatrix} = 0,$$

where the elements in the first and second rows are proportional. This is shown by expansion of the determinant.

(*b*) Let the given determinant be

$$\begin{vmatrix} a_1 & b_1 & c_1 \\ a_2 & b_2 & c_2 \\ a_3 & b_3 & c_3 \end{vmatrix}.$$

We must show that if k is any constant

$$\begin{vmatrix} a_1 & b_1 & c_1 \\ a_2 & b_2 & c_2 \\ a_3+ka_2 & b_3+kb_2 & c_3+kc_2 \end{vmatrix} = \begin{vmatrix} a_1 & b_1 & c_1 \\ a_2 & b_2 & c_2 \\ a_3 & b_3 & c_3 \end{vmatrix}.$$

where we have multiplied each of the elements in the second row of the given determinant by k and added to the corresponding elements in the third row. The result is proved by expanding each of the determinants and showing that they are equal.

28.9 Solve the following systems of equations.

(*a*) $\begin{cases} 2x+y-z=5 \\ 3x-2y+2z=-3 \\ x-3y-3z=-2 \end{cases}$

Here $$D = \begin{vmatrix} 2 & 1 & -1 \\ 3 & -2 & 2 \\ 1 & -3 & -3 \end{vmatrix} = 42 \quad \text{and}$$

$$D_x = \begin{vmatrix} 5 & 1 & -1 \\ -3 & -2 & 2 \\ -2 & -3 & -3 \end{vmatrix} = 42 \qquad D_y = \begin{vmatrix} 2 & 5 & -1 \\ 3 & -3 & 2 \\ 1 & -2 & -3 \end{vmatrix} = 84 \qquad D_z = \begin{vmatrix} 2 & 1 & 5 \\ 3 & -2 & -3 \\ 1 & -3 & -2 \end{vmatrix} = -42$$

$$x = \frac{D_x}{D} = \frac{42}{42} = 1, \qquad y = \frac{D_y}{D} = \frac{84}{42} = 2, \qquad z = \frac{D_z}{D} = \frac{-42}{42} = -1$$

The solution of the system is $(1, 2, -1)$.

(*b*) $\begin{cases} x+2z=7 \\ 3x+y=5 \\ 2y-3z\ -5 \end{cases}$

Write as $$\begin{cases} x+0y+2z=7 \\ 3x+y+0z=5 \\ 0x+2y-3z=-5. \end{cases}$$

Then $$D = \begin{vmatrix} 1 & 0 & 2 \\ 3 & 1 & 0 \\ 0 & 2 & -3 \end{vmatrix} = 9 \quad \text{and}$$

$$D_x = \begin{vmatrix} 7 & 0 & 2 \\ 5 & 1 & 0 \\ -5 & 2 & -3 \end{vmatrix} = 9 \qquad D_x = \begin{vmatrix} 1 & 7 & 2 \\ 3 & 5 & 0 \\ 0 & -5 & -3 \end{vmatrix} = 18 \qquad D_z = \begin{vmatrix} 1 & 0 & 7 \\ 3 & 1 & 5 \\ 0 & 2 & -5 \end{vmatrix} = 27$$

$$x = \frac{D_x}{D} = \frac{9}{9} = 1, \qquad y = \frac{D_y}{D} = \frac{18}{9} = 2, \qquad z = \frac{D_z}{D} = \frac{27}{9} = 3$$

The solution of the system is $(1, 2, 3)$.

28.10 The equations for the currents i_1, i_2, i_3 in a given electrical network are

$$\begin{cases} 3i_1 - 2i_2 + 4i_3 = 2 \\ i_1 + 3i_2 - 6i_3 = 8 \\ 2i_1 - i_2 - 2i_3 = 0. \end{cases}$$

Find i_3.

SOLUTION

$$i_3 = \frac{\begin{vmatrix} 3 & -2 & 2 \\ 1 & 3 & 8 \\ 2 & -1 & 0 \end{vmatrix}}{\begin{vmatrix} 3 & -2 & 4 \\ 1 & 3 & -6 \\ 2 & -1 & -2 \end{vmatrix}} = \frac{-22}{-44} = \frac{1}{2}$$

Supplementary Problems

28.11 Evaluate each of the following determinants.

(a) $\begin{vmatrix} 4 & -3 \\ -1 & 2 \end{vmatrix}$ (b) $\begin{vmatrix} -2 & 4 \\ -3 & 7 \end{vmatrix}$ (c) $\begin{vmatrix} 2 & -1 \\ 4 & 0 \end{vmatrix}$

(d) $\begin{vmatrix} -2x & -3y \\ 4x & -y \end{vmatrix}$ (e) $\begin{vmatrix} a+b & a-b \\ a & -b \end{vmatrix}$ (f) $\begin{vmatrix} 2x-1 & x+1 \\ x+2 & x-2 \end{vmatrix}$

28.12 Show that if the elements of one row (or column) of a second-order determinant are multiplied by the same number, the determinant is multiplied by the number.

28.13 Solve the unknowns in each of the following systems.

(a) $\begin{cases} 5x + 2y = 4 \\ 2x - y = 7 \end{cases}$ (b) $\begin{cases} 3r - 5s = -6 \\ 4r + 2s = 5 \end{cases}$ (c) $\begin{cases} 28 + 4x + 5y = 0 \\ -3x + 4y + 10 = 0 \end{cases}$ (d) $\begin{cases} 5x - 4y = 16 \\ 2x + 3y = -10 \end{cases}$

(e) $\begin{cases} \dfrac{x-3}{3} + \dfrac{y+4}{5} = 7 \\ \dfrac{x+2}{7} - \dfrac{y-6}{2} = -3 \end{cases}$ (f) $\begin{cases} \dfrac{3x+2y+1}{x+y} = 4 \\ \dfrac{5x+6y-7}{x+y} = 2 \end{cases}$ (g) $\begin{cases} \dfrac{4}{x} + \dfrac{1}{y} = \dfrac{2}{5} \\ \dfrac{3}{x} - \dfrac{5}{y} = -\dfrac{1}{12} \end{cases}$ (h) $\begin{cases} \dfrac{4}{3u} - \dfrac{3}{5v} = 1 \\ \dfrac{1}{u} - \dfrac{1}{v} = -\dfrac{1}{6} \end{cases}$

28.14 Evaluate each determinant.

(a) $\begin{vmatrix} -2 & 1 & 2 \\ 3 & -1 & 3 \\ 1 & 3 & -2 \end{vmatrix}$ (b) $\begin{vmatrix} 1 & 0 & -2 \\ 0 & -3 & 4 \\ -4 & 2 & -1 \end{vmatrix}$ (c) $\begin{vmatrix} 3 & -1 & 4 \\ -2 & 1 & -3 \\ 1 & 3 & -2 \end{vmatrix}$

(d) $\begin{vmatrix} x & y & z \\ -2 & 3 & 1 \\ 4 & 1 & 2 \end{vmatrix}$ (e) $\begin{vmatrix} 1 & 1 & 1 \\ a & b & c \\ a^2 & b^2 & c^2 \end{vmatrix}$

28.15 For what value of k does

$$\begin{vmatrix} k+3 & 1 & -2 \\ 3 & -2 & 1 \\ -k & -3 & 3 \end{vmatrix} = 0?$$

28.16 Show that if the elements of one row (or column) of a third-order determinant are multiplied by the same number, the determinant is multiplied by the number.

28.17 Solve for the unknowns in each of the following systems.

(*a*) $\begin{cases} 3x + y - 2z = 1 \\ 2x + 3y - z = 2 \\ x - 2y + 2z = -10 \end{cases}$ (*b*) $\begin{cases} u + 2v - 3w = -7 \\ 2u - v + w = 5 \\ 3u - v + 2w = 8 \end{cases}$ (*c*) $\begin{cases} 2x + 3y = -2 \\ 5y - 2z = 4 \\ 3z + 4x = -7 \end{cases}$

28.18 Solve for the indicated unknown.

(*a*) $\begin{cases} 3i_1 + i_2 + 2i_3 = 0 \\ i_1 + 2i_2 - 3i_3 = 5 \\ 2i_1 - i_2 + i_3 = -1 \end{cases}$ for i_2 (*b*) $\begin{cases} 1/x + 2/y + 1/z = 1/2 \\ 4/x + 2/y - 3/z = 2/3 \\ 3/x - 4/y + 4/z = 1/3 \end{cases}$ for x

ANSWERS TO SUPPLEMENTARY PROBLEMS

28.11 (*a*) 5 (*b*) −2 (*c*) 4 (*d*) $14xy$ (*e*) $-a^2 - b^2$ (*f*) $x^2 - 8x$

28.13 (*a*) $x = 2,\ y = -3;\ (2, -3)$
(*b*) $r = 1/2,\ s = 3/2;\ (1/2, 3/2)$
(*c*) $x = -2,\ y = -4;\ (-2, -4)$
(*d*) $x = 8/23,\ y = -82/23;\ (8/23, -82/23)$
(*e*) $x = 12,\ y = 16;\ (12, 16)$
(*f*) $x = 5y = -2;\ (5, -2)$
(*g*) $x = 12,\ y = 15;\ (12, 15)$
(*h*) $u = 2/3,\ v = 3/5;\ (2/3, 3/5)$

28.14 (*a*) 43 (*b*) 19 (*c*) 0 (*d*) $5x + 8y - 14z$ (*e*) $bc^2 - cb^2 + a^2c - ac^2 + ab^2 - ba^2$

28.15 All values of k.

28.17 (*a*) $x = -2,\ y = 1,\ z = -3;\ (-2, 1, -3)$
(*b*) $u = 1,\ v = -1,\ w = 2;\ (1, -1, 2)$
(*c*) $x = -4,\ y = 2,\ z = 3;\ (-4, 2, 3)$

28.18 (*a*) $i_2 = 0.8$ (*b*) $x = 6$

Chapter 29

Determinants of Order n

29.1 INVERSION

An inversion of the arrangement of the positive integers occurs whenever one integer precedes a smaller integer.

For example, in $4, 3, 1, 5, 2$ the integer 4 precedes 3, 1 and 2, the integer 3 precedes 1 and 2, and 5 precedes 2; hence there are 6 inversions.

Similarly, an inversion of the arrangement of letters in alphabetical order occurs whenever one letter precedes another letter which occurs earlier in alphabetical order. Thus *bdca* has b preceding a, d preceding c and a; and c preceding a; hence there are 4 inversions.

29.2 DETERMINANTS OF ORDER n

The symbol

$$\begin{vmatrix} a_1 & b_1 & c_1 & \cdots & k_1 \\ a_2 & b_2 & c_2 & \cdots & k_2 \\ a_3 & b_3 & c_3 & \cdots & k_3 \\ \vdots & \vdots & \vdots & & \vdots \\ a_n & b_n & c_n & \cdots & k_n \end{vmatrix}$$

consisting of n^2 numbers (called elements) arranged in n rows and n columns, is called a *determinant of order* n. This symbol is an abbreviation for the algebraic sum of all possible products, each consisting of n factors, where:

(1) Each product has as factors one and only one element from each row and each column. There will be $n!$ such products.

(2) Each product has associated with it a plus or minus sign according as the number of inversions of the subscripts is even or odd, after the letters in the product have been written in the order of appearance in the first row of the determinant.

The algebraic sum thus obtained is called the *expansion* or *value* of the determinant. Each product in the expansion with its associated sign is called a *term* in the expansion of the determinant.

Sometimes an nth order determinant is written

$$\begin{vmatrix} a_{11} & a_{12} & a_{13} & \cdots & a_{1n} \\ a_{21} & a_{22} & a_{23} & \cdots & a_{2n} \\ a_{31} & a_{32} & a_{33} & \cdots & a_{3n} \\ \vdots & \vdots & \vdots & & \vdots \\ a_{n1} & a_{n2} & a_{n3} & \cdots & a_{nn} \end{vmatrix}$$

In this notation each element is characterized by two subscripts, the first indicating the *row* in which the element appears, the second indicating the *column* in which the element appears. Thus a_{23} is the element in the 2nd row and 3rd column whereas a_{32} is the element in the 3rd row and 2nd column.

The *principal diagonal* of a determinant consists of the elements in the determinant which lie in a straight line from the upper left-hand corner to the lower right-hand corner.

29.3 PROPERTIES OF DETERMINANTS

I. Interchanging corresponding rows and columns of a determinant does not change the value of the determinant. Thus any theorem proved true for rows holds for columns, and conversely.

EXAMPLE 29.1.

$$\begin{vmatrix} a_1 & b_1 & c_1 \\ a_2 & b_2 & c_2 \\ a_3 & b_3 & c_3 \end{vmatrix} = \begin{vmatrix} a_1 & a_2 & a_3 \\ b_1 & b_2 & b_3 \\ c_1 & c_2 & c_3 \end{vmatrix}$$

II. If each element in a row (or column) is zero, the value of the determinant is zero.

EXAMPLE 29.2.

$$\begin{vmatrix} a_1 & 0 & c_1 \\ a_2 & 0 & c_2 \\ a_3 & 0 & c_3 \end{vmatrix} = 0$$

III. Interchanging any two rows (or columns) reverses the sign of the determinant.

EXAMPLE 29.3.

$$\begin{vmatrix} a_1 & b_1 & c_1 \\ a_2 & b_2 & c_2 \\ a_3 & b_3 & c_3 \end{vmatrix} = - \begin{vmatrix} a_3 & b_3 & c_3 \\ a_2 & b_2 & c_2 \\ a_1 & b_1 & c_1 \end{vmatrix}$$

IV. If two rows (or columns) of a determinant are identical, the value of the determinant is zero.

EXAMPLE 29.4.

$$\begin{vmatrix} a_1 & b_1 & a_1 \\ a_2 & b_2 & a_2 \\ a_3 & b_3 & a_3 \end{vmatrix} = 0$$

V. If each of the elements in a row (or column) of a determinant is multiplied by the same number p, the value of the determinant is multiplied by p.

EXAMPLE 29.5.

$$\begin{vmatrix} pa_1 & b_1 & c_1 \\ pa_2 & b_2 & c_2 \\ pa_3 & b_3 & c_3 \end{vmatrix} = p \begin{vmatrix} a_1 & b_1 & c_1 \\ a_2 & b_2 & c_2 \\ a_3 & b_3 & c_3 \end{vmatrix}$$

VI. If each element of a row (or column) of a determinant is expressed as the sum of two (or more) terms, the determinant can be expressed as the sum of two (or more) determinants.

EXAMPLE 29.6.

$$\begin{vmatrix} a_1 + a_1' & b_1 & c_1 \\ a_2 + a_2' & b_2 & c_2 \\ a_3 + a_3' & b_3 & c_3 \end{vmatrix} = \begin{vmatrix} a_1 & b_1 & c_1 \\ a_2 & b_2 & c_2 \\ a_3 & b_3 & c_3 \end{vmatrix} + \begin{vmatrix} a_1' & b_1 & c_1 \\ a_2' & b_2 & c_2 \\ a_3' & b_3 & c_3 \end{vmatrix}$$

VII. If to each element of a row (or column) of a determinant is added m times the corresponding element of any other row (or column), the value of the determinant is not changed.

EXAMPLE 29.7.

$$\begin{vmatrix} a_1 + mb_1 & b_1 & c_1 \\ a_2 + mb_2 & b_2 & c_2 \\ a_3 + mb_3 & b_3 & c_3 \end{vmatrix} = \begin{vmatrix} a_1 & b_1 & c_1 \\ a_2 & b_2 & c_2 \\ a_3 & b_3 & c_3 \end{vmatrix}$$

These properties may be proved for the special cases of second and third order determinants by using the methods of expansion of Chapter 28. For proofs of the general cases see the solved problems below.

29.4 MINORS

The minor of an element in a determinant of order n is the determinant of order $n-1$ obtained by removing the row and the column which contain the given element.

For example, the minor of b_3 in the 4th order determinant

$$\begin{vmatrix} a_1 & b_1 & c_1 & d_1 \\ a_2 & b_2 & c_2 & d_2 \\ a_3 & b_3 & c_3 & d_3 \\ a_4 & b_4 & c_4 & d_4 \end{vmatrix}$$

is obtained by crossing out the row and column containing b_3, as shown, and writing the resulting determinant of order 3, namely

$$\begin{vmatrix} a_1 & c_1 & d_1 \\ a_2 & c_2 & d_2 \\ a_4 & c_4 & d_4 \end{vmatrix}.$$

The minor of an element is denoted by capital letters. Thus the minor corresponding to the element b_3 is denoted by B_3.

29.5 VALUE OF A DETERMINANT

The value of a determinant may be obtained in terms of minors as follows:

(1) Choose any row (or column).

(2) Multiply each element in the row (or column) by its corresponding minor preceded by a sign which is *plus* or *minus* according as the sum of the column number and row number is *even* or *odd*. The minor of an element with the attached sign is called the *cofactor* of the element.

(3) Add algebraically the products obtained in (2).

For example, let us expand the determinant

$$\begin{vmatrix} a_1 & b_1 & c_1 & d_1 \\ a_2 & b_2 & c_2 & d_2 \\ a_3 & b_3 & c_3 & d_3 \\ a_4 & b_4 & c_4 & d_4 \end{vmatrix}$$

by the elements in the third row. The minors of a_3, b_3, c_3, d_3 are A_3, B_3, C_3, D_3, respectively. The sign corresponding to the element a_3 is + since it appears in the 1st column and 3rd row and $1+3=4$ is even. Similarly, the signs associated with the elements b_3, c_3, d_3 are $-,+,-$ respectively. Thus the value of the determinant is

$$a_3A_3 - b_3B_3 + c_3C_3 - d_3D_3.$$

Property VII is useful in producing zeros in a given row or column. This property coupled with the expansion in terms of minors makes for easy determination of the value of a determinant.

29.6 CRAMER'S RULE

Cramer's Rule for the solution of n simultaneous linear equations in n unknowns is exactly analogous to the rule given in Chapter 28 for the case $n=2$ and $n=3$.

Given n linear equations in n unknowns $x_1, x_2, x_3, \ldots, x_n$

$$\begin{array}{c} a_{11}x_1 + a_{12}x_2 + a_{13}x_3 + \ldots + a_{1n}x_n = r_1 \\ a_{21}x_1 + a_{22}x_2 + a_{23}x_3 + \ldots + a_{2n}x_n = r_2 \\ \vdots \qquad \vdots \qquad \vdots \qquad \qquad \vdots \qquad \vdots \\ a_{n1}x_1 + a_{n2}x_2 + a_{n3}x_3 + \ldots + a_{nn}x_n = r_n. \end{array} \tag{1}$$

Let D be the determinant of the coefficients of $x_1, x_2, x_3, \ldots, x_n$, i.e.,

$$D = \begin{vmatrix} a_{11} & a_{12} & a_{13} & \ldots & a_{1n} \\ a_{21} & a_{22} & a_{23} & \ldots & a_{2n} \\ \vdots & \vdots & \vdots & & \vdots \\ a_{n1} & a_{n2} & a_{n3} & \ldots & a_{nn} \end{vmatrix}$$

Denote by D_k the determinant D with the kth column (which corresponds to the coefficients of the unknown x_k) replaced by the column of the coefficients on the right-hand side of (1). Then

$$x_1 = \frac{D_1}{D}, \qquad x_2 = \frac{D_2}{D}, \qquad x_3 = \frac{D_3}{D}, \ldots \qquad \text{provided } D \neq 0.$$

If $D \neq 0$ there is one and only one solution.

If $D = 0$ the system of equations may or may not have solutions.

Equations having no simultaneous solution are called *inconsistent*, otherwise they are consistent. If $D = 0$ and at least one of the determinants $D_1, D_2, \ldots, D_n \neq 0$, the given system is inconsistent. If $D = D_1 = D_2 = \ldots = D_n = 0$, the system may or may not be consistent.

Equations having an infinite number of simultaneous solutions are called *dependent*. If a system of equations is dependent then $D = 0$ and all of the determinants $D_1, D_2, \ldots, D_n = 0$. The converse, however, is not always true.

29.7 HOMOGENEOUS LINEAR EQUATIONS

If $r_1, r_2, \ldots, r_n$ in equations (1) are all zero, the system is said to be *homogeneous*. In this case $D_1 = D_2 = D_3 = \ldots = D_n = 0$ and the following theorem is true.

> **Theorem:** A necessary and sufficient condition that n homogeneous linear equations in n unknowns have solutions other than the trivial solution (where all the unknowns equal zero) is that the determinant of the coefficients, $D = 0$.

A system of m equations in n unknowns may or may not have simultaneous solutions.

(1) If $m > n$, the unknowns in n of the given equations may be obtained. If these values satisfy the remaining $m - n$ equations the system is consistent, otherwise it is inconsistent.

(2) If $m < n$, then m of the unknowns may be determined in terms of the remaining $n - m$ unknowns.

Solved Problems

29.1 Determine the number of inversions in each of the following groupings.

(*a*) $3, 1, 2$	3 precedes 1 and 2. There are 2 inversions.
(*b*) $4, 2, 3, 1$	4 precedes 2, 3, 1; 2 precedes 1; 3 precedes 1. There are 5 inversions.
(*c*) $5, 1, 4, 3, 2$	5 precedes 1, 4, 3, 2; 4 precedes 3, 2; 3 precedes 2. There are 7 inversions.
(*d*) b, a, c	b precedes a. There is 1 inversion.
(*e*) b, a, e, d, c	b precedes a; e precedes d, c; d precedes c. There are 4 inversions.

29.2 (*a*) What would be the number of inversions in the subscripts of $b_1d_3c_2a_4$ when the letters are in alphabetical order?

(*b*) What would be the number of inversions in the letters of $b_1d_3c_2a_4$ when the subscripts are arranged in natural order?

SOLUTION

(*a*) Write $a_4b_1c_2d_3$. The subscripts 4, 1, 2, 3 have 3 inversions.

(*b*) Write $b_1c_2d_3a_4$. The letters *bcda* have 3 inversions.

The fact that there are 3 inversions in (*a*) and (*b*) is not merely a coincidence. There are the same number of inversions with regard to subscripts when letters are in alphabetical order as inversions with regard to letters when subscripts are in natural order.

29.3 Write the expansion of the determinant

$$\begin{vmatrix} a_1 & b_1 & c_1 \\ a_2 & b_2 & c_2 \\ a_3 & b_3 & c_3 \end{vmatrix}$$

by use of inversions.

SOLUTION

The expansion consists of terms of the form *abc* with all possible arrangements of the subscripts, the sign before the product being determined by the number of inversions in the subscripts, a plus sign if there is an even number of inversions, a minus sign if there is an odd number of inversions.

There are 6 terms in the expansion. These terms are:

$a_1b_2c_3$	0 inversions, sign +	$a_2b_3c_1$	2 inversions, sign +
$a_1b_3c_2$	1 inversion, sign −	$a_3b_2c_1$	3 inversions, sign −
$a_2b_1c_3$	1 inversion, sign −	$a_3b_1c_2$	2 inversions, sign +.

The required expansion is: $a_1b_2c_3 - a_1b_3c_2 - a_2b_1c_3 + a_2b_3c_1 - a_3b_2c_1 + a_3b_1c_2$.

29.4 Determine the signs associated with the terms $d_1a_3c_2b_4$ and $a_3c_2d_4b_1$ in the expansion of the determinant

$$\begin{vmatrix} a_1 & b_1 & c_1 & d_1 \\ a_2 & b_2 & c_2 & d_2 \\ a_3 & b_3 & c_3 & d_3 \\ a_4 & b_4 & c_4 & d_4 \end{vmatrix}$$

SOLUTION

$d_1a_3c_2b_4$ written $a_3b_4c_2d_1$ has 5 inversions in the subscripts. The associated sign is $-$.
$a_3c_2d_4b_1$ written $a_3b_1c_2d_4$ has 2 inversions in the subscripts. The associated sign is $+$.

29.5 Prove Property III: If two rows (or columns) are interchanged, the sign of the determinant is changed.

SOLUTION

Case 1. Rows are adjacent.
Interchanging two adjacent rows results in the interchange of two adjacent subscripts in each term of the expansion. Thus the number of inversions of subscripts is either increased by one or decreased by one. Hence the sign of each term is changed and so the sign of the determinant is changed.

Case 2. Rows are not adjacent.
Suppose there are k rows between the ones to be interchanged. It will then take k interchanges of the adjacent rows to bring the upper row to the row just above the lower row, one more to interchange them, and k more interchanges to bring the lower row up to where the upper row was. This involves a total of $k + 1 + k = 2k + 1$ interchanges. Since $2k + 1$ is odd, there is an odd number of changes in sign and the result is the same as a single change in sign.

29.6 Prove Property IV: If two rows (or columns) are identical, the determinant has value zero.

SOLUTION

Let D be the value of the determinant. By Property III, interchange of the two identical rows should change the value to $-D$. Since the determinants are the same, $D = -D$ or $D = 0$.

29.7 Prove Property V: If each of the elements of a row (or column) are multiplied by the same number p, the value of the determinant is multiplied by p.

SOLUTION

Each term of the determinant contains one and only one element from the row multiplied by p and thus each term has factor p. This factor is therefore common to all the terms of the expansion and so the determinant is multiplied by p.

29.8 Prove Property VI: If each element of a row (or column) of a determinant is expressed as the sum of two (or more) terms, the determinant can be expressed as the sum of two (or more) determinants.

SOLUTION

For the case of third-order determinants we must show that

$$\begin{vmatrix} a_1 + a_1' & b_1 & c_1 \\ a_2 + a_2' & b_2 & c_2 \\ a_3 + a_3' & b_3 & c_3 \end{vmatrix} = \begin{vmatrix} a_1 & b_1 & c_1 \\ a_2 & b_2 & c_2 \\ a_3 & b_3 & c_3 \end{vmatrix} + \begin{vmatrix} a_1' & b_1 & c_1 \\ a_2' & b_2 & c_2 \\ a_3' & b_3 & c_3 \end{vmatrix}.$$

Each term in the expansion of the determinant on the left equals the sum of the two corresponding terms in the determinants on the right, e.g., $(a_2 + a_2')b_3c_1 = a_2b_3c_1 + a_2'b_3c_1$. Thus the property holds for third order determinants. The method of proof holds in the general case.

29.9 Prove Property VII: If to each element of a row (or column) of a determinant is added m times the corresponding element of any other row (or column), the value of the determinant is not changed.

SOLUTION

For the case of a third-order determinant we must show that

$$\begin{vmatrix} a_1 + mb_1 & b_1 & c_1 \\ a_2 + mb_2 & b_2 & c_2 \\ a_3 + mb_3 & b_3 & c_3 \end{vmatrix} = \begin{vmatrix} a_1 & b_1 & c_1 \\ a_2 & b_2 & c_2 \\ a_3 & b_3 & c_3 \end{vmatrix}.$$

By property VI the right-hand side may be written

$$\begin{vmatrix} a_1 & b_1 & c_1 \\ a_2 & b_2 & c_2 \\ a_3 & b_3 & c_3 \end{vmatrix} + \begin{vmatrix} mb_1 & b_1 & c_1 \\ mb_2 & b_2 & c_2 \\ mb_3 & b_3 & c_3 \end{vmatrix}.$$

This last determinant may be written

$$m\begin{vmatrix} b_1 & b_1 & c_1 \\ b_2 & b_2 & c_2 \\ b_3 & b_3 & c_3 \end{vmatrix}$$

which is zero by Property IV.

29.10 Show that

$$\begin{vmatrix} 3 & 2 & 2 & 1 \\ 6 & 5 & 4 & -2 \\ 9 & -3 & 6 & -5 \\ 12 & 2 & 8 & 7 \end{vmatrix} = 0.$$

SOLUTION

The number 3 may be factored from each element in the first column and 2 may be factored from each element in the third column to yield

$$(3)(2)\begin{vmatrix} 1 & 2 & 1 & 1 \\ 2 & 5 & 2 & -2 \\ 3 & -3 & 3 & -5 \\ 4 & 2 & 4 & 7 \end{vmatrix}$$

which equals zero since the first and third columns are identical.

29.11 Use Property VII to transform the determinant

$$\begin{vmatrix} 1 & -2 & 3 \\ 2 & -1 & 4 \\ -2 & 3 & 1 \end{vmatrix}$$

into a determinant of equal value with zeros in the first row, second and third columns.

SOLUTION

Multiply each element in the first column by 2 and add to the corresponding elements in the second column, thus obtaining

$$\begin{vmatrix} 1 & (2)(1)-2 & 3 \\ 2 & (2)(2)-1 & 4 \\ -2 & (2)(-2)+3 & 1 \end{vmatrix} = \begin{vmatrix} 1 & 0 & 3 \\ 2 & 3 & 4 \\ -2 & -1 & 1 \end{vmatrix}.$$

Multiply each element in the first column of the new determinant by -3 and add to the corresponding elements in the third column to obtain

$$\begin{vmatrix} 1 & 0 & (-3)(1)+3 \\ 2 & 3 & (-3)(2)+4 \\ -2 & -1 & (-3)(-2)+1 \end{vmatrix} = \begin{vmatrix} 1 & 0 & 0 \\ 2 & 3 & -2 \\ -2 & -1 & 7 \end{vmatrix}.$$

The result could have been obtained in one step by writing

$$\begin{vmatrix} 1 & (2)(1)-2 & (-3)(1)+3 \\ 2 & (2)(2)-1 & (-3)(2)+4 \\ -2 & (2)(-2)+3 & (-3)(-2)+1 \end{vmatrix} = \begin{vmatrix} 1 & 0 & 0 \\ 2 & 3 & -2 \\ -2 & -1 & 7 \end{vmatrix}.$$

The choice of the numbers 2 and -3 was made in order to obtain zeros in the desired places.

29.12 Use Property VII to transform the determinant

$$\begin{vmatrix} 3 & 6 & 2 & 3 \\ -2 & 1 & -2 & 2 \\ 4 & -5 & 1 & 4 \\ 1 & 3 & 4 & -2 \end{vmatrix}$$

into an equal determinant having three zeros in the 4th row.

SOLUTION

Multiply each element in the 1st column (the *basic* column shown shaded) by -3, -4, $+2$ and add respectively to the corresponding elements in the 2nd, 3rd, 4th columns. The result is

$$\begin{vmatrix} 3 & (-3)(3)+6 & (-4)(3)+2 & (2)(3)+3 \\ -2 & (-3)(-2)+1 & (-4)(-2)-2 & (2)(-2)+2 \\ 4 & (-3)(4)-5 & (-4)(4)+1 & (2)(4)+4 \\ 1 & (-3)(1)+3 & (-4)(1)+4 & (2)(1)-2 \end{vmatrix} = \begin{vmatrix} 3 & -3 & -10 & 9 \\ -2 & 7 & 6 & -2 \\ 4 & -17 & -15 & 12 \\ 1 & 0 & 0 & 0 \end{vmatrix}$$

Note that it is useful to choose a basic row or column containing the element 1.

29.13 Obtain 4 zeros in a row or column of the 5th order determinant

$$\begin{vmatrix} 3 & 5 & 4 & 6 & 2 \\ -2 & 3 & 2 & 3 & 4 \\ 4 & 1 & 3 & -2 & -3 \\ 6 & -3 & 2 & 4 & 3 \\ 2 & 2 & 5 & 3 & -2 \end{vmatrix}.$$

SOLUTION

We shall produce zeros in the 2nd column by use of the basic row shown shaded. Multiply the elements in this basic row by $-5, -3, 3, -2$ and add respectively to the corresponding elements in the 1st, 2nd, 4th, 5th rows to obtain

$$\begin{vmatrix} -17 & 0 & -11 & 16 & 17 \\ -14 & 0 & -7 & 9 & 13 \\ 4 & 1 & 3 & -2 & -3 \\ 18 & 0 & 11 & -2 & -6 \\ -6 & 0 & -1 & 7 & 4 \end{vmatrix}$$

29.14 Obtain 3 zeros in a row or column of the determinant

$$\begin{vmatrix} 3 & 4 & 2 & 3 \\ -2 & 2 & 3 & -2 \\ 2 & -3 & 3 & 4 \\ 4 & 5 & -2 & -2 \end{vmatrix}$$

without changing its value.

SOLUTION

It is convenient to use Property VII to obtain an element 1 in a row or column. For example, by multiplying each of the elements in column 2 by -1 and adding to the corresponding elements in column 3, we obtain

$$\begin{vmatrix} 3 & 4 & -2 & 3 \\ -2 & 2 & 1 & -2 \\ 2 & -3 & 6 & 4 \\ 4 & 5 & -7 & -2 \end{vmatrix}.$$

Using the 3rd column as the basic column, multiply its elements by $2, -2, 2$ and add respectively to the 1st, 2nd, 4th columns to obtain

$$\begin{vmatrix} -1 & 8 & -2 & -1 \\ 0 & 0 & 1 & 0 \\ 14 & -15 & 6 & 16 \\ -10 & 19 & -7 & -16 \end{vmatrix}$$

which equals the given determinant.

29.15 Write the minor and corresponding cofactor of the element in the second row, third column for the determinant

$$\begin{vmatrix} 2 & -2 & 3 & 1 \\ 1 & 3 & 2 & 5 \\ 1 & -2 & 5 & -1 \\ 2 & 1 & 3 & -2 \end{vmatrix}.$$

SOLUTION

Crossing out the row and column containing the element, the minor is given by

$$\begin{vmatrix} 2 & -2 & 1 \\ 1 & -2 & -1 \\ 2 & 1 & -2 \end{vmatrix}.$$

Since the element is in the 2nd row, 3rd column and $2 + 3 = 5$ is an odd number, the associated sign is minus. Thus the cofactor corresponding to the given element is

$$-\begin{vmatrix} 2 & -2 & 1 \\ 1 & -2 & -1 \\ 2 & 1 & -2 \end{vmatrix}.$$

29.16 Write the minors and cofactors of the elements in the 4th row of the determinant

$$\begin{vmatrix} 3 & -2 & 4 & 2 \\ 2 & 1 & 5 & -3 \\ 1 & 5 & -2 & 2 \\ -3 & -2 & -4 & 1 \end{vmatrix}.$$

SOLUTION

The elements in the 4th row are -3, -2, -4, 1.

$$\text{Minor of element } -3 = \begin{vmatrix} -2 & 4 & 2 \\ 1 & 5 & -3 \\ 5 & -2 & 2 \end{vmatrix} \qquad \text{Cofactor} = -\text{Minor}$$

$$\text{Minor of element } -2 = \begin{vmatrix} 3 & 4 & 2 \\ 2 & 5 & -3 \\ 1 & -2 & 2 \end{vmatrix} \qquad \text{Cofactor} = +\text{Minor}$$

$$\text{Minor of element } -4 = \begin{vmatrix} 3 & -2 & 2 \\ 2 & 1 & -3 \\ 1 & 5 & 2 \end{vmatrix} \qquad \text{Cofactor} = -\text{Minor}$$

$$\text{Minor of element } 1 = \begin{vmatrix} 3 & -2 & 4 \\ 2 & 1 & 5 \\ 1 & 5 & -2 \end{vmatrix} \qquad \text{Cofactor} = +\text{Minor}$$

29.17 Express the value of the determinant of Problem 29.16 in terms of minors or cofactors.

SOLUTION

Value of determinant = sum of elements each multiplied by associated cofactor

$$= (-3)\left\{-\begin{vmatrix} -2 & 4 & 2 \\ 1 & 5 & -3 \\ 5 & -2 & 2 \end{vmatrix}\right\} + (-2)\left\{+\begin{vmatrix} 3 & 4 & 2 \\ 2 & 5 & -3 \\ 1 & -2 & 2 \end{vmatrix}\right\}$$

$$+ (-4)\left\{-\begin{vmatrix} 3 & -2 & 2 \\ 2 & 1 & -3 \\ 1 & 5 & 2 \end{vmatrix}\right\} + (1)\left\{+\begin{vmatrix} 3 & -2 & 4 \\ 2 & 1 & 5 \\ 1 & 5 & -2 \end{vmatrix}\right\}$$

Upon evaluating each of the 3rd order determinants the result -53 is obtained.

The method of evaluation here indicated is tedious. However, the labor involved may be considerably reduced by first transforming a given determinant into an equivalent one having zeros in a row or column by use of Property VII as shown in the following problem.

29.18 Evaluate the determinant in Problem 29.16 by first transforming it into one having three zeros in a row or column and then expanding by minors.

SOLUTION

Choosing the basic column indicated,

$$\begin{vmatrix} 3 & -2 & 4 & 2 \\ 2 & 1 & 5 & -3 \\ 1 & 5 & -2 & 2 \\ -3 & -2 & -4 & 1 \end{vmatrix},$$

multiply its elements by -2, -5, 3 and add respectively to the corresponding elements of the 1st, 3rd, 4th colums to obtain

$$\begin{vmatrix} 7 & -2 & 14 & -4 \\ 0 & 1 & 0 & 0 \\ -9 & 5 & -27 & 17 \\ 1 & -2 & 6 & -5 \end{vmatrix}.$$

Expand according to the cofactors of the elements in the second row and obtain

$$(0)(\text{its cofactor}) + (1)(\text{its cofactor}) + (0)(\text{its cofactor}) + (0)(\text{its cofactor})$$

$$= (1)(\text{its cofactor}) = 1\left\{+\begin{vmatrix} 7 & 14 & -4 \\ -9 & -27 & 17 \\ 1 & 6 & -5 \end{vmatrix}\right\}.$$

Expanding this determinant, we obtain the value -53 which agrees with the result of Problem 29.16.

Note that the method of this problem may be employed to evaluate 3rd order determinants in terms of 2nd order determinants.

29.19 Evaluate each of the following determinants.

(*a*) $$\begin{vmatrix} 4 & 1 & -2 & 3 \\ -1 & 2 & 1 & 4 \\ 3 & -1 & 3 & 4 \\ 2 & 3 & -3 & 2 \end{vmatrix}$$

Multiply the elements in the indicated basic row by -2, 1, -3 and add respectively to the corresponding elements in the 2nd, 3rd, 4th rows to obtain

$$\begin{vmatrix} 4 & 1 & -2 & 3 \\ -9 & 0 & 5 & -2 \\ 7 & 0 & 1 & 7 \\ -10 & 0 & 3 & -7 \end{vmatrix} = 1\left\{-\begin{vmatrix} -9 & 5 & -2 \\ 7 & 1 & 7 \\ -10 & 3 & -7 \end{vmatrix}\right\}$$

$$= -\begin{vmatrix} -9 & 5 & -2 \\ 7 & 1 & 7 \\ -10 & 3 & -7 \end{vmatrix}.$$

Multiply the elements in the indicated basic column by -7 and add to the corresponding elements in the 1st and 3rd columns to obtain

$$-\begin{vmatrix} -44 & 5 & -37 \\ 0 & 1 & 0 \\ -31 & 3 & -28 \end{vmatrix} = -(1)\left\{+\begin{vmatrix} -44 & -37 \\ -31 & -28 \end{vmatrix}\right\} = -85.$$

(b)
$$\begin{vmatrix} 1 & -3 & 2 & -3 & 1 \\ -1 & 2 & 1 & 2 & -3 \\ -3 & 1 & -2 & -1 & 4 \\ 2 & -3 & 3 & 4 & -1 \\ 3 & -2 & -4 & 2 & 1 \end{vmatrix}$$

Multiply the elements in the indicated basic column by 3, 2, 1, −4 respectively and add to the corresponding elements in the 1st, 3rd, 4th, 5th columns to obtain

$$\begin{vmatrix} -8 & -3 & -4 & -6 & 13 \\ 5 & 2 & 5 & 4 & -11 \\ 0 & 1 & 0 & 0 & 0 \\ -7 & -3 & -3 & 1 & 11 \\ -3 & -2 & -8 & 0 & 9 \end{vmatrix} = 1\left\{-\begin{vmatrix} -8 & -4 & -6 & 13 \\ 5 & 5 & 4 & -11 \\ -7 & -3 & 1 & 11 \\ -3 & -8 & 0 & 9 \end{vmatrix}\right\} = -\begin{vmatrix} -8 & -4 & -6 & 13 \\ 5 & 5 & 4 & -11 \\ -7 & -3 & 1 & 11 \\ -3 & -8 & 0 & 9 \end{vmatrix}.$$

In the last determinant, multiply the elements in the indicated basic row by 6, −4 and add respectively to the elements in the 1st and 2nd rows to obtain

$$-\begin{vmatrix} -50 & -22 & 0 & 79 \\ 33 & 17 & 0 & -55 \\ -7 & -3 & 1 & 11 \\ -3 & -8 & 0 & 9 \end{vmatrix} = -(1)\left\{+\begin{vmatrix} -50 & -22 & 79 \\ 33 & 17 & -55 \\ -3 & -8 & 9 \end{vmatrix}\right\} = -\begin{vmatrix} -50 & -22 & 79 \\ 33 & 17 & -55 \\ -3 & -8 & 9 \end{vmatrix}.$$

Multiply the elements in the indicated row of the last determinant by 2 and add to the 2nd row to obtain

$$-\begin{vmatrix} -50 & -22 & 79 \\ 27 & 1 & -37 \\ -3 & -8 & 9 \end{vmatrix}.$$

Multiply the elements in the indicated row of the last determinant by 22, 8 and add respectively to the elements in the 1st and 3rd rows to obtain

$$-\begin{vmatrix} 544 & 0 & -735 \\ 27 & 1 & -37 \\ 213 & 0 & -287 \end{vmatrix} = -(1)\left\{+\begin{vmatrix} 544 & -735 \\ 213 & -287 \end{vmatrix}\right\} = -427.$$

29.20 Factor the following determinant.

$$\begin{vmatrix} x & y & 1 \\ x^2 & y^2 & 1 \\ x^3 & y^3 & 1 \end{vmatrix} = xy\begin{vmatrix} 1 & 1 & 1 \\ x & y & 1 \\ x^2 & y^2 & 1 \end{vmatrix}$$

Removing factors x and y from 1st and 2nd columns respectively.

$$= xy\begin{vmatrix} 0 & 0 & 1 \\ x-1 & y-1 & 1 \\ x^2-1 & y^2-1 & 1 \end{vmatrix}$$

Adding −1 times elements in 3rd column to the corresponding elements in 1st and 2nd columns.

$$= xy\begin{vmatrix} x-1 & y-1 \\ x^2-1 & y^2-1 \end{vmatrix}$$

$$= xy(x-1)(y-1)\begin{vmatrix} 1 & 1 \\ x+1 & y+1 \end{vmatrix}$$

Removing factors $(x-1)$ and $(y-1)$ from 1st and 2nd columns respectively.

$$= xy(x-1)(y-1)(y-x).$$

29.21 Solve the system

$$\begin{aligned} 2x + y - z + w &= -4 \\ x + 2y + 2z - 3w &= 6 \\ 3x - y - z + 2w &= 0 \\ 2x + 3y + z + 4w &= -5. \end{aligned}$$

SOLUTION

$$D = \begin{vmatrix} 2 & 1 & -1 & 1 \\ 1 & 2 & 2 & -3 \\ 3 & -1 & -1 & 2 \\ 2 & 3 & 1 & 4 \end{vmatrix} = 86$$

$$D_1 = \begin{vmatrix} -4 & 1 & -1 & 1 \\ 6 & 2 & 2 & -3 \\ 0 & -1 & -1 & 2 \\ -5 & 3 & 1 & 4 \end{vmatrix} = 86 \qquad D_2 = \begin{vmatrix} 2 & -4 & -1 & 1 \\ 1 & 6 & 2 & -3 \\ 3 & 0 & -1 & 2 \\ 2 & -5 & 1 & 4 \end{vmatrix} = -172$$

$$D_3 = \begin{vmatrix} 2 & 1 & -4 & 1 \\ 1 & 2 & 6 & -3 \\ 3 & -1 & 0 & 2 \\ 2 & 3 & -5 & 4 \end{vmatrix} = 258 \qquad D_4 = \begin{vmatrix} 2 & 1 & -1 & -4 \\ 1 & 2 & 2 & 6 \\ 3 & -1 & -1 & 0 \\ 2 & 3 & 1 & -5 \end{vmatrix} = -86$$

Then $\quad x = \dfrac{D_1}{D} = 1, \quad y = \dfrac{D_2}{D} = -2, \quad z = \dfrac{D_3}{D} = 3, \quad w = \dfrac{D_4}{D} = -1.$

29.22 The currents i_1, i_2, i_3, i_4, i_5 (measured in amperes) can be determined from the following set of equations. Find i_3.

$$\begin{aligned} i_1 - 2i_2 + i_3 &= 3 \\ i_2 + 3i_4 - i_5 &= -5 \\ i_1 + i_2 + i_3 - i_5 &= 1 \\ 2i_2 + i_3 - 2i_4 - 2i_5 &= 0 \\ i_1 + i_3 + 2i_4 + i_5 &= 3 \end{aligned}$$

SOLUTION

$$D_3 = \begin{vmatrix} 1 & -2 & 3 & 0 & 0 \\ 0 & 1 & -5 & 3 & -1 \\ 1 & 1 & 1 & 0 & -1 \\ 0 & 2 & 0 & -2 & -2 \\ 1 & 0 & 3 & 2 & 1 \end{vmatrix} = 38, \qquad D = \begin{vmatrix} 1 & -2 & 1 & 0 & 0 \\ 0 & 1 & 0 & 3 & -1 \\ 1 & 1 & 1 & 0 & -1 \\ 0 & 2 & 1 & -2 & -2 \\ 1 & 0 & 1 & 2 & 1 \end{vmatrix} = 19, \qquad i_3 = \frac{D_3}{D} = 2 \text{ amp.}$$

29.23 Determine whether the system

$$\begin{aligned} x - 3y + 2z &= 4 \\ 2x + y - 3z &= -2 \\ 4x - 5y + z &= 5 \end{aligned}$$

is consistent.

SOLUTION

$$D = \begin{vmatrix} 1 & -3 & 2 \\ 2 & 1 & -3 \\ 4 & -5 & 1 \end{vmatrix} = 0.$$

However,
$$D_1 = \begin{vmatrix} 4 & -3 & 2 \\ -2 & 1 & -3 \\ 5 & -5 & 1 \end{vmatrix} = -7.$$

Hence at least one of the determinants $D_1, D_2, D_3 \neq 0$ so that the equations are inconsistent.

This could be seen in another way by multiplying the first equation by 2 and adding to the second equation to obtain $4x - 5y + z = 6$ which is not consistent with the last equation.

29.24 Determine whether the system

$$4x - 2y + 6z = 8$$
$$2x - y + 3z = 5$$
$$2x - y + 3z = 4$$

is consistent.

SOLUTION

$$D = \begin{vmatrix} 4 & -2 & 6 \\ 2 & -1 & 3 \\ 2 & -1 & 3 \end{vmatrix} = 0 \qquad D_1 = \begin{vmatrix} 8 & -2 & 6 \\ 5 & -1 & 3 \\ 4 & -1 & 3 \end{vmatrix} = 0$$

$$D_2 = \begin{vmatrix} 4 & 8 & 6 \\ 2 & 5 & 3 \\ 2 & 4 & 3 \end{vmatrix} = 0 \qquad D_3 = \begin{vmatrix} 4 & -2 & 8 \\ 2 & -1 & 5 \\ 2 & -1 & 4 \end{vmatrix} = 0$$

Nothing can be said about the consistency from these facts. On closer examination of the system is is noticed that the second and third equations are inconsistent. Hence the system is inconsistent.

29.25 Determine whether the system

$$2x + y - 2z = 4$$
$$x - 2y + z = -2$$
$$5x - 5y + z = -2$$

is consistent.

SOLUTION

$D = D_1 = D_2 = D_3 = 0$. Hence nothing can be concluded from these facts.

Solving the first two equations for x and y (in terms of z), $x = \frac{3}{5}(z + 2), y = \frac{4}{5}(z + 2)$. These values are found by substitution to satisfy the third equation. (If they did not satisfy the third equation the system would be inconsistent.)

Hence the values $x = \frac{3}{5}(z + 2), y = \frac{4}{5}(z + 2)$ satisfy the system and there are infinite sets of solutions, obtained by assigning various values to z. Thus if $z = 3$, then $x = 3, y = 4$; if $z = -2$, then $x = 0, y = 0$; etc.

It follows that the given equations are *dependent*. This may be seen in another way by multiplying the second equation by 3 and adding to the first equation to obtain $5x - 5y + z = -2$ which is the third equation.

29.26 Does the system

$$\begin{aligned} 2x - 3y + 4z &= 0 \\ x + y - 2z &= 0 \\ 3x + 2y - 3z &= 0 \end{aligned}$$

possess only the trivial solution $x = y = z = 0$?

SOLUTION

$$D = \begin{vmatrix} 2 & -3 & 4 \\ 1 & 1 & -2 \\ 3 & 2 & -3 \end{vmatrix} = -17 \qquad D_1 = D_2 = D_3 = 0$$

Since $D \neq 0$ and $D_1 = D_2 = D_3 = 0$, the system has only the trivial solution.

29.27 Find non-trivial solutions for the system

$$\begin{aligned} x + 3y - 2z &= 0 \\ 2x - 4y + z &= 0 \\ x + y - z &= 0 \end{aligned}$$

if they exist.

SOLUTION

$$D = \begin{vmatrix} 1 & 3 & -2 \\ 2 & -4 & 1 \\ 1 & 1 & -1 \end{vmatrix} = 0 \qquad D_1 = D_2 = D_3 = 0$$

Hence there are non-trivial solutions.

To determine these non-trivial solutions solve for x and y (in terms of z) from the first two equations (this may not always be possible). We find $x = z/2, y = z/2$. These satisfy the third equation. An infinite number of solutions is obtained by assigning various values to z. For example, if $z = 6$, then $x = 3, y = 3$; if $z = -4$, then $x = -2, y = -2$; etc.

29.28 For what values of k will the system

$$\begin{aligned} x + 2y + kz &= 0 \\ 2x + ky + 2z &= 0 \\ 3x + y + z &= 0 \end{aligned}$$

have non-trivial solutions?

SOLUTION

Non-trivial solutions are obtained when

$$D = \begin{vmatrix} 1 & 2 & k \\ 2 & k & 2 \\ 3 & 1 & 1 \end{vmatrix} = 0.$$

Hence $D = -3k^2 + 3k + 6 = 0$ or $k = -1, 2$.

Supplementary Problems

29.29 Determine the number of inversions in each grouping.

(*a*) 4, 3, 1, 2 (*b*) 3, 1, 5, 4, 2 (*c*) c, a, d, b, e

29.30 For the grouping d_3, b_4, c_1, e_2, a_5 determine

(*a*) the number of inversions in the subscripts when the letters are in alphabetical order,

(*b*) the number of inversions in the letters when the subscripts are in natural order.

29.31 In the expansion of

$$\begin{vmatrix} a_1 & b_1 & c_1 & d_1 \\ a_2 & b_2 & c_2 & d_2 \\ a_3 & b_3 & c_3 & d_3 \\ a_4 & b_4 & c_4 & d_4 \end{vmatrix}$$

determine the signs associated with the terms $d_2b_4c_3a_1$ and $b_3c_2a_4d_1$.

29.32 (*a*) Prove Property I: If the rows and columns of a determinant are interchanged, the value of the determinant is the same.

(*b*) Prove Property II: If each element in a row (or column) is zero, the value of the determinant is zero.

29.33 Show that the determinant

$$\begin{vmatrix} 1 & 2 & 3 & 4 \\ 2 & 4 & 6 & 3 \\ 3 & 8 & 12 & 2 \\ 4 & 16 & 24 & 1 \end{vmatrix}$$

equals zero.

29.34 Transform the determinant

$$\begin{vmatrix} -2 & 4 & 1 & 3 \\ 1 & -2 & 2 & 4 \\ 3 & 1 & -3 & 2 \\ 4 & 3 & -2 & -1 \end{vmatrix}$$

into an equal determinant having three zeros in the 3rd column.

29.35 Without changing the value of the determinant

$$\begin{vmatrix} 4 & -2 & 1 & 3 & 1 \\ -2 & 1 & -3 & -2 & -2 \\ 3 & 4 & 2 & 1 & 3 \\ 1 & -3 & 4 & -1 & -1 \\ 2 & -1 & 2 & 4 & 2 \end{vmatrix}$$

obtain four zeros in the 4th column.

29.36 For the determinant

$$\begin{vmatrix} -1 & 2 & 3 & -2 \\ 4 & -1 & -2 & 2 \\ -3 & 1 & 2 & -1 \\ 2 & 4 & -1 & 3 \end{vmatrix}$$

(*a*) write the minors and cofactors of the elements in the 3rd row,

(*b*) express the value of the determinant in terms of minors or cofactors,

(*c*) find the value of the determinant.

29.37 Transform the determinant

$$\begin{vmatrix} -2 & 1 & 2 & 3 \\ 3 & -2 & -3 & 2 \\ 1 & 2 & 1 & 2 \\ 4 & 3 & -1 & -3 \end{vmatrix}$$

into a determinant having three zeros in a row and then evaluate the determinant by use of expansion by minors.

29.38 Evaluate each determinant.

(*a*) $\begin{vmatrix} 2 & -1 & 3 & 2 \\ -3 & 1 & 2 & 4 \\ 1 & -3 & -1 & 3 \\ -1 & 2 & -2 & -3 \end{vmatrix}$ (*b*) $\begin{vmatrix} 3 & -1 & 2 & 1 \\ 4 & 2 & 0 & -3 \\ -2 & 1 & -3 & 2 \\ 1 & 3 & -1 & 4 \end{vmatrix}$ (*c*) $\begin{vmatrix} 1 & 2 & -1 & 1 \\ -2 & 3 & 2 & -1 \\ 3 & -1 & 1 & -4 \\ -1 & 4 & -3 & 2 \end{vmatrix}$

(*d*) $\begin{vmatrix} 3 & 2 & -1 & 3 & 2 \\ -2 & 0 & 3 & 4 & 3 \\ 1 & -3 & -2 & 1 & 0 \\ 2 & 4 & 1 & 0 & 1 \\ -1 & -1 & 2 & 1 & 0 \end{vmatrix}$

29.39 Factor each determinant:

(*a*) $\begin{vmatrix} a & b & c \\ a^2 & b^2 & c^2 \\ a^3 & b^3 & c^3 \end{vmatrix}$ (*b*) $\begin{vmatrix} 1 & 1 & 1 & 1 \\ 1 & x & y & z \\ 1 & x^2 & y^2 & z^2 \\ 1 & x^3 & y^3 & z^3 \end{vmatrix}$

29.40 Solve each system:

(*a*) $\begin{cases} x - 2y + z - 3w = 4 \\ 2x + 3y - z - 2w = -4 \\ 3x - 4y + 2z - 4w = 12 \\ 2x - y - 3z + 2w = -2 \end{cases}$ (*b*) $\begin{cases} 2x + y - 3z = -5 \\ 3y + 4z + w = 5 \\ 2z - w - 4x = 0 \\ w + 3x - y = 4 \end{cases}$

29.41 Find i_1 and i_4 for the system

$$\begin{cases} 2i_1 - 3i_3 - i_4 = -4 \\ 3i_1 + i_2 - 2i_3 + 2i_4 + 2i_5 = 0 \\ -i_1 - 3i_2 + 2i_4 + 3i_5 = 2 \\ i_1 + 2i_3 - i_5 = 9 \\ 2i_1 + i_2 = 5 \end{cases}$$

29.42 Determine whether each system is consistent.

(a) $\begin{cases} 2x - 3y + z = 1 \\ x + 2y - z = 1 \\ 3x - y + 2z = 6 \end{cases}$ (b) $\begin{cases} 2x - y + z = 2 \\ 3x + 2y - 4z = 1 \\ x - 4y + 6z = 3 \end{cases}$ (c) $\begin{cases} x + 3y - 2z = 2 \\ 3x - y - z = 1 \\ 2x + 6y - 4z = 3 \end{cases}$ (d) $\begin{cases} 2u + v - 3w = 1 \\ u - 2v - w = 2 \\ u + 3v - 2w = -2 \end{cases}$

29.43 Find non-trivial solutions, if they exist, for the system

$$\begin{cases} 3x - 2y + 4z = 0 \\ 2x + y - 3z = 0 \\ x + 3y - 2z = 0 \end{cases}$$

29.44 For what value of k will the system

$$\begin{cases} 2x + ky + z + w = 0 \\ 3x + (k-1)y - 2z - w = 0 \\ x - 2y + 4z + 2w = 0 \\ 2x + y + z + 2w = 0 \end{cases}$$

possess non-trivial solutions?

ANSWERS TO SUPPLEMENTARY PROBLEMS

29.29 (a) 5 (b) 5 (c) 3

29.30 (a) 8 (b) 8

29.31 − and + respectively

29.36 (c) −38

29.37 28

29.38 (a) 38 (b) −143 (c) −108 (d) 88

29.39 (a) $abc(a-b)(b-c)(c-a)$ (b) $(x-1)(y-1)(z-1)(x-y)(y-z)(z-x)$

29.40 (a) $x = 2,\ y = -1,\ z = 3,\ w = 1$ (b) $x = 1,\ y = -1,\ z = 2,\ w = 0$

29.41 $i_1 = 3,\ i_4 = -2$

29.42 (a) consistent (b) dependent (c) inconsistent (d) inconsistent

29.43 Only trivial solution $x = y = z = 0$

29.44 $k = -1$

Chapter 30

Matrices

30.1 DEFINITION OF A MATRIX

A matrix is a rectangular array of numbers. The numbers are the entries or elements of the matrix. The following are examples of matrices.

$$\begin{bmatrix} 1 & -4 \\ 7 & 0 \end{bmatrix}, \quad \begin{bmatrix} 1 & 2 \\ 4 & 7 \\ -1 & 3 \end{bmatrix}, \quad \begin{bmatrix} -5 \\ -3 \\ 8 \end{bmatrix}, \quad \begin{bmatrix} 0 & 8 & -1 \\ 6 & 5 & 3 \end{bmatrix}$$

Matrices are classified by the number of rows and columns. The matrices above are 2×2, 3×2, 3×1, and 2×3 with the first number indicating the number of rows and the second number indicating the number of columns. When a matrix has the same number of rows as columns, it is a square matrix.

$$\mathbf{A} = \begin{bmatrix} a_{11} & a_{12} & a_{13} & \cdots & a_{1n} \\ a_{21} & a_{22} & a_{23} & \cdots & a_{2n} \\ \vdots & \vdots & \vdots & & \vdots \\ a_{m1} & a_{m2} & a_{m3} & \cdots & a_{mn} \end{bmatrix}$$

The matrix **A** is an $m \times n$ matrix. The entries in matrix **A** are double subscripted with the first number indicating the row of the entry and the second number indicating the column of the entry. The general entry for the matrix is denoted by a_{ij}. The matrix **A** can be denoted by $[a_{ij}]$.

30.2 OPERATIONS WITH MATRICES

If matrix **A** and matrix **B** have the same size, same number of rows and same number of columns, and the general entries are of the form a_{ij} and b_{ij} respectively, then the sum $\mathbf{A} + \mathbf{B} = [a_{ij}] + [b_{ij}] = [a_{ij} + b_{ij}] = [c_{ij}] = \mathbf{C}$ for all i and j.

EXAMPLE 30.1. Find the sum of $\mathbf{A} = \begin{bmatrix} 2 & 3 & 4 \\ 6 & 0 & -1 \end{bmatrix}$ and $\mathbf{B} = \begin{bmatrix} 0 & 3 & -2 \\ -1 & 1 & 2 \end{bmatrix}$.

$$\mathbf{A} + \mathbf{B} = \begin{bmatrix} 2 & 3 & 4 \\ 6 & 0 & -1 \end{bmatrix} + \begin{bmatrix} 0 & 3 & -2 \\ -1 & 1 & 2 \end{bmatrix} = \begin{bmatrix} 2+0 & 3+3 & 4+(-2) \\ 6+(-1) & 0+1 & -1+2 \end{bmatrix} = \begin{bmatrix} 2 & 6 & 2 \\ 5 & 1 & 1 \end{bmatrix}$$

The matrix $-\mathbf{A}$ is called the opposite of matrix **A** and each entry in $-\mathbf{A}$ is the opposite of the corresponding entry in **A**.

Thus, for $\mathbf{A} = \begin{bmatrix} -1 & 2 & -3 \\ -2 & 0 & 1 \end{bmatrix}$, $-\mathbf{A} = \begin{bmatrix} 1 & -2 & 3 \\ 2 & 0 & -1 \end{bmatrix}$

Multiplying a matrix by a scalar (real number) results in every entry in the matrix being multiplied by the scalar.

EXAMPLE 30.2. Multiply the matrix $\mathbf{A} = \begin{bmatrix} 2 & 3 & 4 \\ 6 & 0 & -1 \end{bmatrix}$ by -2.

$$\begin{bmatrix} 2 & 3 & 4 \\ 6 & 0 & -1 \end{bmatrix} = \begin{bmatrix} -4 & -6 & -8 \\ -12 & 0 & 2 \end{bmatrix} \qquad -2\mathbf{A} = -2$$

The product **AB** where **A** is an $m \times p$ matrix and **B** is a $p \times n$ matrix is **C**, an $m \times n$ matrix. The entries c_{ij} in matrix **C** are found by the formula $c_{ij} = a_{i1}b_{1j} + a_{i2}b_{2j} + a_{i3}b_{3j} + \cdots + a_{ip}b_{pj}$.

$$\begin{matrix} \mathbf{A} & \times & \mathbf{B} & = & \mathbf{C} \end{matrix}$$

$$\begin{bmatrix} a_{11} & a_{12} & a_{13} & \cdots & a_{1p} \\ a_{21} & a_{22} & a_{23} & \cdots & a_{2p} \\ \vdots & \vdots & \vdots & \vdots & \vdots \\ a_{i1} & a_{i2} & a_{i3} & \cdots & a_{ip} \\ \vdots & \vdots & \vdots & \vdots & \vdots \\ a_{m1} & a_{m2} & a_{m3} & \cdots & a_{mn} \end{bmatrix} \times \begin{bmatrix} b_{11} & b_{12} & \cdots & b_{1j} & \cdots & b_{1n} \\ b_{21} & b_{22} & \cdots & b_{2j} & \cdots & b_{2n} \\ b_{31} & b_{32} & \cdots & b_{3j} & \cdots & b_{3n} \\ \vdots & \vdots & \vdots & \vdots & \vdots & \vdots \\ b_{p1} & b_{p2} & \cdots & b_{pj} & \cdots & b_{pn} \end{bmatrix} = \begin{bmatrix} c_{11} & c_{12} & \cdots & c_{1j} & \cdots & c_{1n} \\ c_{21} & c_{22} & \cdots & c_{2j} & \cdots & c_{2n} \\ \vdots & \vdots & \vdots & \vdots & \vdots & \vdots \\ c_{i1} & c_{i2} & \cdots & c_{ij} & \cdots & c_{in} \\ \vdots & \vdots & \vdots & \vdots & \vdots & \vdots \\ c_{m1} & c_{m2} & \cdots & c_{mj} & \cdots & c_{mn} \end{bmatrix}$$

EXAMPLE 30.3. Find the product **AB** when $\mathbf{A} = \begin{bmatrix} 2 & 4 & 1 \\ 0 & 1 & -2 \end{bmatrix}$ and $\mathbf{B} = \begin{bmatrix} 3 & 0 & 1 & -1 \\ -1 & 3 & 1 & 2 \\ 4 & 0 & 3 & -2 \end{bmatrix}$.

$$\mathbf{AB} = \begin{bmatrix} 2 & 4 & 1 \\ 0 & 1 & -2 \end{bmatrix} \begin{bmatrix} 3 & 0 & 1 & -1 \\ -1 & 3 & 1 & 2 \\ 4 & 0 & 3 & -2 \end{bmatrix}$$

$$\mathbf{AB} = \begin{bmatrix} 2(3)+4(-1)+1(4) & 2(0)+4(3)+1(0) & 2(1)+4(1)+1(3) & 2(-1)+4(2)+1(-2) \\ 0(3)+1(-1)+(-2)(4) & 0(0)+1(3)+(-2)(0) & 0(1)+1(1)+(-2)(3) & 0(-1)+1(2)+(-2)(-2) \end{bmatrix}$$

$$\mathbf{AB} = \begin{bmatrix} 6-4+4 & 0+12+0 & 2+4+3 & -2+8-2 \\ 0-1-8 & 0+3+0 & 0+1-6 & 0+2+4 \end{bmatrix}$$

$$\mathbf{AB} = \begin{bmatrix} 6 & 12 & 9 & 4 \\ -9 & 3 & -5 & 6 \end{bmatrix}$$

EXAMPLE 30.4. Find the products **CD** and **DC** when $\mathbf{C} = \begin{bmatrix} 1 & 2 & 3 \\ -1 & 0 & 4 \end{bmatrix}$ and $\mathbf{D} = \begin{bmatrix} 1 & -3 \\ 0 & 2 \\ 4 & -2 \end{bmatrix}$.

$$\mathbf{CD} = \begin{bmatrix} 1 & 2 & 3 \\ -1 & 0 & 4 \end{bmatrix} \begin{bmatrix} 1 & -3 \\ 0 & 2 \\ 4 & -2 \end{bmatrix} = \begin{bmatrix} 1(1)+2(0)+3(4) & 1(-3)+2(2)+3(-2) \\ -1(1)+0(0)+4(4) & -1(-3)+0(2)+4(-2) \end{bmatrix}$$

$$\mathbf{CD} = \begin{bmatrix} 1+0+12 & -3+4-6 \\ -1+0+16 & 3+0-8 \end{bmatrix} = \begin{bmatrix} 13 & -5 \\ 15 & -5 \end{bmatrix}$$

$$\mathbf{DC} = \begin{bmatrix} 1 & -3 \\ 0 & 2 \\ 4 & -2 \end{bmatrix} \begin{bmatrix} 1 & 2 & 3 \\ -1 & 0 & 4 \end{bmatrix} = \begin{bmatrix} 1(1)+(-3)(-1) & 1(2)+(-3)(0) & 1(3)+(-3)(4) \\ 0(1)+2(-1) & 0(2)+2(0) & 0(3)+2(4) \\ 4(1)+(-2)(-1) & 4(2)+(-2)(0) & 4(3)+(-2)(4) \end{bmatrix}$$

$$\mathbf{DC} = \begin{bmatrix} 1+3 & 2+0 & 3-12 \\ 0-2 & 0+0 & 0+8 \\ 4+2 & 8+0 & 12-8 \end{bmatrix} = \begin{bmatrix} 4 & 2 & -9 \\ -2 & 0 & 8 \\ 6 & 8 & 4 \end{bmatrix}$$

In Example 30.4, note that although both products **CD** and **DC** exist, **CD** ≠ **DC**. Thus, multiplication of matrices is not commutative.

An identity matrix is an $n \times n$ matrix with entries of 1 when the row and column numbers are equal and 0 everywhere else. We denote the $n \times n$ identity matrix by $\mathbf{I}_n$.

For example,

$$\mathbf{I}_2 = \begin{bmatrix} 1 & 0 \\ 0 & 1 \end{bmatrix} \quad \text{and} \quad \mathbf{I}_3 = \begin{bmatrix} 1 & 0 & 0 \\ 0 & 1 & 0 \\ 0 & 0 & 1 \end{bmatrix}.$$

If **A** is a square matrix and **I** is the identity matrix the same size as **A**, then **AI** = **IA** = **A**.

30.3 ELEMENTARY ROW OPERATIONS

Two matrices are said to be row equivalent if one can be obtained from the other by a sequence of elementary row operations.

Elementary Row Operations

(1) Interchange two rows.
(2) Multiplying a row by a nonzero constant.
(3) Add a multiple of a row to another row.

A matrix is said to be in reduced row-echelon form if it has the following properties:

(1) All rows consisting of all zeros occur at the bottom of the matrix.
(2) A row that is not all zeros has a 1 as its first non-zero entry, which is called the leading 1.
(3) For two successive non-zero rows, the leading 1 in the higher row is further to the left than the leading 1 in the lower row.
(4) Every column that contains a leading 1 has zeros in every other position in the column.

EXAMPLE 30.5. Use elementary row operations to put the matrix **A** in reduced row-echelon form when

$$\mathbf{A} = \begin{bmatrix} 2 & 1 & 4 \\ 1 & 3 & 2 \\ 3 & -1 & 6 \end{bmatrix}.$$

$$\mathbf{A} = \begin{bmatrix} 2 & 1 & 4 \\ 1 & 3 & 2 \\ 3 & -1 & 6 \end{bmatrix} \sim \begin{matrix} R_2 \\ R_1 \\ \end{matrix} \begin{bmatrix} 1 & 3 & 2 \\ 2 & 1 & 4 \\ 3 & -1 & 6 \end{bmatrix} \sim \begin{matrix} \\ R_2 - 2R_1 \\ R_3 - 3R_1 \end{matrix} \begin{bmatrix} 1 & 3 & 2 \\ 0 & -5 & 0 \\ 0 & -10 & 0 \end{bmatrix} \sim -\tfrac{1}{5}R_2 \begin{bmatrix} 1 & 3 & 2 \\ 0 & 1 & 0 \\ 0 & -10 & 0 \end{bmatrix}$$

$$\sim \begin{matrix} R_1 - 3R_2 \\ \\ R_3 + 10R_2 \end{matrix} \begin{bmatrix} 1 & 0 & 2 \\ 0 & 1 & 0 \\ 0 & 0 & 0 \end{bmatrix}$$

The reduced row-echelon form of matrix **A** is $\begin{bmatrix} 1 & 0 & 2 \\ 0 & 1 & 0 \\ 0 & 0 & 0 \end{bmatrix}$.

30.4 INVERSE OF A MATRIX

A square matrix **A** has an inverse if there is a matrix $\mathbf{A}^{-1}$ such that $\mathbf{AA}^{-1} = \mathbf{A}^{-1}\mathbf{A} = \mathbf{I}$.

To find the inverse, if it exists, of a square matrix **A** we complete the following procedure.

(1) Form the partitioned matrix $[\mathbf{A}|\mathbf{I}]$, where **A** is the given $n \times n$ matrix and **I** is the $n \times n$ identity matrix.

(2) Perform elementary row operations on $[\mathbf{A}|\mathbf{I}]$ until the partitioned matrix has the form $[\mathbf{I}|\mathbf{B}]$, that is, until the matrix **A** on the left is transformed into the identity matrix. If **A** cannot be transformed into the identity matrix, matrix **A** does not have an inverse.

(3) The matrix **B** is $\mathbf{A}^{-1}$, the inverse of matrix **A**.

EXAMPLE 30.6. Find the inverse of matrix $\mathbf{A} = \begin{bmatrix} 2 & 5 & 4 \\ 1 & 4 & 3 \\ 1 & -3 & -2 \end{bmatrix}$.

$$[\mathbf{A}|\mathbf{I}] = \left[\begin{array}{rrr|rrr} 2 & 5 & 4 & 1 & 0 & 0 \\ 1 & 4 & 3 & 0 & 1 & 0 \\ 1 & -3 & -2 & 0 & 0 & 1 \end{array}\right] \sim \begin{matrix} R_2 \\ R_1 \\ \\ \end{matrix} \left[\begin{array}{rrr|rrr} 1 & 4 & 3 & 0 & 1 & 0 \\ 2 & 5 & 4 & 1 & 0 & 0 \\ 1 & -3 & -2 & 0 & 0 & 1 \end{array}\right]$$

$$\sim \begin{matrix} \\ R_2 - 2R_1 \\ R_3 - R_1 \end{matrix} \left[\begin{array}{rrr|rrr} 1 & 4 & 3 & 0 & 1 & 0 \\ 0 & -3 & -2 & 1 & -2 & 0 \\ 0 & -7 & -5 & 0 & -1 & 1 \end{array}\right] \sim \begin{matrix} \\ -\frac{1}{3}R_2 \\ \\ \end{matrix} \left[\begin{array}{rrr|rrr} 1 & 4 & 3 & 0 & 1 & 0 \\ 0 & 1 & 2/3 & -1/3 & 2/3 & 0 \\ 0 & -7 & -5 & 0 & -1 & 1 \end{array}\right]$$

$$\sim \begin{matrix} R_1 - 4R_2 \\ \\ R_3 + 7R_2 \end{matrix} \left[\begin{array}{rrr|rrr} 1 & 0 & 1/3 & 4/3 & -5/3 & 0 \\ 0 & 1 & 2/3 & -1/3 & 2/3 & 0 \\ 0 & 0 & -1/3 & -7/3 & 11/3 & 1 \end{array}\right] \sim \begin{matrix} \\ \\ -3R_3 \end{matrix} \left[\begin{array}{rrr|rrr} 1 & 0 & 1/3 & 4/3 & -5/3 & 0 \\ 0 & 1 & 2/3 & -1/3 & 2/3 & 0 \\ 0 & 0 & 1 & 7 & -11 & -3 \end{array}\right]$$

$$\sim \begin{matrix} R_1 - (1/3)R_3 \\ R_2 - (2/3)R_3 \\ \\ \end{matrix} \left[\begin{array}{rrr|rrr} 1 & 0 & 0 & -1 & 2 & 1 \\ 0 & 1 & 0 & -5 & 8 & 2 \\ 0 & 0 & 1 & 7 & -11 & -3 \end{array}\right] = [\mathbf{I}|\mathbf{A}^{-1}]$$

$$\mathbf{A}^{-1} = \begin{bmatrix} -1 & 2 & 1 \\ -5 & 8 & 2 \\ 7 & -11 & -3 \end{bmatrix}$$

If the matrix **A** is row equivalent to **I**, then the matrix **A** has an inverse and is said to be invertible. **A** does not have an inverse if it is not row equivalent to **I**.

EXAMPLE 30.7. Find the inverse, if it exists, for matrix $\mathbf{A} = \begin{bmatrix} 1 & 3 & 4 \\ -2 & -5 & -3 \\ 1 & 4 & 9 \end{bmatrix}$.

$$[\mathbf{A}|\mathbf{I}] = \left[\begin{array}{rrr|rrr} 1 & 3 & 4 & 1 & 0 & 0 \\ -2 & -5 & -3 & 0 & 1 & 0 \\ 1 & 4 & 9 & 0 & 0 & 1 \end{array}\right] \sim \begin{matrix} \\ R_2 + 2R_1 \\ R_3 - R_1 \end{matrix} \left[\begin{array}{rrr|rrr} 1 & 3 & 4 & 1 & 0 & 0 \\ 0 & 1 & 5 & 2 & 1 & 0 \\ 0 & 1 & 5 & -1 & 0 & 1 \end{array}\right]$$

$$\sim \begin{matrix} R_1 - 3R_2 \\ \\ R_3 - R_2 \end{matrix} \left[\begin{array}{rrr|rrr} 1 & 0 & -11 & -5 & -3 & 0 \\ 0 & 1 & 5 & 2 & 1 & 0 \\ 0 & 0 & 0 & -3 & -1 & 1 \end{array}\right]$$

The matrix **A** is row equivalent to the matrix on the left. Since the matrix on the left has a row of all zeros, it is not row equivalent to **I**. Thus, the matrix **A** does not have an inverse.

Another way to determine whether the inverse of a matrix **A** exists is that the determinant associated with an invertible matrix is non-zero, that is, det $\mathbf{A} \neq 0$ if $\mathbf{A}^{-1}$ exists.

For 2×2 matrices, the inverse can be found by a special procedure:

$$\text{If } \mathbf{A} = \begin{bmatrix} a_{11} & a_{12} \\ a_{21} & a_{22} \end{bmatrix} \quad \text{then} \quad \mathbf{A}^{-1} = \frac{1}{\det \mathbf{A}} \begin{bmatrix} a_{22} & -a_{12} \\ -a_{21} & a_{11} \end{bmatrix} \text{ where } \det \mathbf{A} \neq 0.$$

(1) Find the value of det **A**. If det $\mathbf{A} \neq 0$, then the inverse exists.
(2) Exchange the entries on the main diagonal, swap a_{11} and a_{22}.
(3) Change the signs of the entries on the off diagonal, replace a_{21} by $-a_{21}$ and a_{12} by $-a_{12}$.
(4) Multiply the new matrix by 1/det **A**. This product is $\mathbf{A}^{-1}$.

30.5 MATRIX EQUATIONS

A matrix equation $\mathbf{AX} = \mathbf{B}$ has a solution if and only if the matrix $\mathbf{A}^{-1}$ exists and the solution is $\mathbf{X} = \mathbf{A}^{-1}\mathbf{B}$.

EXAMPLE 30.8. Solve the matrix equation $\begin{bmatrix} 7 & -5 \\ 2 & -3 \end{bmatrix}\begin{bmatrix} x \\ y \end{bmatrix} = \begin{bmatrix} 12 \\ 6 \end{bmatrix}$.

$$\text{If } \mathbf{A} = \begin{bmatrix} 7 & -5 \\ 2 & -3 \end{bmatrix} \quad \text{then} \quad \mathbf{A}^{-1} = \begin{bmatrix} 3/11 & -5/11 \\ 2/11 & -7/11 \end{bmatrix} \quad \text{or} \quad \frac{-1}{11}\begin{bmatrix} -3 & 5 \\ -2 & 7 \end{bmatrix}$$

$$\frac{-1}{11}\begin{bmatrix} -3 & 5 \\ -2 & 7 \end{bmatrix}\begin{bmatrix} 7 & -5 \\ 2 & -3 \end{bmatrix}\begin{bmatrix} x \\ y \end{bmatrix} = \frac{-1}{11}\begin{bmatrix} -3 & 5 \\ -2 & 7 \end{bmatrix}\begin{bmatrix} 12 \\ 6 \end{bmatrix}$$

$$\frac{-1}{11}\begin{bmatrix} -11 & 0 \\ 0 & -11 \end{bmatrix}\begin{bmatrix} x \\ y \end{bmatrix} = \frac{-1}{11}\begin{bmatrix} -6 \\ 18 \end{bmatrix}$$

$$\begin{bmatrix} 1 & 0 \\ 0 & 1 \end{bmatrix}\begin{bmatrix} x \\ y \end{bmatrix} = \frac{-1}{11}\begin{bmatrix} -6 \\ 18 \end{bmatrix}$$

$$\begin{bmatrix} x \\ y \end{bmatrix} = \begin{bmatrix} 6/11 \\ -18/11 \end{bmatrix}$$

30.6 MATRIX SOLUTION OF A SYSTEM OF EQUATIONS

To solve a system of equations using matrices, we write a partitioned matrix which is the coefficient matrix on the left augmented by the constants matrix on the right.

The augmented matrix associated with the system $\begin{aligned} x + 2y + 3z &= 6 \\ x \qquad - z &= 0 \\ x - y - z &= -4 \end{aligned}$ is

$$\mathbf{A} = \left[\begin{array}{ccc|c} 1 & 2 & 3 & 6 \\ 1 & 0 & -1 & 0 \\ 1 & -1 & -1 & -4 \end{array}\right]$$

EXAMPLE 30.9. Use matrices to solve the system of equations:

$$\begin{aligned} x_2 + x_3 - 2x_4 &= -3 \\ x_1 + 2x_2 - x_3 \quad &= 2 \\ 2x_1 + 4x_2 + x_3 - 3x_4 &= -2 \\ x_1 - 4x_2 - 7x_3 - x_4 &= -19 \end{aligned}$$

Write the augmented matrix for the system.

$$\left[\begin{array}{cccc|c} 0 & 1 & 1 & -2 & -3 \\ 1 & 2 & -1 & 0 & 0 \\ 2 & 4 & 1 & -3 & -2 \\ 1 & -4 & -7 & -1 & -19 \end{array}\right]$$

Put the matrix on the left in reduced row-echelon form.

$$\begin{array}{r} R_2 \\ \sim R_1 \\ \\ \\ \end{array}\left[\begin{array}{cccc|c} 1 & 2 & -1 & 0 & 2 \\ 0 & 1 & 1 & -2 & -3 \\ 2 & 4 & 1 & -3 & -2 \\ 1 & -4 & -7 & -1 & -19 \end{array}\right] \sim \begin{array}{r} \\ \\ R_3 - 2R_1 \\ R_4 - R_1 \end{array}\left[\begin{array}{cccc|c} 1 & 2 & -1 & 0 & 2 \\ 0 & 1 & 1 & -2 & -3 \\ 0 & 0 & 3 & -3 & -6 \\ 0 & -6 & -6 & -1 & -21 \end{array}\right]$$

$$\begin{array}{r} R_1 - 2R_2 \\ \sim \\ (1/3)R_3 \\ R_4 + 6R_2 \end{array}\left[\begin{array}{cccc|c} 1 & 0 & -3 & 4 & 8 \\ 0 & 1 & 1 & -2 & -3 \\ 0 & 0 & 1 & -1 & -2 \\ 0 & 0 & 0 & -13 & -39 \end{array}\right] \begin{array}{r} R_1 + 3R_3 \\ \sim R_2 - R_3 \\ \\ (-1/13)R_4 \end{array}\left[\begin{array}{cccc|c} 1 & 0 & 0 & 1 & 2 \\ 0 & 1 & 0 & -1 & -1 \\ 0 & 0 & 1 & -1 & -2 \\ 0 & 0 & 0 & 1 & 3 \end{array}\right]$$

$$\begin{array}{r} R_1 - R_4 \\ \sim R_2 + R_4 \\ R_3 + R_4 \\ \\ \end{array}\left[\begin{array}{cccc|c} 1 & 0 & 0 & 0 & -1 \\ 0 & 1 & 0 & 0 & 2 \\ 0 & 0 & 1 & 0 & 1 \\ 0 & 0 & 0 & 1 & 3 \end{array}\right]$$

From the reduced row-echelon form of the augmented matrix, we write the equations:

$$x_1 = -1,\ x_2 = 2,\ x_3 = 1,\ \text{and } x_4 = 3.$$

Thus, the solution of the system is $(-1, 2, 1, 3)$.

EXAMPLE 30.10. Solve the system of equations: $\begin{aligned} x_1 + 2x_2 - x_3 &= 0 \\ 3x_1 + 5x_2 \quad &= 1 \end{aligned}$

$$\left[\begin{array}{ccc|c} 1 & 2 & -1 & 0 \\ 3 & 5 & 0 & 1 \end{array}\right] \sim \begin{array}{r} \\ R_2 - 3R_1 \end{array}\left[\begin{array}{ccc|c} 1 & 2 & -1 & 0 \\ 0 & -1 & 3 & 1 \end{array}\right] \sim \begin{array}{r} \\ -R_2 \end{array}\left[\begin{array}{ccc|c} 1 & 2 & -1 & 0 \\ 0 & 1 & -3 & -1 \end{array}\right]$$

$$\sim R_1 - 2R_2\left[\begin{array}{ccc|c} 1 & 0 & 5 & 2 \\ 0 & 1 & -3 & -1 \end{array}\right]$$

$x_1 + 5x_3 = 2$ and $x_2 - 3x_3 = -1$ Thus, $x_1 = 2 - 5x_3$ and $x_2 = -1 + 3x_3$.

The system has infinitely many solutions of the form $(2 - 5x_3,\ -1 + 3x_3,\ x_3)$, where x_3 is a real number.

Solved Problems

30.1 Find (*a*) $\mathbf{A}+\mathbf{B}$, (*b*) $\mathbf{A}-\mathbf{B}$, (*c*) $3\mathbf{A}$, and (*d*) $5\mathbf{A}-2\mathbf{B}$ when

$$\mathbf{A}=\begin{bmatrix} 2 & 1 & 1 \\ -1 & -1 & 4 \end{bmatrix} \quad \text{and} \quad \mathbf{B}=\begin{bmatrix} 2 & -3 & 4 \\ -3 & 1 & -2 \end{bmatrix}$$

SOLUTION

(*a*) $$\mathbf{A}+\mathbf{B}=\begin{bmatrix} 2 & 1 & 1 \\ -1 & -1 & 4 \end{bmatrix}+\begin{bmatrix} 2 & -3 & 4 \\ -3 & 1 & -2 \end{bmatrix}=\begin{bmatrix} 2+2 & 1+(-3) & 1+4 \\ -1+(-3) & -1+1 & 4+(-2) \end{bmatrix}=\begin{bmatrix} 4 & -2 & 5 \\ -4 & 0 & 2 \end{bmatrix}$$

(*b*) $$\mathbf{A}-\mathbf{B}=\begin{bmatrix} 2 & 1 & 1 \\ -1 & -1 & 4 \end{bmatrix}-\begin{bmatrix} 2 & -3 & 4 \\ -3 & 1 & -2 \end{bmatrix}=\begin{bmatrix} 2-2 & 1-(-3) & 1-4 \\ -1-(-3) & -1-1 & 4-(-2) \end{bmatrix}=\begin{bmatrix} 0 & 4 & -3 \\ 2 & -2 & 6 \end{bmatrix}$$

(*c*) $$3\mathbf{A}=3\begin{bmatrix} 2 & 1 & 1 \\ -1 & -1 & 4 \end{bmatrix}=\begin{bmatrix} 3(2) & 3(1) & 3(1) \\ 3(-1) & 3(-1) & 3(4) \end{bmatrix}=\begin{bmatrix} 6 & 3 & 3 \\ -3 & -3 & 12 \end{bmatrix}$$

(*d*) $$5\mathbf{A}-2\mathbf{B}=5\begin{bmatrix} 2 & 1 & 1 \\ -1 & -1 & 4 \end{bmatrix}-2\begin{bmatrix} 2 & -3 & 4 \\ -3 & 1 & -2 \end{bmatrix}=\begin{bmatrix} 5(2)-2(2) & 5(1)-2(-3) & 5(1)-2(4) \\ 5(-1)-2(-3) & 5(-1)-2(1) & 5(4)-2(-2) \end{bmatrix}$$

$$=\begin{bmatrix} 6 & 11 & -3 \\ 1 & -7 & 24 \end{bmatrix}$$

30.2 Find, if they exist, (*a*) $\mathbf{AB}$, (*b*) $\mathbf{BA}$, and (*c*) $\mathbf{A}^2$ when

$$\mathbf{A}=[3\ 2\ 1] \quad \text{and} \quad \mathbf{B}=\begin{bmatrix} 2 \\ 3 \\ 0 \end{bmatrix}$$

SOLUTION

(*a*) $$\mathbf{AB}=[3\ 2\ 1]\begin{bmatrix} 2 \\ 3 \\ 0 \end{bmatrix}=[3(2)+2(3)+1(0)]=[12]$$

(*b*) $$\mathbf{BA}=\begin{bmatrix} 2 \\ 3 \\ 0 \end{bmatrix}[3\ 2\ 1]=\begin{bmatrix} 2(3) & 2(2) & 2(1) \\ 3(3) & 3(2) & 3(1) \\ 0(3) & 0(2) & 0(1) \end{bmatrix}=\begin{bmatrix} 6 & 4 & 2 \\ 9 & 6 & 3 \\ 0 & 0 & 0 \end{bmatrix}$$

(*c*) $\mathbf{A}^2=[3\ 2\ 1][3\ 2\ 1]$; not possible. $\mathbf{A}^n$, $n>1$, exists for square matrices only.

30.3 Find $\mathbf{AB}$, if possible.

(*a*) $$\mathbf{A}=\begin{bmatrix} 2 & 1 \\ -3 & 4 \\ 1 & 6 \end{bmatrix} \quad \text{and} \quad \mathbf{B}=\begin{bmatrix} 0 & -1 & 0 \\ 4 & 0 & 2 \\ 8 & -1 & 7 \end{bmatrix}$$ (*b*) $$\mathbf{A}=\begin{bmatrix} -1 & 3 \\ 4 & -5 \\ 0 & 2 \end{bmatrix} \quad \text{and} \quad \mathbf{B}=\begin{bmatrix} 1 & 2 \\ 0 & 7 \end{bmatrix}$$

SOLUTION

(*a*) $$\mathbf{AB}=\begin{bmatrix} 2 & 1 \\ -3 & 4 \\ 1 & 6 \end{bmatrix}\begin{bmatrix} 0 & -1 & 0 \\ 4 & 0 & 2 \\ 8 & -1 & 7 \end{bmatrix}$$; not possible.

A is a 3×2 matrix and **B** is a 3×3. Since **A** has only two columns it can only multiply matrices having two rows, $2\times k$ matrices.

(*b*) $\mathbf{AB} = \begin{bmatrix} -1 & 3 \\ 4 & -5 \\ 0 & 2 \end{bmatrix} \begin{bmatrix} 1 & 2 \\ 0 & 7 \end{bmatrix} = \begin{bmatrix} -1(1)+3(0) & -1(2)+3(7) \\ 4(1)+(-5)(0) & 4(2)+(-5)(7) \\ 0(1)+2(0) & 0(2)+2(7) \end{bmatrix} = \begin{bmatrix} -1 & 19 \\ 4 & -27 \\ 0 & 14 \end{bmatrix}$

30.4 Write each matrix in reduced row-echelon form.

(*a*) $\begin{bmatrix} 0 & 1 & -3 \\ 2 & 3 & -1 \\ 4 & 5 & -2 \end{bmatrix}$ (*b*) $\begin{bmatrix} 1 & -2 & 1 & -1 & 4 \\ 2 & -3 & 2 & -3 & -1 \\ 3 & -5 & 3 & -4 & 3 \\ -1 & 1 & -1 & 2 & 5 \end{bmatrix}$

SOLUTION

(*a*) $\begin{bmatrix} 0 & 1 & -3 \\ 2 & 3 & -1 \\ 4 & 5 & -2 \end{bmatrix} \sim \begin{matrix} R_2 \\ R_1 \\ \end{matrix} \begin{bmatrix} 2 & 3 & -1 \\ 0 & 1 & -3 \\ 4 & 5 & -2 \end{bmatrix} \sim \begin{matrix} \\ \\ R_3 - 2R_1 \end{matrix} \begin{bmatrix} 2 & 3 & -1 \\ 0 & 1 & -3 \\ 0 & -1 & 0 \end{bmatrix} \sim \begin{matrix} R_1 - 3R_2 \\ \\ R_3 + R_2 \end{matrix} \begin{bmatrix} 2 & 0 & 8 \\ 0 & 1 & -3 \\ 0 & 0 & -3 \end{bmatrix}$

$\sim \begin{matrix} (1/2)R_1 \\ \\ (-1/3)R_3 \end{matrix} \begin{bmatrix} 1 & 0 & 4 \\ 0 & 1 & -3 \\ 0 & 0 & 1 \end{bmatrix} \sim \begin{matrix} R_1 - 4R_3 \\ R_2 + 3R_3 \\ \end{matrix} \begin{bmatrix} 1 & 0 & 0 \\ 0 & 1 & 0 \\ 0 & 0 & 1 \end{bmatrix}$

The reduced row-echelon form of $\begin{bmatrix} 0 & 1 & -3 \\ 2 & 3 & -1 \\ 4 & 5 & -2 \end{bmatrix}$ is $\begin{bmatrix} 1 & 0 & 0 \\ 0 & 1 & 0 \\ 0 & 0 & 0 \end{bmatrix}$.

(*b*) $\begin{bmatrix} 1 & -2 & 1 & -1 & 4 \\ 2 & -3 & 2 & -3 & -1 \\ 3 & -5 & 3 & -4 & 3 \\ -1 & 1 & -1 & 2 & 5 \end{bmatrix} \sim \begin{matrix} \\ R_2 - 2R_1 \\ R_3 - 3R_1 \\ R_4 + R_1 \end{matrix} \begin{bmatrix} 1 & -2 & 1 & -1 & 4 \\ 0 & 1 & 0 & -1 & -9 \\ 0 & 1 & 0 & -1 & -9 \\ 0 & -1 & 0 & 1 & 9 \end{bmatrix} \sim \begin{matrix} R_1 + 2R_2 \\ \\ R_3 - R_2 \\ R_4 + R_2 \end{matrix} \begin{bmatrix} 1 & 0 & 1 & -3 & -14 \\ 0 & 1 & 0 & -1 & -9 \\ 0 & 0 & 0 & 0 & 0 \\ 0 & 0 & 0 & 0 & 0 \end{bmatrix}$

The reduced row-echelon form of $\begin{bmatrix} 1 & -2 & 1 & -1 & 4 \\ 2 & -3 & 2 & -3 & -1 \\ 3 & -5 & 3 & -4 & 3 \\ -1 & 1 & -1 & 2 & 5 \end{bmatrix}$ is $\begin{bmatrix} 1 & 0 & 1 & -3 & -14 \\ 0 & 1 & 0 & -1 & -9 \\ 0 & 0 & 0 & 0 & 0 \\ 0 & 0 & 0 & 0 & 0 \end{bmatrix}$

30.5 Find the inverse, if it exists, for each matrix.

(*a*) $\mathbf{A} = \begin{bmatrix} 2 & 3 \\ 1 & -7 \end{bmatrix}$ (*b*) $\mathbf{B} = \begin{bmatrix} 3 & -6 \\ -1 & 2 \end{bmatrix}$ (*c*) $\mathbf{C} = \begin{bmatrix} 1 & -1 & 0 \\ 1 & 0 & -1 \\ 6 & -2 & -3 \end{bmatrix}$ (*d*) $\mathbf{D} = \begin{bmatrix} 3 & 2 & 1 \\ 1 & 0 & -1 \\ 0 & 1 & 2 \end{bmatrix}$

SOLUTION

(*a*) $\det \mathbf{A} = \begin{vmatrix} 2 & 3 \\ 1 & -7 \end{vmatrix} = -14 - 3 = -17$; $\det \mathbf{A} \neq 0$ so $\mathbf{A}^{-1}$ exists.

$\mathbf{A}^{-1} = \frac{-1}{17} \begin{bmatrix} -7 & -3 \\ -1 & 2 \end{bmatrix}$ or $\mathbf{A}^{-1} = \begin{bmatrix} 7/17 & 3/17 \\ 1/17 & -2/17 \end{bmatrix}$

The first form of the matrix is frequently used because it reduces the amount of computation with fractions that needs to be done. Also, it makes it easier to work with matrices on a graphing calculator.

(*b*) $\det \mathbf{B} = \begin{vmatrix} 3 & -6 \\ -1 & 2 \end{vmatrix} = 6 - 6 = 0$. Since $\det \mathbf{B} = 0$, $\mathbf{B}^{-1}$ does not exist.

(*c*) $$[\mathbf{C}|\mathbf{I}] = \left[\begin{array}{rrr|rrr} 1 & -1 & 0 & 1 & 0 & 0 \\ 1 & 0 & -1 & 0 & 1 & 0 \\ 6 & -2 & -3 & 0 & 0 & 1 \end{array}\right] \sim \left[\begin{array}{rrr|rrr} 1 & -1 & 0 & 1 & 0 & 0 \\ 0 & 1 & -1 & -1 & 1 & 0 \\ 0 & 4 & -3 & -6 & 0 & 1 \end{array}\right] \sim \left[\begin{array}{rrr|rrr} 1 & 0 & -1 & 0 & 1 & 0 \\ 0 & 1 & -1 & -1 & 1 & 0 \\ 0 & 0 & 1 & -2 & -4 & 1 \end{array}\right]$$

$$\sim \left[\begin{array}{rrr|rrr} 1 & 0 & 0 & -2 & -3 & 1 \\ 0 & 1 & 0 & -3 & -3 & 1 \\ 0 & 0 & 1 & -2 & -4 & 1 \end{array}\right] = [\mathbf{I}|\mathbf{C}^{-1}]$$

$$\mathbf{C}^{-1} = \begin{bmatrix} -2 & -3 & 1 \\ -3 & -3 & 1 \\ -2 & -4 & 1 \end{bmatrix}$$

(*d*) $$[\mathbf{D}|\mathbf{I}] = \left[\begin{array}{rrr|rrr} 3 & 2 & 1 & 1 & 0 & 0 \\ 1 & 0 & -1 & 0 & 1 & 0 \\ 0 & 1 & 2 & 0 & 0 & 1 \end{array}\right] \sim \left[\begin{array}{rrr|rrr} 1 & 0 & -1 & 0 & 1 & 0 \\ 3 & 2 & 1 & 1 & 0 & 0 \\ 0 & 1 & 2 & 0 & 0 & 1 \end{array}\right] \sim \left[\begin{array}{rrr|rrr} 1 & 0 & -1 & 0 & 1 & 0 \\ 0 & 2 & 4 & 1 & -3 & 0 \\ 0 & 1 & 2 & 0 & 0 & 1 \end{array}\right]$$

$$\sim \left[\begin{array}{rrr|rrr} 1 & 0 & -1 & 0 & 1 & 0 \\ 0 & 1 & 2 & 0 & 0 & 1 \\ 0 & 2 & 4 & 1 & -3 & 0 \end{array}\right] \sim \left[\begin{array}{rrr|rrr} 1 & 0 & -1 & 0 & 1 & 0 \\ 0 & 1 & 2 & 0 & 0 & 1 \\ 0 & 0 & 0 & 1 & -3 & -2 \end{array}\right]$$

Since the left matrix in the last form is not row equivalent to the identity matrix **I**, **D** does not have an inverse.

30.6 If $\mathbf{A} = \begin{bmatrix} -2 & -1 \\ 1 & 0 \\ 3 & -4 \end{bmatrix}$ and $\mathbf{B} = \begin{bmatrix} 0 & 3 \\ 2 & 0 \\ -4 & -1 \end{bmatrix}$, solve each equation for **X**.

(*a*) $2\mathbf{X} + 3\mathbf{A} = \mathbf{B}$ (*b*) $3\mathbf{A} + 6\mathbf{B} = -3\mathbf{X}$

SOLUTION

(*a*) $2\mathbf{X} + 3\mathbf{A} = \mathbf{B}$. So $2\mathbf{X} = -3\mathbf{A} + \mathbf{B}$ and $\mathbf{X} = -\frac{3}{2}\mathbf{A} + \frac{1}{2}\mathbf{B}$.

$$\mathbf{X} = -\frac{3}{2}\begin{bmatrix} -2 & -1 \\ 1 & 0 \\ 3 & -4 \end{bmatrix} + \frac{1}{2}\begin{bmatrix} 0 & 3 \\ 2 & 0 \\ -4 & -1 \end{bmatrix} = \begin{bmatrix} 3+0 & (3/2)+(3/2) \\ (-3/2)+1 & 0+0 \\ (-9/2)-2 & 6+(-1/2) \end{bmatrix} = \begin{bmatrix} 3 & 3 \\ -1/2 & 0 \\ -13/2 & 11/2 \end{bmatrix}$$

(*b*) $3\mathbf{A} + 6\mathbf{B} = -3\mathbf{X}$. So $-3\mathbf{X} = 3\mathbf{A} + 6\mathbf{B}$ and $\mathbf{X} = -\mathbf{A} - 2\mathbf{B}$.

$$\mathbf{X} = -\begin{bmatrix} -2 & -1 \\ 1 & 0 \\ 3 & -4 \end{bmatrix} - 2\begin{bmatrix} 0 & 3 \\ 2 & 0 \\ -4 & -1 \end{bmatrix} = \begin{bmatrix} 2+0 & 1-6 \\ -1-4 & 0+0 \\ -3+8 & 4+2 \end{bmatrix} = \begin{bmatrix} 2 & -5 \\ -5 & 0 \\ 5 & 6 \end{bmatrix}$$

30.7 Write the matrix equation $\mathbf{AX} = \mathbf{B}$ and use it to solve the system $\begin{array}{r} -x + y = 4 \\ -2x + y = 0. \end{array}$

SOLUTION

$$\mathbf{A} \cdot \mathbf{X} = \mathbf{B}$$

$$\begin{bmatrix} -1 & 1 \\ -2 & 1 \end{bmatrix}\begin{bmatrix} x \\ y \end{bmatrix} = \begin{bmatrix} 4 \\ 0 \end{bmatrix} \qquad \mathbf{A} = \begin{bmatrix} -1 & 1 \\ -2 & 1 \end{bmatrix} \qquad \text{so} \qquad \mathbf{A}^{-1} = \frac{1}{1}\begin{bmatrix} 1 & -1 \\ 2 & -1 \end{bmatrix} = \begin{bmatrix} 1 & -1 \\ 2 & -1 \end{bmatrix}$$

$$\begin{bmatrix} 1 & -1 \\ 2 & -1 \end{bmatrix}\begin{bmatrix} -1 & 1 \\ -2 & 1 \end{bmatrix}\begin{bmatrix} x \\ y \end{bmatrix} = \begin{bmatrix} 1 & -1 \\ 2 & -1 \end{bmatrix}\begin{bmatrix} 4 \\ 0 \end{bmatrix}$$

$$\begin{bmatrix} x \\ y \end{bmatrix} = \begin{bmatrix} 4 \\ 8 \end{bmatrix}$$

The solution to the system is (4, 8).

30.8 Solve each system of equations using matrices.

(*a*) $x - 2y + 3z = 9$; $-x + 3y = -4$; $2x - 5y + 5z = 17$ (*b*) $x + 2y - z = 3$; $3x + y = 4$; $2x - y + z = 2$

SOLUTION

(*a*)
$$\left[\begin{array}{ccc|c} 1 & -2 & 3 & 9 \\ -1 & 3 & 0 & -4 \\ 2 & -5 & 5 & 17 \end{array}\right] \sim \left[\begin{array}{ccc|c} 1 & -2 & 3 & 9 \\ 0 & 1 & 3 & 5 \\ 0 & -1 & -1 & -1 \end{array}\right] \sim \left[\begin{array}{ccc|c} 1 & 0 & 9 & 19 \\ 0 & 1 & 3 & 5 \\ 0 & 0 & 2 & 4 \end{array}\right] \sim \left[\begin{array}{ccc|c} 1 & 0 & 9 & 19 \\ 0 & 1 & 3 & 5 \\ 0 & 0 & 1 & 2 \end{array}\right]$$
$$\sim \left[\begin{array}{ccc|c} 1 & 0 & 0 & 1 \\ 0 & 1 & 0 & -1 \\ 0 & 0 & 1 & 2 \end{array}\right]$$

From the reduced row echelon form of the matrix, we write the equations:

$$x = 1,\ y = -1,\ \text{and } z = 2.$$

The system has the solution (1, −1, 2).

(*b*)
$$\left[\begin{array}{ccc|c} 1 & 2 & -1 & 3 \\ 3 & 1 & 0 & 4 \\ 2 & -1 & 1 & 2 \end{array}\right] \sim \left[\begin{array}{ccc|c} 1 & 2 & -1 & 3 \\ 0 & -5 & 3 & -5 \\ 0 & -5 & 3 & -4 \end{array}\right] \sim \left[\begin{array}{ccc|c} 1 & 2 & -1 & 3 \\ 0 & -5 & 3 & -5 \\ 0 & 0 & 0 & 1 \end{array}\right]$$

Since the last row results in the equation $0z = 1$, which has no solution, the system of equations has no solution.

Supplementary Problems

30.9 $\mathbf{A} = \begin{bmatrix} 2 & -5 \\ 0 & 7 \end{bmatrix}$ $\mathbf{B} = \begin{bmatrix} 3 & 1/2 & 5 \\ 1 & -1 & 3 \end{bmatrix}$ $\mathbf{C} = \begin{bmatrix} 2 & -5/2 & 0 \\ 0 & 2 & -3 \end{bmatrix}$ $\mathbf{D} = [7 \quad 3]$

Perform the indicated operations, if possible.

(*a*) $\mathbf{B} + \mathbf{C}$	(*e*) $3\mathbf{B} + 2\mathbf{C}$	(*i*) $\mathbf{C} - 5\mathbf{A}$	(*m*) $\mathbf{B}^2$
(*b*) $5\mathbf{A}$	(*f*) $\mathbf{DA}$	(*j*) $\mathbf{BC}$	(*n*) $\mathbf{D}(\mathbf{AB})$
(*c*) $2\mathbf{C} - 6\mathbf{B}$	(*g*) $\mathbf{AD}$	(*k*) $(\mathbf{DA})\mathbf{B}$	(*o*) $\mathbf{A}^3$
(*d*) $-6\mathbf{B}$	(*h*) $\mathbf{C} - \mathbf{B}$	(*l*) $\mathbf{A}^2$	(*p*) $\mathbf{DB} + \mathbf{DC}$

30.10 Find the product **AB**, if possible.

(a) $\mathbf{A}=\begin{bmatrix}0 & -1 & 0\\ 4 & 0 & 2\\ 8 & -1 & 7\end{bmatrix}$ and $\mathbf{B}=\begin{bmatrix}2 & 1\\ -3 & 4\\ 1 & 6\end{bmatrix}$

(b) $\mathbf{A}=\begin{bmatrix}1 & -1 & 7\\ 2 & -1 & 8\\ 3 & 1 & -1\end{bmatrix}$ and $\mathbf{B}=\begin{bmatrix}1 & 1 & 2\\ 2 & 1 & 1\\ 1 & -3 & 2\end{bmatrix}$

(c) $\mathbf{A}=\begin{bmatrix}1 & 2 & 3\\ 0 & 5 & 4\\ 3 & -2 & 1\end{bmatrix}$ and $\mathbf{B}=\begin{bmatrix}4 & -6 & 3\\ 5 & 4 & 4\\ -1 & 0 & 1\end{bmatrix}$

(d) $\mathbf{A}=\begin{bmatrix}6\\ -2\\ 1\\ 6\end{bmatrix}$ and $\mathbf{B}=[10 \quad 12]$

30.11 Solve each system of equations using a matrix equation of the form $\mathbf{AX}=\mathbf{B}$.

(a) $\begin{aligned}x-\ y&=0\\ 5x-3y&=10\end{aligned}$ (b) $\begin{aligned}x+2y&=1\\ 5x-4y&=-23\end{aligned}$ (c) $\begin{aligned}1.5x+0.8y&=2.3\\ 0.3x-0.2y&=0.1\end{aligned}$ (d) $\begin{aligned}2x+3y&=40\\ 3x-2y&=8\end{aligned}$

30.12 Write each matrix in reduced row echelon form.

(a) $\begin{bmatrix}2 & -1 & -3 & 1\\ 1 & 0 & -2 & 1\\ -3 & 1 & 1 & 2\end{bmatrix}$

(b) $\begin{bmatrix}1 & 0 & 2 & 4 & 0\\ 1 & 1 & 1 & 5 & 1\\ 1 & 2 & 0 & 6 & 3\\ 1 & 1 & 1 & 5 & 0\end{bmatrix}$

(c) $\begin{bmatrix}4 & -1 & 2\\ 1 & 2 & -1\\ 3 & 0 & 4\\ -1 & 0 & 2\end{bmatrix}$

(d) $\begin{bmatrix}2 & 5 & 3 & 3\\ 3 & 2 & 4 & 9\\ 5 & -3 & -2 & 4\end{bmatrix}$

(e) $\begin{bmatrix}1 & 2 & 0 & -1 & -1\\ -1 & -3 & 1 & 2 & 3\\ 1 & -1 & 3 & 1 & 1\\ 2 & -3 & 7 & 3 & 4\end{bmatrix}$

(f) $\begin{bmatrix}2 & -1 & 3 & 1 & 1\\ -1 & 0 & -2 & 1 & -3\\ 1 & 2 & -1 & -4 & 3\\ 3 & 2 & -2 & -3 & -1\end{bmatrix}$

30.13 Find the inverse, if it exists, of each matrix.

(a) $\begin{bmatrix}9 & 13\\ 2 & 3\end{bmatrix}$

(b) $\begin{bmatrix}3 & -2\\ -1 & 2\end{bmatrix}$

(c) $\begin{bmatrix}2 & 0 & 1 & -1\\ 1 & -1 & 0 & 2\\ 0 & -1 & 2 & 1\\ -2 & 1 & 3 & 0\end{bmatrix}$

(d) $\begin{bmatrix}1 & 3 & -2\\ -2 & 4 & 1\\ 5 & 1 & -3\end{bmatrix}$

(e) $\begin{bmatrix}5 & 3 & 4\\ -3 & 2 & 5\\ 7 & 4 & 6\end{bmatrix}$

(f) $\begin{bmatrix}1 & 2 & -1\\ 2 & 3 & 2\\ 4 & -2 & 3\end{bmatrix}$

(g) $\begin{bmatrix}3 & -2 & 4\\ 5 & 3 & 3\\ 2 & 5 & -2\end{bmatrix}$

(h) $\begin{bmatrix}1 & -2 & 0 & 1\\ 0 & 1 & 2 & -1\\ 2 & -3 & 1 & 3\\ -1 & 3 & -2 & 0\end{bmatrix}$

30.14 Solve each system of equations using matrices.

(a) $x - 2y + 3z = -1$, $-x + 3y = 10$, $2x - 5y + 5z = -7$

(b) $x - 3y + z = 1$, $2x - y - 2z = 2$, $x + 2y - 3z = -1$

(c) $x + y - 3z = -1$, $y - z = 0$, $-x + 2y = 1$

(d) $4x - y + 5z = 11$, $x + 2y - z = 5$, $5x - 8y + 13z = 7$

(e) $x_1 + x_3 = 1$, $5x_2 + 3x_2 = 4$, $3x_2 - 4x_3 = 4$

(f) $4x_1 + 3x_2 + 17x_3 = 0$, $5x_1 + 4x_2 + 22x_3 = 0$, $4x_1 + 2x_2 + 19x_3 = 0$

(g) $x_1 + x_2 + x_3 + x_4 = 6$, $2x_1 + 3x_2 - x_4 = 0$, $-3x_1 + 4x_2 + x_3 + 2x_4 = 4$, $x_1 + 2x_2 - x_3 + x_4 = 0$

(h) $3x_1 - 2x_2 - 6x_3 = -4$, $-3x_1 + 2x_2 + 6x_3 = 1$, $x_1 - x_2 - 5x_3 = -3$

ANSWERS TO SUPPLEMENTARY PROBLEMS

30.9 (a) $\begin{bmatrix} 5 & -2 & 5 \\ 1 & 1 & 0 \end{bmatrix}$ (b) $\begin{bmatrix} 10 & -25 \\ 0 & 35 \end{bmatrix}$ (c) $\begin{bmatrix} -14 & -8 & -30 \\ -6 & 10 & -24 \end{bmatrix}$ (d) $\begin{bmatrix} -18 & -3 & -30 \\ -6 & 6 & -18 \end{bmatrix}$ (e) $\begin{bmatrix} 13 & -7/2 & 15 \\ 3 & 1 & 3 \end{bmatrix}$ (f) $[14 \quad -14]$ (g) not possible (h) $\begin{bmatrix} -1 & -3 & -5 \\ -1 & 3 & -6 \end{bmatrix}$

(i) not possible (j) not possible (k) $[28 \quad 21 \quad 28]$ (l) $\begin{bmatrix} 4 & -45 \\ 0 & 49 \end{bmatrix}$ (m) not possible (n) $[28 \quad 21 \quad 28]$ (o) $\begin{bmatrix} 8 & -335 \\ 0 & 343 \end{bmatrix}$ (p) $[38 \quad -11 \quad 35]$

30.10 (a) $\begin{bmatrix} 3 & -4 \\ 10 & 16 \\ 26 & 46 \end{bmatrix}$ (b) $\begin{bmatrix} 6 & -21 & 15 \\ 8 & -23 & 19 \\ 4 & 7 & 5 \end{bmatrix}$ (c) $\begin{bmatrix} 11 & 2 & 14 \\ 21 & 20 & 24 \\ 1 & -26 & 2 \end{bmatrix}$ (d) $\begin{bmatrix} 60 & 72 \\ -20 & -24 \\ 10 & 12 \\ 60 & 72 \end{bmatrix}$

30.11 (a) $(5, 5)$ (b) $(-3, 2)$ (c) $(1, 1)$ (d) $(8, 8)$

30.12 (a) $\begin{bmatrix} 1 & 0 & 0 & -1 \\ 0 & 1 & 0 & 0 \\ 0 & 0 & 1 & -1 \end{bmatrix}$ (b) $\begin{bmatrix} 1 & 0 & 2 & 4 & 0 \\ 0 & 1 & -1 & 1 & 0 \\ 0 & 0 & 0 & 0 & 1 \\ 0 & 0 & 0 & 0 & 0 \end{bmatrix}$

(c) $\begin{bmatrix} 1 & 0 & 0 \\ 0 & 1 & 0 \\ 0 & 0 & 1 \\ 0 & 0 & 0 \end{bmatrix}$ (d) $\begin{bmatrix} 1 & 0 & 0 & 1 \\ 0 & 1 & 0 & -1 \\ 0 & 0 & 1 & -2 \end{bmatrix}$

(e) $\begin{bmatrix} 1 & 0 & 2 & 0 & -1 \\ 0 & 1 & -1 & 0 & 2 \\ 0 & 0 & 0 & 1 & 4 \\ 0 & 0 & 0 & 0 & 0 \end{bmatrix}$ (f) $\begin{bmatrix} 1 & 0 & 0 & 1/5 & -1 \\ 0 & 1 & 0 & -12/5 & 3 \\ 0 & 0 & 1 & -3/5 & 2 \\ 0 & 0 & 0 & 0 & 0 \end{bmatrix}$

30.13 (a) $\begin{bmatrix} 3 & -13 \\ -2 & 9 \end{bmatrix}$

(b) $\frac{1}{4}\begin{bmatrix} 2 & 2 \\ 1 & 3 \end{bmatrix}$

(c) $\frac{1}{18}\begin{bmatrix} 7 & 6 & -5 & 1 \\ 5 & 12 & -19 & 11 \\ 3 & 0 & 3 & 3 \\ -1 & 12 & -7 & 5 \end{bmatrix}$

(d) $\frac{1}{28}\begin{bmatrix} -13 & 7 & 11 \\ -1 & 7 & 3 \\ -22 & 14 & 10 \end{bmatrix}$

(e) $\frac{1}{15}\begin{bmatrix} -8 & -2 & 7 \\ 53 & 2 & -37 \\ -26 & 1 & 19 \end{bmatrix}$

(f) $\frac{1}{33}\begin{bmatrix} 13 & -4 & 7 \\ 2 & 7 & -4 \\ -16 & 10 & -1 \end{bmatrix}$

(g) $\frac{1}{19}\begin{bmatrix} 21 & -16 & 18 \\ -16 & 14 & -11 \\ -19 & 19 & -19 \end{bmatrix}$

(h) $\frac{1}{6}\begin{bmatrix} 21 & 9 & -4 & 7 \\ 3 & 3 & 0 & 3 \\ -6 & 0 & 2 & -2 \\ -9 & -3 & 4 & -1 \end{bmatrix}$

30.14 (a) $(-1, 3, 2)$
(b) no solution
(c) $(2z-1, z, z)$ where z is a real number
(d) $(-z+3, z+1, z)$ where z is a real number
(e) $(-4, 8, 5)$
(f) $(0, 0, 0)$
(g) $(1, 0, 3, 2)$
(h) no solution

Chapter 31

Mathematical Induction

31.1 PRINCIPLE OF MATHEMATICAL INDUCTION

Some statements are defined on the set of positive integers. To establish the truth of such a statement, we could prove it for each positive integer of interest separately. However, since there are infinitely many positive integers, this case-by-case procedure can never prove that the statement is always true. A procedure called mathematical induction can be used to establish the truth of the state for all positive integers.

Principle of Mathematical Induction

Let $P(n)$ be a statement that is either true or false for each positive integer n. If the following two conditions are satisfied:

(1) $P(1)$ is true. and

(2) Whenever for $n = k\,P(k)$ is true implies $P(k+1)$ is true.

Then $P(n)$ is true for all positive integers n.

31.2 PROOF BY MATHEMATICAL INDUCTION

To prove a theorem or formula by mathematical induction there are two distinct steps in the proof.

(1) Show by actual substitution that the proposed theorem or formula is true for some one positive integer n, as $n = 1$, or $n = 2$, etc.

(2) Assume that the theorem or formula is true for $n = k$. Then prove that it is true for $n = k + 1$.

Once steps (1) and (2) have been completed, then you can conclude the theorem or formula is true for all positive integers greater than or equal to a, the positive integer from step (1).

Solved Problems

31.1 Prove by mathematical induction that, for all positive integers n,

$$1 + 2 + 3 + \cdots + n = \frac{n(n+1)}{2}.$$

SOLUTION

Step 1. The formula is true for $n = 1$, since

$$1 = \frac{1(1+1)}{2} = 1.$$

Step 2. Assume that the formula is true for $n = k$. Then, adding $(k+1)$ to both sides,

$$1 + 2 + 3 + \cdots + k + (k+1) = \frac{k(k+1)}{2} + (k+1) = \frac{(k+1)(k+2)}{2}$$

which is the value of $n(n+1)/2$ when $(k+1)$ is substituted for n.

Hence if the formula is true for $n = k$, we have proved it to be true for $n = k+1$. But the formula holds for $n = 1$; hence it holds for $n = 1 + 1 = 2$. Then, since it holds for $n = 2$, it holds for $n = 2 + 1 = 3$. and so on. Thus the formula is true for all positive integers n.

31.2 Prove by mathematical induction that the sum of n terms of an arithmetic sequence $a, a+d$, $a+2d, \cdots$ is $\left(\frac{n}{2}\right)[2a + (n-1)d]$, that is

$$a + (a+d) + (a+2d) + \cdots + [a + (n-1)d] = \frac{n}{2}[2a + (n-1)d].$$

SOLUTION

Step 1. The formula holds for $n = 1$, since $a = \frac{1}{2}[2a + (1-1)d] = a$.

Step 2. Assume that the formula holds for $n = k$. Then

$$a + (a+d) + (a+2d) + \cdots + [a + (k-1)d] = \frac{k}{2}[2a + (k-1)d].$$

Add the $(k+1)$th term, which is $(a + kd)$, to both sides of the latter equation. Then

$$a + (a+d) + (a+2d) + \cdots + [a + (k-1)d] + (a+kd) = \frac{k}{2}[2a + (k-1)d] + (a+kd).$$

The right-hand side of this equation $= ka + \dfrac{k^2 d}{2} - \dfrac{kd}{2} + a + kd = \dfrac{k^2 d + kd + 2ka + 2a}{2}$

$$= \frac{kd(k+1) + 2a(k+1)}{2} = \frac{k+1}{2}(2a + kd)$$

which is the value of $(n/2)[2a + (n-1)d]$ when n is replaced by $(k+1)$.

Hence if the formula is true for $n = k$, we have proved it to be true for $n = k+1$. But the formula holds for $n = 1$; hence it holds for $n = 1 + 1 = 2$. Then, since it holds for $n = 2$, it holds for $n = 2 + 1 = 3$, and so on. Thus the formula is true for all positive integers n.

31.3 Prove by mathematical induction that, for all positive integers n,

$$1^2 + 2^2 + 3^2 + \cdots + n^2 = \frac{n(n+1)(2n+1)}{6}.$$

SOLUTION

Step 1. The formula is true for $n = 1$, since

$$1^2 = \frac{1(1+1)(2+1)}{6} = 1.$$

Step 2. Assume that the formula is true for $n = k$. Then

$$1^2 + 2^2 + 3^2 + \cdots + k^2 = \frac{k(k+1)(2k+1)}{6}.$$

Add the $(k+1)$th term, which is $(k+1)^2$, to both sides of this equation. Then

$$1^2 + 2^2 + 3^2 + \cdots + k^2 + (k+1)^2 = \frac{k(k+1)(2k+1)}{6} + (k+1)^2.$$

$$\text{The right hand side of this equation} = \frac{k(k+1)(2k+1)+6(k+1)^2}{6}$$

$$= \frac{(k+1)[(2k^2+k)+(6k+6)]}{6} = \frac{(k+1)(k+2)(2k+3)}{6}$$

which is the value of $n(n+1)(2n+1)/6$ when n is replaced by $(k+1)$.

Hence if the formula is true for $n = k$, it is true for $n = k+1$. But the formula holds for $n = 1$; hence it holds for $n = 1+1 = 2$. Then, since it holds for $n = 2$, it holds for $n = 2+1 = 3$, and so on. Thus the formula is true for all positive integers.

31.4 Prove by mathematical induction that, for all positive integers n,

$$\frac{1}{1\cdot 3}+\frac{1}{3\cdot 5}+\frac{1}{5\cdot 7}+\cdots+\frac{1}{(2n-1)(2n+1)} = \frac{n}{2n+1}.$$

SOLUTION

Step 1. The formula is true for $n = 1$, since

$$\frac{1}{(2-1)(2+1)} = \frac{1}{2+1} = \frac{1}{3}.$$

Step 2. Assume that the formula is true for $n = k$. Then

$$\frac{1}{1\cdot 3}+\frac{1}{3\cdot 5}+\frac{1}{5\cdot 7}+\cdots+\frac{1}{(2k-1)(2k+1)} = \frac{k}{2k+1}.$$

Add the $(k+1)$th term, which is

$$\frac{1}{(2k+1)(2k+3)},$$

to both sides of the above equation. Then

$$\frac{1}{1\cdot 3}+\frac{1}{3\cdot 5}+\frac{1}{5\cdot 7}+\cdots+\frac{1}{(2k-1)(2k+1)}+\frac{1}{(2k+1)(2k+3)} = \frac{k}{2k+1}+\frac{1}{(2k+1)(2k+3)}.$$

The right-hand side of this equation is

$$\frac{k(2k+3)+1}{(2k+1)(2k+3)} = \frac{k+1}{2k+3},$$

which is the value of $n/(2n+1)$ when n is replaced by $(k+1)$.

Hence if the formula is true for $n = k$, it is true for $n = k+1$. But the formula holds for $n = 1$; hence it holds for $n = 1+1 = 2$. Then, since it holds for $n = 2$, it holds for $n = 2+1 = 3$, and so on. Thus the formula is true for all positive integers n.

31.5 Prove by mathematical induction that $a^{2n} - b^{2n}$ is divisible by $a + b$ when n is any positive integer.

SOLUTION

Step 1. The theorem is true for $n = 1$, since $a^2 - b^2 = (a+b)(a-b)$.

Step 2. Assume that the theorem is true for $n = k$. Then

$$a^{2k} - b^{2k} \text{ is divisible by } a+b.$$

We must show that $a^{2k+2} - b^{2k+2}$ is divisible by $a + b$. From the identity

$$a^{2k+2} - b^{2k+2} = a^2(a^{2k} - b^{2k}) + b^{2k}(a^2 - b^2)$$

it follows that $a^{2k+2} - b^{2k+2}$ is divisible by $a + b$ if $a^{2k} - b^{2k}$ is.

Hence if the theorem is true for $n = k$, we have proved it to be true for $n = k + 1$. But the theorem holds for $n = 1$; hence it holds for $n = 1 + 1 = 2$. Then, since it holds for $n = 2$, it holds for $n = 2 + 1 = 3$, and so on. Thus the theorem is true for all positive integers n.

31.6 Prove the binomial formula

$$(a + x)^n = a^n + na^{n-1}x + \frac{n(n-1)}{2!}a^{n-2}x^2 + \cdots + \frac{n(n-1)\cdots(n-r+2)}{(r-1)!}a^{n-r+1}x^{r-1} + \cdots + x^n$$

for positive integers n.

SOLUTION

Step 1. The formula is true for $n = 1$.

Step 2. Assume the formula is true for $n = k$. Then

$$(a + x)^k = a^k + na^{k-1}x + \frac{k(k-1)}{2!}a^{k-2}x^2 + \cdots + \frac{k(k-1)\cdots(k-r+2)}{(r-1)!}a^{k-r+1}x^{r-1} + \cdots + x^k$$

Multiply both sides by $a + x$. The multiplication on the right may be written

$$a^{k+1} + ka^k x + \frac{k(k-1)}{2!}a^{k-1}x^2 + \cdots + \frac{k(k-1)\cdots(k-r+2)}{(r-1)!}a^{k-r+2}x^{r-1} + \cdots + ax^k.$$

$$+ a^k x + ka^{k-1}x^2 + \cdots + \frac{k(k-1)\cdots(k-r+3)}{(r-2)!}a^{k-r+2}x^{r-1} + \cdots + x^{k+1}$$

Since

$$\frac{k(k-1)\cdots(k-r+2)}{(r-1)!}a^{k-r+2}x^{r-1} + \frac{k(k-1)\cdots(k-r+3)}{(r-2)!}a^{k-r+2}x^{r-1}.$$

$$= \frac{k(k-1)\cdots(k-r+3)}{(r-2)!}a^{k-r+2}x^{r-1}\left\{\frac{k-r+2}{r-1} + 1\right\}$$

$$= \frac{(k+1)k(k-1)\cdots(k-r+3)}{(r-1)!}a^{k-r+2}x^{r-1},$$

the product may be written

$$(a + x)^{k+1} = a^{k+1} + (k+1)a^k x + \cdots + \frac{(k+1)k(k-1)\cdots(k-r+3)}{(r-1)!}a^{k-r+2}x^{r-1} + \cdots + x^{k+1}$$

which is the binomial formula with n replaced by $k + 1$.

Hence if the formula is true for $n = k$, it is true for $n = k + 1$. But the formula holds for $n = 1$; hence it holds for $n = 1 + 1 = 2$, and so on. Thus the formula is true for all positive integers n.

31.7 Prove by mathematical induction that the sum of the interior angles, $S(n)$, of a convex polygon is $S(n) = (n - 2)180°$, where n is the number of sides on the polygon.

SOLUTION

Step 1. Since a polygon has at least 3 sides, we start with $n = 3$. For $n = 3$, $S(3) = (3 - 2)180° = (1)180° = 180°$. This is true since the sum of the interior angles of a triangle is 180°.

Step 2. Assume that for $n = k$, the formula is true. Then $S(k) = (k - 2)180°$ is true. Now consider a convex polygon with $k + 1$ sides. We can draw in a diagonal that forms a triangle with two of the sides of the polygon. The diagonal also forms a k-sided polygon with the other sides of the original polygon. The sum of the interior angles of the $(k + 1)$-sided polygon, $S(k + 1)$, is equal to the sum of the interior angles of the triangle, $S(3)$, and the sum of the interior angles of the k-sided polygon, $S(k)$.

$$S(k + 1) = S(3) + S(k) = 180° + (k - 2)180° = [1 + (k - 2)]180° = [(k + 1) - 2]180°.$$

Hence, if the formula is true for $n = k$, it is true for $n = k + 1$.

Since the formula is true for $n = 3$, and whenever it is true for $n = k$ it is true for $n = k + 1$, the formula is true for all positive integers $n \geq 3$.

31.8 Prove by mathematical induction that $n^3 + n \geq n^2 + 1$ for all positive integers.

SOLUTION

Step 1. For $n = 1$, $n^3 + 1 = 1^3 + 1 = 1 + 1 = 2$ and $n^2 + n = 1^2 + 1 = 1 + 1 = 2$. So $n^3 + 1 \geq n^2 + n$ is true when $n = 1$.

Step 2. Assume the statement is true for $n = k$. So $k^3 + 1 \geq k^2 + k$ is true.

For $n = k + 1$, $(k+1)^3 + 1 = k^3 + 3k^2 + 3k + 1 + 1 = k^3 + 3k^2 + 3k + 2$

$$= k^3 + 2k^2 + k^2 + 3k + 2 = (k^3 + 2k^2) + (k+1)(k+2)$$
$$= (k^3 + 2k^2) + (k+1)[(k+1) + 1]$$
$$= (k^3 + 2k^2) + [(k+1)^2 + (k+1)]$$

We know $n \geq 1$, so $k \geq 1$ and $k^3 + 2k^2 \geq 3$. Thus $(k+1)^3 + 1 \geq (k+1)^2 + (k+1)$. Hence, when the statement is true for $n = k$, it is true for $n = k + 1$.

Since the statement is true for $n = 1$, and whenever it is true for $n = k$, it is true for $n = k + 1$, the statement is true for all positive integers n.

Supplementary Problems

Prove each of the following by mathematical induction. In each case n is a positive integer.

31.9 $1 + 3 + 5 + \cdots + (2n - 1) = n^2$

31.10 $1 + 3 + 3^2 + \cdots + 3^{n-1} = \dfrac{3^n - 1}{2}$

31.11 $1^3 + 2^3 + 3^3 + \cdots + n^3 = \dfrac{n^2(n+1)^2}{4}$

31.12 $a + ar + ar^2 + \cdots + ar^{n-1} = \dfrac{a(r^n - 1)}{r - 1}, \quad r \neq 1$

31.13 $\dfrac{1}{1\cdot 2} + \dfrac{1}{2\cdot 3} + \dfrac{1}{3\cdot 4} + \cdots + \dfrac{1}{n(n+1)} = \dfrac{n}{n+1}$

31.14 $1\cdot 3 + 2\cdot 3^2 + 3\cdot 3^3 + \cdots + n\cdot 3^n = \dfrac{(2n-1)3^{n+1} + 3}{4}$

31.15 $\dfrac{1}{2\cdot 5} + \dfrac{1}{5\cdot 8} + \dfrac{1}{8\cdot 11} + \cdots + \dfrac{1}{(3n-1)(3n+2)} = \dfrac{n}{6n+4}$

31.16 $\dfrac{1}{1\cdot 2\cdot 3} + \dfrac{1}{2\cdot 3\cdot 4} + \dfrac{1}{3\cdot 4\cdot 5} + \cdots + \dfrac{1}{n(n+1)(n+2)} = \dfrac{n(n+3)}{4(n+1)(n+2)}$

31.17 $a^n - b^n$ is divisible by $a - b$, for n = positive integer.

31.18 $a^{2n-1} + b^{2n-1}$ is divisible by $a + b$, for n = positive integer.

31.19 $1 \cdot 2 \cdot 3 + 2 \cdot 3 \cdot 4 + \cdots + n(n+1)(n+2) = \dfrac{n(n+1)(n+2)(n+3)}{4}$

31.20 $1 + 2 + 2^2 + \cdots + 2^{n-1} = 2^n - 1$

31.21 $(ab)^n = a^n b^n$, for $n = a$ positive integer

31.22 $\left(\dfrac{a}{b}\right)^n = \dfrac{a^n}{b^n}$, for $n = a$ positive integer

31.23 $n^2 + n$ is even

31.24 $n^3 + 5n$ is divisible by 3

31.25 $5^n - 1$ is divisible by 4

31.26 $4^n - 1$ is divisible by 3

31.27 $n(n+1)(n+2)$ is divisible by 6

31.28 $n(n+1)(n+2)(n+3)$ is divisible by 24

31.29 $n^2 + 1 > n$

31.30 $2n \geq n + 1$

Chapter 32

Partial Fractions

32.1 RATIONAL FRACTIONS

A rational fraction in x is the quotient $\dfrac{P(x)}{Q(x)}$ of two polynomials in x.

Thus $\dfrac{3x^2-1}{x^3+7x^2-4}$ is a rational fraction.

32.2 PROPER FRACTIONS

A proper fraction is one in which the degree of the numerator is less than the degree of the denominator.

Thus $\dfrac{2x-3}{x^2+5x+4}$ and $\dfrac{4x^2+1}{x^4-3x}$ are proper fractions.

An improper fraction is one in which the degree of the numerator is greater than or equal to the degree of the denominator.

Thus $\dfrac{2x^3+6x^2-9}{x^2-3x+2}$ is an improper fraction.

By division, an improper fraction may always be written as the sum of a polynomial and a proper fraction.

Thus $$\frac{2x^3+6x^2-9}{x^2-3x+2} = 2x+12+\frac{32x-33}{x^2-3x+2}.$$

32.3 PARTIAL FRACTIONS

A given proper fraction may often be written as the sum of other fractions (called partial fractions) whose denominators are of lower degree than the denominator of the given fraction.

EXAMPLE 32.1.

$$\frac{3x-5}{x^2-3x+2} = \frac{3x-5}{(x-1)(x-2)} = \frac{2}{x-1} + \frac{1}{x-2}.$$

32.4 IDENTICALLY EQUAL POLYNOMIALS

If two polynomials of degree n in the same variable x are equal for more than n values of x, the coefficients of like powers of x are equal and the two polynomials are identically equal. If a term is missing in either of the polynomials, it can be written in with a coefficient of 0.

32.5 FUNDAMENTAL THEOREM

A proper fraction may be written as the sum of partial fractions according to the following rules.

(1) Linear factors none of which are repeated

If a linear factor $ax + b$ occurs once as a factor of the denominator of the given fraction, then corresponding to this factor associate a partial fraction $A/(ax + b)$, where A is a constant $\neq 0$.

EXAMPLE 32.2.

$$\frac{x+4}{(x+7)(2x-1)} = \frac{A}{x+7} + \frac{B}{2x-1}$$

(2) Linear factors some of which are repeated

If a linear factor $ax + b$ occurs p times as a factor of the denominator of the given fraction, then corresponding to this factor associate the p partial fractions

$$\frac{A_1}{ax+b} + \frac{A_2}{(ax+b)^2} + \ldots + \frac{A_p}{(ax+b)^p}$$

where $A_1, A_2, \ldots, A_p$ are constants and $A_p \neq 0$.

EXAMPLES 32.3.

(a) $\dfrac{3x-1}{(x+4)^2} = \dfrac{A}{x+4} + \dfrac{B}{(x+4)^2}$

(b) $\dfrac{5x^2-2}{x^3(x+1)^2} = \dfrac{A}{x^3} + \dfrac{B}{x^2} + \dfrac{C}{x} + \dfrac{D}{(x+1)^2} + \dfrac{E}{x+1}$

(3) Quadratic factors none of which are repeated

If a quadratic factor $ax^2 + bx + c$ occurs once as a factor of the denominator of the given fraction, then corresponding to this factor associate a partial fraction

$$\frac{Ax+B}{ax^2+bx+c}$$

where A and B are constants which are not both zero.

Note. It is assumed that $ax^2 + bx + c$ cannot be factored into two real linear factors with integer coefficients.

EXAMPLES 32.4.

(a) $\dfrac{x^2-3}{(x-2)(x^2+4)} = \dfrac{A}{x-2} + \dfrac{Bx+C}{x^2+4}$

(b) $\dfrac{2x^3-6}{x(2x^2+3x+8)(x^2+x+1)} = \dfrac{A}{x} + \dfrac{Bx+C}{2x^2+3x+8} + \dfrac{Dx+E}{x^2+x+1}$

(4) Quadratic factors some of which are repeated

If a quadratic factor $ax^2 + bx + c$ occurs p times as a factor of the denominator of the given fraction, then corresponding to this factor associate the p partial fractions

$$\frac{A_1x+B_1}{ax^2+bx+ac} + \frac{A_2x+B_2}{(ax^2+bx+c)^2} + \ldots + \frac{A_px+B_p}{(ax^2+bx+c)^p}$$

where $A_1, B_1, A_2, B_2, \ldots, A_p, B_p$ are constants and A_p, B_p are not both zero.

EXAMPLE 32.5.

$$\frac{x^2-4x+1}{(x^2+1)^2(x^2+x+1)}=\frac{Ax+B}{x^2+1}+\frac{Cx+D}{(x^2+1)^2}+\frac{Ex+F}{x^2+x+1}$$

32.6 FINDING THE PARTIAL FRACTION DECOMPOSITION

Once the form of the partial fraction decomposition of a rational fraction has been determined, the next step is to find the system of equations to be solved to get the values of the constants needed in the partial fraction decomposition. The solution of the system of equations can be aided by the use of a graphing calculator, especially when using the matrix methods discussed in Chapter 30.

Although the system of equations usually involves more than three equations, it is often quite easy to determine the value of one or two variables or relationships among the variables that allows the system to be reduced to a size small enough to be solved conveniently by any method. The methods discussed in Chapter 15 and Chapter 28 are the basic procedures used.

EXAMPLE 32.6. Find the partial fraction decomposition of $\dfrac{3x^2+3x+7}{(x-2)^2(x^2+1)}$.

Using Rules (2) and (3) in Section 32.5, the form of the decomposition is:

$$\frac{3x^2+3x+7}{(x-2)^2(x^2+1)}=\frac{A}{x-2}+\frac{B}{(x-2)^2}+\frac{Cx+D}{x^2+1}$$

$$\frac{3x^2+3x+7}{(x-2)^2(x^2+1)}=\frac{A(x-2)(x^2+1)+B(x^2+1)+(Cx+D)(x-2)^2}{(x-2)^2(x^2+1)}$$

$$3x^2+3x+7=Ax^3-2Ax^2+Ax-2A+Bx^2+B+Cx^3-4Cx^2+Dx^2+4Cx-4Dx+4D$$

$$3x^2+3x+7=(A+C)x^3+(-2A+B-4C+D)x^2+(A+4C-4D)x+(-2A+B+4D)$$

Equating the coefficients of the corresponding terms in the two polynomials and setting the others equal to 0, we get the system of equations to solve.

$$\begin{aligned} A+C&=0\\ -2A+B-4C+D&=3\\ A+4C-4D&=3\\ -2A+B+4D&=7 \end{aligned}$$

Solving the system, we get $A=-1$, $B=5$, $C=1$, and $D=0$.

Thus, the partial fraction decomposition is:

$$\frac{3x^2+3x+7}{(x-2)^2(x^2+1)}=\frac{-1}{x-2}+\frac{5}{(x-2)^2}+\frac{x}{x^2+1}$$

Solved Problems

32.1 Resolve into partial fractions

$$\frac{x+2}{2x^2-7x-15} \quad \text{or} \quad \frac{x+2}{(2x+3)(x-5)}.$$

SOLUTION

Let $\dfrac{x+2}{(2x+3)(x-5)} = \dfrac{A}{2x+3} + \dfrac{B}{x-5} = \dfrac{A(x-5)+B(2x+3)}{(2x+3)(x-5)} = \dfrac{(A+2B)x+3B-5A}{(2x+3)(x-5)}.$

We must find the constants A and B such that

$$\frac{x+2}{(2x+3)(x-5)} = \frac{(A+2B)x+3B-5A}{(2x+3)(x-5)} \quad \text{identically}$$

or $$x+2 = (A+2B)x+3B-5A.$$

Equating coefficients of like powers of x, we have $1 = A+2B$ and $2 = 3B-5A$ which when solved simultaneously give $A = -1/13$, $B = 7/13$.

Hence $$\frac{x+2}{2x^2-7x-15} = \frac{-1/13}{2x+3} + \frac{7/13}{x-5} = \frac{-1}{13(2x+3)} + \frac{7}{13(x-5)}.$$

Another method. $x+2 = A(x-5)+B(2x+3)$

To find B, let $x=5$: $5+2 = A(0)+B(10+3)$, $7 = 13B$, $B = 7/13$.

To find A, let $x=-3/2$: $-3/2+2 = A(-3/2-5)+B(0)$, $1/2 = -13A/2$, $A = -1/13$.

32.2 $\dfrac{2x^2+10x-3}{(x+1)(x^2-9)} = \dfrac{A}{x+1} + \dfrac{B}{x+3} + \dfrac{C}{x-3}$

SOLUTION

$$2x^2+10x-3 = A(x^2-9)+B(x+1)(x-3)+C(x+1)(x+3)$$

To find A, let $x=-1$: $2-10-3 = A(1-9)$, $A = 11/8$.

To find B, let $x=-3$: $18-30-3 = B(-3+1)(-3-3)$, $B = -5/4$.

To find C, let $x=3$: $18+30-3 = C(3+1)(3+3)$, $C = 15/8$.

Hence $$\frac{2x^2+10x-3}{(x+1)(x^2-9)} = \frac{11}{8(x+1)} - \frac{5}{4(x+3)} + \frac{15}{8(x-3)}.$$

32.3 $\dfrac{2x^2+7x+23}{(x-1)(x+3)^2} = \dfrac{A}{x-1} + \dfrac{B}{(x+3)^2} + \dfrac{C}{x+3}$

SOLUTION

$$\begin{aligned} 2x^2+7x+23 &= A(x+3)^2+B(x-1)+C(x-1)(x+3) \\ &= A(x^2+6x+9)+B(x-1)+C(x^2+2x-3) \\ &= Ax^2+6Ax+9A+Bx-B+Cx^2+2Cx-3C \\ &= (A+C)x^2+(6A+B+2C)x+9A-B-3C \end{aligned}$$

Equating coefficients of like powers of x, $A+C=2$, $6A+B+2C=7$ and $9A-B-3C=23$. Solving simultaneously, $A=2$, $B=-5$, $C=0$.

Hence $$\frac{2x^2+7x+23}{(x-1)(x+3)^2} = \frac{2}{x-1} - \frac{5}{(x+3)^2}.$$

Another method. $2x^2+7x+23 = A(x+3)^2+B(x-1)+C(x-1)(x+3)$

To find A, let $x=1$: $2+7+23 = A(1+3)^2$, $A=2$.

To find B, let $x=-3$: $18-21+23 = B(-3-1)$, $B=-5$.

To find C, let $x=0$: $23 = 2(3)^2 - 5(-1)+C(-1)(3)$, $C=0$.

32.4 $\dfrac{x^2-6x+2}{x^2(x-2)^2} = \dfrac{A}{x^2} + \dfrac{B}{x} + \dfrac{C}{(x-2)^2} + \dfrac{D}{x-2}$

SOLUTION

$$\begin{aligned} x^2-6x+2 &= A(x-2)^2 + Bx(x-2)^2 + Cx^2 + Dx^2(x-2) \\ &= A(x^2-4x+4) + Bx(x^2-4x+4) + Cx^2 + Dx^2(x-2) \\ &= (B+D)x^3 + (A-4B+C-2D)x^2 + (-4A+4B)x + 4A \end{aligned}$$

Equating coefficients of like powers of x, $B+D=0$, $A-4B+C-2D=1$, $-4A+4B=-6$, $4A=2$. The simultaneous solution of these four equations is $A=1/2$, $B=-1$, $C=-3/2$, $D=1$.

Hence $$\frac{x^2-6x+2}{x^2(x-2)^2} = \frac{1}{2x^2} - \frac{1}{x} - \frac{3}{2(x-2)^2} + \frac{1}{x-2}$$

Another method. $x^2-6x+2 = A(x-2)^2 + Bx(x-2)^2 + Cx^2 + Dx^2(x-2)$

To find A, let $x=0$: $2=4A$, $A=1/2$. To find C, let $x=2$: $4-12+2=4C$, $C=-3/2$.
To find B and D, let $x=$ any values except 0 and 2 (for example, let $x=1$, $x=-1$).

Let $x=1$: $1-6+2 = A(1-2)^2 + B(1-2)^2 + C + D(1-2)$ and (1) $B-D=-2$.
Let $x=-1$: $1+6+2 = A(-1-2)^2 - B(-1-2)^2 + C + D(-1-2)$ and (2) $9B+3D=-6$.

The simultaneous solution of equations (1) and (2) is $B=-1$, $D=1$.

32.5 $\dfrac{x^2-4x-15}{(x+2)^3}$. Let $y=x+2$; then $x=y-2$.

SOLUTION

$$\begin{aligned} \frac{x^2-4x-15}{(x+2)^3} &= \frac{(y-2)^2-4(y-2)-15}{y^3} = \frac{y^2-8y-3}{y^3} \\ &= \frac{1}{y} - \frac{8}{y^2} - \frac{3}{y^3} = \frac{1}{x+2} - \frac{8}{(x+2)^2} - \frac{3}{(x+2)^3} \end{aligned}$$

32.6 $\dfrac{7x^2-25x+6}{(x^2-2x-1)(3x-2)} = \dfrac{Ax+B}{x^2-2x-1} + \dfrac{C}{3x-2}$

SOLUTION

$$\begin{aligned} 7x^2-25x+6 &= (Ax+B)(3x-2) + C(x^2-2x-1) \\ &= (3Ax^2+3Bx-2Ax-2B) + Cx^2 - 2Cx - C \\ &= (3A+C)x^2 + (3B-2A-2C)x + (-2B-C) \end{aligned}$$

Equating coefficients of like powers of x, $3A+C=7$, $3B-2A-2C=-25$, $-2B-C=6$. The simultaneous solution of these three equations is $A=1$, $B=-5$, $C=4$.

Hence $$\frac{7x^2-25x+6}{(x^2-2x-1)(3x-2)} = \frac{x-5}{x^2-2x-1} + \frac{4}{3x-2}.$$

32.7 $\dfrac{4x^2-28}{x^4+x^2-6} = \dfrac{4x^2-28}{(x^2+3)(x^2-2)} = \dfrac{Ax+B}{x^2+3} + \dfrac{Cx+D}{x^2-2}$

SOLUTION

$$\begin{aligned} 4x^2-28 &= (Ax+B)(x^2-2) + (Cx+D)(x^2+3) \\ &= (Ax^3+Bx^2-2Ax-2B) + (Cx^3+Dx^2+3Cx+3D) \\ &= (A+C)x^3 + (B+D)x^2 + (3C-2A)x - 2B + 3D \end{aligned}$$

Equating coefficients of like powers of x,

$$A + C = 0,\ B + D = 4,\ 3C - 2A = 0,\ -2B + 3D = -28.$$

Solving simultaneously, $A = 0,\ B = 8,\ C = 0,\ D = -4.$

Hence $$\frac{4x^2 - 28}{x^4 + x^2 - 6} = \frac{8}{x^2 + 3} - \frac{4}{x^2 - 2}.$$

Supplementary Problems

Find the partial fraction decomposition of each rational fraction.

32.8 $\dfrac{x + 2}{x^2 - 7x + 12}$ **32.9** $\dfrac{12x + 11}{x^2 + x - 6}$ **32.10** $\dfrac{8 - x}{2x^2 + 3x - 2}$

32.11 $\dfrac{5x + 4}{x^2 + 2x}$ **32.12** $\dfrac{x}{x^2 - 3x - 18}$ **32.13** $\dfrac{10x^2 + 9x - 7}{(x + 2)(x^2 - 1)}$

32.14 $\dfrac{x^2 - 9x - 6}{x^3 + x^2 - 6x}$ **32.15** $\dfrac{x^3}{x^2 - 4}$ **32.16** $\dfrac{3x^2 - 8x + 9}{(x - 2)^3}$

32.17 $\dfrac{3x^3 + 10x^2 + 27x + 27}{x^2(x + 3)^2}$ **32.18** $\dfrac{5x^2 + 8x + 21}{(x^2 + x + 6)(x + 1)}$ **32.19** $\dfrac{5x^3 + 4x^2 + 7x + 3}{(x^2 + 2x + 2)(x^2 - x - 1)}$

32.20 $\dfrac{3x}{x^3 - 1}$ **32.21** $\dfrac{7x^3 + 16x^2 + 20x + 5}{(x^2 + 2x + 2)^2}$ **32.22** $\dfrac{7x - 9}{(x + 1)(x - 3)}$

32.23 $\dfrac{x + 10}{x(x - 2)(x + 2)}$ **32.24** $\dfrac{3x - 1}{x^2 - 1}$ **32.25** $\dfrac{7x - 2}{x^3 - x^2 - 2x}$

32.26 $\dfrac{5x^2 + 3x + 1}{(x + 2)(x^2 + 1)}$ **32.27** $\dfrac{-2x + 9}{(2x + 1)(4x^2 + 9)}$ **32.28** $\dfrac{2x^3 - x + 3}{(x^2 + 4)(x^2 + 1)}$

32.29 $\dfrac{x^3}{(x^2 + 4)^2}$ **32.30** $\dfrac{x^4 + 3x^2 + x + 1}{(x + 1)(x^2 + 1)^2}$

ANSWERS TO SUPPLEMENTARY PROBLEMS

32.8 $\dfrac{6}{x - 4} - \dfrac{5}{x - 3}$ **32.9** $\dfrac{7}{x - 2} + \dfrac{5}{x + 3}$ **32.10** $\dfrac{3}{2x - 1} - \dfrac{2}{x + 2}$

32.11 $\dfrac{2}{x} + \dfrac{3}{x + 2}$ **32.12** $\dfrac{2/3}{x - 6} + \dfrac{1/3}{x + 3}$ **32.13** $\dfrac{3}{x + 1} + \dfrac{2}{x - 1} + \dfrac{5}{x + 2}$

32.14 $\dfrac{1}{x} - \dfrac{2}{x - 2} + \dfrac{2}{x + 3}$ **32.15** $x + \dfrac{2}{x - 2} + \dfrac{2}{x + 2}$ **32.16** $\dfrac{3}{x - 2} + \dfrac{4}{(x - 2)^2} + \dfrac{5}{(x - 2)^3}$

32.17 $\dfrac{1}{x} + \dfrac{3}{x^2} + \dfrac{2}{x + 3} - \dfrac{5}{(x + 3)^2}$ **32.18** $\dfrac{2x + 3}{x^2 + x + 6} + \dfrac{3}{x + 1}$ **32.19** $\dfrac{2x - 1}{x^2 + 2x + 2} + \dfrac{3x + 1}{x^2 - x - 1}$

32.20 $\dfrac{1}{x-1}+\dfrac{-x-1}{x^2+x+1}$

32.21 $\dfrac{7x+2}{x^2+2x+2}+\dfrac{2x+1}{(x^2+2x+2)^2}$

32.22 $\dfrac{4}{x+1}+\dfrac{3}{x-3}$

32.23 $\dfrac{-5/2}{x}+\dfrac{3/2}{x-2}+\dfrac{1}{x+2}$

32.24 $\dfrac{1}{x-1}+\dfrac{2}{x+1}$

32.25 $\dfrac{1}{x}+\dfrac{2}{x-2}+\dfrac{-3}{x+1}$

32.26 $\dfrac{3}{x+2}+\dfrac{2x-1}{x^2+1}$

32.27 $\dfrac{1}{2x+1}+\dfrac{-2x}{4x^2+1}$

32.28 $\dfrac{3x-1}{x^2+4}+\dfrac{-x+1}{x^2+1}$

32.29 $\dfrac{x}{x^2+4}+\dfrac{-4x}{(x^2+4)^2}$

32.30 $\dfrac{1}{x+1}+\dfrac{x}{(x^2+1)^2}$

Appendix A

Table of Common Logarithms

N	0	1	2	3	4	5	6	7	8	9
10	0000	0043	0086	0128	0170	0212	0253	0294	0334	0374
11	0414	0453	0492	0531	0569	0607	0645	0682	0719	0755
12	0792	0828	0864	0899	0934	0969	1004	1038	1072	1106
13	1139	1173	1206	1239	1271	1303	1335	1367	1399	1430
14	1461	1492	1523	1553	1584	1614	1644	1673	1703	1732
15	1761	1790	1818	1847	1875	1903	1931	1959	1987	2014
16	2041	2068	2095	2122	2148	2175	2201	2227	2253	2279
17	2304	2330	2355	2380	2405	2430	2455	2480	2504	2529
18	2553	2577	2601	2625	2648	2672	2695	2718	2742	2765
19	2788	2810	2833	2856	2878	2900	2923	2945	2967	2989
20	3010	3032	3054	3075	3096	3118	3139	3160	3181	3201
21	3222	3243	3263	3284	3304	3324	3345	3365	3385	3404
22	3424	3444	3464	3483	3502	3522	3541	3560	3579	3598
23	3617	3636	3655	3674	3692	3711	3729	3747	3766	3784
24	3802	3820	3838	3856	3874	3892	3909	3927	3945	3962
25	3979	3997	4014	4031	4048	4065	4082	4099	4116	4133
26	4150	4166	4183	4200	4216	4232	4249	4265	4281	4298
27	4314	4330	4346	4362	4378	4393	4409	4425	4440	4456
28	4472	4487	4502	4518	4533	4548	4564	4579	4594	4609
29	4624	4639	4654	4669	4683	4698	4713	4728	4742	4757
30	4771	4786	4800	4814	4829	4843	4857	4871	4886	4900
31	4914	4928	4942	4955	4969	4983	4997	5011	5024	5038
32	5051	5065	5079	5092	5105	5119	5132	5145	5159	5172
33	5185	5198	5211	5224	5237	5250	5263	5276	5289	5302
34	5315	5328	5340	5353	5366	5378	5391	5403	5416	5428
35	5441	5453	5465	5478	5490	5502	5514	5527	5539	5551
36	5563	5575	5587	5599	5611	5623	5635	5647	5658	5670
37	5682	5694	5705	5717	5729	5740	5752	5763	5775	5786
38	5798	5809	5821	5832	5843	5855	5866	5877	5888	5899
39	5911	5922	5933	5944	5955	5966	5977	5988	5999	6010
40	6021	6031	6042	6053	6064	6075	6085	6096	6107	6117
41	6128	6138	6149	6160	6170	6180	6191	6201	6212	6222
42	6232	6243	6253	6263	6274	6284	6294	6304	6314	6325
43	6335	6345	6355	6365	6375	6385	6395	6405	6415	6425
44	6435	6444	6454	6464	6474	6484	6493	6503	6513	6522
N	0	1	2	3	4	5	6	7	8	9

N	0	1	2	3	4	5	6	7	8	9
45	6532	6542	6551	6561	6571	6580	6590	6599	6609	6618
46	6628	6637	6646	6656	6665	6675	6684	6693	6702	6712
47	6721	6730	6739	6749	6758	6767	6776	6785	6794	6803
48	6812	6821	6830	6839	6848	6857	6866	6875	6884	6893
49	6902	6911	6920	6928	6937	6946	6955	6964	6972	6981
50	6990	6998	7007	7016	7024	7033	7042	7050	7059	7067
51	7076	7084	7093	7101	7110	7118	7126	7135	7143	7152
52	7160	7168	7177	7185	7193	7202	7210	7218	7226	7235
53	7243	7251	7259	7267	7275	7284	7292	7300	7308	7316
54	7324	7332	7340	7348	7356	7364	7372	7380	7388	7396
55	7404	7412	7419	7427	7435	7443	7451	7459	7466	7474
56	7482	7490	7497	7505	7513	7520	7528	7536	7543	7551
57	7559	7566	7574	7582	7589	7597	7604	7612	7619	7627
58	7634	7642	7649	7657	7664	7672	7679	7686	7694	7701
59	7709	7716	7723	7731	7738	7745	7752	7760	7767	7774
60	7782	7789	7796	7803	7810	7818	7825	7832	7839	7846
61	7853	7860	7868	7875	7882	7889	7896	7903	7910	7917
62	7924	7931	7938	7945	7952	7959	7966	7973	7980	7987
63	7993	8000	8007	8014	8021	8028	8035	8041	8048	8055
64	8062	8069	8075	8082	8089	8096	8102	8109	8116	8122
65	8129	8136	8142	8149	8156	8162	8169	8176	8182	8189
66	8195	8202	8209	8215	8222	8228	8235	8241	8248	8254
67	8261	8267	8274	8280	8287	8293	8299	8306	8312	8319
68	8325	8331	8338	8344	8351	8357	8363	8370	8376	8382
69	8388	8395	8401	8407	8414	8420	8426	8432	8439	8445
70	8451	8457	8463	8470	8476	8482	8488	8494	8500	8506
71	8513	8519	8525	8531	8537	8543	8549	8555	8561	8567
72	8573	8579	8585	8591	8597	8603	8609	8615	8621	8627
73	8633	8639	8645	8651	8657	8663	8669	8675	8681	8686
74	8692	8698	8704	8710	8716	8722	8727	8733	8739	8745
75	8751	8756	8762	8768	8774	8779	8785	8791	8797	8802
76	8808	8814	8820	8825	8831	8837	8842	8848	8854	8859
77	8865	8871	8876	8882	8887	8893	8899	8904	8910	8915
78	8921	8927	8932	8938	8943	8949	8954	8960	8965	8971
79	8976	8982	8987	8993	8998	9004	9009	9015	9020	9025
80	9031	9036	9042	9047	9053	9058	9063	9069	9074	9079
81	9085	9090	9096	9101	9106	9112	9117	9122	9128	9133
82	9138	9143	9149	9154	9159	9165	9170	9175	9180	9186
83	9191	9196	9201	9206	9212	9217	9222	9227	9232	9238
84	9243	9248	9253	9258	9263	9269	9274	9279	9284	9289
N	0	1	2	3	4	5	6	7	8	9

N	0	1	2	3	4	5	6	7	8	9
85	9294	9299	9304	9309	9315	9320	9325	9330	9335	9340
86	9345	9350	9355	9360	9365	9370	9375	9380	9385	9390
87	9395	9400	9405	9410	9415	9420	9425	9430	9435	9440
88	9445	9450	9455	9460	9465	9469	9474	9479	9484	9489
89	9494	9499	9504	9509	9513	9518	9523	9528	9533	9538
90	9542	9547	9552	9557	9562	9566	9571	9576	9581	9586
91	9590	9595	9600	9605	9609	9614	9619	9624	9628	9633
92	9638	9643	9647	9652	9657	9661	9666	9671	9675	9680
93	9685	9689	9694	9699	9703	9708	9713	9717	9722	9727
94	9731	9736	9741	9745	9750	9754	9759	9763	9768	9773
95	9777	9782	9786	9791	9795	9800	9805	9809	9814	9818
96	9823	9827	9832	9836	9841	9845	9850	9854	9859	9863
97	9868	9872	9877	9881	9886	9890	9894	9899	9903	9908
98	9912	9917	9921	9926	9930	9934	9939	9943	9948	9952
99	9956	9961	9965	9969	9974	9978	9983	9987	9991	9996
N	0	1	2	3	4	5	6	7	8	9

Appendix B

Table of Natural Logarithms

N	0.00	0.01	0.02	0.03	0.04	0.05	0.06	0.07	0.08	0.09
1.0	0.0000	0.0100	0.0198	0.0296	0.0392	0.0488	0.0583	0.0677	0.0770	0.0862
1.1	0.0953	0.1044	0.1133	0.1222	0.1310	0.1398	0.1484	0.1570	0.1655	0.1740
1.2	0.1823	0.1906	0.1989	0.2070	0.2151	0.2231	0.2311	0.2390	0.2469	0.2546
1.3	0.2624	0.2700	0.2776	0.2852	0.2927	0.3001	0.3075	0.3148	0.3221	0.3293
1.4	0.3365	0.3436	0.3507	0.3577	0.3646	0.3716	0.3784	0.3853	0.3920	0.3988
1.5	0.4055	0.4121	0.4187	0.4253	0.4318	0.4383	0.4447	0.4511	0.4574	0.4637
1.6	0.4700	0.4762	0.4824	0.4886	0.4947	0.5008	0.5068	0.5128	0.5188	0.5247
1.7	0.5306	0.5365	0.5423	0.5481	0.5539	0.5596	0.5653	0.5710	0.5766	0.5822
1.8	0.5878	0.5933	0.5988	0.6043	0.6098	0.6152	0.6206	0.6259	0.6313	0.6366
1.9	0.6419	0.6471	0.6523	0.6575	0.6627	0.6678	0.6729	0.6780	0.6831	0.6881
2.0	0.6931	0.6981	0.7031	0.7080	0.7130	0.7178	0.7227	0.7275	0.7324	0.7372
2.1	0.7419	0.7467	0.7514	0.7561	0.7608	0.7655	0.7701	0.7747	0.7793	0.7839
2.2	0.7885	0.7930	0.7975	0.8020	0.8065	0.8109	0.8154	0.8198	0.8242	0.8286
2.3	0.8329	0.8372	0.8416	0.8459	0.8502	0.8544	0.8587	0.8629	0.8671	0.8713
2.4	0.8755	0.8796	0.8838	0.8879	0.8920	0.8961	0.9002	0.9042	0.9083	0.9123
2.5	0.9163	0.9203	0.9243	0.9282	0.9322	0.9361	0.9400	0.9439	0.9478	0.9517
2.6	0.9555	0.9594	0.9632	0.9670	0.9708	0.9746	0.9783	0.9821	0.9858	0.9895
2.7	0.9933	0.9969	1.0006	1.0043	1.0080	1.0116	1.0152	1.0188	1.0225	1.0260
2.8	1.0296	1.0332	1.0367	1.0403	1.0438	1.0473	1.0508	1.0543	1.0578	1.0613
2.9	1.0647	1.0682	1.0716	1.0750	1.0784	1.0818	1.0852	1.0886	1.0919	1.0953
3.0	1.0986	1.1019	1.1053	1.1086	1.1119	1.1151	1.1184	1.1217	1.1249	1.1282
3.1	1.1314	1.1346	1.1378	1.1410	1.1442	1.1474	1.1506	1.1537	1.1569	1.1600
3.2	1.1632	1.1663	1.1694	1.1725	1.1756	1.1787	1.1817	1.1848	1.1878	1.1909
3.3	1.1939	1.1970	1.2000	1.2030	1.2060	1.2090	1.2119	1.2149	1.2179	1.2208
3.4	1.2238	1.2267	1.2296	1.2326	1.2355	1.2384	1.2413	1.2442	1.2470	1.2499
3.5	1.2528	1.2556	1.2585	1.2613	1.2641	1.2669	1.2698	1.2726	1.2754	1.2782
3.6	1.2809	1.2837	1.2865	1.2892	1.2920	1.2947	1.2975	1.3002	1.3029	1.3056
3.7	1.3083	1.3110	1.3137	1.3164	1.3191	1.3218	1.3244	1.3271	1.3297	1.3324
3.8	1.3350	1.3376	1.3403	1.3429	1.3455	1.3481	1.3507	1.3533	1.3558	1.3584
3.9	1.3610	1.3635	1.3661	1.3686	1.3712	1.3737	1.3762	1.3788	1.3813	1.3838
4.0	1.3863	1.3888	1.3913	1.3938	1.3962	1.3987	1.4012	1.4036	1.4061	1.4085
4.1	1.4110	1.4134	1.4159	1.4183	1.4207	1.4231	1.4255	1.4279	1.4303	1.4327
4.2	1.4351	1.4375	1.4398	1.4422	1.4446	1.4469	1.4493	1.4516	1.4540	1.4563
4.3	1.4586	1.4609	1.4633	1.4656	1.4679	1.4702	1.4725	1.4748	1.4770	1.4793
4.4	1.4816	1.4839	1.4861	1.4884	1.4907	1.4929	1.4952	1.4974	1.4996	1.5019
N	0.00	0.01	0.02	0.03	0.04	0.05	0.06	0.07	0.08	0.09

N	0.00	0.01	0.02	0.03	0.04	0.05	0.06	0.07	0.08	0.09
4.5	1.5041	1.5063	1.5085	1.5107	1.5129	1.5151	1.5173	1.5195	1.5217	1.5239
4.6	1.5261	1.5282	1.5304	1.5326	1.5347	1.5369	1.5390	1.5412	1.5433	1.5454
4.7	1.5476	1.5497	1.5518	1.5539	1.5560	1.5581	1.5602	1.5623	1.5644	1.5665
4.8	1.5686	1.5707	1.5728	1.5748	1.5769	1.5790	1.5810	1.5831	1.5851	1.5872
4.9	1.5892	1.5913	1.5933	1.5953	1.5974	1.5994	1.6014	1.6034	1.6054	1.6074
5.0	1.6094	1.6114	1.6134	1.6154	1.6174	1.6194	1.6214	1.6233	1.6253	1.6273
5.1	1.6292	1.6312	1.6332	1.6351	1.6371	1.6390	1.6409	1.6429	1.6448	1.6467
5.2	1.6487	1.6506	1.6525	1.6544	1.6563	1.6582	1.6601	1.6620	1.6639	1.6658
5.3	1.6677	1.6696	1.6715	1.6734	1.6752	1.6771	1.6790	1.6808	1.6827	1.6845
5.4	1.6864	1.6882	1.6901	1.6919	1.6938	1.6956	1.6974	1.6993	1.7011	1.7029
5.5	1.7047	1.7066	1.7084	1.7102	1.7120	1.7138	1.7156	1.7174	1.7192	1.7210
5.6	1.7228	1.7246	1.7263	1.7281	1.7299	1.7317	1.7334	1.7352	1.7370	1.7387
5.7	1.7405	1.7422	1.7440	1.7457	1.7475	1.7492	1.7509	1.7527	1.7544	1.7561
5.8	1.7579	1.7596	1.7613	1.7630	1.7647	1.7664	1.7682	1.7699	1.7716	1.7733
5.9	1.7750	1.7766	1.7783	1.7800	1.7817	1.7834	1.7851	1.7867	1.7884	1.7901
6.0	1.7918	1.7934	1.7951	1.7967	1.7984	1.8001	1.8017	1.8034	1.8050	1.8066
6.1	1.8083	1.8099	1.8116	1.8132	1.8148	1.8165	1.8181	1.8197	1.8213	1.8229
6.2	1.8245	1.8262	1.8278	1.8294	1.8310	1.8326	1.8342	1.8358	1.8374	1.8390
6.3	1.8406	1.8421	1.8437	1.8453	1.8469	1.8485	1.8500	1.8516	1.8532	1.8547
6.4	1.8563	1.8579	1.8594	1.8610	1.8625	1.8641	1.8656	1.8672	1.8687	1.8703
6.5	1.8718	1.8733	1.8749	1.8764	1.8779	1.8795	1.8810	1.8825	1.8840	1.8856
6.6	1.8871	1.8886	1.8901	1.8916	1.8931	1.8946	1.8961	1.8976	1.8991	1.9006
6.7	1.9021	1.9036	1.9051	1.9066	1.9081	1.9095	1.9110	1.9125	1.9140	1.9155
6.8	1.9169	1.9184	1.9199	1.9213	1.9228	1.9242	1.9257	1.9272	1.9286	1.9301
6.9	1.9315	1.9330	1.9344	1.9359	1.9373	1.9387	1.9402	1.9416	1.9430	1.9445
7.0	1.9459	1.9473	1.9488	1.9502	1.9516	1.9530	1.9544	1.9559	1.9573	1.9587
7.1	1.9601	1.9615	1.9629	1.9643	1.9657	1.9671	1.9685	1.9699	1.9713	1.9727
7.2	1.9741	1.9755	1.9769	1.9782	1.9796	1.9810	1.9824	1.9838	1.9851	1.9865
7.3	1.9879	1.9892	1.9906	1.9920	1.9933	1.9947	1.9961	1.9974	1.9988	2.0001
7.4	2.0015	2.0028	2.0042	2.0055	2.0069	2.0082	2.0096	2.0109	2.0122	2.0136
7.5	2.0149	2.0162	2.0176	2.0189	2.0202	2.0215	2.0229	2.0242	2.0255	2.0268
7.6	2.0282	2.0295	2.0308	2.0321	2.0334	2.0347	2.0360	2.0373	2.0386	2.0399
7.7	2.0412	2.0425	2.0438	2.0451	2.0464	2.0477	2.0490	2.0503	2.0516	2.0528
7.8	2.0541	2.0554	2.0567	2.0580	2.0592	2.0605	2.0618	2.0631	2.0643	2.0665
7.9	2.0669	2.0681	2.0694	2.0707	2.0719	2.0732	2.0744	2.0757	2.0769	2.0782
8.0	2.0794	2.0807	2.0819	2.0832	2.0844	2.0857	2.0869	2.0882	2.0894	2.0906
8.1	2.0919	2.0931	2.0943	2.0956	2.0968	2.0980	2.0992	2.1005	2.1017	2.1029
8.2	2.1041	2.1054	2.1066	2.1078	2.1090	2.1102	2.1114	2.1126	2.1138	2.1150
8.3	2.1163	2.1175	2.1187	2.1199	2.1211	2.1223	2.1235	2.1247	2.1258	2.1270
8.4	2.1282	2.1294	2.1306	2.1318	2.1330	2.1342	2.1353	2.1365	2.1377	2.1389
N	0.00	0.01	0.02	0.03	0.04	0.05	0.06	0.07	0.08	0.09

N	0.00	0.01	0.02	0.03	0.04	0.05	0.06	0.07	0.08	0.09
8.5	2.1401	2.1412	2.1424	2.1436	2.1448	2.1459	2.1471	2.1483	2.1494	2.1506
8.6	2.1518	2.1529	2.1541	2.1552	2.1564	2.1576	2.1587	2.1599	2.1610	2.1622
8.7	2.1633	2.1645	2.1656	2.1668	2.1679	2.1691	2.1702	2.1713	2.1725	2.1736
8.8	2.1748	2.1759	2.1770	2.1782	2.1793	2.1804	2.1815	2.1827	2.1838	2.1849
8.9	2.1861	2.1872	2.1883	2.1894	2.1905	2.1917	2.1928	2.1939	2.1950	2.1961
9.0	2.1972	2.1983	2.1994	2.2006	2.2017	2.2028	2.2039	2.2050	2.2061	2.2072
9.1	2.2083	2.2094	2.2105	2.2116	2.2127	2.2138	2.2148	2.2159	2.2170	2.2181
9.2	2.2192	2.2203	2.2214	2.2225	2.2235	2.2246	2.2257	2.2268	2.2279	2.2289
9.3	2.2300	2.2311	2.2322	2.2332	2.2343	2.2354	2.2364	2.2375	2.2386	2.2396
9.4	2.2407	2.2418	2.2428	2.2439	2.2450	2.2460	2.2471	2.2481	2.2492	2.2502
9.5	2.2513	2.2523	2.2534	2.2544	2.2555	2.2565	2.2576	2.2586	2.2597	2.2607
9.6	2.2618	2.2628	2.2638	2.2649	2.2659	2.2670	2.2680	2.2690	2.2701	2.2711
9.7	2.2721	2.2732	2.2742	2.2752	2.2762	2.2773	2.2783	2.2793	2.2803	2.2814
9.8	2.2824	2.2834	2.2844	2.2854	2.2865	2.2875	2.2885	2.2895	2.2905	2.2915
9.9	2.2925	2.2935	2.2946	2.2956	2.2966	2.2976	2.2986	2.2996	2.3006	2.3016
N	0.00	0.01	0.02	0.03	0.04	0.05	0.06	0.07	0.08	0.09

If $N \geq 10$, $\ln 10 = 2.3026$ and write N in scientific notation; then use $\ln N = \ln[k \cdot (10^m)] = \ln k + m \ln 10 = \ln k + m\,(2.3026)$, where $1 \leq k < 10$ and m is an integer.

Appendix C

SAMPLE Screens From The Companion *Schaum's Electronic Tutor*

This book has a companion *Schaum's Electronic Tutor* which uses Mathcad® and is designed to help you learn the subject matter more readily. The *Electronic Tutor* uses the LIVE-MATH environment of Mathcad technical calculation software to give you on-screen access to approximately 100 representative solved problems from this book, together with summaries of key theoretical points and electronic cross-referencing and hyperlinking. The following pages reproduce a representative sample of screens from the *Electronic Tutor* and will help you understand the powerful capabilities of this electronic learning tool. Compare these screens with the associated solved problems from this book (the corresponding page numbers are listed at the start of each problem) to see how one complements the other.

In the companion *Schaum's Electronic Tutor*, you'll find all related text, diagrams, and equations for a particular solved problem together on your computer screen. As you can see on the following pages, all the math appears in familiar notation, including units. The format differences you may notice between the printed *Schaum's Outline* and the *Electronic Tutor* are designed to encourage your interaction with the material or show you alternate ways to solve challenging problems.

As you view the following pages, keep in mind that every number, formula, and graph shown ***is completely interactive when viewed on the computer screen***. You can change the starting parameters of a problem and watch as new output graphs are calculated before your eyes; you can change any equation and immediately see the effect of the numerical calculations on the solution. Every equation, graph, and number you see is available for experimentation. Each adapted solved problem becomes a "live" worksheet you can modify to solve dozens of related problems. The companion *Electronic Tutor* thus will help you to learn and retain the material taught in this book and you can also use it as a working problem-solving tool.

The Mathcad icon shown on the right is printed throughout this *Schaum's Outline* to indicate which problems are found in the *Electronic Tutor*.

For more information about the companion *Electronic Tutor*, including system requirements, please see the back cover.

Chapter 1 Fundamental Operations with Numbers

1.7 Operations with Fractions

---

1) Equivalent fractions can be created by multiplying or dividing the numerator and denominator by the same number provided the number is not zero.

To simplify a fraction,

- factor both the numerator and denominator and
- cancel common factors.

2) Note: $\frac{-a}{b} = \frac{a}{-b} = -\left(\frac{a}{b}\right)$ and $\frac{-a}{-b} = \frac{a}{b}$

3) To add or subtract fractions that have a common denominator,

- first add or subtract the numerators
- then write that sum or difference over the common denominator.

4) To add or subtract fractions that do not have a common denominator,

- first rewrite each fraction as equivalent fractions with a common denominator
- then add or subtract the numerators
- then write that sum or difference over the common denominator.

5) To multiply fractions

- first multiply the numerators
- and then the denominators.

6) The reciprocal of a fraction is a fraction
- whose numerator is the denominator of the given fraction and
- whose denominator is the numerator of the given fraction.

7) To divide fractions
- invert the divisor
- then multiply across the numerators and denominators.

Supplemental Problem 1.25 d

Write the sum S, difference D, product, P and Quotient Q of the following pair of numbers:

d) -2/3 , -3/2

Solution

Sum $\frac{-2}{3} + \frac{-3}{2} = \frac{-4}{6} + \frac{-9}{6} = \frac{-13}{6}$

Difference $\frac{-2}{3} - \left(\frac{-3}{2}\right) = \frac{-4}{6} - \frac{(-9)}{6} = \frac{5}{6}$

Product $\frac{-2}{3} \cdot \frac{-3}{2} = 1$

Quotient $\frac{\left(\frac{-2}{3}\right)}{\left(\frac{-3}{2}\right)} = \frac{-2}{3} \cdot \frac{-2}{3} = \frac{4}{9}$

It is easier to assign values to variables than to enter these fractions many times. However, the answers will be reported as decimal numbers.

EQUAL SIGNS

Enter a:-2/3 b:-3/2 See $a := \frac{-2}{3}$ $b := \frac{-3}{2}$

Enter a + b = See a + b = -2.167

CALCULATION ORDER

Chapter 10 Equations in General

10.2 Operations Used in Transforming Equations

A) If equals are added to equals, the results are equal.

B) If equals are subtracted from equals, the results are equal.

C) If equals are multiplied by equals, the results are equal.

D) If equals are divided by equals, the results are equal provided there is no division by zero.

E) The same powers of equals are equal.

F) The same roots of equals are equal.

G) Reciprocals of equals are equal provided the reciprocal of zero does not occur.

Supplemental Problem 10.12 a and b

Use the axioms of equality to solve each equation. Check the solutions obtained.

a) $5\,(x-4)=2\,(x+1)-7$

b) $\frac{2\cdot y}{3}-\frac{y}{6}=2$

Solution

a) $5\,(x-4)=2\,(x+1)-7$

$5\cdot x-20=2\cdot x-5$ Simplify each side (distributive property).

$-2\cdot x+20=-2\cdot x+20$ Collect like terms.

$3\cdot x=15$

$\frac{3\cdot x}{3}=\frac{15}{3}$ Divide by the coefficient of the variable.

$x=5$

SOLVING SYMBOLICALLY

Use Study Works to check your answer: In the original equation, underline the variable with the blue cursor, click on MATH then Solve for variable.

$5\,(x-4)=2\,(x+1)-7$

has solution(s)

5

b) $\frac{2\cdot y}{3}\ \frac{y}{6}=2$ Multiply each side by the common denominator.

$\frac{2\cdot y}{3}\cdot 6-\frac{y}{6}\cdot 6=2\cdot 6$ This will always eliminate the denominators.

$4\cdot y-y=12$ Collect like terms.

$3\cdot y=12$ Divide by the coefficient of the variable.

$y=4$

Check using StudyWorks: The complete check uses the original equation. However, any step along the way can be checked just to see if it is correct.

$$\frac{2\cdot y}{3}\cdot 6-\frac{y}{6}\cdot 6=2\cdot 6$$

has solution(s)

4

Chapter 10 Equations in General

10.3 Equivalent Equations

Equivalent equations are equations having the same solutions. The operations in Section 10.2 may not all yield equations equivalent to the original equations. The use of such operations may yield derived equations with either more or fewer solutions than the original equations.

If the operations yield more solutions, the extra solutions are called extraneous and the derived equation is said to be redundant with respect to the original equation.

If the operations yield fewer solutions than the original, the derived equations is said to be defective with respect to the original equation.

Operations A) and B) always yield equivalent equations.
Operations C) and E) may give rise to redundant equations and extraneous solutions.
Operations D) and F) may give rise to defective equations.

Supplemental Problem 10.12 f and i

Solve for the variable:

f) $\sqrt{3\cdot x - 2} = 4$

i) $(y + 1)^2 = 16$

Solution

f) $\sqrt{3\cdot x-2}=4$ To eliminate the root, square both sides.

$\left(\sqrt{3\cdot x-2}\right)^2=4^2$

$3\cdot x-2=16$ Collect like terms.

$3\cdot x=18$ Divide both sides by the coefficient of the variable.

$x=6$

Check Using Study Works:
(Remember the real check
must be to use the original equation.)

$\sqrt{3\cdot x-2}=4$

has solution(s)

6

i) $(y+1)^2=16$ To eliminate the square, take the square root of each side.

$y+1=\pm 4$ Since there are 2 square roots of 16, both are possibilities.

$y=-1+4$ and $y=-1-4$

$y=3$ $y=-5$ Simplify each

Check Using Study Works: Remember extraneous roots may be introduced when taking the root of both sides. Always check both answers.

$$(y + 1)^2 = 16$$

has solution(s)

$$\begin{bmatrix} 3 \\ -5 \end{bmatrix}$$

Chapter 12 Functions and Graphs

12.4 Function Notation

The notation $y = f(x)$, read as "y equals f of x", is used to designate that y is a function of x.

Thus $y = x^2 - 5x + 2$ may be written $f(x) = x^2 - 5x + 2$.

Then f(2), the value of f(x) when x = 2, is f(2) =
$2^2 - 5(2) + 2 = -4$

Supplemental Problem 12.33

Given $y = 5 + 3x - 2x^2$, find the values of y corresponding to x = -3, -2, -1, 0, 1, 2, 3.

Solution

The y-value corresponding to $x = -3$ is

$$5 + 3\cdot(-3) - 2\cdot(-3)^2 = -22$$

The calculations were obtained from Studyworks

- simply by typing the arithmetic expression resulting from substituting $x = -3$ into the formula,
- and then typing = at the end of the expression. This is done in a math region not a text region. The = causes the expression to be evaluated.

An alternative method in Studyworks uses the assignment operator := which is obtained by typing a colon :.

- first assign a value to x,
- next assign the formula to y,
- and then ask to evaluate y.

$$x := -3 \qquad y := 5 + 3\cdot x - 2\cdot x^2 \qquad y = -22$$

The second alternative has a great advantage.

- Change the -3 to -2 or some other number in the expression $x := -3$, above
- move the cursor,
- and Studyworks will automatically update the evaluated value of y.

Try this now.

A third alternative uses the function notation. Assign the formula to some function notation, and then ask Studyworks to evaluate the function at different numerical values.

$f(x) := 5 + 3 \cdot x - 2 \cdot x^2$ $\qquad f(-3) = -22 \qquad f(-2) = -9$

$f(-1) = 0 \qquad f(0) = 5$

With this tool the problem can be quickly completed.

But there is still another Studyworks device for completing the problem quickly and organizing the results into a table.

RANGE VARIABLES

In Studyworks the notation -3..3 is obtained by typing -3;3 and stands for the list of values -3, -2, -1, 0, 1, 2, 3. Studyworks calls this a range of values.

When this list of values is assigned to x, the evaluation of f(x) will produce the list of corresponding y-values.

The table below was obtained by typing x= and f(x)=, but the = signs do not appear.

$x := -3..3$

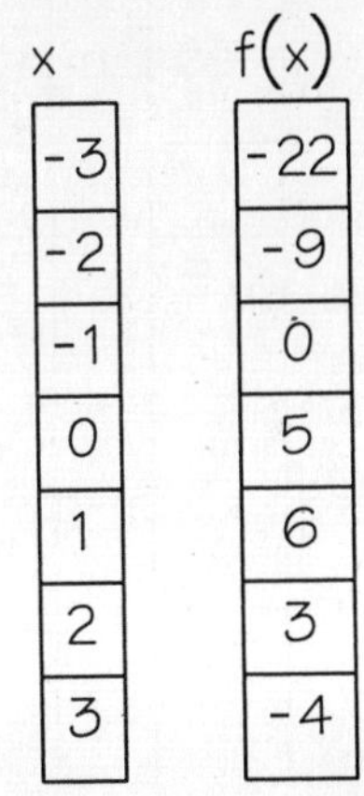

x	f(x)
-3	-22
-2	-9
-1	0
0	5
1	6
2	3
3	-4

Supplemental Problem 12.34

Extend the table of values in Problem 12.33 by finding the values of y which correspond to x = -5/2, -3/2,-1/2, 1/2, 3/2, 5/2.

Solution

Rather than do all the computations by hand use the Studyworks function capabilities by assigning the formula to f(x).

$$f(x) := 5 + 3 \cdot x - 2 \cdot x^2$$

RANGE VARIABLES

Assign the entire list of values to a range variable, and ask Studyworks to evaluate the formula at each value on the list.

x := -3,-2.5..3

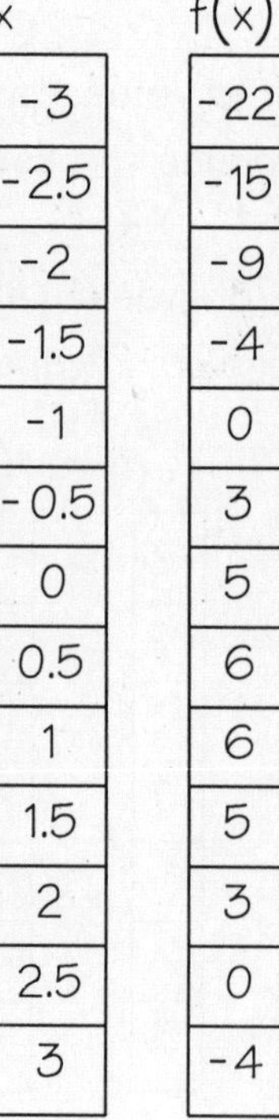

x	f(x)
-3	-22
-2.5	-15
-2	-9
-1.5	-4
-1	0
-0.5	3
0	5
0.5	6
1	6
1.5	5
2	3
2.5	0
3	-4

Note, if these lists are assigned to the horizontal and vertical axes of a graph region, then Studyworks will plot all the ordered pairs of (x, y)-values appearing on the table.

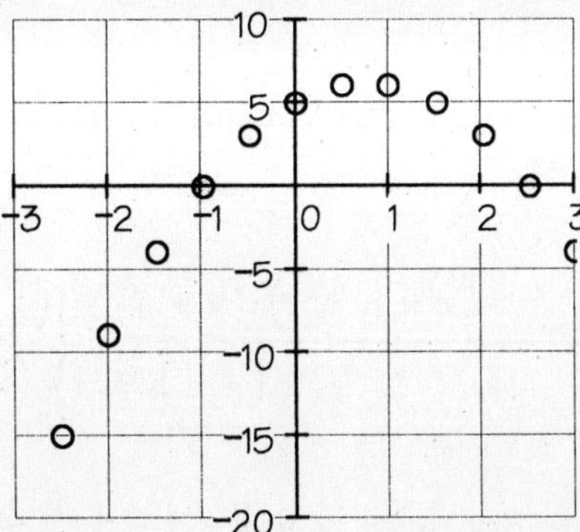

Supplemental Problem 12.39b

If $G(x) := \dfrac{x-1}{x+1}$, find b) $\dfrac{G(x+h) - G(x)}{h}$

Solution

$$G(x+h) = \frac{x+h-1}{x+h+1} \quad \text{so } G(x+h) - G(x) = \frac{x+h-1}{x+h+1} - \frac{x-1}{x+1}.$$

Combine these fractions by using the common denominator $(x+h+1)(x+1)$.

$$G(x+h) - G(x) = \frac{(x+h-1)\cdot(x+1) - (x+h+1)\cdot(x-1)}{(x+h+1)\cdot(x+1)}$$

$$= \frac{x^2 + h\cdot x - x + x + h - 1 - \left(x^2 + h\cdot x + x - x - h - 1\right)}{(x+h+1)\cdot(x+1)}$$

$$= \frac{2\cdot h}{(x+h+1)\cdot(x+1)}$$

Therefore, $\dfrac{G(x+h) - G(x)}{h} = \dfrac{2}{(x+h+1)\cdot(x+1)}$

Chapter 12 Functions and Graphs

12.8 Shifts

Supplemental Problem 12.44

State how the graph of the first equation relates to the graph of the second equation.

d) $y = (x - 1)^3$ and $y = x^3$ e) $y = x^2 - 7$ and $y = x^2$

f) $y = |x| + 1$ and $y = |x|$ g) $y = |x + 5|$ and $y = |x|$.

Solution

d) Let $f(x) := x^3$.

- The second equation is $y = f(x)$.
- The first equation is $y = f(x - 1)$.

This shifts the graph of $y = f(x)$ to the right by 1 unit.

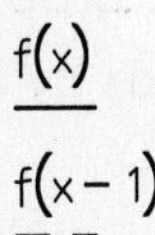

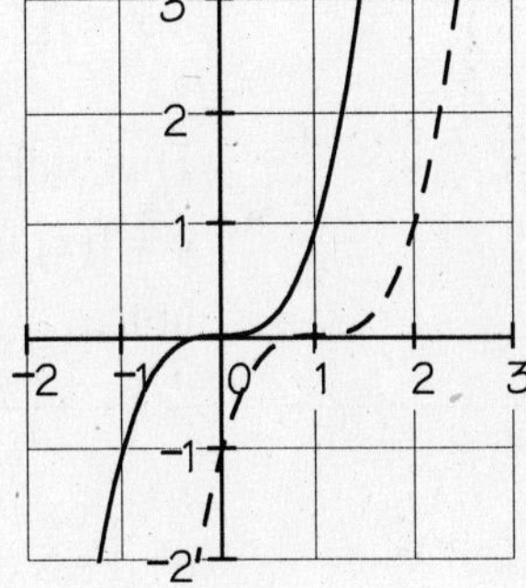

x

e) Let $f(x) := x^2$.

- The second equation is $y = f(x)$.
- The first equation is $y = f(x) - 7$.

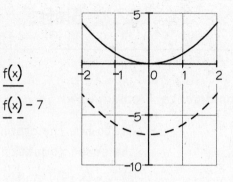

This shifts the graph of $y = f(x)$ down by 7 units.

f(x)

f(x) − 7

5 −2 −1 0 1 2 −5 −10

x

f) Let $f(x) := |x|$.

- The second equation is $y = f(x)$.
- The first equation is $y = f(x) + 1$.

This shifts the graph of $y = f(x)$ upwards by 1 unit.

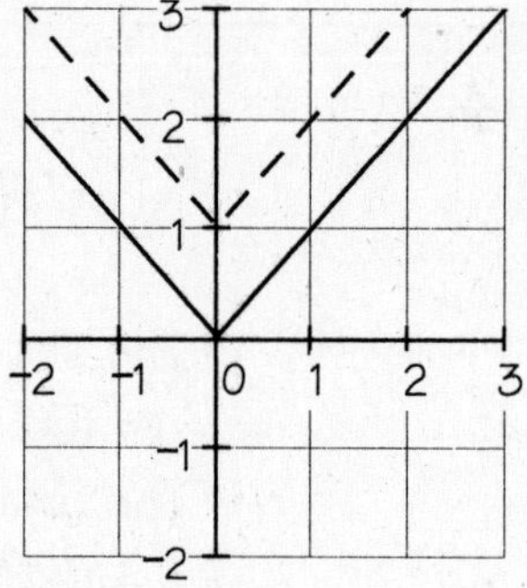

g) Let $f(x) := |x|$.

- The second equation is $y = f(x)$.
- The first equation is $y = f(x + 5)$.

This shifts the graph of $y = f(x)$ to the left by 5 units

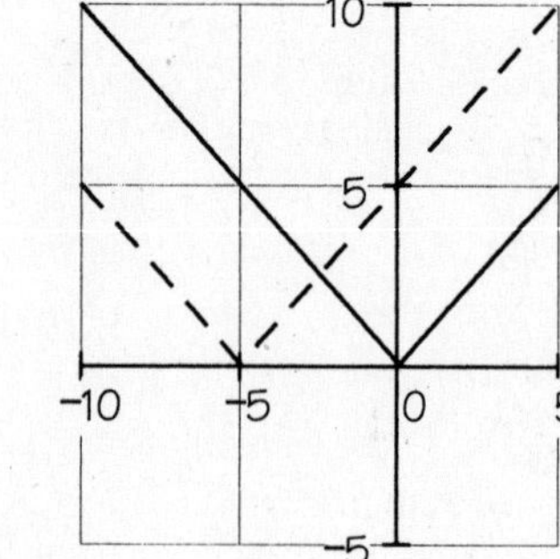

Index

The following license agreement will apply if this Schaum's Outline contains a Schaum's Electronic Tutor.

The McGraw-Hill Companies, Inc. and MathSoft, Inc. - Schaum's Electronic Tutors; MathSoft, Inc--Mathcad Engine License Agreement and Limited Warranty.

Both the program(s) (including all data and information contained therein) and the documentation are protected under applicable copyright laws. Your right to use the program(s) and the documentation are limited to the terms and conditions described herein.

Limited Non-Exclusive License

You May: (a) use the enclosed program(s) on a single computer; (b) physically transfer the program(s) from one computer to another provided that the program(s) are used on only one computer at a time, and that you remove any copies of the program(s) from the computer from which the program(s) are being transferred; (c) make copies of the program(s) solely for backup purposes. You must reproduce and include the copyright notice on a label on any back-up copy.

You May Not: (a) distribute copies of the program(s) or the documentation to others; (b) rent, lease or grant sublicenses or other rights to the program(s); (c) provide use of the program(s) in a computer service business, network, time-sharing, multiple CPU or multiple users arrangement without the prior written consent of MathSoft; (d) translate, decompile, reverse engineer or otherwise alter the program(s) or related documentation without the prior written consent of MathSoft.

Terms

Your license to use the program(s) and documentation will automatically terminate if you fail to comply with the terms of this Agreement. If this license is terminated you agree to destroy all copies of the program(s) and documentation.

Limited Warranty

MathSoft and McGraw-Hill warrant to the original licensee that the disk(s) on which the program(s) are recorded will be free from defects in materials and workmanship under normal use for a period of ninety (90) days from the date of purchase as evidenced by a copy of your receipt. If failure of the disk(s) has resulted from accident, abuse or misapplication of the product, then McGraw-Hill, MathSoft or third party Licensors shall have no responsibility to replace the disk(s) under this Limited Warranty.

This limited warranty and right of replacement is in lieu of, and you hereby waive, any and all other warranties both express and implied, including but not limited to warranties of merchantability and fitness for a particular purpose. The liability of McGraw-Hill, MathSoft or third party Licensors pursuant to this limited warranty shall be limited to the replacement of the defective disk(s), and in no event shall McGraw-Hill, MathSoft or third party Licensors be liable for incidental or consequential damages, including but not limited to loss of use, loss of profits, loss of data or data being rendered inaccurate, or losses sustained by third parties even if McGraw-Hill, MathSoft or third party Licensors have been advised of the possibility of such damages. McGraw-Hill and MathSoft make no representation or warranty that the results obtained will be successful or satisfy licensee's requirements, and McGraw-Hill and MathSoft shall have no liability for any errors or omissions in the programs or in the data and information contain therein. This warranty gives you specific legal rights which may vary from state to state. Some states do not allow the limitation or exclusion of liability for consequential damages, so the above limitation may not apply to you.

This License Agreement shall be governed by the laws of the Commonwealth of Massachusetts and shall inure to the benefit of McGraw-Hill, MathSoft, their respective successors, administrators, heirs and assigns of third party Licensors.